ECE/TRANS/258 (Vol. I)

ECONOMIC COMMISSION FOR EUROPE

Committee on Inland Transport

European Agreement concerning the International Carriage of Dangerous Goods by Inland Waterways (ADN)

including the Annexed Regulations, applicable as from 1 January 2017

Volume I

UNITED NATIONS
New York and Geneva, 2016

NOTE

The designations employed and the presentation of the material in this publication do not imply the expression of any opinion whatsoever on the part of the Secretariat of the United Nations concerning the legal status of any country, territory city or area, or of its authorities, or concerning the delimitation of its frontiers or boundaries.

ECE/TRANS/258 (Vol. I)

UNITED NATIONS PUBLICATIONS
Sales No.: E.16.VIII.3
ISBN 978-92-1-139157-2 *(Complete set of 2 volumes)*
e-ISBN 978-92-1-058306-0

Volumes I and II not to be sold separately.

United Nations Economic Commission for Europe (UNECE)

The United Nations Economic Commission for Europe (UNECE) is one of the five United Nations regional commissions, administered by the Economic and Social Council (ECOSOC). It was established in 1947 with the mandate to help rebuild post-war Europe, develop economic activity and strengthen economic relations among European countries, and between Europe and the rest of the world. During the Cold War, UNECE served as a unique forum for economic dialogue and cooperation between East and West. Despite the complexity of this period, significant achievements were made, with consensus reached on numerous harmonization and standardization agreements.

In the post-Cold War era, UNECE acquired not only many new member States, but also new functions. Since the early 1990s the organization has focused on analyses of the transition process, using its harmonization experience to facilitate the integration of central and eastern European countries into global markets.

UNECE is the forum where the countries of western, central and eastern Europe, Central Asia and North America – 56 countries in all – come together to forge the tools of their cooperation. That cooperation concerns economic cooperation and integration, statistics, environment, transport, trade, sustainable energy, forestry and timber, housing and land management and population. The Commission offers a regional framework for the elaboration and harmonization of conventions, norms and standards. The Commission's experts provide technical assistance to the countries of South-East Europe and the Commonwealth of Independent States. This assistance takes the form of advisory services, training seminars and workshops where countries can share their experiences and best practices.

Transport in UNECE

The UNECE Sustainable Transport Division is the secretariat of the Inland Transport Committee (ITC) and the ECOSOC Committee of Experts on the Transport of Dangerous Goods and on the Globally Harmonized System of Classification and Labelling of Chemicals. The ITC and its 17 working parties, as well as the ECOSOC Committee and its sub-committees are intergovernmental decision-making bodies that work to improve the daily lives of people and businesses around the world, in measurable ways and with concrete actions, to enhance traffic safety, environmental performance, energy efficiency and the competitiveness of the transport sector.

The ECOSOC Committee was set up in 1953 by the Secretary-General of the United Nations at the request of the Economic and Social Council to elaborate recommendations on the transport of dangerous goods. Its mandate was extended to the global (multi-sectoral) harmonization of systems of classification and labelling of chemicals in 1999. It is composed of experts from countries which possess the relevant expertise and experience in the international trade and transport of dangerous goods and chemicals. Its membership is restricted in order to reflect a proper geographical balance between all regions of the world and to ensure adequate participation of developing countries. Although the Committee is a subsidiary body of ECOSOC, the Secretary-General decided in 1963 that the secretariat services would be provided by the UNECE Transport Division.

ITC is a unique intergovernmental forum that was set up in 1947 to support the reconstruction of transport connections in post-war Europe. Over the years, it has specialized in facilitating the harmonized and sustainable development of inland modes of transport. The main results of this persevering and ongoing work are reflected, among other things, (i) in 58 United Nations conventions and many more technical regulations, which are updated on a regular basis and provide an international legal framework for the sustainable development of national and international road, rail, inland water and intermodal transport, including the transport of dangerous goods, as well as the construction and inspection of road motor vehicles; (ii) in the Trans-European North-south Motorway, Trans-European Railway and the Euro-Asia Transport Links projects, that facilitate multi-country coordination of transport infrastructure investment programmes; (iii) in the TIR system, which is a global customs transit facilitation solution; (iv) in the tool called For Future Inland Transport Systems (ForFITS), which can assist national and local governments to monitor carbon dioxide (CO_2) emissions coming from inland transport modes and to select and design climate change mitigation policies, based on their impact and adapted to local conditions; (v) in transport statistics – methods and data – that are internationally agreed on; (vi) in studies and reports that help transport policy development by addressing timely issues, based on cutting-edge research and analysis. ITC also devotes special attention to Intelligent Transport Services (ITS), sustainable urban mobility and city logistics, as well as to increasing the resilience of transport networks and services in response to climate change adaptation and security challenges.

In addition, the UNECE Sustainable Transport and Environment Divisions, together with the World Health Organization (WHO) – Europe, co-service the Transport Health and Environment Pan-European Programme (THE PEP).

Finally, as of 2015, the UNECE Sustainable Transport Division is providing the secretariat services for the Secretary General's Special Envoy for Road Safety, Mr. Jean Todt.

INTRODUCTION

The European Agreement concerning the International Carriage of Dangerous Goods by Inland Waterways (ADN) done at Geneva on 26 May 2000 under the auspices of the United Nations Economic Commission for Europe (UNECE) and the Central Commission for the Navigation of the Rhine (CCNR) entered into force on 28 February 2008.

The Agreement itself and the annexed Regulations, in their original version, were published in 2001 under the symbol ECE/TRANS/150. That publication also contains the Final Act of the Diplomatic Conference held in Geneva from 22 to 26 May 2000 during which the Agreement was adopted as well as the text of a Resolution adopted by the Conference.

At the time of the preparation of the present publication, the Agreement had eighteen Contracting Parties: Austria, Belgium, Bulgaria, Croatia, Czech Republic, France, Germany, Hungary, Luxembourg, Netherlands, Poland, Republic of Moldova, Romania, Russian Federation, Serbia, Slovakia, Switzerland and Ukraine. Other member States of UNECE whose territory contains inland waterways, other than those forming a coastal route, may also become Contracting Parties to the Agreement by acceding to it, on condition that the inland waterways are part of the network of inland waterways of international importance as defined in the European Agreement on Main Inland Waterways of International Importance (AGN).

The Regulations annexed to the ADN contain provisions concerning dangerous substances and articles, provisions concerning their carriage in packages and in bulk on board inland navigation vessels or tank vessels, as well as provisions concerning the construction and operation of such vessels. They also address requirements and procedures for inspections, the issue of certificates of approval, recognition of classification societies, monitoring, and training and examination of experts.

With the exception of the provisions relating to the recognition of classification societies, which have been applicable since the entry into force of the Agreement, the annexed Regulations did not become applicable until twelve months after the entry into force of the Agreement, namely on 28 February 2009 (see Article 11 (1) of the Agreement).

Before the entry into force of the Agreement, updates of the annexed Regulations were carried out regularly by a Joint Meeting of Experts of the UNECE and CCNR. These updates were adopted by the Administrative Committee of the ADN at its first session which was held in Geneva on 19 June 2008 (see document ECE/ADN/2, paragraphs 13 to 16).

Subsequently, the secretariat has published consolidated versions under the symbol ECE/TRANS/203 ("ADN 2009"), ECE/TRANS/220 ("ADN 2011"), ECE/TRANS/231 ("ADN 2013") and ECE/TRANS/243 ("ADN 2015").

At its sixteenth session (Geneva, 29 January 2016), the ADN Administrative Committee requested the secretariat to publish a new consolidated edition of ADN ("ADN 2017") incorporating all agreed corrections and amendments to enter into force on 1 January 2017. The amendments and corrections can be found in the following documents: ECE/ADN/36, ECE/ADN/36/Corr.1, ECE/ADN/36/Add.1, ECE/TRANS/WP.15/AC.2/58, annexes II and III and ECE/TRANS/WP.15/AC.2/60, annex IV.

The annexed Regulations contained in the present publication are the consolidated version which takes account of these updates and which is applicable from 1 January 2017.

It should be noted that, according to Directive 2008/68/EC of the European Parliament and of the Council of 24 September 2008 on the inland transport of dangerous goods, member States of the European Union, have to, with the exclusion of the derogations provided for in Article 1, paragraph 3 of the Directive, apply these annexed Regulations as well as Article 3 (f) and (h) and Article 8, paragraphs 1 and 3 of the ADN to the national and international transport between member States of dangerous goods by inland waterways on their territory.

All requests for information relating to the application of the ADN should be addressed to the relevant competent authority.

Additional information can be found on the website of the UNECE Transport Division at the following address:

http://www.unece.org/trans/danger/publi/adn/adn_e.html

This site, updated on a continuous basis, contains links to the following information:

-       ADN Agreement (excluding the annexed Regulations);

-       Corrections to the ADN Agreement (excluding the annexed Regulations);

-       Status of the Agreement;

-       Depositary notifications;

-       Country information (competent authorities, notifications);

-       Multilateral agreements;

-       Special authorizations;

-       Equivalences and derogations;

-       Classification societies;

-       Accident reports;

-       Catalogue of questions;

-       Harmonized model checklists;

-       Publication details (Corrigenda);

-       ADN 2017 (files);

-       Amendments to ADN 2015;

-       ADN 2015 (files);

-       Previous versions of ADN;

-       Historical information.

TABLE OF CONTENTS

VOLUME I

Page

EUROPEAN AGREEMENT CONCERNING THE INTERNATIONAL CARRIAGE OF DANGEROUS GOODS BY INLAND WATERWAYS (ADN).. xiii

ANNEXED REGULATIONS 1

PART 1   **GENERAL PROVISIONS**.. 3

Chapter 1.1   **Scope and applicability**
1.1.1   Structure.. 5
1.1.2   Scope... 5
1.1.3   Exemptions 6
1.1.4   Applicability of other regulations 10
1.1.5   Application of standards 12

Chapter 1.2   **Definitions and units of measurement**
1.2.1   Definitions 13
1.2.2   Units of measurement 44

Chapter 1.3   **Training of persons involved in the carriage of dangerous goods**
1.3.1   Scope and applicability 47
1.3.2   Nature of the training 47
1.3.3   Documentation.. 48

Chapter 1.4   **Safety obligations of the participants**
1.4.1   General safety measures 49
1.4.2   Obligations of the main participants 49
1.4.3   Obligations of the other participants....................................... 51

Chapter 1.5   **Special rules, derogations**
1.5.1   Bilateral and multilateral agreements 57
1.5.2   Special authorizations concerning transport in tank vessels........... 57
1.5.3   Equivalents and derogations (Article 7, paragraph 3 of ADN)....... 58

Chapter 1.6   **Transitional measures**
1.6.1   General... 59
1.6.2   Pressure receptacles and receptacles for Class 2............................ 61
1.6.3   Fixed tanks (tank-vehicles and tank wagons), demountable tanks, battery vehicles and battery wagons.................................. 61
1.6.4   Tank-containers, portable tanks and MEGCs 61
1.6.5   Vehicles 61
1.6.6   Class 7.. 61
1.6.7   Transitional provisions concerning vessels............................ 61
1.6.8   Transitional provisions concerning training of the crew................. 84
1.6.9   Transitional provisions concerning recognition of classification societies.. 84

Chapter 1.7   **General provisions concerning radioactive material**
1.7.1   Scope and application 85
1.7.2   Radiation protection programme ....

Table of contents (cont'd)

	1.7.3	Management system	88
	1.7.4	Special arrangement	88
	1.7.5	Radioactive material possessing other dangerous properties	88
	1.7.6	Non-compliance	88

Chapter 1.8 **Checks and other support measures to ensure compliance with safety requirements**

	1.8.1	Monitoring compliance with requirements	91
	1.8.2	Administrative assistance during the checking of a foreign vessel	92
	1.8.3	Safety adviser	92
	1.8.4	List of competent authorities and bodies designated by them	99
	1.8.5	Notifications of occurrences involving dangerous goods	99

Chapter 1.9 **Transport restrictions by the competent authorities** ... 105

Chapter 1.10 **Security provisions**

	1.10.1	General provisions	107
	1.10.2	Security training	107
	1.10.3	Provisions for high consequence dangerous goods	107

Chapters 1.11 to 1.14
(Reserved)

Chapter 1.15 **Recognition of classification societies**

	1.15.1	General	115
	1.15.2	Procedure for the recognition of classification societies	115
	1.15.3	Conditions and criteria for the recognition of a classification society applying for recognition under this Agreement	116
	1.15.4	Obligations of recommended classification societies	117

Chapter 1.16 **Procedure for the issue of the certificate of approval**

	1.16.1	Certificate of approval	119
	1.16.2	Issue and recognition of certificates of approval	122
	1.16.3	Inspection procedure	122
	1.16.4	Inspection body	124
	1.16.5	Application for the issue of a certificate of approval	124
	1.16.6	Particulars entered in the certificate of approval and amendments thereto	124
	1.16.7	Presentation of the vessel for inspection	124
	1.16.8	First inspection	125
	1.16.9	Special inspection	125
	1.16.10	Periodic inspection and renewal of the certificate of approval	125
	1.16.11	Extension of the certificate of approval without an inspection	125
	1.16.12	Official inspection	125
	1.16.13	Withholding and return of the certificate of approval	125
	1.16.14	Duplicate copy	127
	1.16.15	Register of certificates of approval	127

PART 2 **CLASSIFICATION** (See Volume II)

PART 3 **DANGEROUS GOODS LIST, SPECIAL PROVISIONS AND EXEMPTIONS RELATED TO LIMITED AND EXCEPTED QUANTITIES** ... 130

Table of contents (cont'd)

Chapter   3.1   **General**...(See Volume II)

Chapter   3.2   **List of dangerous goods**
          3.2.1   Table A: List of dangerous goods in numerical
                               order..(See Volume II)
          3.2.2   Table B: List of dangerous goods in alphabetical
                               order..(See Volume II)
          3.2.3   Table C: List of dangerous goods accepted for
                               carriage in tank vessels in numerical
                               order...135
          3.2.4   Modalities for the application of section 1.5.2 on
                   special authorizations concerning transport in tank
                   vessels 208

Chapter   3.3   **Special provisions applicable to certain articles or
                   substances** ....(See Volume II)

Chapter   3.4   **Dangerous goods packed in limited quantities**..........(See Volume II)

Chapter   3.5   **Dangerous goods packed in excepted quantities**.......(See Volume II)

PART 4   **PROVISIONS CONCERNING THE USE OF PACKAGINGS, TANKS
          AND BULK CARGO TRANSPORT UNITS**.. 225

Chapter   4.1   **General provisions** 227

PART 5   **CONSIGNMENT PROCEDURES** 229

Chapter   5.1   **General provisions**
          5.1.1   Application and general provisions............................. 231
          5.1.2   Use of overpacks... 231
          5.1.3   Empty uncleaned packagings (including IBCs and large
                   packagings), tanks, MEMUs, vehicles, wagons and
                   containers for carriage in bulk 231
          5.1.4   Mixed packing 232
          5.1.5   General provisions for Class 7................................... 232

Chapter   5.2   **Marking and labelling**
          5.2.1   Marking of packages... 241
          5.2.2   Labelling of packages 246

Chapter   5.3   **Placarding and marking of containers, MEGCs, MEMUs,
                   tank-containers, portable tanks, vehicles and wagons**
          5.3.1   Placarding 255
          5.3.2   Orange-coloured plate marking 259
          5.3.3   Mark for elevated temperature substances................... 265
          5.3.4   Marking for carriage in a transport chain
                   including maritime transport.. 266
          5.3.5   (*Reserved*) 267
          5.3.6   Environmentally hazardous substance mark................. 267

Chapter   5.4   **Documentation**
          5.4.0   General.. 269
          5.4.1   Dangerous goods transport document and related information 269

Table of contents (cont'd)

	5.4.2	Large container, vehicle or wagon packing certificate	280
	5.4.3	Instructions in writing	281
	5.4.4	Retention of dangerous goods transport information	287
	5.4.5	Example of a multimodal dangerous goods form	287

Chapter 5.5   **Special provisions**

	5.5.1	*(Deleted)*	291
	5.5.2	Special provisions applicable to fumigated cargo transport units (UN 3359)	291
	5.5.3	Special provisions applicable to packages and vehicles and containers containing substances presenting a risk of asphyxiation when used for cooling or conditioning purposes (such as dry ice (UN 1845) or nitrogen, refrigerated liquid (UN 1977) or argon, refrigerated liquid (UN 1951))	293

PART 6   **REQUIREMENTS FOR THE CONSTRUCTION AND TESTING OF PACKAGINGS (INCLUDING IBCS AND LARGE PACKAGINGS), TANKS AND BULK CARGO TRANSPORT UNITS** ... 297

Chapter 6.1   General requirements ... 299

PART 7   **REQUIREMENTS CONCERNING LOADING, CARRIAGE, UNLOADING AND HANDLING OF CARGO** ... 301

Chapter 7.1   **Dry cargo vessels**

	7.1.0	General requirements	303
	7.1.1	Mode of carriage of goods	303
	7.1.2	Requirements applicable to vessels	304
	7.1.3	General service requirements	305
	7.1.4	Additional requirements concerning loading, carriage, unloading and other handling of the cargo	308
	7.1.5	Additional requirements concerning the operation of vessels	327
	7.1.6	Additional requirements	329

Chapter 7.2   **Tank vessels**

	7.2.0	General requirements	333
	7.2.1	Mode of carriage of goods	333
	7.2.2	Requirements applicable to vessels	334
	7.2.3	General service requirements	335
	7.2.4	Additional requirements concerning loading, carriage, unloading and other handling of cargo	341
	7.2.5	Additional requirements concerning the operation of vessels	353

PART 8   **PROVISIONS FOR VESSEL CREWS, EQUIPMENT, OPERATION AND DOCUMENTATION** ... 355

Chapter 8.1   **General requirements applicable to vessels and equipment**

	8.1.1	*(Reserved)*	357
	8.1.2	Documents	357
	8.1.3	*(Reserved)*	359
	8.1.4	Fire-extinguishing arrangements	359
	8.1.5	Special equipment	360
	8.1.6	Checking and inspection of equipment	360

Table of contents (cont'd)

	8.1.7	Electrical installations	361
	8.1.8	*(Deleted)*	361
	8.1.9	*(Deleted)*	361
	8.1.10	*(Deleted)*	361
	8.1.11	Register of operations during carriage relating to the carriage of UN 1203	361

Chapter 8.2 **Requirements concerning training**

	8.2.1	General requirements concerning training of experts	363
	8.2.2	Special requirements for the training of experts	364

Chapter 8.3 **Miscellaneous requirements to be complied with by the crew of the vessel**

	8.3.1	Persons authorized on board	377
	8.3.2	Portable lamps	377
	8.3.3	Admittance on board	377
	8.3.4	Prohibition on smoking, fire and naked light	377
	8.3.5	Danger caused by work on board	377

Chapter 8.4 *(Reserved)* ... 379

Chapter 8.5 *(Reserved)* ... 381

Chapter 8.6 **Documents**

	8.6.1	Certificate of approval	383
	8.6.2	Certificate of special knowledge of ADN according to 8.2.1.2, 8.2.1.5 or 8.2.1.7	394
	8.6.3	ADN Checklist	395
	8.6.4	*(Deleted)*	399

PART 9 **RULES FOR CONSTRUCTION** ... 401

Chapter 9.1 **Rules for construction of dry cargo vessels**

	9.1.0	Rules for construction applicable to dry cargo vessels	403

Chapter 9.2 **Rules for construction applicable to seagoing vessels which comply with the requirements of the SOLAS 74 Convention, Chapter II-2, Regulation 19 or SOLAS 74, Chapter II-2, Regulation 54** ... 419

Chapter 9.3 **Rules for construction of tank vessels**

	9.3.1	Rules for construction of type G tank vessels	425
	9.3.2	Rules for construction of type C tank vessels	456
	9.3.3	Rules for construction of type N tank vessels	492
	9.3.4	Alternative constructions	528

EUROPEAN AGREEMENT CONCERNING THE INTERNATIONAL CARRIAGE OF DANGEROUS GOODS BY INLAND WATERWAYS (ADN)

THE CONTRACTING PARTIES,

DESIRING to establish by joint agreement uniform principles and rules, for the purposes of:

(a)   increasing the safety of international carriage of dangerous goods by inland waterways;

(b)   contributing effectively to the protection of the environment, by preventing any pollution resulting from accidents or incidents during such carriage; and

(c)   facilitating transport operations and promoting international trade,

CONSIDERING that the best means of achieving this goal is to conclude an agreement to replace the "European Provisions concerning the International Carriage of Dangerous Goods by Inland Waterways" annexed to resolution No. 223 of the Inland Transport Committee of the Economic Commission for Europe, as amended,

HAVE AGREED as follows:

CHAPTER I

GENERAL PROVISIONS

Article 1

Scope

1.   This Agreement shall apply to the international carriage of dangerous goods by vessels on inland waterways.

2.   This Agreement shall not apply to the carriage of dangerous goods by seagoing vessels on maritime waterways forming part of inland waterways.

3.   This Agreement shall not apply to the carriage of dangerous goods by warships or auxiliary warships or to other vessels belonging to or operated by a State, provided such vessels are used by the State exclusively for governmental and non-commercial purposes. However, each Contracting Party shall, by taking appropriate measures which do not impair the operations or operational capacity of such vessels belonging to or operated by it, ensure that such vessels are operated in a manner compatible with this Agreement, where it is reasonable in practice to do so.

Article 2

Regulations annexed to the Agreement

1. The Regulations annexed to this Agreement shall form an integral part thereof. Any reference to this Agreement implies at the same time a reference to the Regulations annexed thereto.

2. The annexed Regulations include:

    (a) Provisions concerning the international carriage of dangerous goods by inland waterways;

    (b) Requirements and procedures concerning inspections, the issue of certificates of approval, recognition of classification societies, derogations, special authorizations, monitoring, training and examination of experts;

    (c) General transitional provisions;

    (d) Supplementary transitional provisions applicable to specific inland waterways.

Article 3

Definitions

For the purposes of this Agreement:

(a) "*vessel*" means an inland waterway or seagoing vessel;

(b) "*dangerous goods*" means substances and articles the international carriage of which is prohibited by, or authorized only on certain conditions by, the annexed Regulations;

(c) "*international carriage of dangerous goods*" means any carriage of dangerous goods performed by a vessel on inland waterways on the territory of at least two Contracting Parties;

(d) "*inland waterways*" means the navigable inland waterways including maritime waterways on the territory of a Contracting Party open to the navigation of vessels under national law;

(e) "*maritime waterways*" means inland waterways linked to the sea, basically used for the traffic of seagoing vessels and designated as such under national law;

(f) "*recognized classification society*" means a classification society which is in conformity with the annexed Regulations and recognized, in accordance with the procedures laid down in these Regulations, by the competent authority of the Contracting Party where the certificate is issued;

(g) "*competent authority*" means the authority or the body designated or recognized as such in each Contracting Party and in each specific case in connection with these provisions;

(h) "*inspection body*" means a body nominated or recognized by the Contracting Party for the purpose of inspecting vessels according to the procedures laid down in the annexed Regulations.

CHAPTER II

TECHNICAL PROVISIONS

Article 4

Prohibitions on carriage, conditions of carriage, monitoring

1. Subject to the provisions of Articles 7 and 8, dangerous goods barred from carriage by the annexed Regulations shall not be accepted for international carriage.

2. Without prejudice to the provisions of Article 6, the international carriage of other dangerous goods shall be authorized, subject to compliance with the conditions laid down in the annexed Regulations.

3. Observance of the prohibitions and the conditions referred to in paragraphs 1 and 2 shall be monitored by the Contracting Parties in accordance with the provisions laid down in the annexed Regulations.

Article 5

Exemptions

This Agreement shall not apply to the carriage of dangerous goods to the extent to which such carriage is exempted in accordance with the annexed Regulations. Exemptions may only be granted when the quantity of the goods exempted, or the nature of the transport operation exempted, or the packagings, ensure that transport is carried out safely.

Article 6

Sovereign right of States

Each Contracting Party shall retain the right to regulate or prohibit the entry of dangerous goods into its territory for reasons other than safety during carriage.

Article 7

Special regulations, derogations

1. The Contracting Parties shall retain the right to arrange, for a limited period established in the annexed Regulations, by special bilateral or multilateral agreements, and provided safety is not impaired:

    (a) that the dangerous goods which under this Agreement are barred from international carriage may, subject to certain conditions, be accepted for international carriage on their inland waterways; or

    (b) that dangerous goods which under this Agreement are accepted for international carriage only on specified conditions may alternatively be accepted for international carriage on their inland waterways under conditions different from those laid down in the annexed Regulations.

    The special bilateral or multilateral agreements referred to in this paragraph shall be communicated immediately to the Executive Secretary of the Economic Commission for Europe, who shall communicate them to the Contracting Parties which are not signatories to the said agreements.

2. Each Contracting Party shall retain the right to issue special authorizations for the international carriage in tank vessels of dangerous substances the carriage of which in tank vessels is not permitted under the provisions concerning carriage in the annexed Regulations, subject to compliance with the procedures relating to special authorizations in the annexed Regulations.

3. The Contracting Parties shall retain the right to authorize, in the following cases, the international carriage of dangerous goods on board vessels which do not comply with conditions established in the annexed Regulations, provided that the procedure established in the annexed Regulations is complied with:

(a) The use on a vessel of materials, installations or equipment or the application on a vessel of certain measures concerning construction or certain provisions other than those prescribed in the annexed Regulations;

(b) Vessel with technical innovations derogating from the provisions of the annexed Regulations.

Article 8

Transitional provisions

1. Certificates of approval and other documents prepared in accordance with the requirements of the Regulations for the Carriage of Dangerous Goods in the Rhine (ADNR), the Regulations for the Carriage of Dangerous Goods on the Danube (ADN-D) or national regulations based on the European Provisions concerning the International Carriage of Dangerous Goods by Inland Waterways as annexed to resolution No. 223 of the Inland Transport Committee of the Economic Commission for Europe or as amended, applicable at the date of application of the annexed Regulations foreseen in Article 11, paragraph 1, shall remain valid until their expiry date, under the same conditions as those prevailing up to the date of such application, including their recognition by other States. In addition, these certificates shall remain valid for a period of one year from the date of application of the annexed Regulations in the event that they would expire during that period. However, the period of validity shall in no case exceed five years beyond the date of application of the annexed Regulations.

2. Vessels which, at the date of application of the annexed Regulations foreseen in Article 11, paragraph 1, are approved for the carriage of dangerous goods on the territory of a Contracting Party and which conform to the requirements of the annexed Regulations, taking into account where necessary, their general transitional provisions, may obtain an ADN certificate of approval under the procedure laid down in the annexed Regulations.

3. In the case of vessels referred to in paragraph 2 to be used exclusively for carriage on inland waterways where ADNR was not applicable under domestic law prior to the date of application of the annexed Regulations foreseen in Article 11, paragraph 1, the supplementary transitional provisions applicable to specific inland waterways may be applied in addition to the general transitional provisions. Such vessels shall obtain an ADN certificate of approval limited to the inland waterways referred to above, or to a portion thereof.

4. If new provisions are added to the annexed Regulations, the Contracting Parties may include new general transitional provisions. These transitional provisions shall indicate the vessels in question and the period for which they are valid.

Article 9

Applicability of other regulations

The transport operations to which this Agreement applies shall remain subject to local, regional or international regulations applicable in general to the carriage of goods by inland waterways.

CHAPTER III

FINAL PROVISIONS

Article 10

Contracting Parties

1. Member States of the Economic Commission for Europe whose territory contains inland waterways, other than those forming a coastal route, which form part of the network of inland waterways of international importance as defined in the European Agreement on Main Inland Waterways of International Importance (AGN) may become Contracting Parties to this Agreement:

   (a) by signing it definitively;

   (b) by depositing an instrument of ratification, acceptance or approval after signing it subject to ratification, acceptance or approval;

   (c) by depositing an instrument of accession.

2. The Agreement shall be open for signature until 31 May 2001 at the Office of the Executive Secretary of the Economic Commission for Europe, Geneva. Thereafter, it shall be open for accession.

3. The instruments of ratification, acceptance, approval or accession shall be deposited with the Secretary-General of the United Nations.

Article 11

Entry into force

1. This Agreement shall enter into force one month after the date on which the number of States mentioned in Article 10, paragraph 1, which have signed it definitively, or have deposited their instruments of ratification, acceptance, approval or accession has reached a total of seven.

   However, the annexed Regulations, except provisions concerning recognition of classification societies, shall not apply until twelve months after the entry into force of the Agreement.

2. For any State signing this Agreement definitively or ratifying, accepting, approving or acceding to it after seven of the States referred to in Article 10, paragraph 1, have signed it definitively or have deposited their instruments of ratification, acceptance, approval or accession, this Agreement shall enter into force one month after the said State has signed it definitively or has deposited its instrument of ratification, acceptance, approval or accession.

   The annexed Regulations shall become applicable on the same date. In the event that the term referred to in paragraph 1 relating to the application of the annexed Regulations has not expired, the annexed Regulations shall become applicable after expiry of the said term.

Article 12

Denunciation

1.   Any Contracting Party may denounce this Agreement by so notifying in writing the Secretary-General of the United Nations.

2.   Denunciation shall take effect twelve months after the date of receipt by the Secretary-General of the written notification of denunciation.

Article 13

Termination

1.   If, after the entry into force of this Agreement, the number of Contracting Parties is less than five during twelve consecutive months, this Agreement shall cease to have effect at the end of the said period of twelve months.

2.   In the event of the conclusion of a world-wide agreement for the regulation of the multimodal transport of dangerous goods, any provision of this Agreement, with the exception of those pertaining exclusively to inland waterways, the construction and equipment of vessels, carriage in bulk or tankers which is contrary to any provision of the said world-wide agreement shall, from the date on which the latter enters into force, automatically cease to apply to relations between the Parties to this Agreement which become parties to the world-wide agreement, and shall automatically be replaced by the relevant provision of the said world-wide agreement.

Article 14

Declarations

1.   Any State may, at the time of signing this Agreement definitively or of depositing its instrument of ratification, acceptance, approval or accession or at any time thereafter, declare by written notification addressed to the Secretary-General of the United Nations that this Agreement shall extend to all or any of the territories for the international relations of which it is responsible. The Agreement shall extend to the territory or territories named in the notification one month after it is received by the Secretary-General.

2.   Any State which has made a declaration under paragraph 1 of this article extending this Agreement to any territory for whose international relations it is responsible may denounce the Agreement in respect of the said territory in accordance with the provisions of Article 12.

3.   (a)   In addition, any State may, at the time of signing this Agreement definitively or of depositing its instrument of ratification, acceptance, approval or accession or at any time thereafter, declare by written notification addressed to the Secretary-General of the United Nations that this Agreement shall not extend to certain inland waterways on its territory, provided that the waterways in question are not part of the network of inland waterways of international importance as defined in the AGN. If this declaration is made subsequent to the time when the State signs this Agreement definitively or when it deposits its instrument of ratification, acceptance, approval or accession, the Agreement shall cease to have effect on the inland waterways in question one month after this notification is received by the Secretary-General.

(b)   However, any State on whose territory there are inland waterways covered by AGN, and which are, at the date of adoption of this Agreement, subject to a mandatory regime under international law concerning the carriage of dangerous goods, may declare that the implementation of this Agreement on these waterways shall be subject to compliance with the procedures set out in the statutes of the said regime. Any declaration of this nature shall be made at the time of signing this Agreement definitively or of depositing its instrument of ratification, acceptance, approval or accession.

4. Any State which has made a declaration under paragraphs 3 (a) or 3 (b) of this article may subsequently declare by means of a written notification to the Secretary-General of the United Nations that this Agreement shall apply to all or part of its inland waterways covered by the declaration made under paragraphs 3 (a) or 3 (b). The Agreement shall apply to the inland waterways mentioned in the notification one month after it is received by the Secretary-General.

Article 15

Disputes

1. Any dispute between two or more Contracting Parties concerning the interpretation or application of this Agreement shall so far as possible be settled by negotiation between the Parties in dispute.

2. Any dispute which is not settled by direct negotiation may be referred by the Contracting Parties in dispute to the Administrative Committee which shall consider it and make recommendations for its settlement.

3. Any dispute which is not settled in accordance with paragraphs 1 or 2 shall be submitted to arbitration if any one of the Contracting Parties in dispute so requests and shall be referred accordingly to one or more arbitrators selected by agreement between the Parties in dispute. If within three months from the date of the request for arbitration the Parties in dispute are unable to agree on the selection of an arbitrator or arbitrators, any of those Parties may request the Secretary-General of the United Nations to nominate a single arbitrator to whom the dispute shall be referred for decision.

4. The decision of the arbitrator or arbitrators appointed under paragraph 3 of this article shall be binding on the Contracting Parties in dispute.

Article 16

Reservations

1. Any State may, at the time of signing this Agreement definitively or of depositing its instrument of ratification, acceptance, approval or accession, declare that it does not consider itself bound by Article 15. Other Contracting Parties shall not be bound by Article 15 in respect of any Contracting Party which has entered such a reservation.

2. Any Contracting State having entered a reservation as provided for in paragraph 1 of this article may at any time withdraw such reservation by notifying in writing the Secretary-General of the United Nations.

3. Reservations other than those provided for in this Agreement are not permitted.

Article 17

Administrative Committee

1. An Administrative Committee shall be established to consider the implementation of this Agreement, to consider any amendments proposed thereto and to consider measures to secure uniformity in the interpretation and application thereof.

2. The Contracting Parties shall be members of the Administrative Committee. The Committee may decide that the States referred to in Article 10, paragraph 1 of this Agreement which are not Contracting Parties, any other Member State of the Economic Commission for Europe or of the United Nations or representatives of international intergovernmental or non-governmental organizations may, for questions which interest them, attend the sessions of the Committee as observers.

3. The Secretary-General of the United Nations and the Secretary-General of the Central Commission for the Navigation of the Rhine shall provide the Administrative Committee with secretariat services.

4. The Administrative Committee shall, at the first session of the year, elect a Chairperson and a Vice-Chairperson.

5. The Executive Secretary of the Economic Commission for Europe shall convene the Administrative Committee annually, or at other intervals decided on by the Committee, and also at the request of at least five Contracting Parties.

6. A quorum consisting of not less than one half of the Contracting Parties shall be required for the purpose of taking decisions.

7. Proposals shall be put to the vote. Each Contracting Party represented at the session shall have one vote. The following rules shall apply:

    (a) Proposed amendments to the Agreement and decisions pertaining thereto shall be adopted in accordance with the provisions of Article 19, paragraph 2;

    (b) Proposed amendments to the annexed Regulations and decisions pertaining thereto shall be adopted in accordance with the provisions of Article 20, paragraph 4;

    (c) Proposals and decisions relating to the recommendation of agreed classification societies, or to the withdrawal of such recommendation, shall be adopted in accordance with the procedure of the provisions of Article 20, paragraph 4;

    (d) Any proposal or decision other than those referred to in paragraphs (a) to (c) above shall be adopted by a majority of the Administrative Committee members present and voting.

8. The Administrative Committee may set up such working groups as it may deem necessary to assist it in carrying out its duties.

9. In the absence of relevant provisions in this Agreement, the Rules of Procedure of the Economic Commission for Europe shall be applicable unless the Administrative Committee decides otherwise.

Article 18

Safety Committee

A Safety Committee shall be established to consider all proposals for the amendment of the Regulations annexed to the Agreement, particularly as regards safety of navigation in relation to the construction, equipment and crews of vessels. The Safety Committee shall function within the framework of the activities of the bodies of the Economic Commission for Europe, of the Central Commission for the Navigation of the Rhine and of the Danube Commission which are competent in the transport of dangerous goods by inland waterways.

Article 19

Procedure for amending the Agreement, excluding the annexed Regulations

1. This Agreement, excluding its annexed Regulations, may be amended upon the proposal of a Contracting Party by the procedure specified in this article.

2. Any proposed amendment to this Agreement, excluding the annexed Regulations, shall be considered by the Administrative Committee. Any such amendment considered or prepared during the meeting of the Administrative Committee and adopted by it by a two-thirds majority of the members present and voting shall be communicated by the Secretary-General of the United Nations to the Contracting Parties for their acceptance.

3. Any proposed amendments communicated for acceptance in accordance with paragraph 2 shall come into force with respect to all Contracting Parties six months after the expiry of a period of twenty-four months following the date of communication of the proposed amendment if, during that period, no objection to the amendment in question has been communicated in writing to the Secretary-General of the United Nations by a Contracting Party.

Article 20

Procedure for amending the annexed Regulations

1. The annexed Regulations may be amended upon the proposal of a Contracting Party.

   The Secretary-General of the United Nations may also propose amendments with a view to bringing the annexed Regulations into line with other international agreements concerning the transport of dangerous goods and the United Nations Recommendations on the Transport of Dangerous Goods, as well as amendments proposed by a subsidiary body of the Economic Commission for Europe with competence in the area of the transport of dangerous goods.

2. Any proposed amendment to the annexed Regulations shall in principle be submitted to the Safety Committee, which shall submit the draft amendments it adopts to the Administrative Committee.

3. At the specific request of a Contracting Party, or if the secretariat of the Administrative Committee considers it appropriate, amendments may also be proposed directly to the Administrative Committee. They shall be examined at a first session and if they are deemed to be acceptable, they shall be reviewed at the following session of the Committee at the same time as any related proposal, unless otherwise decided by the Committee.

4. Decisions on proposed amendments and proposed draft amendments submitted to the Administrative Committee in accordance with paragraphs 2 and 3 shall be made by a majority of the members present and voting. However, a draft amendment shall not be deemed adopted if, immediately after the vote, five members present declare their objection to it. Adopted draft amendments shall be communicated by the Secretary-General of the United Nations to the Contracting Parties for acceptance.

5.	Any draft amendment to the annexed Regulations communicated for acceptance in accordance with paragraph 4 shall be deemed to be accepted unless, within three months from the date on which the Secretary-General circulates it, at least one-third of the Contracting Parties, or five of them if one-third exceeds that figure, have given the Secretary-General written notification of their objection to the proposed amendment. If the amendment is deemed to be accepted, it shall enter into force for all the Contracting Parties, on the expiry of a further period of three months, except in the following cases:

(a)	In cases where similar amendments to other international agreements governing the carriage of dangerous goods have already entered into force, or will enter into force at a different date, the Secretary-General may decide, upon written request by the Executive Secretary of the Economic Commission for Europe, that the amendment shall enter into force on the expiry of a different period so as to allow the simultaneous entry into force of these amendments with those to be made to such other agreements or, if not possible, the quickest entry into force of this amendment after the entry into force of such amendments to other agreements; such period shall not, however, be of less than one month's duration.

(b)	The Administrative Committee may specify, when adopting a draft amendment, for the purpose of entry into force of the amendment, should it be accepted, a period of more than three months' duration.

Article 21

Requests, communications and objections

The Secretary-General of the United Nations shall inform all Contracting Parties and all States referred to in Article 10, paragraph 1 of this Agreement of any request, communication or objection under Articles 19 and 20 above and of the date on which any amendment enters into force.

Article 22

Review conference

1.	Notwithstanding the procedure provided for in Articles 19 and 20, any Contracting Party may, by notification in writing to the Secretary-General of the United Nations, request that a conference be convened for the purpose of reviewing this Agreement.

A review conference to which all Contracting Parties and all States referred to in Article 10, paragraph 1, shall be invited, shall be convened by the Executive Secretary of the Economic Commission for Europe if, within a period of six months following the date of notification by the Secretary-General, not less than one fourth of the Contracting Parties notify him of their concurrence with the request.

2.	Notwithstanding the procedure provided for in Articles 19 and 20, a review conference to which all Contracting Parties and all States referred to in Article 10, paragraph 1, shall be invited, shall also be convened by the Executive Secretary of the Economic Commission for Europe upon notification in writing by the Administrative Committee. The Administrative Committee shall make a request if agreed to by a majority of those present and voting in the Committee.

3.   If a conference is convened in pursuance of paragraphs 1 or 2 of this article, the Executive Secretary of the Economic Commission for Europe shall invite the Contracting Parties to submit, within a period of three months, the proposals which they wish the conference to consider.

4.   The Executive Secretary of the Economic Commission for Europe shall circulate to all the Contracting Parties and to all the States referred to in Article 10, paragraph 1, the provisional agenda for the conference, together with the texts of such proposals, at least six months before the date on which the conference is to meet.

Article 23

Depositary

The Secretary-General of the United Nations shall be the depositary of this Agreement.

IN WITNESS WHEREOF the undersigned, being duly authorized thereto, have signed this Agreement.

DONE at Geneva, this twenty-sixth day of May two thousand, in a single copy, in the English, French, German and Russian languages for the text of the Agreement proper, and in the French language for the annexed Regulations, each text being equally authentic for the Agreement proper.

The Secretary-General of the United Nations is requested to prepare a translation of the annexed Regulations in the English and Russian languages.

The Secretary-General of the Central Commission for the Navigation of the Rhine is requested to prepare a translation of the annexed Regulations in the German language.

ANNEXED REGULATIONS

PART I

General provisions

CHAPTER 1.1

SCOPE AND APPLICABILITY

1.1.1   **Structure**

The Regulations annexed to ADN are grouped into nine parts. Each part is subdivided into chapters and each chapter into sections and subsections (see table of contents). Within each part the number of the part is included with the numbers of the chapters, sections and subsections, for example Part 2, Chapter 2, section 1 is numbered "2.2.1".

1.1.2   **Scope**

1.1.2.1   For the purposes of Article 2 paragraph 2 (a) and Article 4 of ADN, the annexed Regulations specify:

(a)   dangerous goods which are barred from international carriage;

(b)   dangerous goods which are authorized for international carriage and the conditions attaching to them (including exemptions) particularly with regard to:

– classification of goods, including classification criteria and relevant test methods;

– use of packagings (including mixed packing);

– use of tanks (including filling);

– consignment procedures (including marking and labelling of packages and placarding and marking of vehicles and wagons embarked, the marking of vessels as well as documentation and information required);

– provisions concerning the construction, testing and approval of packagings and tanks;

– use of means of transport (including loading, mixed loading and unloading).

1.1.2.2   For the purposes of Article 5 of ADN, section 1.1.3 of this chapter specifies the cases in which the carriage of dangerous goods is partially or totally exempted from the conditions of carriage established by ADN.

1.1.2.3   For the purposes of Article 7 of ADN, Chapter 1.5 of this part specifies the rules concerning the derogations, special authorizations and equivalences for which that article provides.

1.1.2.4   For the purposes of Article 8 of ADN, Chapter 1.6 of this part specifies the transitional measures concerning the application of the Regulations annexed to ADN.

1.1.2.5   The provisions of ADN also apply to empty vessels or vessels which have been unloaded as long as the holds, cargo tanks or receptacles or tanks accepted on board are not free from dangerous substances or gases, except for the exemptions for which section 1.1.3 of these Regulations provides.

1.1.3 **Exemptions**

1.1.3.1 *Exemptions related to the nature of the transport operation*

The provisions laid down in ADN do not apply to:

(a) the carriage of dangerous goods by private individuals where the goods in question are packaged for retail sale and are intended for their personal or domestic use or for their leisure or sporting activities provided that measures have been taken to prevent any leakage of contents in normal conditions of carriage. When these goods are flammable liquids carried in refillable receptacles filled by, or for, a private individual, the total quantity shall not exceed 60 litres per receptacle and 240 litres per cargo transport unit. Dangerous goods in IBCs, large packagings or tanks are not considered to be packaged for retail sale;

(b) the carriage of machinery or equipment not specified in these annexed Regulations and which happen to contain dangerous goods in their internal or operational equipment, provided that measures have been taken to prevent any leakage of contents in normal conditions of carriage;

(c) the carriage undertaken by enterprises which is ancillary to their main activity, such as deliveries to or returns from building or civil engineering sites, or in relation to surveying, repairs and maintenance, in quantities of not more than 450 litres per packaging, including intermediate bulk containers (IBCs) and large packagings, and within the maximum quantities specified in 1.1.3.6. Measures shall be taken to prevent any leakage of contents in normal conditions of carriage. These exemptions do not apply to Class 7.

Carriage undertaken by such enterprises for their supply or external or internal distribution does not fall within the scope of this exemption;

(d) the carriage undertaken by the competent authorities for the emergency response or under their supervision, insofar as such carriage is necessary in relation to the emergency response, in particular carriage undertaken to recover dangerous goods involved in an incident or accident and move them to the nearest appropriate safe place;

(e) emergency transport under the supervision of the competent authorities intended to save human lives or protect the environment provided that all measures are taken to ensure that such transport is carried out in complete safety;

(f) the carriage of uncleaned empty static storage vessels which have contained gases of Class 2, groups A, O or F, substances of Class 3 or Class 9 belonging to packing group II or III or pesticides of Class 6.1 belonging to packing group II or III, subject to the following conditions:

All openings with the exception of pressure relief devices (when fitted) are hermetically closed;

Measures have been taken to prevent any leakage of contents in normal conditions of carriage; and

The load is fixed in cradles or crates or other handling devices or to the vehicle, container or vessel in such a way that they will not become loose or shift during normal conditions of carriage.

This exemption does not apply to static storage vessels which have contained desensitized explosives or substances the carriage of which is prohibited by ADN.

NOTE: *For radioactive material see also 1.7.1.4.*

1.1.3.2   ***Exemptions related to the carriage of gases***

The provisions laid down in ADN do not apply to the carriage of:

(a)   *(Reserved)*;

(b)   *(Reserved)*;

(c)   gases of Groups A and O (according to 2.2.2.1), if the pressure of the gas in the receptacle or tank at a temperature of 20 °C does not exceed 200 kPa (2 bar) and if the gas is not a liquefied or a refrigerated liquefied gas. This includes every kind of receptacle or tank, e.g. also parts of machinery and apparatus;

NOTE: *This exemption does not apply to lamps. For lamps see 1.1.3.10.*

(d)   gases contained in the equipment used for the operation of the vessel (e.g. fire extinguishers), including spare parts;

(e)   *(Reserved)*;

(f)   gases contained in foodstuffs (except UN 1950), including carbonated beverages;

(g)   gases contained in balls intended for use in sports; and

(h)   *(Deleted)*.

1.1.3.3   ***Exemptions related to dangerous goods used for the propulsion of vessels, vehicles, wagons or non-road mobile machinery carried, for the operation of their special equipment, for their upkeep or for their safety***

The requirements of ADN do not apply to substances used

–   for the propulsion of vessels, vehicles, wagons or non-road mobile machinery carried[1],

–   for the upkeep of vessels,

–   for the operation or upkeep of their permanently installed special equipment,

–   for the operation or upkeep of their mobile special equipment used during carriage or intended to be used during carriage, or

–   to ensure safety,

and which are carried on board in the packaging, receptacle or tanks intended for use for this purpose.

---

[1]   *For the definition of non-road mobile machinery see paragraph 2.7 of the Consolidated Resolution on the Construction of Vehicles (R.E.3) (United Nations document ECE/TRANS/WP.29/78/Rev.3) or Article 2 of Directive 97/68/EC of the European Parliament and of the Council of 16 December 1997 on the approximation of the laws of the Member States relating to measures against the emission of gaseous and particulate pollutants from internal combustion engines to be installed in non-road mobile machinery (Official Journal of the European Communities No. L 059 of 27 February 1998).*

1.1.3.4   *Exemptions related to special provisions or to dangerous goods packed in limited or excepted quantities*

NOTE: For radioactive material see also 1.7.1.4.

1.1.3.4.1   Certain special provisions of Chapter 3.3 exempt partially or totally the carriage of specific dangerous goods from the requirements of ADN. The exemption applies when the special provision is referred to in Column (6) of Table A of Chapter 3.2 against the dangerous goods entry concerned.

1.1.3.4.2   Certain dangerous goods may be subject to exemptions provided that the conditions of Chapter 3.4 are met.

1.1.3.4.3   Certain dangerous goods may be subject to exemptions provided that the conditions of Chapter 3.5 are met.

1.1.3.5   *Exemptions related to empty uncleaned packagings*

Empty uncleaned packagings (including IBCs and large packagings) which have contained substances of Classes 2, 3, 4.1, 5.1, 6.1, 8 and 9 are not subject to the conditions of ADN if adequate measures have been taken to nullify any hazards. Hazards are nullified if adequate measures have been taken to nullify all hazards of Classes 1 to 9.

1.1.3.6   *Exemptions related to quantities carried on board vessels*

1.1.3.6.1   (a)   In the event of the carriage of dangerous goods in packages, the provisions of ADN other than those of 1.1.3.6.2 are not applicable when the gross mass of all the dangerous goods carried does not exceed 3,000 kg.

This provision does not apply to the carriage of:

(i)   substances and articles of Class 1;

(ii)   substances of Class 2, groups T, F, TF, TC, TO, TFC or TOC, according to 2.2.2.1.3 and aerosols of groups C, CO, F, FC, T, TF, TC, TO, TFC and TOC according to 2.2.2.1.6;

(iii)   substances of Classes 4.1 or 5.2. for which a danger label of model No. 1 is required in column (5) of Table A of Chapter 3.2;

(iv)   substances of Class 6.2, Category A;

(v)   substances of Class 7 other than UN Nos. 2908, 2909, 2910 and 2911;

(vi)   substances assigned to Packing Group I;

(vii)   substances carried in tanks;

(b)   In the event of the carriage of dangerous goods in packages other than tanks, the provisions of ADN other than those of 1.1.3.6.2 are not applicable to the carriage of:

–   substances of Class 2 of group F in accordance with 2.2.2.1.3 or aerosols of group F according to 2.2.2.1.6; or

–   substances assigned to Packing Group I, except substances of Class 6.1;

–   when the gross mass of these goods does not exceed 300 kg.

1.1.3.6.2 The carriage of exempted quantities according to 1.1.3.6.1 is, however, subject to the following conditions:

(a) The obligation to report in accordance with 1.8.5 remains applicable;

(b) Packages, except vehicles and containers (including swap bodies), shall comply with the requirements for packagings referred to in Parts 4 and 6 of ADR or RID; the provisions of Chapter 5.2 concerning marking and labelling are applicable;

(c) The following documents shall be on board:

– the transport documents (see 5.4.1.1); they shall concern all the dangerous goods carried on board;

– the stowage plan (see 7.1.4.11.1);

(d) The goods shall be stowed in the holds.

This provision does not apply to goods loaded in:

– containers with complete spray-proof walls;

– vehicles with complete spray-proof walls;

(e) Goods of different class shall be separated by a minimum horizontal distance of 3 m. They shall not be stowed on top of each other.

This provision does not apply to:

– containers with complete metal walls;

– vehicles with complete metal walls;

(f) For seagoing and inland navigation vessels, where the latter carry only containers, the above requirements under (d) and (e) shall be considered to have been met if the provisions of the IMDG Code regarding stowage and separation are met and if this particular is recorded in the transport document.

1.1.3.7 *Exemptions related to the carriage of electric energy storage and production systems*

The provisions laid down in ADN do not apply to electric energy storage and production systems (e.g., lithium batteries, electric capacitors, asymmetric capacitors, metal hydride storage systems and fuel cells):

(a) installed in a means of transport, performing a transport operation and destined for its propulsion or for the operation of any of its equipment;

(b) contained in an equipment for the operation of this equipment used or intended for use during carriage (e.g. a laptop computer).

1.1.3.8 *(Reserved)*

1.1.3.9 *Exemptions related to dangerous goods used as a coolant or conditioner during carriage*

When used in vehicles or containers for cooling or conditioning purposes, dangerous goods that are only asphyxiant (which dilute or replace the oxygen normally in the atmosphere) are only subject to the provisions of section 5.5.3.

1.1.3.10   *Exemptions related to the carriage of lamps containing dangerous goods*

The following lamps are not subject to ADN provided that they do not contain radioactive material and do not contain mercury in quantities above those specified in special provision 366 of Chapter 3.3:

(a)   Lamps that are collected directly from individuals and households when carried to a collection or recycling facility;

   NOTE: *This also includes lamps brought by individuals to a first collection point, and then carried to another collection point, intermediate processing or recycling facility.*

(b)   Lamps each containing not more than 1 g of dangerous goods and packaged so that there is not more than 30 g of dangerous goods per package, provided that:

   (i)   the lamps are manufactured according to a certified quality management system;

      NOTE: *ISO 9001 may be used for this purpose.*

   and

   (ii)   each lamp is either individually packed in inner packagings, separated by dividers, or surrounded with cushioning material to protect the lamps and packed into strong outer packagings meeting the general provisions of 4.1.1.1 of ADR and capable of passing a 1.2 m drop test;

(c)   Used, damaged or defective lamps each containing not more than 1 g of dangerous goods with not more than 30 g of dangerous goods per package when carried from a collection or recycling facility. The lamps shall be packed in strong outer packagings sufficient for preventing release of the contents under normal conditions of carriage meeting the general provisions of 4.1.1.1 of ADR and that are capable of passing a drop test of not less than 1.2 m;

(d)   Lamps containing only gases of Groups A and O (according to 2.2.2.1) provided they are packaged so that the projectile effects of any rupture of the lamp will be contained within the package.

NOTE: *Lamps containing radioactive material are addressed in 2.2.7.2.2.2 (b).*

1.1.4   **Applicability of other regulations**

1.1.4.1   *General*

The following requirements are applicable to packages:

(a)   In the case of packagings (including large packagings and intermediate bulk containers (IBCs), the applicable requirements of one of the international regulations shall be met (see also Part 4 and Part 6);

(b)   In the case of containers, tank-containers, portable tanks and multiple element gas containers (MEGCs), the applicable requirements of ADR, RID or the IMDG Code shall be met (see also Part 4 and Part 6);

(c)   In the case of vehicles or wagons, the vehicles or wagons and their load shall meet the applicable requirements of ADR or of RID, as relevant.

NOTE: For the marking, labelling, placarding and orange plate marking, see also Chapters 5.2 and 5.3.

1.1.4.2   *Carriage in a transport chain including maritime, road, rail or air carriage*

1.1.4.2.1   Packages, containers, portable tanks and tank-containers and MEGCs, which do not entirely meet the requirements for packing, mixed packing, marking, labelling of packages or placarding and orange plate marking, of ADN, but are in conformity with the requirements of the IMDG Code or the ICAO Technical Instructions shall be accepted for carriage in a transport chain including maritime or air carriage subject to the following conditions:

(a)   If the packages are not marked and labelled in accordance with ADN, they shall bear marks and danger labels in accordance with the requirements of the IMDG Code or the ICAO Technical Instructions;

(b)   The requirements of the IMDG Code or the ICAO Technical Instructions shall be applicable to mixed packing within a package;

(c)   For carriage in a transport chain including maritime carriage, if the containers, portable tanks, tank-containers or MEGCs are not marked and placarded in accordance with Chapter 5.3 of these Regulations, they shall be marked and placarded in accordance with Chapter 5.3 of the IMDG Code. In such case, only 5.3.2.1.1 of these Regulations is applicable to the marking of the vehicle itself. For empty, uncleaned portable tanks, tank-containers and MEGCs, this requirement shall apply up to and including the subsequent transfer to a cleaning station.

This derogation does not apply in the case of goods classified as dangerous goods in classes 1 to 9 of ADN and considered as non-dangerous goods according to the applicable requirements of the IMDG Code or the ICAO Technical Instructions.

1.1.4.2.2   When a maritime, road, rail or air transport operation follows or precedes carriage by inland waterway, the transport document used or to be used for the maritime, road, rail or air transport operation may be used in place of the transport document prescribed in 5.4.1 provided that the particulars it contains are in conformity with the applicable requirements of the IMDG Code, ADR, RID or the ICAO Technical Instructions, respectively except that, when additional information is required by ADN, it shall be added or entered at the appropriate place.

NOTE: For carriage in accordance with 1.1.4.2.1, see also 5.4.1.1.7. For carriage in containers, see also 5.4.2.

1.1.4.3   *Use of IMO type portable tanks approved for maritime transport*

IMO type portable tanks (types 1, 2, 5 and 7) which do not meet the requirements of Chapters 6.7 or 6.8 of ADR, but which were built and approved before 1 January 2003 in accordance with the provisions of the IMDG Code (Amdt. 29-98) may continue to be used provided that they meet the applicable periodic inspection and test provisions of the IMDG Code[2]. In addition, they shall meet the provisions corresponding to the instructions set out in columns (10) and (11) of Table A in Chapter 3.2 and the provisions of Chapter 4.2 of ADR. See also 4.2.0.1 of the IMDG Code.

1.1.4.4   *(Reserved)*

1.1.4.5   *(Reserved)*

---

[2]   *The International Maritime Organization (IMO) has issued "Guidance on the Continued Use of Existing IMO Type Portable Tanks and Road Tank Vehicles for the Transport of Dangerous Goods" as circular DSC.1/Circ.12 and Corrigenda. The text of this guidance can be found on the IMO website at: www.imo.org.*

1.1.4.6   *Other regulations applicable to carriage by inland waterway*

1.1.4.6.1   In accordance with article 9 of ADN, transport operations shall remain subject to the local, regional or international requirements generally applicable to the carriage of goods by inland waterway.

1.1.4.6.2   Where the requirements of these Regulations are in contradiction with the requirements referred to in 1.1.4.6.1, the requirements referred to in 1.1.4.6.1 shall not apply.

1.1.5   **Application of standards**

Where the application of a standard is required and there is any conflict between the standard and the provisions of ADN, the provisions of ADN take precedence. The requirements of the standard that do not conflict with ADN shall be applied as specified, including the requirements of any other standard, or part of a standard, referenced within that standard as normative.

CHAPTER 1.2

DEFINITIONS AND UNITS OF MEASUREMENT

1.2.1   **Definitions**

NOTE: This section contains all general or specific definitions.

For the purposes of these regulations:

A

Accommodation means spaces intended for the use of persons normally living on board, including galleys, food stores, lavatories, washrooms, bathrooms, laundries, halls, alleyways, etc., but excluding the wheelhouse;

ADR means the European Agreement concerning the International Carriage of Dangerous Goods by Road;

Aerosol, see *Aerosol dispenser*;

Aerosol dispenser means an article consisting of any non-refillable receptacle meeting the requirements of 6.2.6 of ADR made of metal, glass or plastics, and containing a gas, compressed, liquefied or dissolved under pressure, with or without a liquid, paste or powder, and fitted with a release device allowing the contents to be ejected as solid or liquid particles in suspension in a gas, as a foam, paste or powder or in a liquid state or in a gaseous state;

Animal material means animal carcasses, animal body parts, or animal foodstuffs;

Approval

Multilateral approval, for the carriage of radioactive material, means approval by the relevant competent authority of the country of origin of the design or shipment, as applicable, and by the competent authority of each country through or into which the consignment is to be carried;

Unilateral approval, for the carriage of radioactive material, means an approval of a design which is required to be given by the competent authority of the country of origin of the design only. If the country of origin is not a Contracting Party to ADN, the approval shall require validation by the competent authority of a Contracting Party to ADN (see 6.4.22.8 of ADR);

ASTM means the American Society for Testing and Materials (ASTM International, 100 Barr Harbor Drive, PO Box C700, West Conshohocken, PA, 19428-2959, United States of America);

Auto-ignition temperature (EN 13237:2011) means the lowest temperature determined under prescribed test conditions of a hot surface on which a flammable substance in the form of a gas/air or vapour/air mixture ignites.

B

Bag means a flexible packaging made of paper, plastics film, textiles, woven material or other suitable material;

Battery-vehicle means a vehicle containing elements which are linked to each other by a manifold and permanently fixed to this vehicle. The following elements are considered to be elements of a battery-vehicle: cylinders, tubes, bundles of cylinders (also known as frames), pressure drums as well as tanks destined for the carriage of gases as defined in 2.2.2.1.1 with a capacity of more than 450 litres;

Battery-wagon means a wagon containing elements which are linked to each other by a manifold and permanently fixed to a wagon. The following elements are considered to be elements of a battery wagon: cylinders, tubes, bundles of cylinders (also known as frames), pressure drums as well as tanks intended for gases of Class 2 with a capacity greater than 450 litres;

Bilge water means oily water from the engine room bilges, the peak, the cofferdams and the double-hull spaces;

Biological/technical name means a name currently used in scientific and technical handbooks, journals and texts. Trade names shall not be used for this purpose;

Body (for all categories of IBC other than composite IBCs) means the receptacle proper, including openings and closures, but does not include service equipment;

Boil-off means the vapour produced above the surface of a boiling cargo due to evaporation. It is caused by heat ingress or a drop in pressure;

Box means a packaging with complete rectangular or polygonal faces, made of metal, wood, plywood, reconstituted wood, fibreboard, plastics or other suitable material. Small holes for purposes of ease of handling or opening or to meet classification requirements, are permitted as long as they do not compromise the integrity of the packaging during carriage;

Breathing apparatus (ambient air-dependent filter apparatus) means an apparatus which protects the person wearing it when working in a dangerous atmosphere by means of a suitable filter. For such apparatuses, see for example European standard EN 136:1998. For the filters used, see for example European standard EN 14387:2004 + A1:2008;

Breathing apparatus (self-contained) means an apparatus which supplies the person wearing it when working in a dangerous atmosphere with breathing air by means of pressurized air carried with him or by means of an external supply via a tube. For such apparatuses, see for example European standard EN 137:2006 or EN 138:1994;

Bulk container means a containment system (including any liner or coating) intended for the carriage of solid substances which is in direct contact with the containment system. Packagings, intermediate bulk containers (IBCs), large packagings and tanks are not included.

A bulk container is:

− of a permanent character and accordingly strong enough to be suitable for repeated use;

− specially designed to facilitate the carriage of goods by one or more means of transport without intermediate reloading;

− fitted with devices permitting its ready handling;

− of a capacity of not less than 1.0 m³.

Examples of bulk containers are containers, offshore bulk containers, skips, bulk bins, swap bodies, trough-shaped containers, roller containers, load compartments of vehicles or wagons;

NOTE: *This definition only applies to bulk containers meeting the requirements of chapter 6.11 of ADR.*

Closed bulk container means a totally closed bulk container having a rigid roof, sidewalls, end walls and floor (including hopper-type bottoms). The term includes bulk containers with an opening roof, side or end wall that can be closed during carriage. Closed bulk containers may be equipped with openings to allow for the exchange of vapours and gases with air and which prevent under normal conditions of carriage the release of solid contents as well as the penetration of rain and splash water;

Flexible bulk container means a flexible container with a capacity not exceeding 15 m³ and includes liners and attached handling devices and service equipment;

Sheeted bulk container means an open top bulk container with rigid bottom (including hopper-type bottom), side and end walls and a non-rigid covering;

Bulkhead means a metal wall, generally vertical, inside the vessel and which is bounded by the bottom, the side plating, a deck, the hatchway covers or by another bulkhead;

Bulkhead (watertight) means

– In a dry cargo vessel: a bulkhead constructed so that it can withstand water pressure with a head of 1.00 metre above the deck but at least to the top of the hatchway coaming;

– In a tank vessel: a bulkhead constructed to withstand a water pressure of 1.00 metre above the deck;

Bundle of cylinders (frame) means an assembly of cylinders that are fastened together and are interconnected by a manifold and carried as a unit. The total water capacity shall not exceed 3,000 litres except that bundles intended for the carriage of toxic gases of Class 2 (groups starting with letter T according to 2.2.2.1.3) shall be limited to 1,000 litres water capacity.

C

Capacity of shell or shell compartment, for tanks, means the total inner volume of the shell or shell compartment expressed in litres or cubic metres. When it is impossible to completely fill the shell or the shell compartment because of its shape or construction, this reduced capacity shall be used for the determination of the degree of filling and for the marking of the tank;

Cargo area means the whole of the following spaces (see figures below);

Cargo area

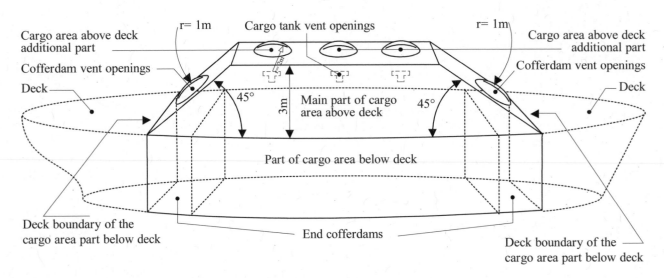

Above deck cargo area for various tank vessel

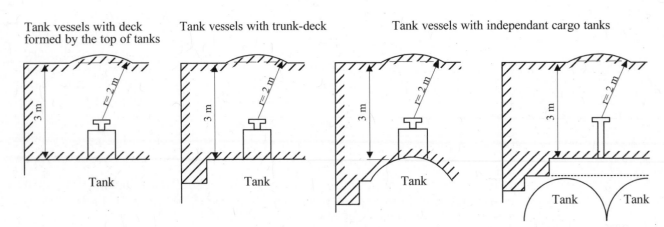

Cargo area (additional part above deck) (when anti-explosion protection is required, comparable to zone 1) means the spaces not included in the main part of the cargo area above deck comprising 1.00 m radius spherical segments centred over the ventilation openings of the cofferdams and the service spaces located in the cargo area part below the deck and 2.00 m spherical segments centred over the ventilation openings of the cargo tanks and the opening of the pump-rooms;

Cargo area (main part above deck) (when anti-explosion protection is required - comparable to zone 1) means the space which is bounded:

–   at the sides, by the shell plating extending upwards from the decks sides;

–   fore and aft, by planes inclined at 45° towards the cargo area, starting at the boundary of the cargo area part below deck;

–   vertically, 3 m above the deck;

Cargo area (part below deck) means the space between two vertical planes perpendicular to the centre-line plane of the vessel, which comprises cargo tanks, hold spaces, cofferdams, double-hull spaces and double bottoms; these planes normally coincide with the outer cofferdam bulkheads or hold end bulkheads. Their intersection line with the deck is referred to as the boundary of the cargo area part below deck;

Cargo piping, see *Piping for loading and unloading*;

Cargo pump-room (when anti-explosion protection is required, comparable to zone 1) means a service space where the cargo pumps and stripping pumps are installed together with their operational equipment;

Cargo residues means liquid cargo which cannot be pumped out of the cargo tanks or piping by means of the stripping system;

Cargo tank (when anti-explosion protection is required, comparable to zone 0) means a tank which is permanently attached to the vessel and intended for the carriage of dangerous goods.

Cargo tank design:

(a)   *Pressure cargo tank* means a cargo tank independent of the vessel's hull, built according to dedicated recognised standards for a working pressure ≥ 400 kPa;

(b)   *Closed cargo tank* means a cargo tank connected to the outside atmosphere through a device preventing unacceptable internal overpressure or underpressure;

(c)   *Open cargo tank with flame arrester* means a cargo tank connected to the outside atmosphere through a device fitted with a flame arrester;

(d)   *Open cargo tank* means a cargo tank in open connection with the outside atmosphere.

Cargo tank type:

(a)   *Independent cargo tank* means a cargo tank which is permanently built in, but which is independent of the vessel's structure;

(b)   *Integral cargo tank* means a cargo tank which is constituted by the vessel's structure itself and bounded by the outer hull or by walls separate from the outer hull;

(c)   *Cargo tank with walls distinct from the outer hull* means an integral cargo tank of which the bottom and side walls do not form the outer hull of the vessel or an independent cargo tank.

Cargo tank (discharged) means a cargo tank which after unloading may contain some residual cargo.

Cargo tank (empty) means a cargo tank which after unloading contains no residual cargo but may not be gas free.

Cargo tank (gas free) means a cargo tank which after unloading does not contain any residual cargo or any measurable concentration of dangerous gases.

Cargo transport unit means a vehicle, a wagon, a container, a tank-container, a portable tank or an MEGC;

Carriage means the change of place of dangerous goods, including stops made necessary by transport conditions and including any period spent by the dangerous goods in vessels, vehicles, wagons, tanks and containers made necessary by traffic conditions before, during and after the change of place.

This definition also covers the intermediate temporary storage of dangerous goods in order to change the mode or means of transport (transshipment). This shall apply provided that transport documents showing the place of dispatch and the place of reception are presented on request and provided that packages and tanks are not opened during intermediate storage, except to be checked by the competent authorities;

Carriage in bulk means the carriage of an unpackaged solid which can be discharged;

NOTE: *Within the meaning of ADN, the carriage in bulk referred to in ADR or RID is considered as carriage in packages.*

Carrier means the enterprise which carries out the transport operation with or without a transport contract;

CDNI means Convention on the Collection, Storage and Reception of Waste Generated during Navigation on the Rhine and Other Inland Waterways;

Certified safe type electrical apparatus means an electrical apparatus which has been tested and approved by the competent authority regarding its safety of operation in an explosive atmosphere, e.g.

– intrinsically safe apparatus;

– flameproof enclosure apparatus;

– apparatus protected by pressurization;

– powder filling apparatus;

– apparatus protected by encapsulation;

– increased safety apparatus.

NOTE: *Limited explosion risk apparatus is not covered by this definition.*

CEVNI means the UNECE European Code for Inland Waterways;

CGA means the Compressed Gas Association (CGA, 14501 George Carter Way, Suite 103, Chantilly, VA 20151, United States of America);

CIM means the Uniform Rules Concerning the Contract of International Carriage of Goods by Rail (Appendix B to the Convention concerning International Carriage by Rail (COTIF)), as amended;

Classification society (recognized) means a classification society which is recognized by the competent authorities in accordance with Chapter 1.15;

Classification of zones (see Directive 1999/92/CE[*]*)*

Zone 0:   areas in which dangerous explosive atmospheres of gases, vapours or sprays exist permanently or during long periods;

Zone 1:   areas in which dangerous explosive atmospheres of gases, vapours or sprays are likely to occur occasionally;

Zone 2:   areas in which dangerous explosive atmospheres of gases, vapours or sprays are likely to occur rarely and if so for short periods only;

---

[*] *Official Journal of the European Communities No. L 23 of 28 January 2000, p.57.*

Closed bulk container, see *Bulk container;*

Closed container, see *Container;*

Closed-type sampling device means a device penetrating through the boundary of the cargo tank or through the piping for loading and unloading but constituting a part of a closed system designed so that during sampling no gas or liquid may escape from the cargo tank. The device shall be of a type approved by the competent authority for this purpose;

Closed vehicle means a vehicle having a body capable of being closed;

Closed wagon means a wagon with sides and a fixed or movable roof.

Closure means a device which closes an opening in a receptacle;

CMNI means the Convention on the Contract for the Carriage of Goods by Inland Waterway (Budapest, 22 June 2001).

CMR means the Convention on the Contract for the International Carriage of Goods by Road (Geneva, 19 May 1956), as amended;

Cofferdam (when anti-explosion protection is required, comparable to zone 1) means an athwartship compartment which is bounded by watertight bulkheads and which can be inspected. The cofferdam shall extend over the whole area of the end bulkheads of the cargo tanks. The bulkhead not facing the cargo area shall extend from one side of the vessel to the other and from the bottom to the deck in one frame plane;

Collective entry means an entry for a defined group of substances or articles (see 2.1.1.2, B, C and D);

Combination packaging means a combination of packagings for carriage purposes, consisting of one or more inner packagings secured in an outer packaging in accordance with 4.1.1.5 of ADR;

NOTE: *The term "inner packaging" used for combination packagings shall not be confused with the term "inner receptacle" used for composite packagings.*

Competent authority means the authority or authorities or any other body or bodies designated as such in each State and in each specific case in accordance with domestic law;

Compliance assurance (radioactive material) means a systematic programme of measures applied by a competent authority which is aimed at ensuring that the requirements of ADN are met in practice;

Composite IBC with plastics inner receptacle means an IBC comprising structural equipment in the form of a rigid outer casing encasing a plastics inner receptacle together with any service or other structural equipment. It is so constructed that the inner receptacle and outer casing once assembled form, and are used as, an integrated single unit to be filled, stored, transported or emptied as such;

NOTE: *Plastics material, when used in connection with inner receptacles for composite IBCs, is taken to include other polymeric materials such as rubber.*

Composite packaging means a packaging consisting of an outer packaging and an inner receptacle so constructed that the inner receptacle and the outer packaging form an integral packaging. Once assembled it remains thereafter an integrated single unit; it is filled, stored, carried and emptied as such;

NOTE: The term "inner receptacle" used for composite packagings shall not be confused with the term "inner packaging" used for combination packagings. For example, the inner of a 6HA1 composite packaging (plastics material) is such an inner receptacle since it is normally not designed to perform a containment function without its outer packaging and is not therefore an inner packaging.

Where a material is mentioned in brackets after the term "composite packaging", it refers to the inner receptacle.

Compressed natural gas (CNG) means a compressed gas composed of natural gas with a high methane content assigned to UN No. 1971;

Confinement system, for the carriage of radioactive material, means the assembly of fissile material and packaging components specified by the designer and agreed to by the competent authority as intended to preserve criticality safety;

Connection for a sampling device means a connection allowing the installation of a closed-type or partly closed-type sampling device. The connection shall be fitted with a lockable mechanism resistant to the internal pressure of the cargo tank. The connection shall be of a type approved by the competent authority for the intended use;

Consignee means the consignee according to the contract for carriage. If the consignee designates a third party in accordance with the provisions applicable to the contract for carriage, this person shall be deemed to be the consignee within the meaning of ADN. If the transport operation takes place without a contract for carriage, the enterprise which takes charge of the dangerous goods on arrival shall be deemed to be the consignee;

Consignment means any package or packages, or load of dangerous goods, presented by a consignor for carriage;

Consignor means the enterprise which consigns dangerous goods either on its own behalf or for a third party. If the transport operation is carried out under a contract for carriage, consignor means the consignor according to the contract for carriage. In the case of a tank vessel, when the cargo tanks are empty or have just been unloaded, the master is considered to be the consignor for the purpose of the transport document;

Containment system, for the carriage of radioactive material, means the assembly of components of the packaging specified by the designer as intended to retain the radioactive material during carriage;

Container means an article of transport equipment (lift van or other similar structure):

- of a permanent character and accordingly strong enough to be suitable for repeated use;

- specially designed to facilitate the carriage of goods, by one or more means of transport, without breakage of load;

- fitted with devices permitting its ready stowage and handling, particularly when being transloaded from one means of transport to another;

- so designed as to be easy to fill and empty;

- having an internal volume of not less than 1 m³, except for containers for the carriage of radioactive material.

In addition:

Closed container means a totally enclosed container having a rigid roof, rigid side walls, rigid end walls and a floor. The term includes containers with an opening roof where the roof can be closed during transport;

Large container means:

(a)   a container which does not meet the definition of a small container;

(b)   in the meaning of the CSC, a container of a size such that the area enclosed by the four outer bottom corners is either

   (i)   at least 14 m^2 (150 square feet) or

   (ii)   at least 7 m^2 (75 square feet) if fitted with top corner fittings;

Open container means an open top container or a platform based container;

Sheeted container means an open container equipped with a sheet to protect the goods loaded;

Small container means a container which has an internal volume of not more than 3 m^3;

A swap body is a container which, in accordance with European Standard EN 283 (1991 edition) has the following characteristics:

–   from the point of view of mechanical strength, it is only built for carriage on a wagon or a vehicle on land or by roll-on roll-off ship;

–   it cannot be stacked;

–   it can be removed from vehicles by means of equipment on board the vehicle and on its own supports, and can be reloaded;

NOTE: *The term "container" does not cover conventional packagings, IBCs, tank-containers, vehicles or wagons. Nevertheless, a container may be used as a packaging for the carriage of radioactive material.*

Control temperature means the maximum temperature at which an organic peroxide or a self-reactive substance can be safely carried;

Conveyance means, with respect to the carriage by inland waterway, any vessel, hold or defined deck area of any vessel; for carriage by road or by rail, it means a vehicle or a wagon;

Crate means an outer packaging with incomplete surfaces;

Criticality safety index (CSI) assigned to a package, overpack or container containing fissile material, for the carriage of radioactive material, means a number which is used to provide control over the accumulation of packages, overpacks or containers containing fissile material;

Critical temperature means the temperature above which the substance cannot exist in the liquid state;

Cryogenic receptacle means a transportable thermally insulated receptacle for refrigerated liquefied gases of a water capacity of not more than 1,000 litres (see also *Open cryogenic receptacle*);

CSC means the International Convention for Safe Containers (Geneva, 1972) as amended and published by the International Maritime Organization (IMO), London;

Cylinder means a transportable pressure receptacle of a water capacity not exceeding 150 litres (see also *Bundle of cylinders (frame)*);

D

Damage control plan means the plan indicating the boundaries of the watertight compartments serving as the basis for the stability calculations, in the event of a leak, the trimming arrangements for the correction of any list due to flooding and the means of closure which are to be kept closed when the vessel is under way;

Dangerous goods means those substances and articles the carriage of which is prohibited by ADN, or authorized only under the conditions prescribed therein;

Dangerous reaction means:

(a)   combustion or evolution of considerable heat;

(b)   evolution of flammable, asphyxiate, oxidizing or toxic gases;

(c)   the formation of corrosive substances;

(d)   the formation of unstable substances; or

(e)   dangerous rise in pressure (for tanks and cargo tanks only);

Deflagration means an explosion which propagates at subsonic speed (see EN 13237:2011);

Demountable tank means a tank, other than a fixed tank, a portable tank, a tank-container or an element of a battery-vehicle or a MEGC which has a capacity of more than 450 litres, is not designed for the carriage of goods without breakage of load, and normally can only be handled when it is empty; or a tank designed to fit the special apparatus of a wagon but which can only be removed from it after dismantling the means of attachment;

Design, for the carriage of radioactive material, means the description of fissile material excepted under 2.2.7.2.3.5 (f), special form radioactive material, low dispersible radioactive material, package or packaging which enables such an item to be fully identified. The description may include specifications, engineering drawings, reports demonstrating compliance with regulatory requirements, and other relevant documentation;

Design life, for composite cylinders and tubes, means the maximum life (in number of years) for which the cylinder or tube is designed and approved in accordance with the applicable standard;

Design pressure means the pressure on the basis of which the cargo tank or the residual cargo tank has been designed and built;

Detonation means an explosion which propagates at supersonic speed and is characterized by a shock-wave (see EN 13237:2011);

Drum means a flat-ended or convex-ended cylindrical packaging made out of metal, fibreboard, plastics, plywood or other suitable materials. This definition also includes packagings of other shapes, e.g. round, taper-necked packagings or pail-shaped packagings. *Wooden barrels* and *jerricans* are not covered by this definition.

E

EC Directive means provisions decided by the competent institutions of the European Community and which are binding, as to the result to be achieved, upon each Member State to which it is addressed, but shall leave to the national authorities the choice of form and methods;

Emergency temperature means the temperature at which emergency procedures shall be implemented in the event of loss of temperature control;

Electrical apparatus protected against water jets means an electrical apparatus so designed that water, projected by a nozzle on the enclosure from any direction, has no damaging effects. The test conditions are specified in the IEC publication 60529, minimum degree of protection IP55;

EN (standard) means a European standard published by the European Committee for Standardization (CEN) (CEN – Avenue Marnix 17, B-1000 Brussels);

Enterprise means any natural person, any legal person, whether profit-making or not, any association or group of persons without legal personality, whether profit-making or not, or any official body, whether it has legal personality itself or is dependent upon an authority that has such personality;

Escape boat means a specially designed directly accessible boat designed to withstand all identified hazards of the cargo and to evacuate the people in danger;

Escape device (suitable) means a respiratory protection device, designed to cover the wearer's mouth, nose and eyes, which can be easily put on and which serves to escape from a danger area. For such devices, see for example European standard EN 13794:2002, EN 402: 2003, EN 403: 2004 or EN 1146:2005;

Escape route means a safe route from danger towards safety or to another means of evacuation;

Evacuation boat means a manned and specially equipped boat called in for rescuing people in danger or evacuating them within the minimum safe period of time provided by a safe haven or a safe area;

Exclusive use, for the carriage of radioactive material, means the sole use, by a single consignor, of a conveyance or of a large container, in respect of which all initial, intermediate and final loading and unloading and shipment are carried out in accordance with the directions of the consignor or consignee where so required by ADN;

Explosion means a sudden reaction of oxidation or decomposition with an increase in temperature or in pressure or both simultaneously (see EN 13237:2011);

Explosion danger areas means areas in which an explosive atmosphere may occur of such a scale that special protection measures are necessary to ensure the safety and health of the persons affected (see Directive 1999/92/EC[*]);

Explosion group means a grouping of flammable gases and vapours according to their maximum experimental safe gaps (standard gap width, determined in accordance with

---

[*]   *Official Journal of the European Communities No. L 23 of 28 January 2000, p.57.*

specified conditions) and minimum ignition currents, and of electrical apparatus intended to be used in a potentially explosive atmosphere (see EN IEC 60079-0:2012);

Explosive atmosphere means a mixture of air with gases, vapours or mists flammable in atmospheric conditions, in which the combustion process spreads after ignition to the entire unconsumed mixture (see EN 13237:2011);

F

Fibreboard IBC means a fibreboard body with or without separate top and bottom caps, if necessary an inner liner (but no inner packagings), and appropriate service and structural equipment;

Filler means any enterprise

(a)   which fills dangerous goods into a tank (tank-vehicle, tank wagon, demountable tank, portable tank or tank-container) or into a battery-vehicle, battery-wagon or MEGC; or

(b)   which fills dangerous goods into a cargo tank; or

(c)   which fills dangerous goods into a vessel, a vehicle, a wagon, a large container or small container for carriage in bulk;

Filling pressure means the maximum pressure actually built up in the tank when it is being filled under pressure; (see also *Calculation pressure, Discharge pressure, Maximum working pressure (gauge pressure)* and *Test pressure*);

Filling ratio means the ratio of the mass of gas to the mass of water at 15° C that would fill completely a pressure receptacle fitted ready for use (capacity);

Filling ratio (cargo tank): Where a filling ratio is given for a cargo tank, it refers to the percentage of the volume of the cargo tank which may be filled with liquid during loading;

Fixed tank means a tank having a capacity of more than 1,000 litres which is permanently attached to a vehicle (which then becomes a tank-vehicle) or to a wagon (which then becomes a tank-wagon) or is an integral part of the frame of such vehicle or wagon;

Flame arrester means a device mounted in the vent of part of an installation or in the interconnecting piping of a system of installations, the purpose of which is to permit flow but prevent the propagation of a flame front. This device shall be tested according to the European standard EN ISO 16852:2010;

Flame arrester plate stack means the part of the flame arrester the main purpose of which is to prevent the passage of a flame front;

Flame arrester housing means the part of a flame arrester the main purpose of which is to form a suitable casing for the flame arrester plate stack and ensure a mechanical connection with other systems;

Flammable component (for aerosols) means flammable liquids, flammable solids or flammable gases and gas mixtures as defined in Notes 1 to 3 of sub-section 31.1.3 of Part III of the Manual of Tests and Criteria. This designation does not cover pyrophoric, self-heating or water-reactive substances. The chemical heat of combustion shall be determined by one of the following methods ASTM D 240, ISO/FDIS 13943: 1999 (E/F) 86.1 to 86.3 or NFPA 30B;

Flammable gas detector means a device allowing measuring of any significant concentration of flammable gases given off by the cargo below the lower explosive limit and which clearly indicates the presence of higher concentrations of such gases. Flammable gas detectors may be designed for measuring flammable gases only but also for measuring both flammable gases and oxygen.

This device shall be so designed that measurements are possible without the necessity of entering the spaces to be checked;

Flash-point means the lowest temperature of a liquid at which its vapours form a flammable mixture with air;

Flexible bulk container, see *Bulk container*,

Flexible IBC means a body constituted of film, woven fabric or any other flexible material or combinations thereof, and if necessary, an inner coating or liner, together with any appropriate service equipment and handling devices;

Frame (Class 2), see *Bundle of cylinders*;

Fuel cell means an electrochemical device that converts the chemical energy of a fuel to electrical energy, heat and reaction products;

Fuel cell engine means a device used to power equipment and which consists of a fuel cell and its fuel supply, whether integrated with or separate from the fuel cell, and includes all appurtenances necessary to fulfil its function;

Full load means any load originating from one consignor for which the use of a vehicle, of a wagon or of a large container is exclusively reserved and all operations for the loading and unloading of which are carried out in conformity with the instructions of the consignor or of the consignee;

NOTE: *The corresponding term for radioactive material is "exclusive use".*

G

Gas (for the purposes of Class 2) means a substance which:

(a)   at 50° C has a vapour pressure greater than 300 kPa (3 bar); or

(b)   is completely gaseous at 20° C under standard pressure of 101.3 kPa;

Otherwise, *Gases* means gases or vapours;

Gas cartridge, see *Small receptacle containing gas*;

Gas detection system means a fixed system capable of detecting in time significant concentrations of flammable gases given off by the cargoes at concentrations below the lower explosion limit and capable of activating the alarms;

GESAMP means the Joint Group of Experts on the Scientific Aspects of Marine Environmental Protection. IMO publication: "The Revised GESAMP Hazard Evaluation Procedure for Chemical Substances Carried by Ships", GESAMP Reports and Studies No. 64, IMO, London, 2002.

In applying the GESAMP model for the purposes of the present Regulations, the reference temperature for the relative density, vapour pressure and water solubility is 20°C. The reference relative density to be used to differentiate between floating substances ("floater") and substances that sink ("sinker") is 1,000 (corresponding to the water density in inland waterways of 1000 kg/m^3);

GHS means the sixth revised edition of the Globally Harmonized System of Classification and Labelling of Chemicals, published by the United Nations as document ST/SG/AC.10/30/Rev.6;

H

Handling device (for flexible IBCs) means any sling, loop, eye or frame attached to the body of the IBC or formed from the continuation of the IBC body material;

Hermetically closed tank means a tank intended for the carriage of liquid substances with a calculation pressure of at least 4 bar or intended for the carriage of solid substances (powdery or granular) regardless of its calculation pressure, the openings of which are hermetically closed and which:

- is not equipped with safety valves, bursting discs, other similar safety devices or vacuum valves; or

- is not equipped with safety valves, bursting discs or other similar safety devices, but is equipped with vacuum valves, in accordance with the requirements of 6.8.2.2.3 of ADR; or

- is equipped with safety valves preceded by a bursting disc according to 6.8.2.2.10 of ADR, but is not equipped with vacuum valves; or

- is equipped with safety valves preceded by a bursting disc according to 6.8.2.2.10 of ADR and vacuum valves, in accordance with the requirements of 6.8.2.2.3 of ADR;

Highest class may be assigned to a vessel when:

- the hull, inclusive of rudder and steering gear and equipment of anchors and chains, complies with the rules and regulations of a recognized classification society and has been built and tested under its supervision;

- the propulsion plant, together with the essential auxiliary engines, mechanical and electrical installations, have been made and tested in conformity with the rules and regulations of this classification society, and the installation has been carried out under its supervision, and the complete plant was tested to its satisfaction on completion;

High-velocity vent valve means a pressure relief valve designed to have nominal flow velocities which exceed the flame velocity of the flammable mixture, thus preventing flame transmission. This type of installation shall be tested in accordance with standard EN ISO 16852:2010;

Hold (when anti-explosion protection is required, comparable to zone 1 - see *Classification of zones*) means a part of the vessel which, whether covered by hatchway covers or not, is bounded fore and aft by bulkheads and which is intended to carry goods in packages or in bulk. The upper boundary of the hold is the upper edge of the hatchway coaming. Cargo extending above the hatchway coaming shall be considered as loaded on deck;

Hold (discharged) means a hold which after unloading may contain some dry cargo remains;

Hold (empty) means a hold which after unloading contains no dry cargo remains (swept clean);

Hold space (when anti-explosion protection is required, comparable to zone 1) means an enclosed part of the vessel which is bounded fore and aft by watertight bulkheads and which is intended only to carry cargo tanks independent of the vessel's hull.

Holding time means the time that will elapse from the establishment of the initial filling condition until the pressure has risen due to heat influx to the lowest set pressure of the pressure limiting devices (s) of tanks intended for the carriage of refrigerated liquefied gases;

NOTE: *For portable tanks, see 6.7.4.1 of ADR.*

Hose assemblies means hoses, which are integrated or welded on both sides into hose fittings; hose fittings shall be integrated so that it is only possible to loosen them with a tool.

Hose fittings means couplings and connection elements of hoses.

Hoses means flexible tubular semi-finished products of elastomers, thermoplastics or stainless steel composed of one or several coatings and liners.

I

IAEA means the International Atomic Energy Agency (IAEA), (IAEA, P.O. Box 100 – A-1400 Vienna);

IBC see *Intermediate bulk container*;

IBC Code means the International Code for the Construction and Equipment of Ships carrying Dangerous Chemicals in Bulk, published by the International Maritime Organization (IMO);

ICAO means the International Civil Aviation Organization (ICAO, 999 University Street, Montreal, Quebec H3C 5H7, Canada);

ICAO Technical Instructions means the Technical Instructions for the Safe Transport of Dangerous Goods by Air, which complement Annex 18 to the Chicago Convention on International Civil Aviation (Chicago 1944) published by the International Civil Aviation Organization (ICAO) in Montreal;

Identification number means the number for identifying a substance to which no UN number has been assigned or which cannot be classified under a collective entry with a UN number.

These numbers have four figures beginning with 9;

IEC means the International Electrotechnical Commission;

IMDG Code means the International Maritime Dangerous Goods Code, for the implementation of Chapter VII, Part A, of the International Convention for the Safety of Life at Sea, 1974 (SOLAS Convention), published by the International Maritime Organization (IMO), London;

IMO means the International Maritime Organization (IMO, 4 Albert Embankment, London SE1 7SR, United Kingdom);

IMSBC Code means the International Maritime Solid Bulk Cargoes Code of the International Maritime Organization (IMO);

Inner packaging means a packaging for which an outer packaging is required for carriage;

Inner receptacle means a receptacle which requires an outer packaging in order to perform its containment function;

Inspection body means an independent monitoring and verification body certified by the competent authority;

Instruction means transmitting know-how or teaching how to do something or how to act. This transmission or teaching may be dispensed internally by the personnel;

Intermediate bulk container (IBC) means a rigid, or flexible portable packaging, other than those specified in Chapter 6.1 of ADR, that:

(a)   has a capacity of:

      (i)   not more than 3 m^3 for solids and liquids of packing groups II and III;

      (ii)   not more than 1.5 m^3 for solids of packing group I when packed in flexible, rigid plastics, composite, fibreboard and wooden IBCs;

      (iii)   not more than 3 m^3 for solids of packing group I when packed in metal IBCs;

      (iv)   not more than 3 m^3 for radioactive material of Class 7;

(b)   is designed for mechanical handling;

(c)   is resistant to the stresses produced in handling and transport as determined by the tests specified in Chapter 6.5 of ADR;

(see also *Composite IBC with plastics inner receptacle*, *Fibreboard IBC*, *Flexible IBC*, *Metal IBC*, *Rigid plastics IBC* and *Wooden IBC*)

NOTE 1: *Portable tanks or tank-containers that meet the requirements of Chapter 6.7 or 6.8 of ADR respectively are not considered to be intermediate bulk containers (IBCs).*

NOTE 2: *Intermediate bulk containers (IBCs) which meet the requirements of Chapter 6.5 of ADR are not considered to be containers for the purposes of ADN.*

Intermediate packaging means a packaging placed between inner packagings or articles and an outer packaging;

International regulations means ADR, ICAO-TI, IMDG Code, IMSBC Code or RID.

ISO (standard) means an international standard published by the International Organization for Standardization (ISO) (ISO, 1, rue de Varembé, CH-1204, Geneva 20);

J

Jerrican means a metal or plastics packaging of rectangular or polygonal cross-section with one or more orifices.

L

Large container, see *Container*;

Large packaging means a packaging consisting of an outer packaging which contains articles or inner packagings and which:

(a)   is designed for mechanical handling;

(b)   exceeds 400 kg net mass or 450 litres capacity but has a volume of not more than 3 m^3;

Remanufactured large packaging means a metal or rigid plastics large packaging that:

(a)   Is produced as a UN type from a non-UN type; or

(b)   Is converted from one UN design type to another UN design type.

Remanufactured large packagings are subject to the same requirements of ADR that apply to new large packagings of the same type (see also design type definition in 6.6.5.1.2 of ADR);

Reused large packaging means a large packaging to be refilled which has been examined and found free of defects affecting the ability to withstand the performance tests; the term includes those which are refilled with the same or similar compatible contents and are carried within distribution chains controlled by the consignor of the product;

Large salvage packaging means a special packaging which

(a)   is designed for mechanical handling; and

(b)   exceeds 400 kg net mass or 450 litres capacity but has a volume of not more than 3 m³;

into which damaged, defective, leaking or non-conforming dangerous goods packages, or dangerous goods that have spilled or leaked are placed for purposes of carriage for recovery or disposal;

Life boat (i.e. ship's boat) means an onboard boat in transport, rescue, salvage and work duties;

Light-gauge metal packaging means a packaging of circular, elliptical, rectangular or polygonal cross-section (also conical) and taper-necked and pail-shaped packaging made of metal, having a wall thickness of less than 0.5 mm (e.g. tinplate), flat or convex bottomed and with one or more orifices, which is not covered by the definitions for drums or jerricans;

Limited explosion risk electrical apparatus means an electrical apparatus which, during normal operation, does not cause sparks or exhibits surface temperatures which are above the required temperature class, including e.g.:

–   three-phase squirrel cage rotor motors;

–   brushless generators with contactless excitation;

–   fuses with an enclosed fuse element;

–   contactless electronic apparatus;

or means an electrical apparatus with an enclosure protected against water jets (degree of protection IP55) which during normal operation does not exhibit surface temperatures which are above the required temperature class;

Liner means a tube or bag inserted into a packaging, including large packagings or IBCs, but not forming an integral part of it, including the closures of its openings;

Liquefied natural gas (LNG) means a refrigerated liquefied gas composed of natural gas with a high methane content assigned to UN No. 1972;

Liquefied petroleum gas (LPG) means a low pressure liquefied gas composed of one or more light hydrocarbons which are assigned to UN 1011, UN 1075, UN 1965, UN 1969 or

UN 1978 only and which consists mainly of propane, propene, butane, butane isomers, butene with traces of other hydrocarbon gases.

NOTE 1: Flammable gases assigned to other UN numbers shall not be regarded as LPG.

NOTE 2: For UN No. 1075 see NOTE 2 under 2F, UN No. 1965, in the table for liquefied gases in 2.2.2.3.

Liquid means a substance which at 50° C has a vapour pressure of not more than 300 kPa (3 bar) which is not completely gaseous at 20° C and 101.3 kPa, and which:

(a)   has a melting point or initial melting point of 20° C or less at a pressure of 101.3 kPa, or

(b)   is liquid according to the ASTM D 4359-90 test method or

(c)   is not pasty according to the criteria applicable to the test for determining fluidity (penetrometer test) described in 2.3.4;

NOTE: "Carriage in the liquid state" for the purpose of tank requirements means:

–   Carriage of liquids according to the above definition, or

–   Solids handed over for carriage in the molten state;

Loader means any enterprise which:

(a)   Loads packaged dangerous goods, small containers or portable tanks into or onto a conveyance or a container; or

(b)   Loads a container, bulk-container, MEGC, tank-container or portable tank onto a conveyance; or

(c)   Loads a vehicle or a wagon into or onto a vessel;

Loading means all actions carried out by the loader, in accordance with the definition of loader;

Loading instrument: A loading instrument consists of a computer (hardware) and a programme (software) and offers the possibility of ensuring that in every ballast or loading case:

–   the permissible values concerning longitudinal strength as well as the maximum permissible draught are not exceeded; and

–   the stability of the vessel complies with the requirements applicable to the vessel. For this purpose intact stability and damage stability shall be calculated.

M

Management system, for the carriage of radioactive material, means a set of interrelated or interacting elements (system) for establishing policies and objectives and enabling the objectives to be achieved in an efficient and effective manner;

Manual of Tests and Criteria means the sixth revised edition of the Recommendations on the Transport of Dangerous Goods, Manual of Tests and Criteria, published by the United Nations (ST/SG/AC.10/11/Rev.6);

Mass density shall be expressed in kg/m^3. In the event of repetition, the number alone shall be used;

Mass of package means gross mass of the package unless otherwise stated. The mass of containers, tanks, vehicles and wagons used for the carriage of goods is not included in the gross mass;

Master means a person as defined in Article 1.02 of the European Code for Inland Waterways (CEVNI);

Maximum capacity means the maximum inner volume of receptacles or packagings including intermediate bulk containers (IBCs) and large packagings expressed in cubic metres or litres;

Maximum net mass means the maximum net mass of contents in a single packaging or maximum combined mass of inner packagings and the contents thereof expressed in kilograms;

Maximum normal operating pressure, for the carriage of radioactive material, means the maximum pressure above atmospheric pressure at mean sea-level that would develop in the containment system in a period of one year under the conditions of temperature and solar radiation corresponding to environmental conditions in the absence of venting, external cooling by an ancillary system, or operational controls during carriage;

Maximum permissible gross mass, means

(a)   (for IBCs) the mass of the IBC and any service or structural equipment together with the maximum net mass;

(b)   (for tanks) the tare of the tank and the heaviest load authorized for carriage;

NOTE: *For portable tanks, see Chapter 6.7 of ADR.*

Maximum working pressure means the maximum pressure occurring in a cargo tank or a residual cargo tank during operation. This pressure equals the opening pressure of high velocity vent valves or pressure relief valves;

Means of evacuation means any means that can be used by people to move from danger to safety as follows:

Dangers that have to be taken into account are:

–   For class 3, packing group III, UN 1202, second and third entry and for classes 4.1, 8 and 9 on tank vessels: leakage at the manifold;

–   For other substances of class 3 and class 2 and for flammable substances of class 8 on tank vessels: fire in the area of the manifold on the deck and burning liquid on the water;

–   For class 5.1 on tank vessels: oxidizing substances in combination with flammable liquids may cause an explosion;

–   For class 6.1 on tank vessels: toxic gases around the manifold and in the direction of the wind;

–   For dangerous goods on dry cargo vessels: dangers emanating from the goods in the cargo holds;

MEGC, see *Multiple-element gas container;*

MEMU, see *Mobile explosives manufacturing unit;*

Metal hydride storage system means a single complete hydrogen storage system, including a receptacle, metal hydride, pressure relief device, shut-off valve, service equipment and internal components used for the carriage of hydrogen only;

Metal IBC means a metal body together with appropriate service and structural equipment;

Mobile explosives manufacturing unit (MEMU) means a unit, or a vehicle mounted with a unit, for manufacturing and charging explosives from dangerous goods that are not explosives. The unit consists of various tanks and bulk containers and process equipment as well as pumps and related equipment. The MEMU may have special compartments for packaged explosives;

NOTE: *Even though the definition of MEMU includes the expression "manufacturing and charging explosives" the requirements for MEMUs apply only to carriage and not to manufacturing and charging of explosives.*

Multiple-element gas container (MEGC) means a unit containing elements which are linked to each other by a manifold and mounted on a frame. The following elements are considered to be elements of a multiple-element gas container: cylinders, tubes, pressure drums or bundles of cylinders as well as tanks for the carriage of gases as defined in 2.2.2.1.1 having a capacity of more than 450 litres.

NOTE: *For UN MEGCs, see Chapter 6.7 of ADR.*

N

Naked light means a source of light using a flame which is not enclosed in a flameproof enclosure.

Net explosive mass (NEM) means the total mass of the explosive substances, without the packagings, casings, etc. *(Net explosive quantity (NEQ), net explosive contents (NEC), net explosive weight (NEW)* or *net mass of explosive contents* are often used to convey the same meaning.);

Neutron radiation detector means a device that detects neutron radiation. In such a device, a gas may be contained in a hermetically sealed electron tube transducer that converts neutron radiation into a measureable electric signal;

N.O.S. entry (not otherwise specified entry) means a collective entry to which substances, mixtures, solutions or articles may be assigned if they:

(a)   are not mentioned by name in Table A of Chapter 3.2, and

(b)   exhibit chemical, physical and/or dangerous properties corresponding to the Class, classification code, packing group and the name and description of the n.o.s. entry;

Not readily flammable means a material which is not in itself readily flammable or whose outer surface at least is not readily flammable and limits the propagation of a fire to an appropriate degree.

In order to determine flammability, the IMO procedure, Resolution A.653(16), or any equivalent requirements of a Contracting State are recognized;

O

Offshore bulk container means a bulk container specially designed for repeated use for carriage to, from and between offshore facilities. An offshore bulk container is designed and constructed in accordance with the guidelines for the approval of offshore containers handled in open seas specified by the International Maritime Organization (IMO) in document MSC/Circ.860;

Oil separator vessel means an open type N tank-vessel with a dead weight of up to 300 tonnes, constructed and fitted to accept and carry oily and greasy wastes from the operation of vessels. Vessels without cargo tanks are considered to be subject to Chapters 9.1 or 9.2;

Oily and greasy wastes from the operation of the vessel means used oils, bilge water and other oily or greasy wastes, such as used grease, used filters, used rags, and receptacles and packagings for such wastes;

Open container, see *Container;*

Open cryogenic receptacle means a transportable thermally insulated receptacle for refrigerated liquefied gases maintained at atmospheric pressure by continuous venting of the refrigerated liquefied gas;

Open vehicle means a vehicle the platform of which has no superstructure or is merely provided with side boards and a tailboard;

Open wagon means a wagon with or without side boards and a tailboard, the loading surfaces of which are open.

Opening pressure means the pressure referred to in a list of substances in Chapter 3.2, Table C at which the high velocity vent valves open. For pressure tanks the opening pressure of the safety valve shall be established in accordance with the requirements of the competent authority or a recognized classification society;

OTIF means Intergovernmental Organisation for International Carriage by Rail (OTIF, Gryphenhübeliweg 30, CH-3006 Bern);

Outer packaging means the outer protection of the composite or combination packaging together with any absorbent materials, cushioning and any other components necessary to contain and protect inner receptacles or inner packagings;

Overpack means an enclosure used (by a single consignor in the case of radioactive material) to contain one or more packages, consolidated into a single unit easier to handle and stow during carriage;

Examples of overpacks:

(a)   a loading tray such as a pallet, on which several packages are placed or stacked and secured by a plastics strip, shrink or stretch wrapping or other appropriate means; or

(b)   an outer protective packaging such as a box or a crate;

Oxygen meter means a device allowing measuring of any significant reduction of the oxygen content of the air. Oxygen meters may either be a device for measuring oxygen only or part of a combination device for measuring both flammable gas and oxygen.

This device shall be so designed that measurements are possible without the necessity of entering the spaces to be checked.

P

Package means the complete product of the packing operation, consisting of the packaging or large packaging or IBC and its contents prepared for dispatch. Except for the carriage of radioactive material, the term includes receptacles for gases as defined in this section as well as articles which, because of their size, mass or configuration may be carried unpackaged or carried in cradles, crates or handling devices.

The term does not apply to goods which are carried in bulk in the holds of vessels, nor to substances carried in tanks in tank vessels.

On board vessels, the term also includes vehicles, wagons, containers (including swap bodies), tank-containers, portable tanks, battery-vehicles, battery-wagons, tank vehicles, tank wagons and multiple element gas containers (MECGs).

NOTE: For radioactive material, see 2.2.7.2., 4.1.9.1.1 and Chapter 6.4 of ADR.

Packaging means one or more receptacles and any other components or materials necessary for the receptacles to perform their containment and other safety functions (see also *Combination packaging, Composite packaging, Inner packaging, Intermediate bulk container (IBC), Intermediate packaging, Large packaging, Light-gauge metal packaging, Outer packaging, Reconditioned packaging, Remanufactured packaging, Reused packaging, Salvage packaging and Sift-proof packaging)*;

Packer means any enterprise which puts dangerous goods into packagings, including large packagings and intermediate bulk containers (IBCs) and, where necessary, prepares packages for carriage;

Packing group means a group to which, for packing purposes, certain substances may be assigned in accordance with their degree of danger. The packing groups have the following meanings which are explained more fully in Part 2:

Packing group I   :   Substances presenting high danger;

Packing group II :   Substances presenting medium danger; and

Packing group III :   Substances presenting low danger;

NOTE: Certain articles containing dangerous goods are assigned to a packing group.

Partly closed-type sampling device means a device penetrating through the boundary of the cargo tank or through the piping for loading and unloading such that during sampling only a small quantity of gaseous or liquid cargo can escape into the open air. As long as the device is not used it shall be closed completely. The device shall be of a type approved by the competent authority for this purpose;

Piping for loading and unloading (cargo piping) means all piping which may contain liquid or gaseous cargo, including pipes, hose assemblies, connected pumps, filters and closure devices.

Portable tank means a multimodal tank having, when used for the carriage of gases as defined in 2.2.2.1.1, a capacity of more than 450 litres in accordance with the definitions in Chapter 6.7 of ADR or the IMDG Code and indicated by a portable tank instruction (T-Code) in Column (10) of Table A of Chapter 3.2 of ADR;

Portable tank operator, see *Tank-container/portable tank operator*;

Possibility of cargo heating means a cargo heating installation in the cargo tanks using a heat insulator. The heat insulator may be heated by means of a boiler on board the tank vessel (cargo heating system in accordance with 9.3.2.42 or 9.3.3.42) or from shore;

Pressure drum means a welded, transportable pressure receptacle of a water capacity exceeding 150 litres and of not more than 1,000 litres (e.g. cylindrical receptacles equipped with rolling hoops, spheres on skids);

Pressure relief device means a spring-loaded device which is activated automatically by pressure the purpose of which is to protect the cargo tank against unacceptable excess internal pressure;

Pressure receptacle means a collective term that includes cylinders, tubes, pressure drums, closed cryogenic receptacles, metal hydride storage systems, bundles of cylinders and salvage pressure receptacles;

Pressures means for tanks, all kinds of pressures (e.g. working pressure, opening pressure of the high velocity vent valves, test pressure) shall be expressed as gauge pressures in kPa (bar); the vapour pressure of substances, however, shall be expressed as an absolute pressure in kPa (bar);

Pressurized gas cartridge, see *Aerosol dispenser*;

Protected area means

(a)   the hold or holds (when anti-explosion protection is required, comparable to zone 1);

(b)   the space situated above the deck (when anti-explosion protection is required, comparable to zone 2), bounded:

   (i)   athwartships, by vertical planes corresponding to the side plating;

   (ii)   fore and aft, by vertical planes corresponding to the end bulkheads of the hold; and

   (iii) upwards, by a horizontal plane 2.00 m above the upper level of the load, but at least by a horizontal plane 3.00 m above the deck.

Protected IBC (for metal IBCs) means an IBC provided with additional protection against impact, the protection taking the form of, for example, a multi-layer (sandwich) or double-wall construction, or a frame with a metal lattice-work casing.

Protective gloves means gloves which protect the wearer's hands during work in a danger area. The choice of appropriate gloves shall correspond to the dangers likely to arise. For protective gloves, see for example European standard EN 374-1:2003, EN 374-2:2003 or EN 374-3:2003 + AC:2006;

Protective goggles, protective masks means goggles or face protection which protects the wearer's eyes or face during work in a danger area. The choice of appropriate goggles or masks shall correspond to the dangers likely to arise. For protective goggles or masks, see for example European standard EN 166:2001;

Protective shoes (or protective boots) means shoes or boots which protect the wearer's feet during work in a danger area. The choice of appropriate protective shoes or boots shall correspond to the dangers likely to arise. For protective shoes or boots, see for example European standard EN ISO 20346:2014;

Protective suit means a suit which protects the wearer's body during work in a danger area. The choice of appropriate suit shall correspond to the dangers likely to arise. For protective suits, see for example European standard EN 340:2003;

Q

Quality assurance means a systematic programme of controls and inspections applied by any organization or body which is aimed at providing confidence that the safety prescriptions in ADN are met in practice.

R

Radiation detection system means an apparatus that contains radiation detectors as components;

Radiation level, for the carriage of radioactive material, means the corresponding dose rate expressed in millisieverts per hour or microsieverts per hour;

Radioactive contents, for the carriage of radioactive material, mean the radioactive material together with any contaminated or activated solids, liquids, and gases within the packaging;

Receptacle (Class 1) includes boxes, cylinders, cans, drums, jars and tubes, including any means of closure used in the inner or intermediate packaging;

Receptacle means a containment vessel for receiving and holding substances or articles, including any means of closing. This definition does not apply to shells (see also *Cryogenic receptacle, Inner receptacle, Rigid inner receptacle* and *Gas cartridge*);

Receptacle for residual products means a tank, intermediate bulk container or tank-container or portable tank intended to collect residual cargo, washing water, cargo residues or slops which are suitable for pumping;

Receptacle for slops means a steel drum intended to collect slops which are unsuitable for pumping;

Recycled plastics material means material recovered from used industrial packagings that has been cleaned and prepared for processing into new packagings;

Reel (Class 1) means a device made of plastics, wood, fibreboard, metal or other suitable material comprising a central spindle with, or without, side walls at each end of the spindle. Articles and substances can be wound on to the spindle and may be retained by side walls;

Relative density (or specific density) describes the ratio of the density of a substance to the density of pure water at 3.98 °C (1000 kg/m³) and is dimensionless;

Remanufactured large packaging see *Large packaging*;

Rescue winch means a device for hoisting persons from spaces such as cargo tanks, cofferdams and double-hull spaces. The device shall be operable by one person;

Residual cargo means liquid cargo remaining in the cargo tank or cargo piping after unloading without the use of the stripping system;

Reused large packaging see *Large packaging*;

RID means Regulations concerning the International Carriage of Dangerous Goods by Rail, Appendix C of COTIF (Convention concerning International Carriage by Rail);

Rigid inner receptacle (for composite IBCs) means a receptacle which retains its general shape when empty without its closures in place and without benefit of the outer casing. Any inner receptacle that is not rigid is considered to be flexible;

Rigid plastics IBC means a rigid plastics body, which may have structural equipment together with appropriate service equipment;

S

Safe area means a designated, recognisable area outside the cargo area which can be readily accessed by all persons on board. The safe area provides protection against the identified hazards of the cargo by a water spray system for at least 60 minutes. The safe area can be evacuated during an incident. A safe area is not acceptable when the identified danger is explosion;

Safe haven means a designated, recognisable, readily accessible module (fixed or floating) capable of protecting all persons on board against the identified hazards of the cargo for at least sixty minutes during which communication to the emergency and rescue services is possible. A safe haven can be integrated into the wheelhouse or into the accommodation. A safe haven can be evacuated during an incident. A safe haven on board is not acceptable when the identified danger is explosion. A safe haven on board and a floating safe haven outside the ship are certified by a recognized classification society. A safe haven on land is constructed according to local law;

Safety adviser means a person who, in an undertaking the activities of which include the carriage, or the related packing, loading, filling or unloading, of dangerous goods by inland waterways, is responsible for helping to prevent the risks inherent in the carriage of dangerous goods;

Safety valve means a spring-loaded device which is activated automatically by pressure the purpose of which is to protect the cargo tank against unacceptable excess internal pressure or negative internal pressure (see also, *High velocity vent valve, Pressure-relief device* and *Vacuum valve*);

SADT see *Self-accelerating decomposition temperature*;

Salvage packaging means a special packaging into which damaged, defective, leaking or non-conforming dangerous goods packages, or dangerous goods that have spilled or leaked are placed for purposes of carriage for recovery or disposal;

Salvage pressure receptacle means a pressure receptacle with a water capacity not exceeding 3 000 litres into which are placed damaged, defective, leaking or non-conforming pressure receptacle(s) for the purpose of carriage e.g. for recovery or disposal;

Sampling opening means an opening with a diameter of not more than 0.30 m. When the list of substances on the vessel according to 1.16.1.2.5 contains substances for which protection against explosion is required in column (17) of Table C of Chapter 3.2, it shall be fitted with a flame arrester plate stack, capable of withstanding steady burning and so designed that the opening period will be as short as possible and that the flame arrester plate stack cannot remain open without external intervention. The flame arrester plate stack shall be of a type approved by the competent authority for this purpose;

SAPT see Self-accelerating polymerization temperature;

Self-accelerating decomposition temperature (SADT) means the lowest temperature at which self-accelerating decomposition may occur with the substance in the packaging as used during carriage. Provisions for determining the SADT and the effects of heating under confinement are contained in Part II of the *Manual of Tests and Criteria;*

Self-accelerating polymerization temperature (SAPT) means the lowest temperature at which polymerization may occur with a substance in the packaging, IBC or tank as offered for carriage. The SAPT shall be determined in accordance with the test procedures established for the self-accelerating decomposition temperature for self-reactive substances in accordance with Part II, section 28 of the Manual of Tests and Criteria;

Service life, for composite cylinders and tubes, means the number of years the cylinder or tube is permitted to be in service;

Service space means a space which is accessible during the operation of the vessel and which is neither part of the accommodation nor of the cargo tanks, with the exception of the forepeak and after peak, provided no machinery has been installed in these latter spaces;

Settled pressure means the pressure of the contents of a pressure receptacle in thermal and diffusive equilibrium;

Sheeted bulk container, see *Bulk container;*

Sheeted container, see Container;

Sheeted vehicle means an open vehicle provided with a sheet to protect the load;

Sheeted wagon means an open wagon provided with a sheet to protect the load;

Sift-proof packaging means a packaging impermeable to dry contents, including fine solid material produced during carriage;

Slops means a mixture of cargo residues and washing water, rust or sludge which is either suitable or not suitable for pumping;

Small container, see *Container;*

Small receptacle containing gas (gas cartridge) means a non-refillable receptacle having a water capacity not exceeding 1000 ml for receptacles made of metal and not exceeding 500 ml for receptacles made of synthetic material or glass, containing, under pressure, a gas or a mixture of gases. It may be fitted with a valve;

SOLAS means the International Convention for the Safety of Life at Sea, 1974, as amended;

Solid means:

(a)   a substance with a melting point or initial melting point of more than 20 °C at a pressure of 101.3 kPa; or

(b)   a substance which is not liquid according to the ASTM D 4359-90 test method or which is pasty according to the criteria applicable to the test for determining fluidity (penetrometer test) described in 2.3.4;

STCW means the International Convention on Standards of Training, Certification and Watchkeeping for Seafarers, 1978, as amended.

Steady burning means combustion stabilized for an indeterminate period (see EN ISO 16852:2010);

Stripping system (efficient) means a system according to Annex II of CDNI for complete draining, if possible, of the cargo tanks and stripping the cargo piping except for the cargo residues;

Supply installation (bunkering system) means an installation for the supply of vessels with liquid fuels;

Supply vessel means an open type N tank vessel with a dead weight of up to 300 tonnes, constructed and fitted for the carriage and delivery to other vessels of products intended for the operation of vessels;

Swap-body, see *Container*.

T

Tank means a shell, including its service and structural equipment. When used alone, the term tank means a tank-container, portable tank, demountable tank, fixed tank or tank wagon as defined in this section, including tanks forming elements of battery-vehicles, battery wagons or MEGCs (see also *Demountable tank, Fixed tank, Portable tank and Multiple-element gas container)*;

NOTE: For portable tanks, see 6.7.4.1 of ADR.

Tank-container means an article of transport equipment meeting the definition of a container, and comprising a shell and items of equipment, including the equipment to facilitate movement of the tank-container without significant change of attitude, used for the carriage of gases, liquid, powdery or granular substances and, when used for the carriage of gases as defined in 2.2.2.1.1 having a capacity of more than 0.45 m^3 (450 litres);

NOTE: IBCs which meet the requirements of Chapter 6.5 of ADR are not considered to be tank-containers.

Tank-container/portable tank operator means any enterprise in whose name the tank-container/portable tank is registered;

Tank for residual products means a permanently built-in tank intended to collect residual cargo, washing water, cargo residues or slops which are suitable for pumping;

Tank record means a file containing all the important technical information concerning a tank, a battery-vehicle, a battery wagon or an MEGC, such as certificates referred to in 6.8.2.3, 6.8.2.4 and 6.8.3.4 of ADR;

Tank swap body is considered to be a tank-container;

Tank-vehicle means a vehicle built to carry liquids, gases or powdery or granular substances and comprising one or more fixed tanks. In addition to the vehicle proper, or the units of running gear used in its stead, a tank-vehicle comprises one or more shells, their items of equipment and the fittings for attaching them to the vehicle or to the running-gear units;

Tank vessel means a vessel intended for the carriage of substances in cargo tanks;

Tank wagon means a wagon intended for the carriage of liquids, gases, powdery or granular substances, comprising a superstructure, consisting of one or more tanks and their equipment and an underframe fitted with its own items of equipment (running gear, suspension, buffing, traction, braking gear and inscriptions).

NOTE: Tank wagon also includes wagons with demountable tanks.

Technical name means a recognized chemical name, or a recognized biological name where relevant, or another name currently used in scientific and technical handbooks, journals and texts (see 3.1.2.8.1.1);

Temperature class means a grouping of flammable gases and vapours of flammable liquids according to their ignition temperature; and of the electrical apparatus intended to be used in the corresponding potentially explosive atmosphere according to their maximum surface temperature (*see EN 13237:2011*);

Test pressure means the pressure at which a cargo tank, a residual cargo tank, a cofferdam or the loading and unloading piping shall be tested prior to being brought into service for the first time and subsequently regularly within prescribed times;

Through or into, for the carriage of radioactive material, means through or into the countries in which a consignment is carried but specifically excludes countries "over" which a consignment is carried by air provided that there are no scheduled stops in those countries;

Toximeter means a device allowing measuring of any significant concentration of toxic gases given off by the cargo.

This device shall be so designed that such measurements are possible without the necessity of entering the spaced to be checked.

Training means teaching instruction, courses or apprenticeships dispensed by an organizer approved by the competent authority;

Transport index (TI) assigned to a package, overpack or container, or to unpackaged LSA-I or SCO-I, for the carriage of radioactive material, means a number which is used to provide control over radiation exposure;

Transport unit means a motor vehicle without an attached trailer, or a combination consisting of a motor vehicle and an attached trailer;

Tray (Class 1) means a sheet of metal, plastics, fibreboard or other suitable material which is placed in the inner, intermediate or outer packaging and achieves a close-fit in such packaging. The surface of the tray may be shaped so that packagings or articles can be inserted, held secure and separated from each other;

Tube means a transportable pressure receptacle of seamless or composite construction having a water capacity exceeding 150 litres and of not more than 3,000 litres;

Types of protection: (see IEC 60079-0:2011)

EEx (d):	flameproof enclosure	(IEC 60079-1:2007);
EEx (e):	increased safety	(IEC 60079-7:2015);
EEx (ia) and EEx (ib):	intrinsic safety	(IEC 60079-11:2011);
EEx (m):	encapsulation	(IEC 60079-18:2009);
EEx (p):	pressurized apparatus	(IEC 60079-2:2007);
EEx (q):	powder filling	(IEC 60079-5:2007);

Type of vessel

Type G   :   means a tank vessel intended for the carriage of gases. Carriage may be under pressure or under refrigeration.

Type C   :   means a tank vessel intended for the carriage of liquids. The vessel shall be of the flush-deck/double-hull type with double-hull spaces, double bottoms, but without trunk. The cargo tanks may be formed by the vessel's inner hull or may be installed in the hold spaces as independent tanks.

Type N: means a tank vessel intended for the carriage of liquids.

Closed Type N: a tank vessel intended for the carriage of liquids in closed cargo tanks.

Open type N: a tank vessel intended for the carriage of liquids in open cargo tanks.

Open Type N: with flame arrester :a tank vessel intended for the carriage of liquids in open cargo tanks whose openings to the atmosphere are equipped with a flame arrester capable of withstanding steady burning.

Sketches (as example)

Type G :

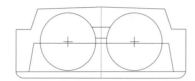

Type G   Cargo tank design 1,
         Type of cargo tank 1
         (also by flush-deck)

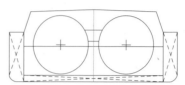

Type G   Cargo tank design 1,
         Type of cargo tank 1
         (also by flush-deck)

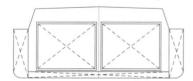

Type G   Cargo tank design 2,
         Type of cargo tank 1
         (also by flush-deck)

Type C :

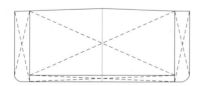

Type C   Cargo tank design 2,
         Type of cargo tank 2

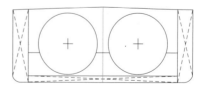

Type C   Cargo tank design 1,
         Type of cargo tank 1

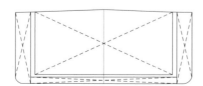

Type C   Cargo tank design 2
         Type of cargo tank 1

Type N :

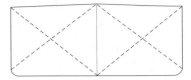

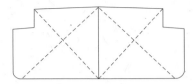

Type N Cargo tank design 2, 3 or 4
Type of cargo tank 2

Type N Cargo tank design 2, 3 or 4
Type of cargo tank 2

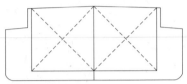

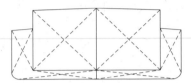

Type N Cargo tank design 2, 3 or 4
Type of cargo tanks 1
(also by flush-deck)

Type N Cargo tank design 2, 3 or 4
Type of cargo tank 3
(also by flush-deck)

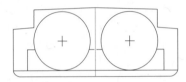

Type N Cargo tank design 2, 3 or 4
Type of cargo tank 1
(also by flush-deck)

U

UIC means the International Union of Railways (UIC, 16 rue Jean Rey, F-75015 Paris, France);

Undertaking, see *Enterprise*;

UNECE means the United Nations Economic Commission for Europe (UNECE, Palais des Nations, 8-14 avenue de la Paix, CH-1211 Geneva 10, Switzerland);

Unloader means any enterprise which:

(a)   Removes a container, bulk-container, MEGC, tank-container or portable tank from a conveyance; or

(b)   Unloads packaged dangerous goods, small containers or portable tanks out of or from a conveyance or a container; or

(c)   Discharges dangerous goods from a cargo tank, tank-vehicle, demountable tank, portable tank or tank-container; or from a battery-wagon, battery-vehicle, MEMU or MEGC; or from a conveyance for carriage in bulk, a large container or small container for carriage in bulk or a bulk container;

(d)   Removes a vehicle or a wagon from a vessel;

Unloading means all actions carried out by the unloader, in accordance with the definition of unloader;

UN Model Regulations means the Model Regulations annexed to the nineteenth revised edition of the Recommendations on the Transport of Dangerous Goods published by the United Nations (ST/SG/AC.10/1/Rev.19);

UN number means the four-figure identification number of the substance or article taken from the United Nations Model Regulations.

V

Vacuum design pressure means the vacuum pressure on the basis of which the cargo tank or the residual cargo tank has been designed and built;

Vacuum-operated waste tank means a fixed or demountable tank primarily used for the carriage of dangerous wastes, with special constructional features and/or equipment to facilitate the filling and discharging of wastes as specified in Chapter 6.10 of ADR. A tank which fully complies with the requirements of Chapter 6.7 or 6.8 of ADR is not considered to be a vacuum-operated waste tank;

Vacuum valve means a spring-loaded device which is activated automatically by pressure the purpose of which is to protect the cargo tank against unacceptable negative internal pressure;

Vapour return piping (on shore) means a pipe of the shore facility which is connected during loading or unloading to the vessel's venting piping. This pipe is designed so as to protect the vessel against detonations or the passage of flames from the shore side;

Vehicle means any vehicle covered by the definition of the term vehicle in the ADR (see *Battery-vehicle, Closed vehicle, Open vehicle, Sheeted vehicle* and *Tank-vehicle)*;

Venting piping (on board) means a pipe of the vessel's installation connecting one or more cargo tanks to the vapour return piping during loading or unloading. This pipe is fitted with safety valves protecting the cargo tank(s) against unacceptable internal overpressure or vacuums;

Vessel means an inland navigation vessel or a seagoing vessel.

Vessel record means a file containing all the important technical information concerning a vessel or a barge such as construction plans and documents about the equipment;

W

Wagon means a rail vehicle without its own means of propulsion that runs on its own wheels on railway tracks and is used for the carriage of goods (see also *battery-wagon, closed wagon, open wagon, sheeted wagon and tank wagon*);

Wastes means substances, solutions, mixtures or articles for which no direct use is envisaged but which are transported for reprocessing, dumping, elimination by incineration or other methods of disposal;

Water film means a deluge of water for protection against brittle fracture;

Water spray system means an on-board installation that, by means of a uniform distribution of water, is capable of protecting all the vertical external surfaces of the ship's hull fore and aft, all vertical surfaces of superstructures and deckhouses and deck surfaces above the superstructures, engine rooms and spaces in which combustible materials may be stored. The capacity of the water spray system for the area to be protected should be at least 10 l/m^2 per minute. The water spray system shall be designed for full-year use. The spray system should be operable from the wheelhouse and the safe area;

Watertight means a structural component or device so fitted as to prevent any ingress of water;

Weathertight means a structural component or device so fitted that in normal conditions it allows only a negligible quantity of water to penetrate;

Wooden barrel means a packaging made of natural wood, of round cross-section, having convex walls, consisting of staves and heads and fitted with hoops;

Wooden IBC means a rigid or collapsible wooden body, together with an inner liner (but no inner packaging) and appropriate service and structural equipment;

Working pressure means the settled pressure of a compressed gas at a reference temperature of 15° C in a full pressure receptacle.

NOTE: *For tanks, see Maximum working pressure.*

1.2.2   Units of measurement

1.2.2.1   The following units of measurement [a] are applicable in ADN:

Measurement of	SI Unit[b]	Acceptable alternative unit	Relationship between units
Length	m (metre)	-	-
Area	m^2 (square metre)	-	-
Volume	m^3 (cubic metre)	l [c] (litre)	$1\,l = 10^{-3}\,m^3$
Time	s (second)	min. (minute)	1 min. = 60 s
		h (hour)	1 h = 3 600 s
		d (day)	1 d = 86 400 s
Mass	kg (kilogram)	g (gramme)	$1\,g = 10^{-3}\,kg$
		t (ton)	$1\,t = 10^3\,kg$
Mass density	kg/m^3	kg/l	$1\,kg/l = 10^3\,kg/m^3$
Temperature	K (kelvin)	°C (degree Celsius)	0° C = 273.15 K
Temperature difference	K (kelvin)	°C (degree Celsius)	1° C = 1 K
Force	N (newton)	-	$1\,N = 1\,kg.m/s^2$
Pressure	Pa (pascal)		$1\,Pa = 1\,N/m^2$
		bar (bar)	$1\,bar = 10^5\,Pa$
Stress	N/m^2	N/mm^2	$1\,N/mm^2 = 1\,MPa$
Work		kWh (kilowatt hours)	1 kWh = 3.6 MJ
Energy	J (joule)		1 J = 1 N.m = 1 W.s
Quantity of heat		eV (electronvolt)	$1\,eV = 0.1602\,H\,10^{-18}\,J$
Power	W (watt)	-	1 W = 1 J/s = 1 N.m/s
Kinematic viscosity	m^2/s	mm^2/s	$1\,mm^2/s = 10^{-6}\,m^2/s$
Dynamic viscosity	Pa.s	mPa.s	$1\,mPa.s = 10^{-3}\,Pa.s$
Activity	Bq (becquerel)		
Dose equivalent	Sv (sievert)		

[a] *The following round figures are applicable for the conversion of the units hitherto used into SI Units.*

Force			*Stress*		
1 kg	=	*9.807 N*	*1 kg/mm²*	=	*9.807 N/mm²*
1 N	=	*0.102 kg*	*1 N/mm²*	=	*0.102 kg/mm²*

Pressure

1 Pa	=	*1 N/m²*	=	*10⁻⁵ bar*	=	*1.02 H 10⁻⁵ kg/cm²*	=	*0.75 H 10⁻² torr*
1 bar	=	*10⁵ Pa*	=	*1.02 kg/cm²*	=	*750 torr*		
1 kg/cm²	=	*9.807 H 10⁴ Pa*	=	*0.9807 bar*	=	*736 torr*		
1 torr	=	*1.33 H 10² Pa*	=	*1.33 H 10⁻³ bar*	=	*1.36 H 10⁻³ kg/cm²*		

Energy, Work, Quantity of heat

1 J	=	1 N.m	=	0.278 × 10^{-6} kWh	=	0.102 kgm	= 0.239 × 10^{-3} kcal
1 kWh	=	3.6 × 10^6 J	=	367 × 10^3 kgm	=	860 kcal	
1 kgm	=	9.807 J	=	2.72 × 10^{-6} kWh	=	2.34 × 10^{-3} kcal	
1 kcal	=	4.19 × 10^3 J	=	1.16 × 10^{-3} kWh	=	427 kgm	

Power

1 W	=	0.102 kgm/s	=	0.86 kcal/h
1 kgm/s	=	9.807 W	=	8.43 kcal/h
1 kcal/h	=	1.16 W	=	0.119 kgm/s

Kinematic viscosity

1 m^2/s	=	10^4 St (Stokes)
1 St	=	10^{-4} m^2/s

Dynamic viscosity

1 Pa.s	=	1 N.s/m^2	=	10 P (poise)	=	0.102 kg.s/m^2
1 P	=	0.1 Pa.s	=	0.1 N.s/m^2	=	1.02 × 10^{-2} kg.s/m^2
1 kg.s/m^2	=	9.807 Pa.s	=	9.807 N.s/m^2	=	98.07 P

[b] *The International System of Units (SI) is the result of decisions taken at the General Conference on Weights and Measures (Address: Pavillon de Breteuil, Parc de St-Cloud, F-92 310 Sèvres).*

[c] *The abbreviation "L" for litre may also be used in place of the abbreviation "l" when a typewriter cannot distinguish between figure "1" and letter "l".*

The decimal multiples and sub-multiples of a unit may be formed by prefixes or symbols, having the following meanings, placed before the name or symbol of the unit:

Factor		Prefix		Symbol
1 000 000 000 000 000 000	= 10^{18}	quintillion	exa	E
1 000 000 000 000 000	= 10^{15}	quadrillion	peta	P
1 000 000 000 000	= 10^{12}	trillion	tera	T
1 000 000 000	= 10^9	billion	giga	G
1 000 000	= 10^6	million	mega	M
1 000	= 10^3	thousand	kilo	k
100	= 10^2	hundred	hecto	h
10	= 10^1	ten	deca	da
0.1	= 10^{-1}	tenth	deci	d
0.01	= 10^{-2}	hundredth	centi	c
0.001	= 10^{-3}	thousandth	milli	m
0.000 001	= 10^{-6}	millionth	micro	μ
0.000 000 001	= 10^{-9}	billionth	nano	n
0.000 000 000 001	= 10^{-12}	trillionth	pico	p
0.000 000 000 000 001	= 10^{-15}	quadrillionth	femto	f
0.000 000 000 000 000 001	= 10^{-18}	quintillionth	atto	a

NOTE: 10^9 = 1 billion is United Nations usage in English. By analogy, so is 10^{-9} = 1 billionth.

1.2.2.2   Unless expressly stated otherwise, the sign "%" in ADN represents:

(a)   In the case of mixtures of solids or of liquids, and also in the case of solutions and of solids wetted by a liquid, a percentage mass based on the total mass of the mixture, the solution or the wetted solid;

(b)   In the case of mixtures of compressed gases, when filled by pressure, the proportion of the volume indicated as a percentage of the total volume of the gaseous mixture, or, when filled by mass, the proportion of the mass indicated as a percentage of the total mass of the mixture;

(c)   In the case of mixtures of liquefied gases and dissolved gases, the proportion of the mass indicated as a percentage of the total mass of the mixture.

1.2.2.3   Pressures of all kinds relating to receptacles (such as test pressure, internal pressure, safety valve opening pressure) are always indicated in gauge pressure (pressure in excess of atmospheric pressure); however, the vapour pressure of substances is always expressed in absolute pressure.

1.2.2.4   Where ADN specifies a degree of filling for receptacles, this is always related to a reference temperature of the substances of 15° C, unless some other temperature is indicated.

CHAPTER 1.3

TRAINING OF PERSONS INVOLVED IN THE CARRIAGE
OF DANGEROUS GOODS

1.3.1   **Scope and applicability**

Persons employed by the participants referred to in Chapter 1.4, whose duties concern the carriage of dangerous goods, shall be trained in the requirements governing the carriage of such goods appropriate to their responsibilities and duties. Employees shall be trained in accordance with 1.3.2 before assuming responsibilities and shall only perform functions, for which required training has not yet been provided, under the direct supervision of a trained person. Training requirements specific to security of dangerous goods in Chapter 1.10 shall also be addressed.

NOTE 1: With regard to the training for the safety adviser, see 1.8.3 instead of this section.

NOTE 2: With regard to expert training, see Chapter 8.2 instead of this section.

NOTE 3: For training with regard to Class 7, see also 1.7.2.5.

1.3.2   **Nature of the training**

The training shall take the following form, appropriate to the responsibility and duties of the individual concerned.

1.3.2.1   *General awareness training*

Personnel shall be familiar with the general requirements of the provisions for the carriage of dangerous goods.

1.3.2.2   *Function-specific training*

1.3.2.2.1   Personnel shall be trained, commensurate directly with their duties and responsibilities in the requirements of the regulations concerning the carriage of dangerous goods. Where the carriage of dangerous goods involves a multimodal transport operation, the personnel shall be aware of the requirements concerning other transport modes.

1.3.2.2.2   The crew shall be familiarized with the handling of fire-extinguishing systems and fire-extinguishers.

1.3.2.2.3   The crew shall be familiarized with the handling of the special equipment referred to in 8.1.5.

1.3.2.2.4   Persons wearing self-contained breathing apparatus shall be physically able to bear the additional constraints.

They shall:

–   in the case of devices operating with pressurized air, be trained in their handling and maintenance;

–   in the case of devices supplied with pressurized air through a hose, be instructed in their handling and maintenance. The instruction shall be supplemented by practical exercises.

1.3.2.2.5   The master shall bring the instructions in writing referred to in 5.4.3 to the attention of the other persons on board to ensure that they are capable of applying them.

1.3.2.3   *Safety training*

Commensurate with the degree of risk of injury or exposure arising from an incident involving the carriage of dangerous goods, including loading and unloading, personnel shall be trained in the hazards and dangers presented by dangerous goods.

The training provided shall aim to make personnel aware of the safe handling and emergency response procedures.

1.3.2.4   The training shall be periodically supplemented with refresher training to take account of changes in regulations.

1.3.3   **Documentation**

Records of training received according to this Chapter shall be kept by the employer and made available to the employee or competent authority, upon request. Records shall be kept by the employer for a period of time established by the competent authority. Records of training shall be verified upon commencing a new employment.

CHAPTER 1.4

SAFETY OBLIGATIONS OF THE PARTICIPANTS

1.4.1 **General safety measures**

1.4.1.1   The participants in the carriage of dangerous goods shall take appropriate measures according to the nature and the extent of foreseeable dangers, so as to avoid damage or injury and, if necessary, to minimize their effects. They shall, in all events, comply with the requirements of ADN in their respective fields.

1.4.1.2   When there is an immediate risk that public safety may be jeopardized, the participants shall immediately notify the emergency services and shall make available to them the information they require to take action.

1.4.1.3   ADN may specify certain of the obligations falling to the various participants.

If a Contracting Party considers that no lessening of safety is involved, it may in its domestic legislation transfer the obligations falling to a specific participant to one or several other participants, provided that the obligations of 1.4.2 and 1.4.3 are met. These derogations shall be communicated by the Contracting Party to the secretariat of the United Nations Economic Commission for Europe which will bring them to the attention of the Contracting Parties.

The requirements of 1.2.1, 1.4.2 and 1.4.3 concerning the definitions of participants and their respective obligations shall not affect the provisions of domestic law concerning the legal consequences (criminal nature, liability, etc.) stemming from the fact that the participant in question is e.g. a legal entity, a self-employed worker, an employer or an employee.

1.4.2 **Obligations of the main participants**

NOTE 1: Several participants to which safety obligations are assigned in this section may be one and the same enterprise. Also, the activities and the corresponding safety obligations of a participant can be assumed by several enterprises.

NOTE 2: For radioactive material see also 1.7.6.

1.4.2.1 *Consignor*

1.4.2.1.1   The consignor of dangerous goods is required to hand over for carriage only consignments which conform to the requirements of ADN. In the context of 1.4.1, he shall in particular:

(a)   ascertain that the dangerous goods are classified and authorized for carriage in accordance with ADN;

(b)   furnish the carrier with information and data in a traceable form and, if necessary, the required transport documents and accompanying documents (authorizations, approvals, notifications, certificates, etc.), taking into account in particular the requirements of Chapter 5.4 and of the tables in Part 3;

(c)   use only packagings, large packagings, intermediate bulk containers (IBCs) and tanks (tank-vehicles, demountable tanks, battery-vehicles, MEGCs, portable tanks, tank-containers, tank wagons and battery wagons) approved for and suited to the carriage of the substances concerned and bearing the marks prescribed by one of the international Regulations, and use only approved vessels or tank-vessels suitable for the carriage of the goods in question;

(d)   comply with the requirements on the means of dispatch and on forwarding restrictions;

(e) ensure that even empty uncleaned and non-degassed tanks (tank-vehicles, demountable tanks, battery-vehicles, MEGCs, portable tanks, tank-containers, tank wagons and tank vehicles) or empty uncleaned vehicles and bulk containers are placarded, marked and labelled in accordance with Chapter 5.3 and that empty uncleaned tanks are closed and present the same degree of leakproofness as if they were full.

1.4.2.1.2   If the consignor uses the services of other participants (packer, loader, filler, etc.), he shall take appropriate measures to ensure that the consignment meets the requirements of ADN. He may, however, in the case of 1.4.2.1.1 (a), (b), (c) and (e), rely on the information and data made available to him by other participants.

1.4.2.1.3   When the consignor acts on behalf of a third party, the latter shall inform the consignor in writing that dangerous goods are involved and make available to him all the information and documents he needs to perform his obligations.

1.4.2.2   *Carrier*

1.4.2.2.1   In the context of 1.4.1, where appropriate, the carrier shall in particular:

(a) ascertain that the dangerous goods to be carried are authorized for carriage in accordance with ADN;

(b) ascertain that all information prescribed in ADN related to the dangerous goods to be carried has been provided by the consignor before carriage, that the prescribed documentation is on board the vessel or if electronic data processing (EDP) or electronic data interchange (EDI) techniques are used instead of paper documentation, that data is available during transport in a manner at least equivalent to that of paper documentation;

(c) ascertain visually that the vessels and loads have no obvious defects, leakages or cracks, missing equipment, etc.;

(d) ascertain that a second means of evacuation in the event of an emergency from the vessel side is available, when the landside installation is not equipped with a second necessary means of evacuation;

NOTE: Before loading and unloading, the carrier shall consult the administration of the landside installation on the availability of means of evacuation.

(e) verify that the vessels are not overloaded;

(f) *(Reserved)*;

(g) provide the master with the required instructions in writing and ascertain that the prescribed equipment is on board the vessel;

(h) ascertain that the marking requirements for the vessel have been met;

(i) ascertain that during loading, carriage, unloading and any other handling of the dangerous goods in the holds or cargo tanks, special requirements are complied with;

(j) ascertain that the vessel substance list in accordance with 1.16.1.2.5 complies with Table C of chapter 3.2 including the modifications made to it.

Where appropriate, this shall be done on the basis of the transport documents and accompanying documents, by a visual inspection of the vessel or the containers and, where appropriate, the load.

1.4.2.2.2	The carrier may, however, in the case of 1.4.2.2.1 (a) and (b), rely on information and data made available to him by other participants.
1.4.2.2.3	If the carrier observes an infringement of the requirements of ADN, in accordance with 1.4.2.2.1, he shall not forward the consignment until the matter has been rectified.
1.4.2.2.4	*(Reserved)*
1.4.2.2.5	*(Reserved)*

1.4.2.3 *Consignee*

1.4.2.3.1 The consignee has the obligation not to defer acceptance of the goods without compelling reasons and to verify, before, during or after unloading, that the requirements of ADN concerning him have been complied with.

In the context of 1.4.1, he shall in particular:

(a) *(Deleted)*;

(b) carry out in the cases provided for by ADN the prescribed cleaning and decontamination of the vessels;

(c) *(Deleted)*;

(d) *(Deleted)*;

(e) *(Deleted)*;

(f) *(Deleted)*;

(g) *(Deleted)*;

(h) *(Deleted)*.

1.4.2.3.2 *(Deleted)*

1.4.2.3.3 *(Deleted)*

1.4.3 **Obligations of the other participants**

A non-exhaustive list of the other participants and their respective obligations is given below. The obligations of the other participants flow from section 1.4.1 above insofar as they know or should have known that their duties are performed as part of a transport operation subject to AND.

1.4.3.1 *Loader*

1.4.3.1.1 In the context of 1.4.1, the loader has the following obligations in particular:

(a) He shall hand the dangerous goods over to the carrier only if they are authorized for carriage in accordance with ADN;

(b) He shall, when handing over for carriage packed dangerous goods or uncleaned empty packagings, check whether the packaging is damaged. He shall not hand over a package the packaging of which is damaged, especially if it is not leakproof, and there are leakages or the possibility of leakages of the dangerous substance, until the damage has been repaired; this obligation also applies to empty uncleaned packagings;

(c)   He shall comply with the special requirements concerning loading and handling;

(d)   He shall, after loading dangerous goods into a container comply with the requirements concerning placarding, marking and orange-coloured plates conforming to Chapter 5.3;

(e)   He shall, when loading packages, comply with the prohibitions on mixed loading taking into account dangerous goods already in the vessel, vehicle, wagon or large container and requirements concerning the separation of foodstuffs, other articles of consumption or animal feedstuffs;

(f)   He shall ascertain that the landside installation is equipped with one or two means of evacuation from the vessel in the event of an emergency;

(g)   (*Reserved*).

1.4.3.1.2   The loader may, however, in the case of 1.4.3.1.1 (a), (d) and (e), rely on information and data made available to him by other participants.

1.4.3.2   *Packer*

In the context of 1.4.1, the packer shall comply with in particular:

(a)   the requirements concerning packing conditions, or mixed packing conditions; and

(b)   when he prepares packages for carriage, the requirements concerning marking and labelling of the packages.

1.4.3.3   *Filler*

In the context of 1.4.1, the filler has the following obligations in particular:

Obligations concerning the filling of tanks (tank-vehicles, battery-vehicles, demountable tanks, portable tanks, tank-containers, MEGCs, tank wagons and battery wagons):

(a)   He shall ascertain prior to the filling of tanks that both they and their equipment are technically in a satisfactory condition;

(b)   He shall ascertain that the date of the next test for tanks has not expired;

(c)   He shall only fill tanks with the dangerous goods authorized for carriage in those tanks;

(d)   He shall, in filling the tank, comply with the requirements concerning dangerous goods in adjoining compartments;

(e)   He shall, during the filling of the tank, observe the maximum permissible degree of filling or the maximum permissible mass of contents per litre of capacity for the substance being filled;

(f)   He shall, after filling the tank, ensure that all closures are in a closed position and that there is no leakage;

(g)   He shall ensure that no dangerous residue of the filling substance adheres to the outside of the tanks filled by him;

(h)   He shall, in preparing the dangerous goods for carriage, ensure that the placards, marks, orange-coloured plates and labels are affixed in accordance with Chapter 5.3.

Obligations concerning the bulk loading of dangerous solids in vehicles, wagons or containers:

(i)   He shall ascertain, prior to loading, that the vehicles, wagons and containers, and if necessary their equipment, are technically in a satisfactory condition and that the carriage in bulk of the dangerous goods in question is authorized in these vehicles, wagons or containers;

(j)   He shall ensure after loading that the orange plates and placards or labels prescribed are affixed in accordance with the requirements of Chapter 5.3 applicable to such vehicles, wagons or containers;

(k)   He shall, when filling vehicles, wagons or containers with dangerous goods in bulk, ascertain that the relevant provisions of Chapter 7.3 of RID or ADR are complied with.

Obligations concerning the filling of cargo tanks:

(l)   (*Reserved*);

(m)   He shall complete his section of the checklist referred to in 7.2.4.10 prior to the loading of the cargo tanks of a tank vessel;

(n)   He shall only fill cargo tanks with the dangerous goods accepted in such tanks;

(o)   He shall, when necessary, issue a heating instruction in the case of the carriage of substances whose melting point is 0 °C or higher;

(p)   He shall ascertain that during loading the trigger for the automatic device for the prevention of overfilling switches off the electric line established and supplied by the on-shore installation and that he can take steps against overfilling;

(q)   He shall ascertain that the landside installation is equipped with one or two means of evacuation from the vessel in the event of an emergency;

(r)   He shall ascertain that, when prescribed in 7.2.4.25.5, there is a flame-arrester in the vapour return piping to protect the vessel against detonations and flame-fronts from the landward side;

(s)   He shall ascertain that the loading flows conform to the loading and unloading instructions referred to in 9.3.2.25.9 or 9.3.3.25.9 and that the pressure at the crossing-point of the gas discharge pipe or the compensation pipe is not greater than the opening pressure of the high velocity vent valve;

(t)   He shall ascertain that the joints provided by him for the connecting flange of the ship/shore connections of the loading and unloading piping consist of a material which is not susceptible to be damaged by the cargo or causes a decomposition of the cargo nor forms harmful or dangerous components with it;

(u)   He shall ascertain that during the entire duration of loading a permanent and appropriate supervision is assured.

Obligations concerning the bulk loading of dangerous solids in vessels:

(v)   When special provision 803 applies, shall guarantee and document, using an appropriate procedure, that the maximum permissible temperature of the cargo is not exceeded and shall provide instructions to the master in a traceable form;

(w)   He shall only load the vessel with dangerous goods the bulk carriage of which is authorized in that vessel;

(x)   He shall ascertain that the landside installation is equipped with one or two means of evacuation from the vessel in the event of an emergency.

1.4.3.4 **_Tank-container/portable tank operator_**

In the context of 1.4.1, the tank-container/portable tank operator shall in particular:

(a)   ensure compliance with the requirements for construction, equipment, tests and marking;

(b)   ensure that the maintenance of shells and their equipment is carried out in such a way as to ensure that, under normal operating conditions, the tank-container/portable tank satisfies the requirements of ADR, RID or the IMDG Code until the next inspection;

(c)   have an exceptional check made when the safety of the shell or its equipment is liable to be impaired by a repair, an alteration or an accident.

1.4.3.5   *(Reserved)*

1.4.3.6   *(Reserved)*

1.4.3.7 **_Unloader_**

1.4.3.7.1   In the context of 1.4.1, the unloader shall in particular:

(a)   Ascertain that the correct goods are unloaded by comparing the relevant information on the transport document with the information on the package, container, tank, MEMU, MEGC or conveyance;

(b)   Before and during unloading, check whether the packagings, the tank, the conveyance or container have been damaged to an extent which would endanger the unloading operation. If this is the case, ascertain that unloading is not carried out until appropriate measures have been taken;

(c)   Comply with all relevant requirements concerning unloading and handling;

(d)   Immediately following the unloading of the tank, conveyance or container:

(i)   Ensure the removal of any dangerous residues which have adhered to the outside of the tank, conveyance or container during the process of unloading; and

(ii)   By unloading of packages, ensure the closure of valves and inspection openings;

(e)   Ensure that the prescribed cleaning and decontamination of the conveyances or containers is carried out;

(f)   Ensure that the containers, vehicles and wagons, once completely unloaded, cleaned and decontaminated, no longer display the placards, marks and orange-coloured plates that had been displayed in accordance with Chapter 5.3;

(g)   Ascertain that the landside installation is equipped with one or two means of evacuation from the vessel in the event of an emergency;

Additional obligations concerning the unloading of cargo tanks:

(h)   Complete his section of the checklist referred to in 7.2.4.10 prior to the unloading of the cargo tanks of a tank vessel;

(i)   Ascertain that, when prescribed in 7.2.4.25.5, there is a flame-arrester in the vapour return piping to protect the vessel against detonations and flame-fronts from the landward side;

(j)   Ascertain that the unloading flows conform to the instructions on loading and unloading flows referred to in 9.3.2.25.9 or 9.3.3.25.9 and that the pressure at the connecting-point of the gas discharge pipe or the gas return pipe does not exceed the opening pressure of the high velocity vent valve;

(k)   Ascertain that the gaskets provided by him for the connecting flange of the ship/shore connections of the loading and unloading piping consist of a material which will not be damaged by the cargo nor causes a decomposition of the cargo nor forms harmful or dangerous components with it;

(l)   Ascertain that during the entire duration of unloading a permanent and appropriate supervision is assured;

(m)   Ascertain that, during unloading by means of the on-board pump, it is possible for the shore facility to switch it off;

1.4.3.7.2   If the unloader makes use of the services of other participants (cleaner, decontamination facility, etc.) he shall take appropriate measures to ensure that the requirements of ADN have been complied with.

CHAPTER 1.5

SPECIAL RULES, DEROGATIONS

1.5.1   **Bilateral and multilateral agreements**

1.5.1.1   In accordance with Article 7, paragraph 1 of ADN, the competent authorities of the Contracting Parties may agree directly among themselves to authorize certain transport operations in their territories by temporary derogation from the requirements of ADN, provided that safety is not compromised thereby. The authority which has taken the initiative with respect to the temporary derogation shall notify such derogations to the Secretariat of the United Nations Economic Commission for Europe which shall bring them to the attention of the Contracting Parties.

     NOTE: *"Special arrangement" in accordance with 1.7.4 is not considered to be a temporary derogation in accordance with this section.*

1.5.1.2   The period of validity of the temporary derogation shall not be more than five years from the date of its entry into force. The temporary derogation shall automatically cease as from the date of the entry into force of a relevant amendment to these annexed Regulations.

1.5.1.3   Transport operations on the basis of these agreements shall constitute transport operations in the sense of ADN.

1.5.2   **Special authorizations concerning transport in tank vessels**

1.5.2.1   *Special authorizations*

1.5.2.1.1   In accordance with paragraph 2 of Article 7 of ADN, the competent authority shall have the right to issue special authorizations to a carrier or a consignor for the international carriage in tank vessels of dangerous substances, including mixtures, the carriage of which in tank vessels is not authorized under these Regulations, in accordance with the procedure set out below.

1.5.2.1.2   The special authorization shall be valid, due account being taken of the restrictions specified therein, for the Contracting Parties and on whose territory the transport operation will take place, for not more than two years unless it is repealed at an earlier date. With the approval of the competent authorities of these Contracting Parties, the special authorization may be renewed for a period of not more than one year.

1.5.2.1.3   The special authorization shall include a statement concerning its repeal at an earlier date and shall conform to the model contained in subsection 3.2.4.1.

1.5.2.2   *Procedure*

1.5.2.2.1   The carrier or the consignor shall apply to the competent authority of a Contracting Party on whose territory the transport operation takes place for the issue of a special authorization.

     The application shall conform to the model contained in subsection 3.2.4.2. The applicant shall be responsible for the accuracy of the particulars.

1.5.2.2.2	The competent authority shall consider the application from the technical and safety point of view. If it has no reservations, it shall draw up a special authorization in accordance with the criteria contained in subsection 3.2.4.3 and immediately inform the other competent authorities involved in the carriage in question. The special authorization shall be issued only when the authorities concerned agree to it or have not expressed opposition within a period of two months after receiving the information. The applicant shall receive the original of the special authorization and keep a copy of it on board the vessel(s) involved in the carriage in question. The competent authorities shall immediately communicate to the Administrative Committee the applications for special authorizations, the applications rejected and the special authorizations granted.
1.5.2.2.3	If the special authorization is not issued because doubts or opposition have been expressed, the Administrative Committee shall decide whether or not to issue a special authorization.

1.5.2.3   *Update of the list of substances authorized for carriage in tank vessels*

1.5.2.3.1	The Administrative Committee shall consider all the special authorizations and applications communicated to it and decide whether the substance is to be included in the list of substances in these Regulations, authorized for carriage in tank vessels.
1.5.2.3.2	If the Administrative Committee enters technical or safety reservations concerning the inclusion of the substance in the list of substances of these Regulations authorized for carriage in tank vessels or concerning certain conditions, the competent authority shall be so informed. The competent authority shall immediately withdraw or, if necessary, modify the special authorization.

1.5.3   **Equivalents and derogations (Article 7, paragraph 3 of ADN)**

1.5.3.1   *Procedure for equivalents*

When the provisions of these Regulations prescribe for a vessel the use or the presence on board of certain materials, installations or equipment or the adoption of certain construction measures or certain fixtures, the competent authority may agree to the use or the presence on board of other materials, installations or equipment or the adoption of other construction measures or other fixtures for this vessel if, in line with recommendations established by the Administrative Committee, they are accepted as equivalent.

1.5.3.2   *Derogations on a trial basis*

The competent authority may, on the basis of a recommendation by the Administrative Committee, issue a trial certificate of approval for a limited period for a specific vessel having new technical characteristics departing from the requirements of these Regulations, provided that these characteristics are sufficiently safe.

1.5.3.3   *Particulars of equivalents and derogations*

The equivalents and derogations referred to in 1.5.3.1 and 1.5.3.2 shall be entered in the certificate of approval.

CHAPTER 1.6

TRANSITIONAL MEASURES

1.6.1   **General**

1.6.1.1   Unless otherwise provided, the substances and articles of ADN may be carried until 30 June 2017 in accordance with the requirements of ADN applicable up to 31 December 2016.

1.6.1.2   (*Deleted*)

1.6.1.3   The transitional measures of 1.6.1.3 and 1.6.1.4 of ADR and RID, or falling within the scope of 4.1.5.19 of the IMDG Code, concerning the packaging of substances and articles of Class 1, are also valid for carriage subject to ADN.

1.6.1.4   (*Deleted*)

1.6.1.5 to 1.6.1.7   (*Reserved*)

1.6.1.8   Existing orange-coloured plates which meet the requirements of sub-section 5.3.2.2 applicable up to 31 December 2004 may continue to be used provided that the requirements in 5.3.2.2.1 and 5.3.2.2.2 that the plate, numbers and letters shall remain affixed irrespective of the orientation of the vehicle or wagon are met.

1.6.1.9   (*Reserved*)

1.6.1.10   (*Deleted*)

1.6.1.11 and 1.6.1.12   (*Reserved*)

1.6.1.13   (*Deleted*)

1.6.1.14   IBCs manufactured before 1 January 2011 and conforming to a design type which has not passed the vibration test of 6.5.6.13 of ADR or which was not required to meet the criteria of 6.5.6.9.5 (d) of ADR at the time it was subjected to the drop test, may still be used.

1.6.1.15   IBCs manufactured, remanufactured or repaired before 1 January 2011 need not be marked with the maximum permitted stacking load in accordance with 6.5.2.2.2 of ADR. Such IBCs, not marked in accordance with 6.5.2.2.2 of ADR, may still be used after 31 December 2010 but must be marked in accordance with 6.5.2.2.2 of ADR if they are remanufactured or repaired after that date. IBCs manufactured, remanufactured or repaired between 1 January 2011 and 31 December 2016 and marked with the maximum permitted stacking load in accordance with 6.5.2.2.2 of ADR in force up to 31 December 2014 may continue to be used.

1.6.1.16 to 1.6.1.20   (*Deleted*)

1.6.1.21 to 1.6.1.23   (*Reserved*)

1.6.1.24   (*Deleted*)

1.6.1.25   Cylinders of 60 litres water capacity or less marked with a UN number in accordance with the provisions of ADN applicable up to 31 December 2012 and which do not conform to the requirements of 5.2.1.1 regarding the size of the UN number and of the letters "UN" applicable as from 1 January 2013 may continue to be used until the next periodic inspection but no later than 30 June 2018.

1.6.1.26   Large packagings manufactured or remanufactured before 1 January 2014 and which do not conform to the requirements of 6.6.3.1 of ADR regarding the height of letters, numerals and symbols applicable as from 1 January 2013 may continue to be used. Those manufactured or remanufactured before 1 January 2015 need not be marked with the maximum permitted stacking load in accordance with 6.6.3.3 of ADR. Such large packagings not marked in accordance with 6.6.3.3 of ADR may still be used after 31 December 2014 but must be marked in accordance with 6.6.3.3 of ADR if they are remanufactured after that date. Large packagings manufactured or remanufactured between 1 January 2011 and 31 December 2016 and marked with the maximum permitted stacking load in accordance with 6.6.3.3 of ADR in force up to 31 December 2014 may continue to be used.

1.6.1.27   Means of containment integral to equipment or machinery containing liquid fuels of UN Nos. 1202, 1203, 1223, 1268, 1863 and 3475 constructed before 1 July 2013, which do not conform to the requirements of paragraph (a) of special provision 363 of chapter 3.3 applicable as from 1 January 2013, may still be used.

1.6.1.28   *(Deleted)*

1.6.1.29   Lithium cells and batteries manufactured according to a type meeting the requirements of sub-section 38.3 of the Manual of Tests and Criteria, Revision 3, Amendment 1 or any subsequent revision and amendment applicable at the date of the type testing may continue to be carried, unless otherwise provided in ADN.

Lithium cells and batteries manufactured before 1 July 2003 meeting the requirements of the Manual of Tests and Criteria, Revision 3, may continue to be carried if all other applicable requirements are fulfilled.

1.6.1.30   Labels which meet the requirements of 5.2.2.2.1.1 applicable up to 31 December 2014, may continue to be used until 30 June 2019.

1.6.1.31 and 1.6.1.32   *(Deleted)*

1.6.1.33   Electric double layer capacitors of UN No. 3499, manufactured before 1 January 2014, need not be marked with the energy storage capacity in Wh as required by sub-paragraph (e) of special provision 361 of Chapter 3.3.

1.6.1.34   Asymmetric capacitors of UN No. 3508, manufactured before 1 January 2016, need not be marked with the energy storage capacity in Wh as required by sub-paragraph (c) of special provision 372 of Chapter 3.3.

1.6.1.35 to 1.6.1.37   *(Reserved)*

1.6.1.38   Contracting Parties may continue to issue training certificates for dangerous goods safety advisers conforming to the model applicable until 31 December 2016, instead of those conforming to the requirements of 1.8.3.18 applicable from 1 January 2017, until 31 December 2018. Such certificates may continue in use to the end of their five year validity.

1.6.1.39   Notwithstanding the requirements of special provision 188 of Chapter 3.3 applicable as from 1 January 2017, packages containing lithium cells or batteries may continue to be marked until 31 December 2018 in accordance with the requirements of special provision 188 of Chapter 3.3 in force up to 31 December 2016.

1.6.1.40   Notwithstanding the requirements of ADN applicable as from 1 January 2017, articles of UN Nos. 0015, 0016 and 0303 containing smoke-producing substance(s) toxic by inhalation according to the criteria for Class 6.1 manufactured before 31 December 2016 may be carried until 31 December 2018 without a "TOXIC" subsidiary risk label (model No. 6.1, see 5.2.2.2.2).

1.6.1.41	Notwithstanding the requirements of ADN applicable as from 1 January 2017, large packagings conforming to the packing group III performance level in accordance with special packing provision L2 of packing instruction LP02 of 4.1.4.3 of ADR applicable until 31 December 2016 may continue to be used until 31 December 2022 for UN No. 1950.
1.6.1.42	Notwithstanding the requirements of column (5) of Table A of Chapter 3.2 applicable as from 1 January 2017 to UN Nos. 3090, 3091, 3480 and 3481, the Class 9 label (model No 9, see 5.2.2.2.2) may continue to be used for these UN numbers until 31 December 2018.
1.6.1.43	Vehicles registered or brought into service before 1 July 2017, as defined in special provisions 240, 385 and 669 of Chapter 3.3, and their equipment intended for use during carriage, which conform to the requirements of ADN applicable until 31 December 2016 but containing lithium cells and batteries which do not conform to the requirement of 2.2.9.1.7 may continue to be carried as a load in accordance with the requirements of special provision 666 of Chapter 3.3.

1.6.2 Pressure receptacles and receptacles for Class 2

The transitional measures of sections 1.6.2 of ADR and RID are also valid for transport operations subject to ADN.

1.6.3 Fixed tanks (tank-vehicles and tank wagons), demountable tanks, battery vehicles and battery wagons

The transitional measures of sections 1.6.3 of ADR and RID are also valid for transport operations subject to ADN.

1.6.4 Tank-containers, portable tanks and MEGCs

The transitional measures of sections 1.6.4 of ADR and RID or of section 4.2.0 of the IMDG Code, depending on the case, are also valid for transport operations subject to ADN.

1.6.5 Vehicles

The transitional measures of section 1.6.5 of ADR are also valid for transport operations subject to ADN.

1.6.6 Class 7

The transitional measures of sections 1.6.6 of ADR and RID or of section 6.4.24 of the IMDG Code are also valid for transport operations subject to ADN.

1.6.7 Transitional provisions concerning vessels

1.6.7.1 *General*

1.6.7.1.1 For the purposes of Article 8 of ADN, section 1.6.7 sets out general transitional provisions in 1.6.7.2 (see Article 8, paragraphs 1, 2 and 4) and supplementary transitional provisions in 1.6.7.3 (see Article 8, paragraph 3).

1.6.7.1.2 In this section:

(a) "Vessel in service" means

– A vessel according to Article 8, paragraph 2, of ADN;

– A vessel for which a certificate of approval has already been issued according to 8.6.1.1 to 8.6.1.4;

In both cases vessels that, as from 31 December 2014, have been without a valid certificate of approval for more than twelve months shall be excluded;

(b) "N.R.M." means that the requirement does not apply to vessels in service except where the parts concerned are replaced or modified, i.e. it applies only to vessels which are **n**ew (as from the date indicated), or to parts which are **r**eplaced or **m**odified after the date indicated; the date of presentation for first inspection for obtaining a certificate of approval shall be decisive for nomination as a new vessel; where existing parts are replaced by spare or replacement parts of the same type and manufacture, this shall not be considered a replacement 'R' as defined in these transitional provisions.

Modification shall also be taken to mean the conversion of an existing type of tank vessel, a type of cargo tank or a cargo tank design to another type or design at a higher level.

When in the general transitional provisions in 1.6.7.2 no date is specified after 'N.R.M.', it refers to N.R.M. after 26 May 2000. When in the supplementary transitional provisions in 1.6.7.3, no date is specified, it refers to N.R.M. after 26 May 2000.

(c) "Renewal of the certificate of approval after the ..." means that when a vessel has benefitted from the transitional measure in paragraph (b) the requirement shall be met at the next renewal of the certificate of approval following the date indicated. If the certificate of approval expires during the first year after the date of application of these Regulations, the requirement shall be mandatory only after the expiry of this first year.

(d) Requirements of chapter 1.6.7 applicable on board vessels in service are only valid if N.R.M. is not applicable.

1.6.7.2   ***General transitional provisions***

1.6.7.2.1   *General transitional provisions for dry cargo vessels*

1.6.7.2.1.1   Vessels in service shall meet:

(a) the requirements of paragraphs mentioned in the table below within the period established therein;

(b) the requirements of paragraphs not mentioned in the table below at the date of application of these Regulations.

The construction and equipment of vessels in service shall be maintained at least at the previous standard of safety.

1.6.7.2.1.1 Table of general transitional provisions: Dry cargo		
Paragraphs	Subject	Time limit and comments
1.16.1.4 and 1.16.2.5	Annex to certificate of approval and provisional certificate of approval	Renewal of the certificate of approval after 31 December 2014
9.1.0.12.1	Ventilation of holds	N.R.M. Renewal of the certificate of approval after 31 December 2018 Until then, the following requirements apply on board vessels in service: Each hold shall have appropriate natural or artificial ventilation; for the carriage of substances of Class 4.3, each hold shall be equipped with forced-air ventilation; the appliances used for this purpose must be so constructed that water cannot enter the hold.
9.1.0.12.3	Ventilation of service spaces	N.R.M. Renewal of the certificate of approval after 31 December 2018
9.1.0.17.2	Gas-tight openings facing holds	N.R.M. Renewal of the certificate of approval after 31 December 2018 Until then, the following requirements apply on board vessels in service: Openings of accommodation and the wheelhouse facing the holds must be capable of being tightly closed.
9.1.0.17.3	Entrances and openings in the protected area	N.R.M. Renewal of the certificate of approval after 31 December 2018 Until then, the following requirements apply on board vessels in service: Openings of engine rooms and service spaces facing the holds must be capable of being tightly closed.
9.1.0.31.2	Air intakes of engines	N.R.M. Renewal of the certificate of approval after 31 December 2034
9.1.0.32.2	Air pipes 50 cm above the deck	N.R.M. Renewal of the certificate of approval after 31 December 2018
9.1.0.34.1	Position of exhaust pipes	N.R.M. Renewal of the certificate of approval after 31 December 2018

1.6.7.2.1.1 Table of general transitional provisions: Dry cargo		
Paragraphs	Subject	Time limit and comments
9.1.0.35	Stripping pumps in the protected area	N.R.M. Renewal of the certificate of approval after 31 December 2018 Until then, the following requirements apply on board vessels in service: In the event of the carriage of substances of Class 4.1, UN No. 3175, of all substances of Class 4.3 in bulk or unpackaged and polymeric beads, expandable, of Class 9, UN No. 2211, the stripping of the holds may only be effected using a stripping installation located in the protected area. The stripping installation located above the engine room must be clamped.
9.1.0.40.1	Fire extinguishers, two pumps, etc.	N.R.M. Renewal of the certificate of approval after 31 December 2018
9.1.0.40.2	Fire extinguishing systems permanently fixed in engine rooms	N.R.M. Renewal of the certificate of approval after 31 December 2034
9.1.0.41 in conjunction with 7.1.3.41	Fire and naked light	N.R.M. Renewal of the certificate of approval after 31 December 2018 Until then, the following requirements apply on board vessels in service: Outlets of funnels shall be located not less than 2 m from the nearest point on hold hatchways. Heating and cooking appliances shall be permitted only in metal-based accommodation and wheelhouses. However: - Heating appliances fuelled with liquid fuels having a flashpoint above 55 °C shall be permitted in engine rooms; - Central-heating boilers fuelled with solid fuels shall be permitted in spaces situated below deck and accessible only from the deck.
9.2.0.31.2	Air intakes of engines	N.R.M. Renewal of the certificate of approval after 31 December 2034
9.2.0.34.1	Position of exhaust pipes	N.R.M. Renewal of the certificate of approval after 31 December 2018

1.6.7.2.1.1 Table of general transitional provisions: Dry cargo		
Paragraphs	Subject	Time limit and comments
9.2.0.41 in conjunction with 7.1.3.41	Fire and naked light	N.R.M. Renewal of the certificate of approval after 31 December 2018 Until then, the following requirements apply on board vessels in service: Outlets of funnels shall be located not less than 2 m from the nearest point on hold hatchways. Heating and cooking appliances shall be permitted only in metal-based accommodation and wheelhouses. However: - Heating appliances fuelled with liquid fuels having a flashpoint above 55 °C shall be permitted in engine rooms; - Central-heating boilers fuelled with solid fuels shall be permitted in spaces situated below deck and accessible only from the deck.

1.6.7.2.1.2   (*Deleted*)

1.6.7.2.1.3   By way of derogation from 7.1.4.1, transport in bulk of UN Nos. 1690, 1812 and 2505, may be carried out with single hull vessels until 31.12.2018.

1.6.7.2.1.4   For a vessel or a barge whose keel was laid before 1 January July 2017 and which does not conform to the requirements of 9.0.X.1 concerning the vessel record, the retention of files for the vessel record shall start at the latest at the next renewal of the certificate of approval.

1.6.7.2.2   *General transitional provisions for tank vessels*

1.6.7.2.2.1   Vessels in service shall meet:

(a)   the requirements of paragraphs mentioned in the table below within the period established therein;

(b)   the requirements of paragraphs not mentioned in the table below at the date of application of these Regulations.

The construction and equipment of vessels in service shall be maintained at least at the previous standard of safety.

1.6.7.2.2.2   Table of general transitional provisions for tank vessels

1.6.7.2.2.2 Table of general transitional provisions: Tank vessels		
Paragraphs	Subject	Time limit and comments
1.2.1	Limited explosion risk electrical apparatus	N.R.M. Renewal of the certificate of approval after 31 December 2034 Until then, the following requirements apply on board vessels in service: Limited explosion risk electrical apparatus is: - Electrical apparatus which, during normal operation, does not cause sparks or exhibit surface temperatures exceeding 200 °C; or - Electrical apparatus with a spray-water protected housing which, during normal operation, does not exhibit surface temperatures above 200 °C.
1.2.1	Hold spaces	N.R.M. For Type N open vessels whose hold spaces contain auxiliary appliances and which are carrying only substances of Class 8, with remark 30 in column (20) of Table C of Chapter 3.2. Renewal of the certificate of approval after 31 December 2038.
1.2.1	Flame arrester Test according to standard EN ISO 16852:2010	N.R.M. from 1 January 2001 Renewal of the certificate of approval after 31 December 2034 Until then, the following requirements are applicable on board vessels in service: Flame arresters shall conform to the standard EN 12874:1999 on board vessels built or modified from 1 January 2001 or if they have been replaced from 1 January 2001. In other cases, they shall be of a type approved by the competent authority for the use prescribed.
1.2.1	High velocity vent valve Test according to standard EN ISO 16852:2010	N.R.M. from 1 January 2015 Renewal of the certificate of approval after 31 December 2034 Until then, the following requirements are applicable on board vessels in service: High velocity vent valves shall conform to the standard EN 12874:1999 on board vessels built or modified from 1 January 2001 or if they have been replaced from 1 January 2001. In other cases, they shall be of a type approved by the competent authority for the use prescribed.
1.16.1.4 and 1.16.2.5	Annex to certificate of approval and provisional certificate of approval	Renewal of the certificate of approval after 31 December 2014
7.2.2.6	Approved gas detection system	N.R.M. Renewal of the certificate of approval after 31 December 2010

1.6.7.2.2.2 Table of general transitional provisions: Tank vessels		
Paragraphs	Subject	Time limit and comments
7.2.2.19.3	Vessels used for propulsion	N.R.M. Renewal of the certificate of approval after 31 December 2044
7.2.3.20.1	Ballast water Prohibition against filling cofferdams with water	N.R.M. Renewal of the certificate of approval after 31 December 2038 Until then, the following requirements apply on board vessels in service: Cofferdams may be filled with water during unloading to provide trim and to permit residue-free drainage as far as possible. When the vessel is underway, cofferdams may be filled with ballast water only when cargo tanks are empty.
7.2.3.20.1	Proof of stability in the event of a leak connected with ballast water	N.R.M. for Type G and Type N vessels. Renewal of the certificate of approval after 31 December 2044.
7.2.3.20.1	Fitting of ballast tanks and compartments with level indicators	N.R.M. after 1 January 2013 for Type C and Type G tank vessels and Type N double hull tank vessels. Renewal of the certificate of approval after 31 December 2012.
7.2.3.31.2	Motor vehicles only outside the cargo area	N.R.M. for Type N vessels. Renewal of the certificate of approval after 31 December 2034 Until then, the following requirement applies on board vessels in service: the vehicle shall not be started on board.
7.2.4.22.3	Sampling from other openings	N.R.M. for Type N open vessels. Renewal of the certificate of approval after 31 December 2018 Until then, on board vessels in service, cargo tank covers may be opened during loading for control and sampling.
8.1.6.2.	Hose assemblies	Hose assemblies of previous standards EN 12115:1999, EN 13765:2003 or EN ISO 10380:2003 may be used until 31 December 2018.
9.3.2.0.1 (c) 9.3.3.0.1 (c)	Protection of venting piping against corrosion	N.R.M. from 1 January 2001 Renewal of the certificate of approval after 31 December 2034
9.3.1.0.3 (d) 9.3.2.0.3 (d) 9.3.3.0.3 (d)	Fire-resistant materials of accommodation and wheelhouse	N.R.M. Renewal of the certificate of approval after 31 December 2034

1.6.7.2.2.2 Table of general transitional provisions: Tank vessels		
Paragraphs	Subject	Time limit and comments
9.3.3.8.1	Continuation of class	N.R.M. for Type N open vessels with flame arresters and Type N open vessels. Renewal of the certificate of approval after 31 December 2044. Until then, the following requirements apply on board vessels in service: Except where otherwise provided, the type of construction, the strength, the subdivision, the equipment and the gear of the vessel shall conform or be equivalent to the construction requirements for classification in the highest class of a recognized classification society.
9.3.1.10.2 9.3.2.10.2 9.3.3.10.2	Door coamings, etc.	N.R.M. Renewal of the certificate of approval after 31 December 2034 Until then, the following requirements apply on board vessels in service, with the exception of Type N open vessels: This requirement may be met by fitting vertical protection walls not less than 0.50 m in height. Until then, on board vessels in service less than 50.00 m long, the height of 0.50 m may be reduced to 0.30 m in passageways leading to the deck.
9.3.1.10.3 9.3.2.10.3 9.3.3.10.3	Height of sills of hatches and openings above the deck	N.R.M. from 1 January 2005 Renewal of the certificate of approval after 31 December 2010
9.3.1.11.1 (b)	Ratio of length to diameter of pressure cargo tanks	N.R.M. Renewal of the certificate of approval after 31 December 2044
9.3.3.11.1 (d)	Limitation of length of cargo tanks	N.R.M. Renewal of the certificate of approval after 31 December 2044
9.3.1.11.2 (a)	Arrangement of cargo tanks Distance between cargo tanks and side walls Height of saddles	N.R.M. for Type G vessels whose keels were laid before 1 January 1977. Renewal of the certificate of approval after 31 December 2044

1.6.7.2.2.2 Table of general transitional provisions: Tank vessels		
Paragraphs	Subject	Time limit and comments

9.3.1.11.2 (a)	Arrangement of cargo tanks
	Distance between cargo tanks and side walls
	Height of saddles
		N.R.M.
		Renewal of the certificate of approval after 31 December 2044
		Until then, the following requirements apply on board vessels in service whose keels were laid after 31 December 1976:
		Where tank volume is more than 200 m^3 or where the ratio of length to diameter is less than 7 but more than 5, the hull in the tank area shall be such that, in the event of a collision, the tanks remain intact as far as possible. This requirement shall be considered as having been met where, in the tank area, the vessel:
		- is double-hulled with a distance of at least 80 cm between the side plating and the longitudinal bulkhead
		- or is designed as follows:
		  (a) Between the gangboard and the top of the floorplates there shall be side stringers at regular intervals of not more than 60 cm;
		  (b) The side stringers shall be supported by web frames spaced at intervals of not more than 2.00 m. The height of the web frames shall be not less than 10% of the depth and in any event not less than 30 cm. They shall be fitted with a face plate made of flat steel having a cross section of not less than 15 cm^2;
		  (c) The side stringers referred to in (a) shall have the same height as the web frames and be fitted with a face plate made of flat steel having a cross section of not less than 7.5 cm^2.

9.3.1.11.2 (a)	Distance between suction wells and floor plates	N.R.M. Renewal of the certificate of approval after 31 December 2044
9.3.1.11.2 (b) 9.3.2.11.2 (b) 9.3.3.11.2 (a)	Cargo tank fastenings	N.R.M. Renewal of the certificate of approval after 31 December 2044
9.3.1.11.2 (c) 9.3.2.11.2 (c) 9.3.3.11.2 (b)	Capacity of suction well	N.R.M. Renewal of the certificate of approval after 31 December 2044
9.3.1.11.2 (d) 9.3.2.11.2 (d)	Side struts between the hull and the cargo tanks	N.R.M. from 1 January 2001 Renewal of the certificate of approval after 31 December 2044
9.3.1.11.3 (a)	End bulkheads of cargo area with "A-60" insulation. Distance of 0.50 m from cargo tanks to end bulkheads	N.R.M. Renewal of the certificate of approval after 31 December 2044

1.6.7.2.2.2 Table of general transitional provisions: Tank vessels		
Paragraphs	Subject	Time limit and comments
9.3.2.11.3 (a) 9.3.3.11.3 (a)	Width of cofferdams of 0.60 m Hold spaces with cofferdams or "A-60" insulated bulkheads Distance of 0.50 m from cargo tanks in hold spaces	N.R.M. Renewal of the certificate of approval after 31 December 2044 Until then, the following requirements apply on board vessels in service: Type C: minimum width of cofferdams: 0.50 m; Type N: minimum width of cofferdams: 0.50 m; on board vessels with a deadweight of up to 150 t: 0.40 m;
		Type N open: cofferdams shall not be required on board vessels with a deadweight up to 150 t and oil separator vessels: The distance between cargo tanks and end bulkheads of hold spaces shall be at least 0.40m.
9.3.3.11.4	Penetrations through the end bulkheads of hold spaces	N.R.M. from 1 January 2005 for Type N open vessels whose keels were laid before 1 January 1977. Renewal of the certificate of approval after 31 December 2044.
9.3.3.11.4	Distance of piping in relation to the bottom	N.R.M. from 1 January 2005 Renewal of the certificate of approval after 31 December 2038
9.3.3.11.4	Shut-off devices of the loading and unloading piping in the cargo tank from which they come	N.R.M. from 1 January 2005 Renewal of the certificate of approval after 31 December 2018
9.3.3.11.6 (a)	Form of cofferdam arranged as a pump room	N.R.M for Type N vessels whose keels were laid before 1 January 1977. Renewal of the certificate of approval after 31 December 2044.
9.3.3.11.7	Distance between the cargo tanks and the outer wall of the vessel	N.R.M. after 1 January 2001 Renewal of the certificate of approval after 31 December 2038
9.3.3.11.7	Width of double hull	N.R.M. after 1 January 2007 Renewal of the certificate of approval after 31 December 2038
9.3.1.11.7	Distance between the suction well and the bottom spaces	N.R.M. after 1 January 2003 Renewal of the certificate of approval after 31 December 2038
9.3.3.11.8	Arrangement of service spaces located in the cargo area below decks	N.R.M. for Type N open vessels. Renewal of the certificate of approval after 31 December 2038.
9.3.1.11.8 9.3.3.11.9	Dimensions of openings for access to spaces within the cargo area	N.R.M. Renewal of the certificate of approval after 31 December 2018
9.3.1.11.8 9.3.2.11.10 9.3.3.11.9	Interval between reinforcing elements	N.R.M. Renewal of the certificate of approval after 31 December 2044

1.6.7.2.2.2 Table of general transitional provisions: Tank vessels		
Paragraphs	Subject	Time limit and comments
9.3.2.12.1 9.3.3.12.1	Ventilation openings in hold spaces	N.R.M. from 1 January 2003 Renewal of the certificate of approval after 31 December 2018
9.3.1.12.2 9.3.3.12.2	Ventilation systems in double-hull spaces and double bottoms	N.R.M. Renewal of the certificate of approval after 31 December 2018
9.3.1.12.3 9.3.2.12.3 9.3.3.12.3	Height above the deck of the air intake for service spaces located below deck	N.R.M. Renewal of the certificate of approval after 31 December 2018
9.3.1.12.6 9.3.2.12.6 9.3.3.12.6	Distance of ventilation inlets from cargo area	N.R.M. from 1 January 2003 Renewal of the certificate of approval after 31 December 2044
9.3.1.12.6 9.3.2.12.6 9.3.3.12.6	Permanently installed flame screens	N.R.M. from 1 January 2003 Renewal of the certificate of approval after 31 December 2018
9.3.3.12.7	Approval of flame arresters	N.R.M for Type N vessels whose keels were laid before 1 January 1977. Renewal of the certificate of approval after 31 December 2018.
9.3.1.13 9.3.3.13	Stability (general)	N.R.M. Renewal of the certificate of approval after 31 December 2044
9.3.3.13.3 paragraph 2	Stability (general)	N.R.M. from 1 January 2007 Renewal of the certificate of approval after 31 December 2044
9.3.1.14 9.3.3.14	Stability (intact)	N.R.M. Renewal of the certificate of approval after 31 December 2044
9.3.1.15	Stability (damaged condition)	N.R.M. Renewal of the certificate of approval after 31 December 2044
9.3.3.15	Stability (damaged condition)	N.R.M. after 1 January 2007 Renewal of the certificate of approval after 31 December 2044
9.3.1.16.1 9.3.3.16.1	Distance of openings of engine rooms from the cargo area	N.R.M. Renewal of the certificate of approval after 31 December 2044
9.3.3.16.1	Internal combustion engines outside the cargo area	N.R.M. for Type N open vessels. Renewal of the certificate of approval after 31 December 2034.
9.3.1.16.2 9.3.3.16.2	Hinges of doors facing the cargo area	N.R.M. for vessels whose keels were laid before 1 January 1977 where alterations would obstruct other major openings. Renewal of the certificate of approval after 31 December 2034.

1.6.7.2.2.2 Table of general transitional provisions: Tank vessels		
Paragraphs	Subject	Time limit and comments
9.3.3.16.2	Engine rooms accessible from the deck	N.R.M. for Type N open vessels. Renewal of the certificate of approval after 31 December 2034.
9.3.1.17.1 9.3.3.17.1	Accommodation and wheelhouse outside the cargo area	N.R.M. for vessels whose keels were laid before 1 January 1977, provided that there is no connection between the wheelhouse and other enclosed spaces. Renewal of the certificate of approval after 31 December 2044. Renewal of the certificate of approval after 31 December 2044 for vessels up to 50 m in length whose keels were laid before 1 January 1977 and whose wheelhouses are located in the cargo area even if it provides access to another enclosed space, provided that safety is ensured by appropriate service requirements of the competent authority.
9.3.3.17.1	Accommodation and wheelhouse outside the cargo area	N.R.M. for Type N open vessels. Renewal of the certificate of approval after 31 December 2044.
9.3.1.17.2 9.3.2.17.2 9.3.3.17.2	Arrangement of entrances and openings of forward superstructures	N.R.M. Renewal of the certificate of approval after 31 December 2044
9.3.1.17.2 9.3.2.17.2 9.3.3.17.2	Entrances facing the cargo area	N.R.M. for vessels up to 50 m in length whose keels were laid before 1 January 1977, provided that gas screens are installed. Renewal of the certificate of approval after 31 December 2044.
9.3.3.17.2	Entrances and openings	N.R.M. for Type N open vessels. Renewal of the certificate of approval after 31 December 2044.
9.3.1.17.4 9.3.3.17.4	Distance of openings from the cargo area	N.R.M. for Type N open vessels. Renewal of the certificate of approval after 31 December 2018.
9.3.3.17.5 (b), (c)	Approval of shaft passages and displaying of instructions	N.R.M. for Type N open vessels. Renewal of the certificate of approval after 31 December 2018.

9.3.1.17.6
9.3.3.17.6
	Pump-room below deck	N.R.M.
		Renewal of the certificate of approval after 31 December 2018
		Until then, the following requirements apply on board vessels in service:
		Pump-rooms below deck shall
		- meet the requirements for service spaces:
		  - for Type G vessels: 9.3.1.12.3
		  - for Type N vessels: 9.3.3.12.3
		- be equipped with a gas detection system referred to in 9.3.1.17.6 or 9.3.3.17.6

1.6.7.2.2.2 Table of general transitional provisions: Tank vessels		
Paragraphs	Subject	Time limit and comments
9.3.2.20.1 9.3.3.20.1	Access to cofferdams or cofferdam compartments	N.R.M. from 1 January 2015 Renewal of the certificate of approval after 31 December 2034
9.3.2.20.2 9.3.3.20.2	Intake valve	N.R.M. Renewal of the certificate of approval after 31 December 2018
9.3.3.20.2	Filling of cofferdams with pump	N.R.M. for Type N open vessels. Renewal of the certificate of approval after 31 December 2018.
9.3.2.20.2 9.3.3.20.2	Filling of cofferdams within 30 minutes	N.R.M. Renewal of the certificate of approval after 31 December 2018
9.3.1.21.3 9.3.2.21.3 9.3.3.21.3	Marking on each level gauge of all permissible maximum filling levels of cargo tanks	N.R.M. from 1 January 2015 Renewal of the certificate of approval after 31 December 2018
9.3.3.21.1 (b)	Liquid level gauge	N.R.M. from 1 January 2005 for vessels of Type N open with flame-arresters and those of Type N open. Renewal of the certificate of approval after 31 December 2018. Until then, on board vessels in service fitted with gauging openings, such openings shall: • Be arranged so that the degree of filling can be measured using a sounding rod; • Be fitted with an automatically-closing cover.
9.3.3.21.1 (g)	Sampling opening	N.R.M. for Type N open vessels. Renewal of the certificate of approval after 31 December 2018.
9.3.1.21.4 9.3.2.21.4 9.3.3.21.4	Liquid-level alarm device independent from the liquid-level gauge	N.R.M. Renewal of the certificate of approval after 31 December 2018
9.3.1.21.5 (a) 9.3.2.21.5 (a) 9.3.3.21.5 (a)	Socket close to the shore connections of the loading and unloading piping and switching off of vessel's pump	N.R.M. Renewal of the certificate of approval after 31 December 2018
9.3.1.21.5 (b) 9.3.2.21.5 (b) 9.3.3.21.5 (d)	Installation of on-board pump switch-off from the shore	N.R.M. Renewal of the certificate of approval after 31 December 2006
9.3.2.21.5 (c)	Device for rapid shutting off of refuelling	N.R.M. Renewal of the certificate of approval after 31 December 2008

1.6.7.2.2.2 Table of general transitional provisions: Tank vessels		
Paragraphs	Subject	Time limit and comments
9.3.1.21.7 9.3.2.21.7 9.3.3.21.7	Vacuum or over-pressure alarms in cargo tanks for the carriage of substances without remark 5 in column (20) of Table C of Chapter 3.2	N.R.M. from 1 January 2001 Renewal of the certificate of approval after 31 December 2018
9.3.1.21.7 9.3.2.21.7 9.3.3.21.7	Temperature alarms in cargo tanks	N.R.M. from 1 January 2001 Renewal of the certificate of approval after 31 December 2018
9.3.1.22.1 (b)	Height of cargo tank openings above the deck	N.R.M. Renewal of the certificate of approval after 31 December 2044
9.3.3.22.1 (b)	Cargo tank openings 0.50 m above the deck	N.R.M. Renewal of the certificate of approval after 31 December 2044 for vessels whose keels were laid before 1 January 1977.
9.3.1.22.4	Prevention of spark-formation by closure devices	N.R.M. from 1 January 2003 Renewal of the certificate of approval after 31 December 2018
9.3.1.22.3 9.3.2.22.4 (b) 9.3.3.22.4 (b)	Position of outlets of valves above the deck	N.R.M. Renewal of the certificate of approval after 31 December 2018
9.3.2.22.4 (b) 9.3.3.22.4 (b)	Pressure setting of high velocity vent valves	N.R.M. Renewal of the certificate of approval after 31 December 2018
9.3.3.23.2	Test pressure for cargo tanks	N.R.M. for vessels whose keels were laid before 1 January 1977, for which a test pressure of 15 kPa (0.15 bar) is required. Renewal of the certificate of approval after 31 December 2044. Until then, a test pressure of 10 kPa (0.10 bar) shall be sufficient.
9.3.3.23.2	Test pressure for cargo tanks	N.R.M. for oil-separator vessels in service before 1 January 1999. Renewal of the certificate of approval after 31 December 2044. Until then, a test pressure of 5 kPa (0.05 bar) is sufficient.
9.3.3.23.3	Test pressure for piping for loading and unloading	N.R.M. for oil-separator vessels in service before 1 January 1999. Renewal of the certificate of approval at the latest by 1 January 2039. Until then, a test pressure of 400 kPa (4 bar) is sufficient.
9.3.2.25.1 9.3.3.25.1	Shut-down of cargo pumps	N.R.M. Renewal of the certificate of approval after 31 December 2018
9.3.1.25.1 9.3.2.25.1 9.3.3.25.1	Distance of pumps, etc. from accommodation, etc.	N.R.M. Renewal of the certificate of approval after 31 December 2044

1.6.7.2.2.2 Table of general transitional provisions: Tank vessels		
Paragraphs	Subject	Time limit and comments
9.3.1.25.2 (d) 9.3.2.25.2 (d)	Position of loading and unloading piping on deck	N.R.M. Renewal of the certificate of approval after 31 December 2044
9.3.1.25.2 (e) 9.3.2.25.2 (e) 9.3.3.25.2 (e)	Distance of shore connections from accommodation, etc.	N.R.M. Renewal of the certificate of approval after 31 December 2034
9.3.2.25.2 (i)	Piping for loading and unloading, and venting piping, shall not have flexible connections fitted with sliding seals.	N.R.M. from 1 January 2009 Vessels in service having connections with sliding seals may no longer transport substances with toxic or corrosive properties (see column (5) of Table C of Chapter 3.2, hazards 6.1 and 8) following the renewal of the certificate of approval after 31 December 2008. Vessels in service shall not have flexible connections fitted with sliding seals following the renewal of the certificate of approval after 31 December 2018
9.3.3.25.2 (h)	Piping for loading and unloading, and venting piping, shall not have flexible connections fitted with sliding seals	N.R.M. from 1 January 2009 Vessels in service having connections with sliding seals may no longer transport substances with corrosive properties (see column (5) of Table C of Chapter 3.2, hazard 8) following the renewal of the certificate of approval after 31 December 2008. Vessels in service shall not have flexible connections with sliding seals following the renewal of the certificate of approval after 31 December 2018.
9.3.2.25.8 (a)	Ballasting suction pipes located within the cargo area but outside the cargo tanks	N.R.M. Renewal of the certificate of approval after 31 December 2018
9.3.2.25.9 9.3.3.25.9	Loading and unloading flow	N.R.M. from 1 January 2003 Renewal of the certificate of approval after 31 December 2018
9.3.3.25.12	9.3.3.25.1 (a) and (c), 9.3.3.25.2 (e), 9.3.3.25.3 and 9.3.3.25.4 (a) are not applicable for Type N open with the exception of Type N open carrying corrosive substances (see Chapter 3.2, Table C, column (5), hazard 8)	N.R.M. Renewal of the certificate of approval after 31 December 2018 This time limit concerns only Type N open vessels carrying corrosive substances (see Chapter 3.2, Table C, column (5), hazard 8).
9.3.1.31.2 9.3.2.31.2 9.3.3.31.2	Distance of engine air intakes from the cargo area	N.R.M. Renewal of the certificate of approval after 31 December 2044

1.6.7.2.2.2 Table of general transitional provisions: Tank vessels		
Paragraphs	Subject	Time limit and comments
9.3.1.31.4 9.3.2.31.4 9.3.3.31.4	Temperature of outer parts of engines, etc.	N.R.M. Renewal of the certificate of approval after 31 December 2018 Until then, the following requirements apply on board vessels in service: The temperature of outer parts shall not exceed 300 °C.
9.3.1.31.5 9.3.2.31.5 9.3.3.31.5	Temperature in the engine room	N.R.M. Renewal of the certificate of approval after 31 December 2018 Until then, the following requirements apply on board vessels in service: The temperature in the engine room shall not exceed 45 °C.
9.3.1.32.2 9.3.2.32.2 9.3.3.32.2	Openings of air pipes 0.50 m above the deck	N.R.M. Renewal of the certificate of approval after 31 December 2010
9.3.3.34.1	Exhaust pipes	N.R.M. Renewal of the certificate of approval after 31 December 2018
9.3.1.35.1 9.3.3.35.1	Stripping and ballast pumps in the cargo area	N.R.M. Renewal of the certificate of approval after 31 December 2034
9.3.3.35.3	Suction pipes for ballasting located within the cargo area but outside the cargo tanks	N.R.M. Renewal of the certificate of approval after 31 December 2018
9.3.1.35.4	Stripping installation of the pump-room outside the pump-room	N.R.M. from 1 January 2003 Renewal of the certificate of approval after 31 December 2018
9.3.1.40.1 9.3.2.40.1 9.3.3.40.1	Fire extinguishing systems, two pumps, etc.	N.R.M. Renewal of the certificate of approval after 31 December 2018
9.3.1.40.2 9.3.2.40.2 9.3.3.40.2	Fixed fire extinguishing system in engine room	N.R.M. Renewal of the certificate of approval after 31 December 2034
9.3.1.41.1 9.3.3.41.1	Outlets of funnels located not less than 2 m from the cargo area	N.R.M. Renewal of the certificate of approval after 31 December 2044 for vessels whose keels were laid before 1 January 1977.
9.3.3.41.1	Outlets of funnels	N.R.M. at the latest by 1 January 2039 for oil-separator vessels
9.3.1.41.2 9.3.2.41.2 9.3.3.41.2 in conjunction with 7.2.3.41	Heating, cooking and refrigerating appliances	N.R.M. Renewal of the certificate of approval after 31 December 2010

1.6.7.2.2.2 Table of general transitional provisions: Tank vessels		
Paragraphs	Subject	Time limit and comments
9.3.3.42.2	Cargo heating system	N.R.M for Type N open vessels. Renewal of the certificate of approval after 31 December 2034. Until then, the following requirements apply on board vessels in service: This can be achieved by one oil separator fitted to the condensed water return pipe.
9.3.1.51.2 9.3.2.51.2 9.3.3.51.2	Visual and audible alarm	N.R.M. Renewal of the certificate of approval after 31 December 2034
9.3.1.51.3 9.3.2.51.3 9.3.3.51.3	Temperature class and explosion group	N.R.M. Renewal of the certificate of approval after 31 December 2034
9.3.3.52.1 (b), (c), (d) and (e)	Electrical installations	N.R.M. for Type N open vessels. Renewal of the certificate of approval after 31 December 2034.
9.3.1.52.1 (e) 9.3.3.52.1 (e)	Electrical installations of the "certified safe" type in the cargo area	N.R.M. for vessels whose keels were laid before 1 January 1977. Renewal of the certificate of approval after 31 December 2034. Until then, the following conditions shall be met during loading, unloading and gas freeing on board vessels having non-gastight wheelhouse openings (e.g. doors, windows, etc.) in the cargo area: (a) All electrical installations designed to be used shall be of a limited explosion-risk type, i.e. they shall be so designed that there is no sparking under normal operating conditions and the temperature of their outer surface does not rise above 200°C, or be of a type protected against water spray the temperature of whose outer surfaces does not exceed 200°C under normal operating conditions; (b) Electrical installations which do not meet the requirements of (a) above shall be marked in red and it shall be possible to switch them off by means of a central switch.
9.3.3.52.2	Accumulators located outside the cargo area	N.R.M. for Type N open vessels. Renewal of the certificate of approval after 31 December 2034.

1.6.7.2.2.2 Table of general transitional provisions: Tank vessels		
Paragraphs	Subject	Time limit and comments
9.3.1.52.3 (a) 9.3.1.52.3 (b) 9.3.3.52.3 (a) 9.3.3.52.3 (b)	Electrical installations used during loading, unloading or gas-freeing	N.R.M. for vessels whose keels were laid before 1 January 1977. Renewal of the certificate of approval after 31 December 2034 for the following installations: • Lighting installations in accommodation, with the exception of switches near the entrances to accommodation; • Radio telephone installations in accommodation and wheelhouses and combustion engine control appliances. Until then, all other electrical installations shall meet the following requirements: a) Generators, engine, etc. IP 13 protection mode; b) Control panels, lamps, etc. IP 23 protection mode; c) Appliances, etc. IP 55 protection mode.
9.3.3.52.3 (a) 9.3.3.52.3 (b)	Electrical installations used during loading, unloading or gas-freeing	N.R.M. for Type N open vessels. Renewal of the certificate of approval after 31 December 2034
9.3.1.52.3 (b) 9.3.2.52.3 (b) 9.3.3.52.3 (b) in conjunction with 3 (a)	Electrical installations used during loading, unloading or gas-freeing	N.R.M. Renewal of the certificate of approval after 31 December 2034 Until then, on board vessels in service, paragraph (3) (a) shall not apply to: - Lighting installations in accommodation, with the exception of switches near entrances to accommodation; - Radio telephone installations in accommodation and wheelhouses.
9.3.1.52.4 9.3.2.52.4 9.3.3.52.4 last sentence	Disconnection of such installations from a centralized location	N.R.M. Renewal of the certificate of approval after 31 December 2034
9.3.3.52.4	Red mark on electrical installations	N.R.M. for Type N open vessels Renewal of the certificate of approval after 31 December 2034
9.3.3.52.5	Shutting down switch for continuously driven generator	N.R.M. for Type N open vessels Renewal of the certificate of approval after 31 December 2034
9.3.3.52.6	Permanently fitted sockets	N.R.M. for Type N open vessels Renewal of the certificate of approval after 31 December 2034
9.3.1.56.1 9.3.3.56.1	Metallic sheaths for all cables in the cargo area	N.R.M. for vessels whose keels were laid before 1 January 1977. Renewal of the certificate of approval after 31 December 2034.

1.6.7.2.2.2 Table of general transitional provisions: Tank vessels		
Paragraphs	Subject	Time limit and comments
9.3.3.56.1	Metallic sheath for all cables in the cargo area	N.R.M. by 1 January 2039 at the latest for oil-separator vessels.

1.6.7.2.2.3   Transitional provisions concerning the application of the requirements of Table C of Chapter 3.2 to the carriage of goods in tank vessels.

1.6.7.2.2.3.1   The goods for which Type N closed with a minimum valve setting of 10 kPa (0.10 bar) is required in Table C of Chapter 3.2, may be carried in tank-vessels in service of Type N closed with a minimum valve setting of 6 kPa (0.06 bar) (cargo tank test pressure of 10 kPa (0.10 bar)). This transitional provision is valid until 31 December 2018.

1.6.7.2.2.3.2 and 1.6.7.2.2.3.3   (*Deleted*)

1.6.7.2.2.4   (*Deleted*)

1.6.7.2.2.5   For a vessel or a barge whose keel was laid before 1 January July 2017 and which does not conform to the requirements of 9.3.X.1 concerning the vessel record, the retention of files for the vessel record shall start at the latest at the next renewal of the certificate of approval.

1.6.7.3   ***Supplementary transitional provisions applicable to specific inland waterways***

Vessels in service to which the transitional provisions of this subsection are applied shall meet:

–   the requirements of paragraphs and subparagraphs mentioned in the table below and in the table of general transitional provisions (see 1.6.7.2.1.1 and 1.6.7.2.2.1) within the period established therein;

–   the requirements of paragraphs and subparagraphs not mentioned in the table below or in the table of general transitional provisions at the date of application of these Regulations.

The construction and equipment of vessels in service shall be maintained at least at the previous standard of safety.

Table of supplementary transitional provisions		
Paragraph	Subject	Time limit and comments
9.1.0.11.1 (b)	Holds, common bulkheads with oil fuel tanks	N.R.M. The following requirements apply on board vessels in service: Holds may share a common bulkhead with the oil fuel tanks, provided that the cargo or its packaging does not react chemically with the fuel.
9.1.0.92	Emergency exit	N.R.M. The following requirements apply on board vessels in service: Spaces the entrances or exits of which are partly or fully immersed in damaged condition shall be provided with an emergency exit not less than 0.075 m above the damage waterline.

Table of supplementary transitional provisions		
Paragraph	Subject	Time limit and comments
9.1.0.95.1 (c)	Height of openings above damage waterline	N.R.M. The following requirements apply on board vessels in service: The lower edge of any non-watertight openings (e.g. doors, windows, access hatchways) shall, at the final stage of flooding, be not less than 0.075 m above the damage waterline.
9.1.0.95.2 9.3.2.15.2	Extent of the stability diagram (damaged condition)	N.R.M. The following requirements apply on board vessels in service: At the final stage of flooding the angle of heel shall not exceed: 20° before measures to right the vessel; 12° following measures to right the vessel.
9.3.3.8.1	Classification	N.R.M. for Type N open vessels with flame arresters and Type N open vessels. Renewal of the certificate of approval after 31 December 2044.
9.3.1.11.1 (a) 9.3.2.11.1 (a) 9.3.3.11.1 (a)	Maximum capacity of cargo tanks	N.R.M. The following requirements apply on board vessels in service: The maximum permissible capacity of a cargo tank shall be 760 m³.
9.3.2.11.1 (d)	Length of cargo tanks	N.R.M. The following requirements apply on board vessels in service: The length of a cargo tank may exceed 10 m and 0.2 L.
9.3.1.12.3 9.3.2.12.3 9.3.3.12.3	Position of air inlets	N.R.M. The following requirements apply on board vessels in service: The air inlets to be positioned at least 5.00 m from the safety-valve outlets
9.3.2.15.1 (c)	Height of openings above damage waterline	N.R.M. The following requirements apply on board vessels in service: The lower edge of any non-watertight openings (e.g. doors, windows, access hatchways) shall, at the final stage of flooding, be not less than 0.075 m above the damage waterline.
9.3.2.20.2 9.3.3.20.2	Filling of cofferdams with water	N.R.M. The following requirements apply on board vessels in service: Cofferdams shall be fitted with a system for filling with water or inert gas.

Table of supplementary transitional provisions		
Paragraph	Subject	Time limit and comments
9.3.1.92 9.3.2.92	Emergency exit	N.R.M. The following requirements apply on board vessels in service: Spaces the entrances or exits of which are partly or fully immersed in damaged condition shall be provided with an emergency exit not less than 0.075 m above the damage waterline.

1.6.7.4 *Transitional provisions concerning the transport of substances hazardous to the environment or to health*

1.6.7.4.1 *Transitional provisions: vessels*

Single-hull tank vessels in service on 1 January 2009 with a dead weight on 1 January 2007 of less than 1,000 tonnes may continue to transport the substances they were authorized to carry on 31 December 2008 until 31 December 2018.

Supply vessels and oil separator vessels in service on 1 January 2009 with a dead weight on 1 January 2007 of less than 300 tonnes may continue to transport the substances they were authorized to carry on 31 December 2008 until 31 December 2038.

1.6.7.4.2 *Transitional periods applicable to substances*

By way of derogation from Part 3, Table C, the substances listed below may be transported in accordance with the requirements referred to in the following tables until the date specified.

Table 1. Until 31 December 2012   (*Deleted*)

Table 2. Until 31 December 2015   (*Deleted*)

3. Until 31 December 2018

(1) UN No. or substance identification No.	(2) Name and description	(3a) Class	(3b) Classification code	(4) Packing group	(5) Dangers	(6) Type of tank vessel	(7) Cargo tank design	(8) Cargo tank type	(9) Cargo tank equipment	(10) Opening pressure of the high-velocity vent valve in kPa	(11) Maximum degree of filling in %	(12) Relative density at 20 °C	(13) Type of sampling device	(14) Pump room below deck permitted	(15) Temperature class	(16) Explosion group	(17) Anti-explosion protection required	(18) Equipment required	(19) Number of blue cones/lights	(20) Additional requirements/Remarks
1202	GAS OIL or DIESEL FUEL or HEATING OIL (LIGHT) (flash-point not more than 60 °C)	3	F1	III	3+(N1, N2, N3, CMR, F)	N	4	2			97	<0,85	3	yes			non	*	0	*see 3.2.3.3
1202	GAS OIL complying with standard EN 590: 2004 or DIESEL FUEL or HEATING OIL (LIGHT) with flash-point as specified in EN 590:2009 + A1:2010	3	F1	III	3+N2+F	N	4	2			97	0,82 - 0,85	3	yes			non	PP	0	
1202	GAS OIL or DIESEL FUEL or HEATING OIL (LIGHT) (flash-point more than 60 °C but not more than 100 °C)	3	F1	III	3+(N1, N2, N3, CMR, F or S)	N	4	2			97	< 1,1	3	yes			non	*	0	*see 3.2.3.3
1223	KEROSENE	3	F1	III	3+N2+F	N	3	2			97	≤0,83	3	yes	T3	II A[7]	yes	PP, EX, A	0	14
1300	TURPENTINE SUBSTITUTE	3	F1	III	3+N2+F	N	3	2			97	0,78	3	yes	T3	II B[4]	yes	PP, EX, A	0	
1863	FUEL, AVIATION, TURBINE ENGINE vp50 > 175 kPa	3	F1	I	3+(N1, N2, N3, CMR, F)	N	1	1	1		97		1	yes	T4[3]	II B[4]	yes	PP, EX, A	1	14; 29
1863	FUEL, AVIATION, TURBINE ENGINE vp50 > 175 kPa	3	F1	I	3+(N1, N2, N3, CMR, F)	N	2	2		50	97		2	yes	T4[3]	II B[4]	yes	PP, EX, A	1	14; 29
1863	FUEL, AVIATION, TURBINE ENGINE 110 kPa < vp50 ≤ 175 kPa	3	F1	II	3+(N1, N2, N3, CMR, F)	N	2	2		50	97		3	yes	T4[3]	II B[4]	yes	PP, EX, A	1	14; 29
1863	FUEL, AVIATION, TURBINE ENGINE 110 kPa < vp50 ≤ 150 kPa	3	F1	II	3+(N1, N2, N3, CMR, F)	N	2	2	3	10	97		3	yes	T4[3]	II B[4]	yes	PP, EX, A	1	14; 29
1863	FUEL, AVIATION, TURBINE ENGINE vp50 ≤ 110 kPa	3	F1	II	3+(N1, N2, N3, CMR, F)	N	2	2		10	97		3	yes	T4[3]	II B[4]	yes	PP, EX, A	1	14; 29
1863	FUEL, AVIATION, TURBINE ENGINE	3	F1	III	3+(N1, N2, N3, CMR, F)	N	3	2			97		3	yes	T4[3]	II B[4]	yes	*	0	14 *see 3.2.3.3

1.6.7.5   *Transitional provisions concerning the modification of tank vessels*

1.6.7.5.1   The modification of the cargo area of a vessel in order to achieve a Type N double-hull vessel is admissible until 31 December 2018 under the following conditions:

(a)   The modified or new cargo area shall comply with the provisions of ADN. Transitional provisions under 1.6.7.2.2 may not be applied for the cargo area;

(b)   The vessel parts outside of the cargo area shall comply with the provisions of ADN. Moreover, the following transitional provisions under 1.6.7.2.2 may be applied: 1.2.1, 9.3.3.0.3 (d), 9.3.3.51.3 and 9.3.3.52.4 last sentence;

(c)   If goods which require explosion protection are entered in the list according to 1.16.1.2.5, accommodation and wheelhouses shall be equipped with a fire alarm system according to 9.3.3.40.2.3;

(d)   The application of this sub-section shall be entered in the certificate of approval under No. 12 (Additional observations).

1.6.7.5.2   Modified vessels may continue to be operated beyond 31 December 2018. The time limits stipulated in the applied transitional provisions under 1.6.7.2.2 shall be observed.

1.6.7.6   *Transitional provisions concerning the transport of gases in tank vessels*

Tank vessels in service on 1 January 2011 with a pump room below deck may continue to transport the substances listed in the following table until the renewal of the certificate of approval after 1 January 2045.

UN No. or ID No.	Class and classification code	Name and description
1005	2, 2TC	AMMONIA, ANHYDROUS
1010	2, 2F	1,2-BUTADIENE, STABILIZED
1010	2, 2F	1,3-BUTADIENE, STABILIZED
1010	2, 2F	BUTADIENE STABILIZED or BUTADIENES AND HYDROCARBON MIXTURE, STABILIZED, having a vapour pressure at 70 °C not exceeding 1.1 MPa (11 bar) and a density at 50 °C not lower than 0.525 kg/l
1011	2, 2F	BUTANE
1012	2, 2F	1-BUTYLENE
1020	2,2A	CHLOROPENTAFLUOROETHANE (REFRIGERANT GAS R 115)
1030	2,2F	1,1-DIFLUOROETHANE (REFRIGERANT GAS R 152a)
1033	2,2F	DIMETHYL ETHER
1040	2,2TF	ETHYLENE OXIDE WITH NITROGEN up to a total pressure of 1 MPa (10 bar) at 50 °C
1055	2,2F	ISOBUTYLENE
1063	2,2F	METHYL CHLORIDE (REFRIGERANT GAS R 40)
1077	2,2F	PROPYLENE
1083	2,2F	TRIMETHYLAMINE, ANHYDROUS
1086	2,2F	VINYL CHLORIDE, STABILIZED
1912	2,2F	METHYL CHLORIDE AND METHYLENE CHLORIDE MIXTURE
1965	2,2F	HYDROCARBON GAS MIXTURE, LIQUEFIED, N.O.S., (MIXTURE A)

UN No. or ID No.	Class and classification code	Name and description
1965	2,2F	HYDROCARBON GAS MIXTURE, LIQUEFIED, N.O.S., (MIXTURE A0)
1965	2,2F	HYDROCARBON GAS MIXTURE, LIQUEFIED, N.O.S., (MIXTURE A01)
1965	2,2F	HYDROCARBON GAS MIXTURE, LIQUEFIED, N.O.S., (MIXTURE A02)
1965	2,2F	HYDROCARBON GAS MIXTURE, LIQUEFIED, N.O.S., (MIXTURE A1)
1965	2,2F	HYDROCARBON GAS MIXTURE, LIQUEFIED, N.O.S., (MIXTURE B)
1965	2,2F	HYDROCARBON GAS MIXTURE, LIQUEFIED, N.O.S., (MIXTURE B1)
1965	2,2F	HYDROCARBON GAS MIXTURE, LIQUEFIED, N.O.S., (MIXTURE B2)
1965	2,2F	HYDROCARBON GAS MIXTURE, LIQUEFIED, N.O.S., (MIXTURE C)
1969	2,2F	ISOBUTANE
1978	2,2F	PROPANE
9000		AMMONIA, ANHYDROUS, DEEPLY REFRIGERATED

1.6.8 Transitional provisions concerning training of the crew

The responsible master and the person responsible for the loading or unloading of a barge shall be in possession of a certificate of special knowledge with the entry "The holder of this certificate has participated in an 8-lesson stability training" before 31 December 2019.

The condition for this entry is participation in a basic course required by the Regulations in force after 1 January 2013 or participation in a basic refresher course that, in derogation from 8.2.2.5, comprises 24 lessons of 45 minutes, including eight lessons devoted to the subject of stability.

Until 31 December 2018, the expert on the carriage of gases (as referred to in 8.2.1.5) does not have to be the responsible master (as referred to in 7.2.3.15) but can be any member of the crew when the type G tank vessel is only carrying UN No. 1972. In this case, the responsible master shall have attended the specialization course on gases and shall also have followed an additional training on the carriage of liquefied natural gas (LNG) in accordance with 1.3.2.2.

1.6.9 Transitional provisions concerning recognition of classification societies

1.6.9.1 The provisions of 1.15.3.8 concerning the maintenance of an effective system of internal quality by the recommended classification societies still applicable on 31 December 2015 may continue to be applied until 14 September 2018.

CHAPTER 1.7

GENERAL PROVISIONS CONCERNING RADIOACTIVE MATERIAL

1.7.1   **Scope and application**

NOTE 1: In the event of accidents or incidents during the carriage of radioactive material, emergency provisions, as established by relevant national and/or international organizations, shall be observed to protect persons, property and the environment. Appropriate guidelines for such provisions are contained in "Planning and Preparing for Emergency Response to Transport Accidents Involving Radioactive Material", IAEA Safety Standard Series No. TS-G-1.2 (ST-3), IAEA, Vienna (2002).

NOTE 2: Emergency procedures shall take into account the formation of other dangerous substances that may result from the reaction between the contents of a consignment and the environment in the event of an accident.

1.7.1.1   ADN establishes standards of safety which provide an acceptable level of control of the radiation, criticality and thermal hazards to persons, property and the environment that are associated with the carriage of radioactive material. These standards are based on the IAEA Regulations for the Safe Transport of Radioactive Material, 2012 Edition, IAEA Safety Standards Series No. SSR–6, IAEA, Vienna (2012). Explanatory material can be found in "Advisory material for the IAEA Regulations for the Safe Transport of Radioactive Material (2012 Edition), IAEA Safety Standards Series No. SSG-26, IAEA, Vienna (2014).

1.7.1.2   The objective of ADN is to establish requirements that shall be satisfied to ensure safety and to protect persons, property and the environment from the effects of radiation in the carriage of radioactive material. This protection is achieved by requiring:

(a)   Containment of the radioactive contents;

(b)   Control of external radiation levels;

(c)   Prevention of criticality; and

(d)   Prevention of damage caused by heat.

These requirements are satisfied firstly by applying a graded approach to contents limits for packages and vehicles and to performance standards applied to package designs depending upon the hazard of the radioactive contents. Secondly, they are satisfied by imposing conditions on the design and operation of packages and on the maintenance of packagings, including a consideration of the nature of the radioactive contents. Finally, they are satisfied by requiring administrative controls including, where appropriate, approval by competent authorities.

1.7.1.3　　　　ADN applies to the carriage of radioactive material by inland waterways including carriage which is incidental to the use of the radioactive material. Carriage comprises all operations and conditions associated with and involved in the movement of radioactive material; these include the design, manufacture, maintenance and repair of packaging, and the preparation, consigning, loading, carriage including in-transit storage, unloading and receipt at the final destination of loads of radioactive material and packages. A graded approach is applied to the performance standards in ADN that are characterized by three general severity levels:

(a)　　Routine conditions of carriage (incident free);

(b)　　Normal conditions of carriage (minor mishaps);

(c)　　Accident conditions of carriage.

1.7.1.4　　　　The provisions laid down in ADN do not apply to any of the following:

(a)　　Radioactive material that is an integral part of the means of transport;

(b)　　Radioactive material moved within an establishment which is subject to appropriate safety regulations in force in the establishment and where the movement does not involve public roads or railways;

(c)　　Radioactive material implanted or incorporated into a person or live animal for diagnosis or treatment;

(d)　　Radioactive material in or on a person who is to be transported for medical treatment because the person has been subject to accidental or deliberate intake of radioactive material or to contamination;

(e)　　Radioactive material in consumer products which have received regulatory approval, following their sale to the end user;

(f)　　Natural material and ores containing naturally occurring radionuclides (which may have been processed), provided the activity concentration of the material does not exceed 10 times the values specified in Table 2.2.7.2.2.1, or calculated in accordance with 2.2.7.2.2.2 (a) and 2.2.7.2.2.3 to 2.2.7.2.2.6. For natural materials and ores containing naturally occurring radionuclides that are not in secular equilibrium the calculation of the activity concentration shall be performed in accordance with 2.2.7.2.2.4;

(g)　　Non-radioactive solid objects with radioactive substances present on any surfaces in quantities not in excess of the limit set out in the definition for "contamination" in 2.2.7.1.2.

1.7.1.5　　　*Specific provisions for the carriage of excepted packages*

1.7.1.5.1　　　Excepted packages which may contain radioactive material in limited quantities, instruments, manufactured articles or empty packagings as specified in 2.2.7.2.4.1 shall be subject only to the following provisions of Parts 5 to 7:

(a)　　The applicable provisions specified in 5.1.2.1, 5.1.3.2, 5.1.5.2.2, 5.1.5.4, 5.2.1.10, 7.1.4.14.7.3.1, 7.1.4.14.7.5.1 to 7.1.4.14.7.5.4 and 7.1.4.14.7.7; and

(b)　　The requirements for excepted packages specified in 6.4.4 of ADR;

except when the radioactive material possesses other hazardous properties and has to be classified in a class other than Class 7 in accordance with special provision 290 or 369 of Chapter 3.3, where the provisions listed in (a) and (b) above apply only as relevant and in addition to those relating to the main class.

1.7.1.5.2   Excepted packages are subject to the relevant provisions of all other parts of ADN. If the excepted package contains fissile material, one of the fissile exceptions provided by 2.2.7.2.3.5 shall apply and the requirements of 7.1.4.14.7.4.3 shall be met.

1.7.2   **Radiation protection programme**

1.7.2.1   The carriage of radioactive material shall be subject to a radiation protection programme which shall consist of systematic arrangements aimed at providing adequate consideration of radiation protection measures.

1.7.2.2   Doses to persons shall be below the relevant dose limits. Protection and safety shall be optimized in order that the magnitude of individual doses, the number of persons exposed and the likelihood of incurring exposure shall be kept as low as reasonably achievable, economic and social factors being taken into account within the restriction that the doses to individuals be subject to dose constraints. A structured and systematic approach shall be adopted and shall include consideration of the interfaces between carriage and other activities.

1.7.2.3   The nature and extent of the measures to be employed in the programme shall be related to the magnitude and likelihood of radiation exposures. The programme shall incorporate the requirements in 1.7.2.2, 1.7.2.4, 1.7.2.5 and 7.5.11 CV33 (1.1) of ADR. Programme documents shall be available, on request, for inspection by the relevant competent authority.

1.7.2.4   For occupational exposures arising from transport activities, where it is assessed that the effective dose either:

(a)   is likely to be between 1 mSv and 6 mSv in a year, a dose assessment programme via work place monitoring or individual monitoring shall be conducted; or

(b)   is likely to exceed 6 mSv in a year, individual monitoring shall be conducted.

When individual monitoring or work place monitoring is conducted, appropriate records shall be kept.

NOTE: For occupational exposures arising from transport activities, where it is assessed that the effective dose is most unlikely to exceed 1mSv in a year, no special work patterns, detailed monitoring, dose assessment programmes or individual record keeping need be required.

1.7.2.5   Workers (see 7.1.4.14.7, NOTE 3) shall be appropriately trained in radiation protection including the precautions to be observed in order to restrict their occupational exposure and the exposure of other persons who might be affected by their actions.

1.7.3   **Management system**

1.7.3.1   A management system based on international, national or other standards acceptable to the competent authority shall be established and implemented for all activities within the scope of ADN, as identified in 1.7.1.3, to ensure compliance with the relevant provisions of ADN. Certification that the design specification has been fully implemented shall be available to the competent authority. The manufacturer, consignor or user shall be prepared:

(a)   To provide facilities for inspection during manufacture and use; and

(b)   To demonstrate compliance with ADN to the competent authority.

Where competent authority approval is required, such approval shall take into account and be contingent upon the adequacy of the management system.

1.7.4   **Special arrangement**

1.7.4.1   Special arrangement shall mean those provisions, approved by the competent authority, under which consignments which do not satisfy all the requirements of ADN applicable to radioactive material may be transported.

NOTE: Special arrangement is not considered to be a temporary derogation in accordance with 1.5.1.

1.7.4.2   Consignments for which conformity with any provision applicable to radioactive material is impracticable shall not be transported except under special arrangement. Provided the competent authority is satisfied that conformity with the radioactive material provisions of ADN is impracticable and that the requisite standards of safety established by ADN have been demonstrated through alternative means the competent authority may approve special arrangement transport operations for single or a planned series of multiple consignments. The overall level of safety in carriage shall be at least equivalent to that which would be provided if all the applicable requirements had been met. For international consignments of this type, multilateral approval shall be required.

1.7.5   **Radioactive material possessing other dangerous properties**

In addition to the radioactive and fissile properties, any subsidiary risk of the contents of the package, such as explosiveness, flammability, pyrophoricity, chemical toxicity and corrosiveness, shall also be taken into account in the documentation, packing, labelling, marking, placarding, stowage, segregation and carriage, in order to be in compliance with all relevant provisions for dangerous goods of ADN.

1.7.6   **Non-compliance**

1.7.6.1   In the event of non-compliance with any limit in ADN applicable to radiation level or contamination,

(a)   The consignor, consignee, carrier and any organization involved during carriage who may be affected, as appropriate, shall be informed of the non-compliance by:

(i)   by the carrier if the non-compliance is identified during carriage; or

(ii)   by the consignee if the non-compliance is identified at receipt;

(b)   The carrier, consignor or consignee, as appropriate shall:

(i)   take immediate steps to mitigate the consequences of the non-compliance;

(ii)   investigate the non-compliance and its causes, circumstances and consequences;

(iii) take appropriate action to remedy the causes and circumstances that led to the non-compliance and to prevent a recurrence of similar circumstances that led to the non-compliance; and

(iv)   communicate to the competent authority(ies) on the causes of the non-compliance and on corrective or preventive actions taken or to be taken;

(c)   The communication of the non-compliance to the consignor and competent authority(ies), respectively, shall be made as soon as practicable and it shall be immediate whenever an emergency exposure situation has developed or is developing.

CHAPTER 1.8

CHECKS AND OTHER SUPPORT MEASURES TO ENSURE COMPLIANCE

WITH SAFETY REQUIREMENTS

1.8.1 **Monitoring compliance with requirements**

1.8.1.1 *General*

1.8.1.1.1 In accordance with Article 4, paragraph 3 of ADN, Contracting Parties shall ensure that a representative proportion of consignments of dangerous goods carried by inland waterways is subject to monitoring in accordance with the provisions of this Chapter, and including the requirements of 1.10.1.5.

1.8.1.1.2 Participants in the carriage of dangerous goods (see Chapter 1.4) shall, without delay, in the context of their respective obligations, provide the competent authorities and their agents with the necessary information for carrying out the checks.

1.8.1.2 *Monitoring procedure*

1.8.1.2.1 In order to carry out the checks provided for in Article 4, paragraph 3 of ADN, the Contracting Parties shall use the checklist developed by the Administrative Committee.[*] A copy of this checklist shall be given to the master of the vessel. Competent authorities of other Contracting Parties may decide to simplify or refrain from conducting subsequent checks if a copy of the checklist is presented to them. This paragraph shall not prejudice the right of Contracting Parties to carry out specific measures or more detailed checks.

1.8.1.2.2 The checks shall be random and shall as far as possible cover an extensive portion of the inland waterway network.

1.8.1.2.3 When exercising the right to monitor, the authorities shall make all possible efforts to avoid unduly detaining or delaying a vessel.

1.8.1.3 *Infringements of the requirements*

Without prejudice to other penalties which may be imposed, vessels in respect of which one or more infringements of the rules on the transport of dangerous goods by inland waterways are established may be detained at a place designated for this purpose by the authorities carrying out the check and required to be brought into conformity before continuing their journey or may be subject to other appropriate measures, depending on the circumstances or the requirements of safety.

1.8.1.4 *Checks in companies and at places of loading and unloading*

1.8.1.4.1 Checks may be carried out at the premises of undertakings, as a preventive measure or where infringements which jeopardize safety in the transport of dangerous goods have been recorded during the voyage.

1.8.1.4.2 The purpose of such checks shall be to ensure that safety conditions for the transport of dangerous goods by inland waterways comply with the relevant laws.

---

[*]   *Note by the secretariat: The model of the checklist can be found on the United Nations Economic Commission for Europe website (http://www.unece.org/trans/danger/danger.html).*

1.8.1.4.3   *Sampling*

Where appropriate and provided that this does not constitute a safety hazard, samples of the goods transported may be taken for examination by laboratories recognized by the competent authority.

1.8.1.4.4   *Cooperation of the competent authorities*

1.8.1.4.4.1   Contracting Parties shall assist one another in order to give proper effect to these requirements.

1.8.1.4.4.2   Serious or repeated infringements jeopardizing the safety of the transport of dangerous goods committed by a foreign vessel or undertaking shall be reported to the competent authority in the Contracting Party where the certificate of approval of the vessel was issued or where the undertaking is established.

1.8.1.4.4.3   The competent authority of the Contracting Party where serious or repeated infringements have been recorded may ask the competent authority of the Contracting Party where the certificate of approval of the vessel was issued or where the undertaking is established for appropriate measures to be taken with regard to the offender or offenders.

1.8.1.4.4.4   The latter competent authority shall notify the competent authorities of the Contracting Party where the infringements were recorded of any measures taken with regard to the offender or offenders.

1.8.2   Administrative assistance during the checking of a foreign vessel

If the findings of a check on a foreign vessel give grounds for believing that serious or repeated infringements have been committed which cannot be detected in the course of that check in the absence of the necessary data, the competent authorities of the Contracting Parties concerned shall assist one another in order to clarify the situation.

1.8.3   Safety adviser

1.8.3.1   Each undertaking, the activities of which include the carriage, or the related packing, loading, filling or unloading, of dangerous goods by inland waterways shall appoint one or more safety advisers, hereinafter referred to as "advisers", for the carriage of dangerous goods, responsible for helping to prevent the risks inherent in such activities with regard to persons, property and the environment.

1.8.3.2   The competent authorities of the Contracting Parties may provide that these requirements shall not apply to undertakings:

(a)   the activities of which concern:

(i)   The carriage of dangerous goods fully or partially exempted according to the provisions of 1.7.1.4 or of chapters 3.3, 3.4 or 3.5;

(ii)   Quantities per transport unit, wagon or container smaller than those referred to in 1.1.3.6 of ADR or RID;

(iii)   When (ii) above is not relevant, quantities per vessel smaller than those referred to in 1.1.3.6 of these Regulations.

(b)   the main or secondary activities of which are not the carriage or the related packing, filling, loading or unloading of dangerous goods but which occasionally engage in the national carriage or the related packing, filling, loading or unloading of dangerous goods posing little danger or risk of pollution.

1.8.3.3   The main task of the adviser shall be, under the responsibility of the head of the undertaking, to seek by all appropriate means and by all appropriate action, within the limits of the relevant activities of that undertaking, to facilitate the conduct of those activities in accordance with the requirements applicable and in the safest possible way.

With regard to the undertaking's activities, the adviser has the following duties in particular:

–   monitoring compliance with the requirements governing the carriage of dangerous goods;

–   advising his undertaking on the carriage of dangerous goods;

–   preparing an annual report to the management of his undertaking or a local public authority, as appropriate, on the undertaking's activities in the carriage of dangerous goods. Such annual reports shall be preserved for five years and made available to the national authorities at their request.

The adviser's duties also include monitoring the following practices and procedures relating to the relevant activities of the undertaking:

–   the procedures for compliance with the requirements governing the identification of dangerous goods being transported;

–   the undertaking's practice in taking account, when purchasing means of transport, of any special requirements in connection with the dangerous goods being transported;

–   the procedures for checking the equipment used in connection with the carriage, packing, filling, loading or unloading of dangerous goods;

–   the proper training of the undertaking's employees, including on the changes to the Regulations, and the maintenance of records of such training;

–   the implementation of proper emergency procedures in the event of any accident or incident that may affect safety during the carriage, packing, filling, loading or unloading of dangerous goods;

–   investigating and, where appropriate, preparing reports on serious accidents, incidents or serious infringements recorded during the carriage, packing, filling, loading or unloading of dangerous goods;

–   the implementation of appropriate measures to avoid the recurrence of accidents, incidents or serious infringements;

–   the account taken of the legal prescriptions and special requirements associated with the carriage of dangerous goods in the choice and use of sub-contractors or third parties;

–   verification that employees involved in the carriage, packing, filling, loading or unloading of dangerous goods have detailed operational procedures and instructions,

–   the introduction of measures to increase awareness of the risks inherent in the carriage, packing, filling, loading and unloading of dangerous goods;

–   the implementation of verification procedures to ensure the presence on board means of transport of the documents and safety equipment which must accompany transport and the compliance of such documents and equipment with the regulations;

–   the implementation of verification procedures to ensure compliance with the requirements governing packing, filling, loading and unloading;

–   the existence of the security plan indicated in 1.10.3.2.

1.8.3.4   The safety adviser may also be the head of the undertaking, a person with other duties in the undertaking, or a person not directly employed by that undertaking, provided that that person is capable of performing the duties of adviser.

1.8.3.5   Each undertaking concerned shall, on request, inform the competent authority or the body designated for that purpose by each Contracting Party of the identity of its adviser.

1.8.3.6   Whenever an accident affects persons, property or the environment or results in damage to property or the environment during carriage, packing, filling, loading or unloading carried out by the undertaking concerned, the safety adviser shall, after collecting all the relevant information, prepare an accident report to the management of the undertaking or to a local public authority, as appropriate. That report shall not replace any report by the management of the undertaking which might be required under any other international or national legislation.

1.8.3.7   A safety adviser shall hold a vocational training certificate, valid for transport by inland waterways. That certificate shall be issued by the competent authority or the body designated for that purpose by each Contracting Party.

1.8.3.8   To obtain a certificate, a candidate shall undergo training and pass an examination approved by the competent authority of the Contracting Party.

1.8.3.9   The main aims of the training shall be to provide candidates with sufficient knowledge of the risks inherent in the carriage packing, filling, loading or unloading of dangerous goods, of the applicable laws, regulations and administrative provisions and of the duties listed in 1.8.3.3.

1.8.3.10   The examination shall be organized by the competent authority or by an examining body designated by the competent authority. The examining body shall not be a training provider.

The examining body shall be designated in writing. This approval may be of limited duration and shall be based on the following criteria:

–   competence of the examining body;

–   specifications of the form of the examinations the examining body is proposing, including, if necessary, the infrastructure and organisation of electronic examinations in accordance with 1.8.3.12.5, if these are to be carried out;

–   measures intended to ensure that examinations are impartial;

–   independence of the body from all natural or legal persons employing safety advisers.

1.8.3.11   The aim of the examination is to ascertain whether candidates possess the necessary level of knowledge to carry out the duties incumbent upon a safety adviser as listed in 1.8.3.3, for the purpose of obtaining the certificate prescribed in subsection 1.8.3.7, and it shall cover at least the following subjects:

(a)   Knowledge of the types of consequences which may be caused by an accident involving dangerous goods and knowledge of the main causes of accidents;

(b) Requirements under national law, international conventions and agreements, with regard to the following in particular:

– classification of dangerous goods (procedure for classifying solutions and mixtures, structure of the list of substances, classes of dangerous goods and principles for their classification, nature of dangerous goods transported, physical, chemical and toxicological properties of dangerous goods);

– general packing provisions, provisions for tanks and tank-containers (types, code, marking, construction, initial and periodic inspection and testing);

– marking and labelling, placarding and orange-coloured plate marking (marking and labelling of packages, placing and removal of placards and orange-coloured plates);

– particulars in transport documents (information required);

– method of consignment and restrictions on dispatch (full load, carriage in bulk, carriage in intermediate bulk containers, carriage in containers, carriage in fixed or demountable tanks);

– transport of passengers;

– prohibitions and precautions relating to mixed loading;

– segregation of goods;

– limitation of the quantities carried and quantities exempted;

– handling and stowage (packing, filling, loading and unloading - filling ratios - stowage and segregation);

– cleaning and/or degassing before packing, filling, loading and after unloading;

– crews, vocational training;

– vehicle documents (transport documents, instructions in writing, vessel approval certificate, ADN dangerous goods training certificate, copies of any derogations, other documents);

– instructions in writing (implementation of the instructions and crew protection equipment);

– supervision requirements (berthing);

– traffic regulations and restrictions;

– operational discharges or accidental leaks of pollutants;

– requirements relating to equipment for transport (vessel).

1.8.3.12   *Examinations*

1.8.3.12.1   The examination shall consist of a written test which may be supplemented by an oral examination.

1.8.3.12.2   The competent authority or an examining body designated by the competent authority shall invigilate every examination. Any manipulation and deception shall be ruled out as far as possible. Authentication of the candidate shall be ensured. The use in the written test of documentation other than international or national regulations is not permitted. All examination documents shall be recorded and kept as a print-out or electronically as a file.

1.8.3.12.3   Electronic media may be used only if provided by the examining body. There shall be no means of a candidate introducing further data to the electronic media provided; the candidate may only answer to the questions posed.

1.8.3.12.4   The written test shall consist of two parts:

(a)   Candidates shall receive a questionnaire. It shall include at least 20 open questions covering at least the subjects mentioned in the list in 1.8.3.11. However, multiple choice questions may be used. In this case, two multiple choice questions count as one open question. Amongst these subjects particular attention shall be paid to the following subjects:

–   general preventive and safety measures;

–   classification of dangerous goods;

–   general packing provisions, including tanks, tank-containers, tank-vehicles, etc.;

–   danger marking, labelling and placarding;

–   information in the transport document;

–   handling and stowage;

–   crew, vocational training;

–   vehicle documents and transport certificates;

–   instructions in writing;

–   requirements concerning equipment for transport by vessel;

(b)   Candidates shall undertake a case study in keeping with the duties of the adviser referred to in 1.8.3.3, in order to demonstrate that they have the necessary qualifications to fulfil the task of adviser.

1.8.3.12.5   Written examinations may be performed, in whole or in part, as electronic examinations, where the answers are recorded and evaluated using electronic data processing (EDP) processes, provided the following conditions are met:

(a)   The hardware and software shall be checked and accepted by the competent authority or by an examining body designated by the competent authority;

(b)   Proper technical functioning shall be ensured. Arrangements as to whether and how the examination can be continued shall be made for a failure of the devices and applications. No aids shall be available on the input devices (e.g. electronic search function), the equipment provided according to 1.8.3.12.3 shall not allow the candidates to communicate with any other device during the examination;

(c)   Final inputs of each candidate shall be logged. The determination of the results shall be transparent.

1.8.3.13   The Contracting Parties may decide that candidates who intend working for undertakings specializing in the carriage of certain types of dangerous goods need only be questioned on the substances relating to their activities. These types of goods are:

–   Class 1;

–   Class 2;

–   Class 7;

–   Classes 3, 4.1, 4.2, 4.3, 5.1, 5.2, 6.1, 6.2, 8 and 9;

–   UN Nos. 1202, 1203, 1223, 3475, and aviation fuel classified under UN Nos. 1268 or 1863.

The certificate prescribed in 1.8.3.7 shall clearly indicate that it is only valid for one type of the dangerous goods referred to in this subsection and on which the adviser has been questioned under the conditions defined in 1.8.3.12.

1.8.3.14   The competent authority or the examining body shall keep a running list of the questions that have been included in the examination.

1.8.3.15   The certificate prescribed in 1.8.3.7 shall take the form laid down in 1.8.3.18 and shall be recognized by all Contracting Parties.

1.8.3.16   *Validity and renewal of certificates*

1.8.3.16.1   The certificate shall be valid for five years. The period of validity of a certificate shall be extended from the date of its expiry for five years at a time where, during the year before its expiry, its holder has passed an examination. The examination shall be approved by the competent authority.

1.8.3.16.2   The aim of the examination is to ascertain that the holder has the necessary knowledge to carry out the duties set out in 1.8.3.3. The knowledge required is set out in 1.8.3.11 (b) and shall include the amendments to the Regulations introduced since the award of the last certificate. The examination shall be held and supervised on the same basis as in 1.8.3.10 and 1.8.3.12 to 1.8.3.14. However, holders need not undertake the case study specified in 1.8.3.12.4 (b).

1.8.3.17   The requirements set out in 1.8.3.1 to 1.8.3.16 shall be considered to have been fulfilled if the relevant conditions of Council Directive 96/35/EC of 3 June 1996 on the appointment and vocational qualification of safety advisers for the transport of dangerous goods by road, rail and inland waterway[1] and of Directive 2000/18/EC of the European Parliament and of the Council of 17 April 2000 on minimum examination requirements for safety advisers for the transport of dangerous goods by road, rail or inland waterway[2] are applied.

---

[1]   *Official Journal of the European Communities, No. L145 of 19 June 1996, page 10.*

[2]   *Official Journal of the European Communities, No. L118 of 19 May 2000, page 41.*

1.8.3.18   *Form of certificate*

Certificate of training as safety adviser for the transport of dangerous goods

Certificate No: ....

Distinguishing sign of the State issuing the certificate: ....

Surname: ....

Forename(s): ....

Date and place of birth: ....

Nationality: ....

Signature of holder: ....

Valid until .... for undertakings which transport dangerous goods and for undertakings which carry out related packing, filling, loading or unloading:

☐ by road   ☐ by rail   ☐ by inland waterway

Issued by: ....

Date: ....   Signature: ....

1.8.4   **List of competent authorities and bodies designated by them**

The Contracting Parties shall communicate to the secretariat of the United Nations Economic Commission for Europe the addresses of the authorities and bodies designated by them which are competent in accordance with national law to implement ADN, referring in each case to the relevant requirement of ADN and giving the addresses to which the relevant applications should be made.

The secretariat of the United Nations Economic Commission for Europe shall establish a list on the basis of the information received and shall keep it up-to-date. It shall communicate this list and the amendments thereto to the Contracting Parties.

1.8.5   **Notifications of occurrences involving dangerous goods**

1.8.5.1   If a serious accident or incident takes place during loading, filling, carriage or unloading of dangerous goods on the territory of a Contracting Party, the loader, filler, carrier or consignee, respectively, shall ascertain that a report conforming to the model prescribed in 1.8.5.4 is made to the competent authority of the Contracting Party concerned at the latest one month after the occurrence.

1.8.5.2   The Contracting Party shall in turn, if necessary, make a report to the secretariat of the United Nations Economic Commission for Europe with a view to informing the other Contracting Parties.

1.8.5.3   *An occurrence subject to report* in accordance with 1.8.5.1 has occurred if dangerous goods were released or if there was an imminent risk of loss of product, if personal injury, material or environmental damage occurred, or if the authorities were involved and one or more of the following criteria has/have been met:

Personal injury means an occurrence in which death or injury directly relating to the dangerous goods carried has occurred, and where the injury

(a)   requires intensive medical treatment,

(b)   requires a stay in hospital of at least one day, or

(c)   results in the inability to work for at least three consecutive days.

Loss of product means the release of dangerous goods of:

(a)   Classes 1 or 2 or packing group I or other substances not assigned to a packing group in quantities of 50 kg or 50 litres or more;

(b)   Packing group II in quantities of 333 kg or 333 litres or more; or

(c)   Packing group III in quantities of 1,000 kg or 1,000 litres or more.

The loss of product criterion also applies if there was an imminent risk of loss of product in the above-mentioned quantities. As a rule, this has to be assumed if, owing to structural damage, the means of containment is no longer suitable for further carriage or if, for any other reason, a sufficient level of safety is no longer ensured (e.g. owing to distortion of tanks or containers, overturning of a tank or fire in the immediate vicinity).

If dangerous goods of Class 6.2 are involved, the obligation to report applies without quantity limitation.

In occurrences involving radioactive material, the criteria for loss of product are:

(a)   Any release of radioactive material from the packages;

(b)   Exposure leading to a breach of the limits set out in the regulations for protection of workers and members of the public against ionizing radiation (Schedule II of IAEA Safety Series No. 115 – "International Basic Safety Standards for Protection Against Ionizing Radiation and for Safety of Radiation Sources"); or

(c)   Where there is reason to believe that there has been a significant degradation in any package safety function (containment, shielding, thermal protection or criticality) that may have rendered the package unsuitable for continued carriage without additional safety measures.

NOTE: *See the provisions of 7.1.4.14.7.7 for undeliverable consignments.*

Material damage or *environmental damage* means the release of dangerous goods, irrespective of the quantity, where the estimated amount of damage exceeds 50,000 Euros. Damage to any directly involved means of carriage containing dangerous goods and to the modal infrastructure shall not be taken into account for this purpose.

Involvement of authorities means the direct involvement of the authorities or emergency services during the occurrence involving dangerous goods and the evacuation of persons or closure of public traffic routes (roads/railways/inland waterways) for at least three hours owing to the danger posed by the dangerous goods.

If necessary, the competent authority may request further relevant information.

1.8.5.4   *Model report on occurrences during the carriage of dangerous goods*

Report on occurrences during the carriage of dangerous goods in accordance with ADN, section 1.8.5

Report No.:

Carrier/Filler/Consignee/Loader: ....

Official number of vessel: ....

Dry cargo vessel (single-hull, double-hull): ....

Tank vessel (type): ....

Address:

Contact name: .... Telephone: ....

Fax/e-mail: ....

(The competent authority shall remove this cover sheet before forwarding the report)

1.	**Mode**	
	Inland waterway	Official number of vessel/name of vessel (optional)
	....	....

2.	**Date and location of occurrence**
	Year: .... Month: .... Day: .... Time: ....

		Comments concerning description of location:
☐	Port	....
☐	Loading/unloading/transhipment facility	....
	Location/Country:	
	or	
☐	Free sector	
	Name of sector:	
	Kilometre point:	
	or	
☐	Structure such as bridge or guide wall	

3.	**Conditions of inland waterway**
	Water level (reference gauge): ....
	Estimated speed through water: ....
☐	High water
☐	Low water

4.	**Particular weather conditions**
☐	Rain
☐	Snow
☐	Fog
☐	Thunderstorm
☐	Storm
	Temperature: .... °C

5.	**Description of occurrence**
☐	Collision with bank, structure or berthing installation
☐	Collision with another cargo vessel (collision/impact)
☐	Collision with a passenger vessel (collision/impact)
☐	Contact with the waterway bed, whether or not vessel has run aground
☐	Fire
☐	Explosion
☐	Leak/Location and extent of damage (with additional description)
☐	Shipwreck
☐	Capsizing
☐	Technical fault (optional)
☐	Human error (optional)
	Additional description of occurrence:
	....
	....
	....
	....

6. Dangerous goods involved

UN Number[1] or Identification number	Class	Packing group if known	Estimated quantity of loss of products (kg or l)[2]	Means of containment in accordance with ADN, 1.2.1[3]	Means of containment material	Type of failure of means of containment[4]

[1] For dangerous goods assigned to collective entries to which special provision 274 applies, also the technical name shall be indicated.	[2] For class 7, indicate values according to the criteria in 1.8.5.3.

[3] Indicate the appropriate number: 1 Packaging 2 IBC 3 Large packaging 4 Small container 5 Wagon 6 Vehicle 7 Tank-wagon 8 Tank-vehicle 9 Battery-wagon 10 Battery-vehicle 11 Wagon with demountable tanks 12 Demountable tank 13 Large container	[4] Indicate the appropriate number: 1 Loss 2 Fire 3 Explosion 4 Structural failure

14 Tank container 15 MEGC 16 Portable tank 17 Dry cargo vessel (single-hull, double-hull) 18 Tank vessel (type)	

7. Cause of occurrence (if clearly known) (optional)

☐   Technical fault
☐   Faulty load securing
☐   Operational cause
☐   Other: ....
             ....
             ....
             ....

8. Consequences of occurrence

Personal injury in connection with the dangerous goods involved:

☐   Deaths (number: ....)
☐   Injured (number: ....)

Loss of product:

☐   Yes
☐   No
☐   Imminent risk of loss of product

Material/Environment damage:

☐   Estimated level of damage $\leq$ 50 000 Euros
☐   Estimated level of damage $>$ 50 000 Euros

Involvement of authorities:

☐   Yes   ☐   Evacuation of persons for a duration of at least three hours caused by the dangerous goods involved
                                ☐   Closure of public traffic routes for a duration of at least three hours caused by the dangerous goods involved

☐   No

If necessary, the competent authority may request further relevant information.

CHAPTER 1.9

TRANSPORT RESTRICTIONS BY THE COMPETENT AUTHORITIES

1.9.1   In accordance with Article 6, paragraph 1 of ADN, the entry of dangerous goods into the territory of Contracting Parties may be subject to regulations or prohibitions imposed for reasons other than safety during carriage. Such regulations or prohibitions shall be published in an appropriate form.

1.9.2   Subject to the provisions of 1.9.3, a Contracting Party may apply to vessels engaged in the international carriage of dangerous goods by inland waterways on its territory certain additional provisions not included in ADN, provided that those provisions do not conflict with Article 4, paragraph 2 of ADN, and are contained in its domestic legislation applying equally to vessels engaged in the domestic carriage of dangerous goods by inland waterways on the territory of that Contracting Party.

1.9.3   Additional provisions falling within the scope of 1.9.2 are as follows:

(a)   Additional safety requirements or restrictions concerning vessels using certain structures such as bridges or tunnels, or vessels entering or leaving ports or other transport terminals;

(b)   Requirements for vessels to follow prescribed routes to avoid commercial or residential areas, environmentally sensitive areas, industrial zones containing hazardous installations or inland waterways presenting severe physical hazards;

(c)   Emergency requirements regarding routing or parking of vessels carrying dangerous goods resulting from extreme weather conditions, earthquake, accident, industrial action, civil disorder or military hostilities;

(d)   Restrictions on movement of vessels carrying dangerous goods on certain days of the week or year.

1.9.4   The competent authority of the Contracting Party applying on its territory any additional provisions within the scope of 1.9.3 (a) and (d) above shall notify the secretariat of the United Nations Economic Commission for Europe of the additional provisions, which secretariat shall bring them to the attention of the Contracting Parties.

CHAPTER 1.10

SECURITY PROVISIONS

NOTE: *For the purposes of this Chapter, "security" means measures or precautions to be taken to minimise theft or misuse of dangerous goods that may endanger persons, property or the environment.*

1.10.1   General provisions

1.10.1.1   All persons engaged in the carriage of dangerous goods shall consider the security requirements set out in this Chapter commensurate with their responsibilities.

1.10.1.2   Dangerous goods shall only be offered for carriage to carriers that have been appropriately identified.

1.10.1.3   Holding areas in trans-shipment zones for dangerous goods shall be secured, well lit and, where possible and appropriate, not accessible to the general public.

1.10.1.4   For each crew member of a vessel carrying dangerous goods, means of identification, which includes a photograph, shall be on board during carriage.

1.10.1.5   Safety checks in accordance with 1.8.1 shall also concern the implementation of security measures.

1.10.1.6   The competent authority shall maintain up-to-date registers of all valid certificates for experts stipulated in 8.2.1 issued by it or by any recognized organization.

1.10.2   Security training

1.10.2.1   The training and the refresher training specified in Chapter 1.3 shall also include elements of security awareness. The security refresher training need not be linked to regulatory changes only.

1.10.2.2   Security awareness training shall address the nature of security risks, recognising security risks, methods to address and reduce such risks and actions to be taken in the event of a security breach. It shall include awareness of security plans (if appropriate) commensurate with the responsibilities and duties of individuals and their part in implementing security plans.

1.10.2.3   Such training shall be provided or verified upon employment in a position involving dangerous goods transport and shall be periodically supplemented with refresher training.

1.10.2.4   Records of all security training received shall be kept by the employer and made available to the employee or competent authority, upon request. Records shall be kept by the employer for a period of time established by the competent authority.

1.10.3   Provisions for high consequence dangerous goods

1.10.3.1   *Definition of high consequence dangerous goods*

1.10.3.1.1   High consequence dangerous goods are those which have the potential for misuse in a terrorist event and which may, as a result, produce serious consequences such as mass casualties, mass destruction or, particularly for Class 7, mass socio-economic disruption.

1.10.3.1.2   High consequence dangerous goods in classes other than Class 7 are those listed in Table 1.10.3.1.2 below and carried in quantities greater than those indicated therein.

Table 1.10.3.1.2: List of high consequence dangerous goods

Class	Division	Substance or article	Tank or cargo tank (litres)^c	Bulk$^{*/}$ (kg) ^{d}	Goods in packages (kg)
1	1.1	Explosives	a	a	0
	1.2	Explosives	a	a	0
	1.3	Compatibility group C explosives	a	a	0
	1.5	Explosives	0	a	0
1	1.4	Explosives of UN Nos. 0104, 0237, 0255, 0267, 0289, 0361, 0365, 0366, 0440, 0441, 0455, 0456 and 0500	a	a	0
2		Flammable gases (classification codes including only letter F)	3000	a	b
		Toxic gases (classification codes including letter(s) T, TF, TC, TO, TFC or TOC) excluding aerosols	0	a	0
3		Flammable liquids of packing groups I and II	3000	a	b
		Desensitized explosives	0	a	0
4.1		Desensitized explosives	a	a	0
4.2		Packing group I substances	3000	a	b
4.3		Packing group I substances	3000	a	b
5.1		Oxidizing liquids of packing group I	3000	a	b
		Perchlorates, ammonium nitrate, ammonium nitrate fertilisers and ammonium nitrate emulsions or suspensions or gels	3000	3000	b
6.1		Toxic substances of packing group I	0	a	0
6.2		Infectious substances of Category A (UN Nos. 2814 and 2900, except for animal material)	a	0	0
8		Corrosive substances of packing group I	3000	a	b

*/   Bulk means bulk in the vessel, or bulk in a vehicle or a container.

a   Not relevant.

b   The provisions of 1.10.3 do not apply, whatever the quantity is.

c   A value indicated in this column is applicable only if carriage in tanks is authorized according to chapter 3.2, table A, column (10) or (12) of ADR or RID or if letter "T" is indicated in chapter 3.2, table A, column (8) of ADN. For substances which are not authorized for carriage in tanks, the instruction in this column is not relevant.

d   A value indicated in this column is applicable only if carriage in bulk is authorized according to chapter 3.2, table A, column (10) or (17) of ADR or RID, or if letter "B" is indicated in chapter 3.2, table A, column (8) of ADN. For substances which are not authorized for carriage in bulk, the instruction in this column is not relevant.

1.10.3.1.3   For dangerous goods of Class 7, high consequence radioactive material is that with an activity equal to or greater than a transport security threshold of 3 000 A_2 per single package (see also 2.2.7.2.2.1) except for the following radionuclides where the transport security threshold is given in Table 1.10.3.1.3 below.

Table 1.10.3.1.3: Transport security thresholds for specific radionuclides

Element	Radionuclide	Transport security threshold (TBq)
Americium	Am-241	0.6
Gold	Au-198	2
Cadmium	Cd-109	200
Caesium	Cs-137	1
Californium	Cf-252	0.2

Element	Radionuclide	Transport security threshold (TBq)
Curium	Cm-244	0.5
Cobalt	Co-57	7
Cobalt	Co-60	0.3
Iron	Fe-55	8000
Germanium	Ge-68	7
Gadolinium	Gd-153	10
Iridium	Ir-192	0.8
Nickel	Ni-63	600
Palladium	Pd-103	900
Promethium	Pm-147	400
Polonium	Po-210	0.6
Plutonium	Pu-238	0.6
Plutonium	Pu-239	0.6
Radium	Ra-226	0.4
Ruthenium	Ru-106	3
Selenium	Se-75	2
Strontium	Sr-90	10
Thallium	Tl-204	200
Thulium	Tm-170	200
Ytterbium	Yb-169	3

1.10.3.1.4   For mixtures of radionuclides, determination of whether or not the transport security threshold has been met or exceeded can be calculated by summing the ratios of activity present for each radionuclide divided by the transport security threshold for that radionuclide. If the sum of the fractions is less than 1, then the radioactivity threshold for the mixture has not been met nor exceeded.

This calculation can be made with the formula:

$$\sum_i \frac{A_i}{T_i} < 1$$

Where:

A_i = activity of radionuclide i that is present in a package (TBq)

T_i = transport security threshold for radionuclide i (TBq).

1.10.3.1.5   When radioactive material possess subsidiary risks of other classes, the criteria of Table 1.10.3.1.2 shall also be taken into account (see also 1.7.5).

1.10.3.2   *Security plans*

1.10.3.2.1   Carriers, consignors and other participants specified in 1.4.2 and 1.4.3 engaged in the carriage of high consequence dangerous goods (see Table 1.10.3.1.2) or high consequence radioactive material (see 1.10.3.1.3) shall adopt, implement and comply with a security plan that addresses at least the elements specified in 1.10.3.2.2.

1.10.3.2.2   The security plan shall comprise at least the following elements:

(a)   specific allocation of responsibilities for security to competent and qualified persons with appropriate authority to carry out their responsibilities;

(b)   records of dangerous goods or types of dangerous goods concerned;

(c)   review of current operations and assessment of security risks, including any stops necessary to the transport operation, the keeping of dangerous goods in the vessel, tank or container before, during and after the journey and the intermediate temporary storage of dangerous goods during the course of intermodal transfer or transshipment between units;

(d)   clear statement of measures that are to be taken to reduce security risks, commensurate with the responsibilities and duties of the participant, including:

-     training;

-     security policies (e.g. response to higher threat conditions, new employee/employment verification, etc.);

-     operating practices (e.g. choice/use of routes where known, access to dangerous goods in intermediate temporary storage (as defined in (c)), proximity to vulnerable infrastructure etc.);

-     equipment and resources that are to be used to reduce risks;

(e)   effective and up to date procedures for reporting and dealing with security threats, breaches of security or security incidents;

(f)   procedures for the evaluation and testing of security plans and procedures for periodic review and update of the plans;

(g)   measures to ensure the physical security of transport information contained in the security plan; and

(h)   measures to ensure that the distribution of information relating to the transport operation contained in the security plan is limited to those who need to have it. Such measures shall not preclude the provision of information required elsewhere in ADN.

NOTE: Carriers, consignors and consignees should co-operate with each other and with competent authorities to exchange threat information, apply appropriate security measures and respond to security incidents.

1.10.3.3   Operational or technical measures shall be taken on vessels carrying high consequence dangerous goods (see Table 1.10.3.1.2) or high consequence radioactive material (see 1.10.3.1.3) in order to prevent the improper use of the vessel and of the dangerous goods. The application of these protective measures shall not jeopardize emergency response.

NOTE: When appropriate and already fitted, the use of transport telemetry or other tracking methods or devices should be used to monitor the movement of high consequence dangerous goods (see Table 1.10.3.1.2 or 1.10.3.1.3).

1.10.4   Except for radioactive material, the requirements of 1.10.1, 1.10.2 and 1.10.3 do not apply when the quantities carried in packages on a vessel do not exceed those referred to in 1.1.3.6.1. In addition the provisions of this Chapter do not apply to the carriage of UN No. 2912 RADIOACTIVE MATERIAL, LOW SPECIFIC ACTIVITY (LSA-I) and UN No. 2913 RADIOACTIVE MATERIAL, SURFACE CONTAMINATED OBJECTS (SCO-I).

1.10.5   For radioactive material, the provisions of this Chapter are deemed to be complied with when the provisions of the Convention on Physical Protection of Nuclear Material[1] and the IAEA circular on "The Physical Protection of Nuclear Material and Nuclear Facilities"[2] are applied.

---

[1]   *IAEACIRC/274/Rev.1, IAEA, Vienna (1980).*
[2]   *IAEACIRC/225/Rev.4 (Corrected), IAEA, Vienna (1999.*

CHAPTERS 1.11 to 1.14

(Reserved)

CHAPTER 1.15

RECOGNITION OF CLASSIFICATION SOCIETIES

1.15.1   **General**

In the event of the conclusion of an international agreement concerning more general regulations or the navigation of vessels on inland waterways and containing provisions relating to the full range of activities of classification societies and their recognition, any provision of this Chapter in contradiction with any of the provisions of the said international agreement would, in the relations among Parties to this Agreement which had become parties to the international agreement and as from the day of the entry into force of the latter, automatically be deleted and replaced ipso facto by the relevant provision of the international agreement. This Chapter would become null and void once the international agreement came into force if all Parties to this Agreement became Parties to the international agreement.

1.15.2   **Procedure for the recognition of classification societies**

1.15.2.1   A classification society which wishes to be recommended for recognition under this Agreement shall submit its application for recognition, in accordance with the provisions of this Chapter, to the competent authority of a Contracting Party.

The classification society shall prepare the relevant information in accordance with the provisions of this Chapter. It shall produce it in, at least, an official language of the State where the application is submitted and in English.

The Contracting Party shall forward the application to the Administrative Committee unless in its opinion the conditions and criteria referred to in 1.15.3 have manifestly not been met.

1.15.2.2   The Administrative Committee shall appoint a Committee of Experts and determine its composition and its rules of procedure. This Committee of Experts shall consider the proposal; it shall determine whether the classification society meets the criteria set out in 1.15.3 and shall make a recommendation to the Administrative Committee within a period of six months.

1.15.2.3   The Administrative Committee shall examine the report of the experts. It shall decide in accordance with the procedure set out in Article 17, 7(c), within one year maximum, whether or not to recommend to the Contracting Parties that they should recognize the classification society in question. The Administrative Committee shall establish a list of the classification societies recommended for recognition by the Contracting Parties.

1.15.2.4   Each Contracting Party may or may not decide to recognize the classification societies in question, only on the basis of the list referred to in 1.15.2.3. The Contracting Party shall inform the Administrative Committee and the other Contracting Parties of its decision.

The Administrative Committee shall update the list of recognitions issued by Contracting Parties.

1.15.2.5   If a Contracting Party considers that a classification society no longer meets the conditions and criteria set out in 1.15.3, it may submit a proposal to the Administrative Committee for withdrawal from the list of recommended societies. Such a proposal shall be substantiated by convincing evidence of a failure to meet the conditions and criteria.

1.15.2.6   The Administrative Committee shall set up a new Committee of Experts following the procedure set out under 1.15.2.2 which shall report to the Administrative Committee within a period of six months. The classification society shall be informed and invited by the Committee of Experts to comment on the findings.

1.15.2.7   The Administrative Committee may decide, in case of a failure(s) to meet the conditions and criteria in 1.15.3, that the classification society shall have the opportunity to present a plan to address the identified failure(s) within a deadline of six months and to avoid any reoccurrence or, in accordance with Article 17, 7 (c), to withdraw the name of the society in question from the list of societies recommended for recognition.

1.15.3   Conditions and criteria for the recognition of a classification society applying for recognition

A classification society applying for recognition under this Agreement shall meet all the following conditions and criteria:

1.15.3.1   A classification society shall be able to demonstrate extensive knowledge of and experience in the assessment of the design and construction of inland navigation vessels. The society should have comprehensive rules and regulations for the design, construction and periodical inspection of vessels. These rules and regulations shall be published and continuously updated and improved through research and development programmes.

1.15.3.2   Registers of the vessels classified by the classification society shall be published annually.

1.15.3.3   The classification society shall not be controlled by shipowners or shipbuilders, or by others engaged commercially in the manufacture, fitting out, repair or operation of ships. The classification society shall not be substantially dependent on a single commercial enterprise for its revenue.

1.15.3.4   The headquarters or a branch of the classification society authorized and entitled to give a ruling and to act in all areas incumbent on it under the regulations governing inland navigation shall be located in one of the Contracting Parties.

1.15.3.5   The classification society and its experts shall have a good reputation in inland navigation; the experts shall be able to provide proof of their professional abilities.

1.15.3.6   The classification society:

–   shall have sufficient professional staff and engineers for the technical tasks of monitoring and inspection and for the tasks of management, support and research, in proportion to the tasks and the number of vessels classified and sufficient to keep regulations up to date and develop them in the light of quality requirements;

–   shall have experts in at least two Contracting Parties.

1.15.3.7   The classification society shall be governed by a code of ethics.

1.15.3.8   The classification society shall have prepared and implemented and shall maintain an effective system of internal quality based on the relevant aspects of internationally recognized quality standards and conforming to the standards EN ISO/IEC 17020:2012 (except clause 8.1.3) (inspection bodies) and ISO 9001 or EN ISO 9001:2015. The classification society is subject to certification of its quality system by an independent body of auditors recognized by the administration of the State in which it is located.

1.15.4 Obligations of recommended classification societies

1.15.4.1   Recommended classification societies shall undertake to cooperate with each other so as to guarantee equivalence from the point of view of safety of their technical standards which are relevant to the implementation of the provisions of the present Agreement.

1.15.4.2   They shall exchange experiences in joint meetings at least once a year. They shall report annually to the Safety Committee. The secretariat of the Safety Committee shall be informed of those meetings. The opportunity will be given to Contracting Parties to attend the meetings as observers.

1.15.4.3   Recommended classification societies shall undertake to apply the present and future provisions of the Agreement taking into account the date of their entry into force. In response to requests from the competent authority, recommended classification societies shall provide all relevant information regarding their technical requirements.

CHAPTER 1.16

PROCEDURE FOR THE ISSUE OF THE CERTIFICATE OF APPROVAL

1.16.0   For the purposes of this Chapter, "owner" means "the owner or his designated representative or, if the vessel is chartered by an operator, the operator or his designated representative".

1.16.1   Certificate of approval

1.16.1.1   *General*

1.16.1.1.1   Dry cargo vessels carrying dangerous goods in quantities greater than exempted quantities, the vessels referred to in 7.1.2.19.1, tank vessels carrying dangerous goods and the vessels referred to in 7.2.2.19.3 shall be provided with an appropriate certificate of approval.

1.16.1.1.2   The certificate of approval shall be valid for not more than five years, subject to the provisions of 1.16.11.

1.16.1.2   *Format of the certificate of approval, particulars to be included*

1.16.1.2.1   The certificate of approval shall conform to the model 8.6.1.1 or 8.6.1.3 with regard to content, form and layout and include the required particulars, as appropriate. It shall include the date of expiry of the period of validity.

Its dimensions are 210 mm x 297 mm (A4). Front and back pages may be used.

It shall be drawn up in the language or one of the languages of the issuing country. If this language is not English, French or German, the title of the certificate and each entry under items 5, 9 and 10 in the certificate of approval for dry cargo vessels (8.6.1.1) and under items 12, 16 and 17 in the certificate of approval for tank vessels (8.6.1.3) shall also be provided in English, French or German.

1.16.1.2.2   The certificate of approval shall attest that the vessel has been inspected and that its construction and equipment comply completely with the applicable requirements of this Regulation.

1.16.1.2.3   All particulars for amendments to the certificate of approval provided for in these Regulations and in the other regulations drawn up by mutual agreement by the Contracting Parties may be entered in the certificate by the competent authority.

1.16.1.2.4   The competent authority shall include the following particulars in the certificate of approval of double-hull vessels meeting the additional requirements of 9.1.0.80 to 9.1.0.95 or 9.2.0.80 to 9.2.0.95:

"The vessel meets the additional requirements for double-hull vessels of 9.1.0.80 to 9.1.0.95" or "The vessel meets the additional requirements for double-hull vessels of 9.2.0.80 to 9.2.0.95."

1.16.1.2.5 For tank vessels, the certificate of approval shall be supplemented by a list of all the dangerous goods accepted for carriage in the tank vessel, drawn up by the recognized classification society which has classified the vessel (vessel substance list). To the extent required for safe carriage, the list shall contain reservations for certain dangerous goods regarding:

– the criteria for strength and stability of the vessel; and

– the compatibility of the accepted dangerous goods with all the construction materials of the vessel, including installations and equipment, which come into contact with the cargo.

Classification societies shall update the vessel substance list at each renewal of the class of a vessel on the basis of the annexed Regulations in force at the time. Classification societies shall inform the owner of the vessel about amendments to Table C of chapter 3.2 which have become relevant in the meantime. If these amendments require an update of the vessel substance list, the owner of the vessel shall request this from a recognized classification society. This updated vessel substance list shall be issued within the period referred to in 1.6.1.1.

The entire vessel substance list shall be withdrawn by the recognized classification society within the period referred to in 1.6.1.1 if, due to amendments to these Regulations or due to changes in classification, goods contained in it are no longer permitted to be carried in the vessel.

The recognized classification society shall without delay, after the delivery to the holder of the certificate of approval, transmit a copy of the vessel substance list to the authority responsible for issuing the certificate of approval and without delay inform it about amendments or withdrawal.

NOTE: When the substance list is available electronically, see 5.4.0.2.

1.16.1.2.6 (*Deleted*)

1.16.1.3 **_Provisional certificate of approval_**

1.16.1.3.1 For a vessel which is not provided with a certificate of approval, a provisional certificate of approval of limited duration may be issued in the following cases, subject to the following conditions:

(a) The vessel complies with the applicable requirements of these Regulations, but the normal certificate of approval could not be issued in time. The provisional certificate of approval shall be valid for an appropriate period but not exceeding three months;

(b) The vessel does not comply with every applicable requirement of these Regulations, but the safety of carriage is not impaired according to the appraisal of the competent authority.

The one-off provisional certificate of approval shall be valid for an appropriate period to bring the vessel into compliance with the applicable provisions, but not exceeding three months.

The competent authority may request additional reports in addition to the inspection report and may require additional conditions.

NOTE: For the issuance of the final certificate of approval according to 1.16.1.2 a new inspection report according to 1.16.3.1 shall be prepared, which confirms conformity also with all hitherto unfulfilled requirements of these Regulations.

(c)   The vessel does not comply with every applicable provision of these Regulations after sustaining damage. In this case the provisional certificate of approval shall be valid only for a single specified voyage and for a specified cargo. The competent authority may impose additional conditions.

1.16.1.3.2   The provisional certificate of approval shall conform to the model in 8.6.1.2 or 8.6.1.4 with regard to content, form and layout or a single model certificate combining a provisional certificate of inspection and the provisional certificate of approval provided that the single model certificate contains the same information as the model in 8.6.1.2 or 8.6.1.4 and is approved by the competent authority. Its dimensions are 210 mm x 297 mm (A4). Front and back pages may be used.

It shall be drawn up in the language or one of the languages of the issuing country. If this language is not English, French or German, the title of the certificate and each entry under item 5 in the provisional certificate of approval for dry cargo vessels (8.6.1.2) and under item 12 in the provisional certificate of approval for tank vessels (8.6.1.4) shall also be provided in English, French or German.

1.16.1.3.3   For tank vessels, the relief pressure of the safety valves or of the high-velocity vent valves shall be entered in the certificate of approval.

If a vessel has cargo tanks with different valve opening pressures, the opening pressure of each tank shall be entered in the certificate of approval.

1.16.1.4   *Annex to the certificate of approval*

1.16.1.4.1   The certificate of approval and the provisional certificate of approval according to 1.16.1.3.1 (a) shall be complemented by an annex in accordance with the model under 8.6.1.5.

1.16.1.4.2   The annex to the certificate of approval shall include the date from which the transitional provisions according to 1.6.7 may be applied. This date shall be:

(a)   For vessels according to Article 8, paragraph 2 of ADN for which evidence can be provided that they were already approved for the carriage of dangerous goods on the territory of a Contracting Party before 26 May 2000, 26 May 2000;

(b)   For vessels according to Article 8, paragraph 2, of ADN for which evidence cannot be provided that they were already approved for the carriage of dangerous goods on the territory of a Contracting Party before 26 May 2000, the proven date of the first inspection for the issue of an approval for the carriage of dangerous goods on the territory of a Contracting Party or, if this date is not known, the date of issue of the first proven approval for the carriage of dangerous goods on the territory of a Contracting Party;

(c)   For all other vessels, the proven date of the first inspection for the issue of a certificate of approval in the sense of ADN or, if this date is not known, the date of issue of the first certificate of approval in the sense of ADN;

(d)   In derogation to (a) to (c) above, the date of a renewed first inspection according to 1.16.8 if the vessel no longer had a valid certificate of approval as from 31 December 2014 for more than twelve months.

1.16.1.4.3   All approvals for the carriage of dangerous goods issued on the territory of a Contracting Party which are valid as from the date under 1.16.1.4.2 and all ADN certificates of approval and provisional certificates of approval according to 1.16.1.3.1 (a) shall be entered in the annex to the certificate of approval.

Certificates of approval issued before the issuance of the annex to the certificate of approval shall be recorded by the competent authority that issues the annex to the certificate of approval.

1.16.2 Issue and recognition of certificates of approval

1.16.2.1 The certificate of approval referred to in 1.16.1 shall be issued by the competent authority of the Contracting Party where the vessel is registered, or in its absence, of the Contracting Party where it has its home port or, in its absence, of the Contracting Party where the owner is domiciled or in its absence, by the competent authority selected by the owner.

The other Contracting Parties shall recognize such certificates of approval.

The Contracting Parties shall communicate to the secretariat of the United Nations Economic Commission for Europe (UNECE) the contact information of the authorities and bodies designated by them which are competent in accordance with national law for the issuance of certificates of approval.

The UNECE secretariat shall bring them to the attention of the Contracting Parties through its website.

1.16.2.2 The competent authority of any of the Contracting Parties may request the competent authority of any other Contracting Party to issue a certificate of approval in its stead.

1.16.2.3 The competent authority of any of the Contracting Parties may delegate the authority to issue the certificate of approval to an inspection body as defined in 1.16.4.

1.16.2.4 The provisional certificate of approval referred to in 1.16.1.3 shall be issued by the competent authority of one of the Contracting Parties for the cases and under the conditions referred to in these Regulations.

The other Contracting Parties shall recognize such provisional certificates of approval.

1.16.2.5 The annex to the certificate of approval shall be issued by the competent authority of a Contracting Party. The Contracting Parties shall assist one another at the time of issuance. They shall recognize this annex to the certificate of approval. Each new certificate of approval or provisional certificate of approval issued in accordance with 1.16.1.3.1 (a) shall be entered in the annex to the certificate of approval. Should the annex to the certificate of approval be replaced (e.g. in case of damage or loss), all existing entries shall be transferred.

1.16.2.6 The annex to the certificate of approval shall be withdrawn and a new annex to the certificate of approval shall be issued if according to 1.16.8 a renewed first inspection takes place, as the validity of the certificate of approval expired, as from 31 December 2014, more than twelve months previously.

The valid date is the date on which the application was received by the competent authority. In this case, only such certificates of approval which have been issued after the renewed first inspection shall be recorded.

1.16.3 Inspection procedure

1.16.3.1 The competent authority of the Contracting Party shall supervise the inspection of the vessel. Under this procedure, the inspection may be performed by an inspection body designated by the Contracting Party or by a recognized classification society according to Chapter 1.15. The inspection body or the recognized classification society shall issue an inspection report certifying that the vessel conforms partially or completely to the applicable requirements of these Regulations related to the construction and equipment of the vessel.

1.16.3.2 This inspection report shall contain:

– Name and address of the Inspection Body or the recognized classification society that carried out the inspection;

– Applicant of the inspection;

– Date and place of the inspection;

– Type of the inspected vessel;

– Identification of the vessel (name, vessel number, ENI number, etc.);

– Declaration that the vessel conforms partially or completely to the applicable requirements of ADN on the construction and equipment of the vessel (in the version applicable on the date of the inspection or, if later, on the estimated date of issuance of the certificate of approval);

– Indication (list, description and references in ADN) of any non-conformities;

– Used transitional provisions;

– Used equivalents and derogations from the regulations applicable to the vessel with reference to the relevant recommendation of the ADN Administrative Committee;

– Date of issuance of the inspection report;

– Signature and official seal of the inspection body or recognized classification society.

If the inspection report does not ensure that all the applicable requirements referred to in 1.16.3.1 are fulfilled, the competent authority may require any additional information in order to issue a provisional certificate of approval according to 1.16.1.3.1 (b).

The authority which is issuing the certificate of approval may request information about the name of the office and surveyor(s) which carried out the inspection including email and phone number, but this information will not become part of the vessel record.

1.16.3.3 The inspection report shall be drawn up in a language accepted by the competent authority and shall contain all the necessary information to enable the certificate to be drawn up.

1.16.3.4 The provisions of 1.16.3.1, 1.16.3.2 and 1.16.3.3 apply to the first inspection referred to in 1.16.8, to the special inspection referred to in 1.16.9 and to the periodic inspection referred to in 1.16.10.

1.16.3.5 Where the inspection report is issued by a recognized classification society, the inspection report may include the certificate referred to in 9.1.0.88.1, 9.2.0.88.1, 9.3.1.8.1, 9.3.2.8.1 or 9.3.3.8.1.

The presence on board of the certificates issued by the recognized classification society for the purposes of 8.1.2.3 (f) and 8.1.2.3 (o) remains mandatory.

1.16.4   **Inspection body**

1.16.4.1   Inspection bodies shall be subject to recognition by the Contracting Party administration as expert bodies on the construction and inspection of inland navigation vessels and as expert bodies on the transport of dangerous goods by inland waterway. They shall meet the following criteria:

– Compliance by the body with the requirements of impartiality;

– Existence of a structure and personnel that provide objective evidence of the professional ability and experience of the body;

– Compliance with the material contents of standard EN ISO/IEC 17020:2012 (except clause 8.1.3) supported by detailed inspection procedures.

1.16.4.2   Inspection bodies may be assisted by experts (e.g. an expert in electrical installations) or specialized bodies according to the national provisions applicable (e.g. classification societies).

1.16.4.3   The Administrative Committee shall maintain an up-to-date list of the inspection bodies appointed.

1.16.5   **Application for the issue of a certificate of approval**

The owner of a vessel shall deposit an application for a certificate of approval with the competent authority referred to in 1.16.2.1. The competent authority shall specify the documents to be submitted to it. In order to obtain a certificate of approval, at least a valid vessel certificate, the inspection report referred to in 1.16.3.1 and the certificate referred to in 9.1.0.88.1, 9.2.0.88.1, 9.3.1.8.1, 9.3.2.8.1 or 9.3.3.8.1 shall accompany the request.

1.16.6   **Particulars entered in the certificate of approval and amendments thereto**

1.16.6.1   The owner of a vessel shall inform the competent authority of any change in the name of the vessel or change of official number or registration number and shall transmit to it the certificate of approval for amendment.

1.16.6.2   All amendments to the certificate of approval provided for in these Regulations and in the other regulations drawn up by mutual agreement by the Contracting Parties may be entered in the certificate by the competent authority.

1.16.6.3   When the owner of the vessel has the vessel registered in another Contracting Party, he shall request a new certificate of approval from the competent authority of that Contracting Party. The competent authority may issue the new certificate for the remaining period of validity of the existing certificate without making a new inspection of the vessel, provided that the state and the technical specifications of the vessel have not undergone any modification.

1.16.6.4   In cases of the transfer of responsibility to another competent authority according to 1.16.6.3, the competent authority to which the last certificate of approval was returned shall submit on request the annex to the certificate according to 1.16.1.4 to the competent authority that will issue the new certificate of approval.

1.16.7   **Presentation of the vessel for inspection**

1.16.7.1   The owner shall present the vessel for inspection unladen, cleaned and equipped; he shall be required to provide such assistance as may be necessary for the inspection, such as providing a suitable launch and personnel, and uncovering those parts of the hull or installations which are not directly accessible or visible.

| 1.16.7.2 | In the case of a first, special or periodical inspection, the inspection body or the recognized classification society may require a dry-land inspection. |

1.16.8 First inspection

If a vessel does not yet have a certificate of approval or if the validity of the certificate of approval expired more than twelve months ago, the vessel shall undergo a first inspection.

1.16.9 Special inspection

If the vessel's hull or equipment has undergone alterations liable to diminish safety in respect of the carriage of dangerous goods, or has sustained damage affecting such safety, the vessel shall be presented without delay by the owner for further inspection.

1.16.10 Periodic inspection and renewal of the certificate of approval

1.16.10.1 To renew the certificate of approval, the owner of the vessel shall present the vessel for a periodic inspection. The owner of the vessel may request an inspection at any time.

1.16.10.2 If the request for a periodic inspection is made during the last year preceding the expiry of the validity of the certificate of approval, the period of validity of the new certificate shall commence when the validity of the preceding certificate of approval expires.

1.16.10.3 A periodic inspection may also be requested during a period of twelve months after the expiry of the certificate of approval. After this period of time, the vessel shall undergo a first inspection in accordance with 1.16.8.

1.16.10.4 The competent authority shall establish the period of validity of the new certificate of approval on the basis of the results of the periodic inspection.

1.16.11 Extension of the certificate of approval without an inspection

By derogation from 1.16.10, at the substantiated request of the owner, the competent authority that has issued the certificate of approval may grant an extension of the validity of the certificate of approval of not more than one year without an inspection. This extension shall be granted in writing and shall be kept on board the vessel. Such extensions may be granted only once every two validity periods.

1.16.12 Official inspection

1.16.12.1 If the competent authority of a Contracting Party has reason to assume that a vessel which is in its territory may constitute a danger in relation to the transport of dangerous goods, for the persons on board or for shipping or for the environment, it may order an inspection of the vessel in accordance with 1.16.3.

1.16.12.2 When exercising this right to inspect, the authorities will make all possible efforts to avoid unduly detaining or delaying a vessel. Nothing in this Agreement affects rights relating to compensation for undue detention or delay. In any instance of alleged undue detention or delay the burden of proof shall lie with the owner of the vessel.

1.16.13 Withdrawal, withholding and return of the certificate of approval

1.16.13.1 The certificate of approval may be withdrawn if the vessel is not properly maintained or if the vessel's construction or equipment no longer complies with the applicable provisions of these Regulations, or if the vessel's highest class according to 9.2.0.88.1, 9.3.1.8.1, 9.3.2.8.1 or 9.3.3.8.1 is not valid.

1.16.13.2   The certificate of approval may only be withdrawn by the authority by which it has been issued.

Nevertheless, in the cases referred to in 1.16.9 and 1.16.13.1 above, the competent authority of the State in which the vessel is staying may prohibit its use for the carriage of those dangerous goods for which the certificate is required. For this purpose it may withdraw the certificate until such time as the vessel again complies with the applicable provisions of these Regulations. In that case it shall notify the competent authority which issued the certificate.

1.16.13.3   Notwithstanding 1.16.2.2 above, any competent authority may amend or withdraw the certificate of approval at the request of the vessel's owner, provided that it so notifies the competent authority which issued the certificate.

1.16.13.4   When an inspection body or a recognized classification society observes, in the course of an inspection, that a vessel or its equipment suffers from serious defects in relation to dangerous goods which might jeopardize the safety of the persons on board or the safety of shipping, or constitute a hazard for the environment, or when the vessel's highest class is not valid, it shall immediately notify the competent authority on behalf of which it acts with a view to a decision to withhold the certificate.

If this authority which decided to withdraw the certificate is not the authority which issued the certificate, it shall immediately inform the latter and, where necessary, return the certificate to it if it presumes that the defects cannot be eliminated in the near future.

1.16.13.5   When the inspection body or the recognized classification society referred to in 1.16.13.4 above ascertains, by means of a special inspection according to 1.16.9, that these defects have been remedied, the certificate of approval shall be returned by the competent authority to the owner.

This inspection may be made at the request of the owner by another inspection body or another recognized classification society. In this case, the certificate of approval shall be returned through the competent authority to which the inspection body or the recognized classification society answers.

1.16.13.6   When a vessel is finally immobilized or scrapped, the owner shall send the certificate of approval back to the competent authority which issued it.

1.16.14   **Duplicate copy**

In the event of the loss, theft or destruction of the certificate of approval or when it becomes unusable for other reasons, an application for a duplicate copy, accompanied by appropriate supporting documents, shall be made to the competent authority which issued the certificate.

This authority shall issue a duplicate copy of the certificate of approval, which shall be designated as such.

1.16.15   **Register of certificates of approval**

1.16.15.1   The competent authorities shall assign a serial number to the certificates of approval which they issue. They shall keep a register of all the certificates issued.

1.16.15.2   The competent authorities shall keep copies of all the certificates which they have issued, as well as of the associated vessel substance lists of the recognised classification societies and of all amendments, withdrawals, new issuances and declarations of cancellation of these documents.

PART 2

Classification

(See Volume II)

PART 3

Dangerous goods list, special provisions and exemptions related to limited and excepted quantities

CHAPTER 3.1

GENERAL

(See Volume II)

CHAPTER 3.2

LIST OF DANGEROUS GOODS

3.2.1   **Table A: List of dangerous goods in numerical order**

See Volume II

3.2.2   **Table B: List of dangerous goods in alphabetical order**

See Volume II

3.2.3   **Table C: List of dangerous goods accepted for carriage in tank vessels in numerical order**

3.2.3.1   *Explanations concerning Table C:*

As a rule, each row of Table C of this Chapter deals with the substance(s) covered by a specific UN number or identification number. However, when substances belonging to the same UN number or identification number have different chemical properties, physical properties and/or carriage conditions, several consecutive rows may be used for that UN number or identification number.

Each column of Table C is dedicated to a specific subject as indicated in the explanatory notes below. The intersection of columns and rows (cell) contains information concerning the subject treated in that column, for the substance(s) of that row:

–   The first four cells identify the substance(s) belonging to that row;

–   The following cells give the applicable special provisions, either in the form of complete information or in coded form. The codes cross-refer to detailed information that is to be found in the numbers indicated in the explanatory notes below. An empty cell means either that there is no special provision and that only the general requirements apply, or that the carriage restriction indicated in the explanatory notes is in force.

–   If a cell contains an asterisk, "*", the applicable requirements should be determined in accordance with 3.2.3.3.

The applicable general requirements are not referred to in the corresponding cells.

Explanatory notes for each column:

Column (1)   "UN number/identification number"

Contains the UN number or identification number:

–   of the dangerous substance if the substance has been assigned its own specific UN number or identification number, or

–   of the generic or n.o.s. entry to which the dangerous substances not mentioned by name shall be assigned in accordance with the criteria ("decision trees") of Part 2.

Column (2)	"Name and description"

Contains, in upper case characters, the name of the substance, if the substance has been assigned its own specific UN number or identification number or of the generic or n.o.s. entry to which the dangerous substances have been assigned in accordance with the criteria ("decision trees") of Part 2. This name shall be used as the proper shipping name or, when applicable, as part of the proper shipping name (see 3.1.2 for further details on the proper shipping name).

A descriptive text in lower case characters is added after the proper shipping name to clarify the scope of the entry if the classification or carriage conditions of the substance may be different under certain conditions.

Column (3a)	"Class"

Contains the number of the Class, whose heading covers the dangerous substance. This Class number is assigned in accordance with the procedures and criteria of Part 2.

Column (3b)	"Classification code"

Contains the classification code of the dangerous substance.

- For dangerous substances of Class 2, the code consists of a number and one or more letters representing the hazardous property group, which are explained in 2.2.2.1.2 and 2.2.2.1.3.

- For dangerous substances or articles of Classes 3, 4.1, 6.1, 8 and 9, the codes are explained in 2.2.x.1.2.[1]

Column (4)	"Packing group"

Contains the packing group number(s) (I, II or III) assigned to the dangerous substance. These packing group numbers are assigned on the basis of the procedures and criteria of Part 2. Certain substances are not assigned to packing groups.

Column (5)	"Dangers"

This column contains information concerning the hazards inherent in the dangerous substance. These hazards are included on the basis of the danger labels of Table A, column (5).

In the case of a chemically unstable substance, the code 'unst.' is added to the information.

In the case of a substance or mixture hazardous to the aquatic environment, the code 'N1', 'N2' or 'N3' is added to the information.

In the case of a substance or mixture with CMR properties, the code 'CMR' is added to the information.

---

[1] x = the Class number of the dangerous substance or article, without dividing point if applicable.

In the case of a substance or mixture that floats on the water surface, does not evaporate and is not readily soluble in water or that sinks to the bottom of the water and is not readily soluble, the code 'F' (standing for 'Floater') or 'S' (standing for 'Sinker'), respectively, is added to the information.

Where the information is shown in brackets, only the relevant codes for the substance carried should be used.

Column (6)   "Type of tank vessel"

Contains the type of tank vessel: G, C or N.

Column (7)   "Cargo tank design"

Contains information concerning the design of the cargo tank:

1   Pressure cargo tank

2   Closed cargo tank

3   Open cargo tank with flame arrester

4   Open cargo tank

Column (8)   "Cargo tank type"

Contains information concerning the cargo tank type.

1   Independent cargo tank

2   Integral cargo tank

3   Cargo tank with walls distinct from the outer hull

Column (9)   "Cargo tank equipment"

Contains information concerning the cargo tank equipment.

1   Refrigeration system

2   Possibility of cargo heating

3   Water-spray system

4   Cargo heating system on board

Column (10)   "Opening pressure of the high-velocity vent valve in kPa"

Contains information concerning the opening pressure of the high-velocity vent valve in kPa.

Column (11)   "Maximum degree of filling (%)"

Contains information concerning the maximum degree of filling of cargo tanks as a percentage.

Column (12)   "Relative density at 20 °C"

Contains information concerning the relative density of the substance at 20 °C. Data concerning the density are for information only.

Column (13)   "Type of sampling device"

Contains information concerning the prescribed type of sampling device.

1   Closed-type sampling device

2   Partly closed-type sampling device

3   Sampling opening

Column (14)   "Pump room below deck permitted"

Contains an indication of whether a pump room is permitted below deck.

Yes   pump room below deck permitted

No   pump room below deck not permitted

Column (15)   "Temperature class"

Contains the temperature class of the substance.

Column (16)   "Explosion group"

Contains the explosion group of the substance.

Values between square brackets indicate the explosion group II B subgroups to be used in selecting the relevant self-contained protection systems (flame arresters, pressure/vacuum relief valves with integrated backfire-prevention device, and high velocity vent valves).

NOTE:

Where self-contained protection systems for explosion group II B are in place, products in explosion group II A or II B, including subgroups II B3, II B2 and II B1, may be transported.

Where self-contained protection systems for explosion group II B3 are in place, products in explosion subgroups II B3, II B2 and II B1, or in explosion group II A, may be transported.

Where self-contained protection systems for explosion group II B2 are in place, products in explosion subgroups II B2 and II B1, or in explosion group II A, may be transported.

Where self-contained protection systems for explosion group II B1 are in place, products in explosion subgroup II B1 or in explosion group II A may be transported.

Column (17)	"Anti-explosion protection required"

Contains a code referring to protection against explosions.

Yes   anti-explosion protection required

No   anti-explosion protection not required

Column (18)	"Equipment required"

This column contains the alphanumeric codes for the equipment required for the carriage of the dangerous substance (see 8.1.5).

Column (19)	"Number of cones/blue lights"

This column contains the number of cones/blue lights which should constitute the marking of the vessel during the carriage of this dangerous substance.

Column (20)	"Additional requirements/Remarks"

This column contains the additional requirements or remarks applicable to the vessel.

These additional requirements or remarks are:

1.   Anhydrous ammonia is liable to cause stress crack corrosion in cargo tanks and cooling systems constructed of carbon-manganese steel or nickel steel.

In order to minimize the risk of stress crack corrosion the following measures shall be taken:

(a)   Where carbon-manganese steel is used, cargo tanks, pressure vessels of cargo refrigeration systems and cargo piping shall be constructed of fine-grained steel having a specified minimum yield stress of not more than 355 N/mm^2. The actual yield stress shall not exceed 440 N/mm^2. In addition, one of the following construction or operational measures shall be taken:

.1   Material with a low tensile strength

($R_m < 410$ N/mm^2) shall be used; or

.2   Cargo tanks, etc., shall undergo a post-weld heat treatment for the purpose of stress relieving; or

.3   The transport temperature shall preferably be maintained close to the evaporation temperature of the cargo of -33° C, but in no case above -20° C; or

.4   Ammonia shall contain not less than 0.1 % water, by mass.

(b)   When carbon-manganese steel with yield stress values higher than those referred to in (a) above is used, the completed tanks, pipe sections, etc., shall undergo a post-weld heat treatment for the purpose of stress relieving.

(c) Pressure vessels of the cargo refrigeration systems and the piping systems of the condenser of the cargo refrigeration system constructed of carbon-manganese steel or nickel steel shall undergo a post-weld heat treatment for the purpose of stress relieving.

(d) The yield stress and the tensile strength of welding consumables may exceed only by the smallest value possible the corresponding values of the tank and piping material.

(e) Nickel steels containing more than 5 % nickel and carbon-manganese steel which are not in compliance with the requirements of (a) and (b) above may not be used for cargo tanks and piping systems intended for the transport of this substance.

(f) Nickel steels containing not more than 5 % nickel may be used if the transport temperature is within the limits referred to in (a) above.

(g) The concentration of oxygen dissolved in the ammonia shall not exceed the values given in the table below:

t in °C	O_2 in %
-30 and below	0.90
-20	0.50
-10	0.28
0	0.16
10	0.10
20	0.05
30	0.03

2. Before loading, air shall be removed and subsequently kept away to a sufficient extent from the cargo tanks and the accessory cargo piping by the means of inert gas (see also 7.2.4.18).

3. Arrangements shall be made to ensure that the cargo is sufficiently stabilized in order to prevent a reaction at any time during carriage. The transport document shall contain the following additional particulars:

(a) Name and amount of inhibitor added;

(b) Date on which inhibitor was added and expected duration of effectiveness under normal conditions;

(c) Any temperature limits having an effect on the inhibitor.

When stabilization is ensured solely by blanketing with an inert gas it is sufficient to mention the name of the inert gas used in the transport document.

When stabilization is ensured by another measurement, e.g. the special purity of the substance, this measurement shall be mentioned in the transport document.

4. The substance shall not be allowed to solidify; the transport temperature shall be maintained above the melting point. In instances where cargo heating installations are required, they must be so designed that polymerisation through heating is not possible in any part of the cargo tank. Where the temperature of steam-heated coils could give rise to overheating, lower-temperature indirect heating systems shall be provided.

5. This substance is liable to clog the venting piping and its fittings. Careful surveillance should be ensured. If a closed-type tank vessel is required for the carriage of this substance the venting piping shall conform to 9.3.2.22.5 (a) (i), (ii), (iv), (b), (c) or (d) or to 9.3.3.22.5 (a) (i), (ii), (iv), (b), (c) or (d). This requirement does not apply when the cargo tanks and the corresponding piping are inerted in accordance with 7.2.4.18 nor when protection against explosions is not required in column (17) and when flame-arresters have not been installed.

6. When external temperatures are below or equal to that indicated in column (20), the substance may only be carried in tank vessels equipped with a possibility of heating the cargo.

In addition, in the event of carriage in a closed-type vessel, if the tank vessel:

– is fitted out in accordance with 9.3.2.22.5 (a) (i) or (d) or 9.3.3.22.5 (a) (i) or (d), it shall be equipped with pressure/vacuum valves capable of being heated; or

– is fitted out in accordance with 9.3.2.22.5 (a) (ii), (v), (b) or (c) or 9.3.3.22.5 (a) (ii), (v), (b) or (c), it shall be equipped with heatable venting piping and heatable pressure/vacuum valves; or

– is fitted out in accordance with 9.3.2.22.5 (a) (iii) or (iv) or 9.3.3.22.5 (a) (iii) or (iv), it shall be equipped with heatable venting piping and with heatable pressure/vacuum valves and heatable flame-arresters.

The temperature of the venting piping, pressure/vacuum valves and flame-arresters shall be kept at least above the melting point of the substance.

7. If a closed-type tank vessel is required to carry this substance or if the substance is carried in a closed-type tank vessel, if this vessel:

– is fitted out in accordance with 9.3.2.22.5 (a) (i) or (d) or 9.3.3.22.5 (a) (i) or (d), it shall be equipped with heatable pressure/vacuum valves, or

– is fitted out in accordance with 9.3.2.22.5 (a) (ii), (v), (b) or (c) or 9.3.3.22.5 (a) (ii), (v), (b) or (c), it shall be equipped with heatable venting piping and heatable pressure/vacuum valves, or

– is fitted out in accordance with 9.3.2.22.5 (a) (iii) or (iv) or 9.3.3.22.5 (a) (iii) or (iv), it shall be equipped with heatable venting piping and with heatable pressure/vacuum valves and heatable flame-arresters.

The temperature of the venting piping, pressure/vacuum valves and flame-arresters shall be kept at least above the melting point of the substance.

8. Double-hull spaces, double bottoms and heating coils shall not contain any water.

9. (a) While the vessel is underway, an inert-gas pad shall be maintained in the ullage space above the liquid level.

    (b) Cargo piping and vent lines shall be independent of the corresponding piping used for other cargoes.

    (c) Safety valves shall be made of stainless steel.

10. *(Reserved)*

11. (a) Stainless steel of type 416 or 442 and cast iron shall not be used for cargo tanks and piping for loading and unloading.

    (b) The cargo may be discharged only by deep-well pumps or pressure inert gas displacement. Each cargo pump shall be arranged to ensure that the substance does not heat significantly if the pressure discharge line from the pump is shut off or otherwise blocked.

    (c) The cargo shall be cooled and maintained at temperatures below 30° C.

    (d) The safety valves shall be set at a pressure of not less than 550 kPa (5.5 bar) gauge pressure. Special authorization is required for the maximum setting pressure.

    (e) While the vessel is underway, a nitrogen pad shall be maintained in the ullage space above the cargo (see also 7.2.4.18). An automatic nitrogen supply system shall be installed to prevent the pressure from falling below 7 kPa (0.07 bar) gauge within the cargo tank in the event of a cargo temperature fall due to ambient temperature conditions or to some other reason. In order to satisfy the demand of the automatic pressure control a sufficient amount of nitrogen shall be available on board. Nitrogen of a commercially pure quality of 99.9 %, by volume, shall be used for padding. A battery of nitrogen cylinders connected to the cargo tanks through a pressure reduction valve satisfies the intention of the expression "automatic" in this context.

    The required nitrogen pad shall be such that the nitrogen concentration in the vapour space of the cargo tank is not less than 45 % at any time.

    (f) Before loading and while the cargo tank contains this substance in a liquid or gaseous form, it and the corresponding piping shall be inerted with nitrogen.

(g) The water-spray system shall be fitted with remote-control devices which can be operated from the wheelhouse or from the control station, if any.

(h) Transfer arrangements shall be provided for emergency transfer of ethylene oxide in the event of an uncontrollable self-reaction.

12. (a) The substance shall be acetylene free.

(b) Cargo tanks which have not undergone appropriate cleaning shall not be used for the carriage of these substances if one of the previous three cargoes consisted of a substance known to promote polymerisation, such as:

.1 mineral acids (e.g. sulphuric acid, hydrochloric acid, nitric acid);

.2 carboxylic acids and anhydrides (e.g. formic acid, acetic acid);

.3 halogenated carboxylic acids (e.g. chloroacetic acid);

.4 sulphonic acids (e.g. benzene sulphonic acid);

.5 caustic alkalis (e.g. sodium hydroxide, potassium hydroxide);

.6 ammonia and ammonia solutions;

.7 amines and amine solutions;

.8 oxidizing substances.

(c) Before loading, cargo tanks and their piping shall be efficiently and thoroughly cleaned so as to eliminate all traces of previous cargoes, except when the last cargo was constituted of propylene oxide or a mixture of ethylene oxide and propylene oxide. Special precautions shall be taken in the case of ammonia in cargo tanks built of steel other than stainless steel.

(d) In all cases the efficiency of the cleaning of cargo tanks and their piping shall be monitored by means of appropriate tests or inspections to check that no trace of acid or alkaline substance remains that could present a danger in the presence of these substances.

(e) The cargo tanks shall be entered and inspected prior to each loading of these substances to ensure freedom from contamination, heavy rust deposits or visible structural defects.

When these cargo tanks are in continuous service for these substances, such inspections shall be performed at intervals of not more than two and a half years.

(f)   Cargo tanks which have contained these substances may be reused for other cargoes once they and their piping have been thoroughly cleaned by washing and flushing with an inert gas.

(g)   Substances shall be loaded and unloaded in such a way that there is no release of gas into the atmosphere. If gas is returned to the shore installation during loading, the gas return system connected to the tank containing that substance shall be independent from all other cargo tanks.

(h)   During discharge operations, the pressure in the cargo tanks shall be maintained above 7 kPa (0.07 bar) gauge.

(i)   The cargo shall be discharged only by deep-well pumps, hydraulically operated submerged pumps or pressure inert gas displacement. Each cargo pump shall be arranged to ensure that the substance does not heat significantly if the pressure discharge line from the pump is shut off or otherwise blocked.

(j)   Each cargo tank carrying these substances shall be ventilated by a system independent from the ventilation systems of other cargo tanks carrying other substances.

(k)   Hose assemblies for loading and unloading shall be marked as follows:

"To be used only for the transfer of alkylene oxide."

(l)   (*Reserved*)

(m)   No air shall be allowed to enter the cargo pumps and cargo piping system while these substances are contained within the system.

(n)   Before the shore connections are disconnected, piping containing liquids or gas shall be depressurised at the shore link by means of appropriate devices.

(o)   The piping system for cargo tanks to be loaded with these substances shall be separate from the piping system for all other cargo tanks, including empty cargo tanks. If the piping system for the cargo tanks to be loaded is not independent, separation shall be accomplished by the removal of spool pieces, shut-off valves, other pipe sections and by fitting blank flanges at these locations. The required separation applies to all liquid pipes and vapour vent lines and any other connections which may exist such as common inert gas supply lines.

(p)   These substances may be carried only in accordance with cargo handling plans that have been approved by a competent authority.

Each loading arrangement shall be shown on a separate cargo handling plan. Cargo handling plans shall show the entire cargo piping system and the locations for installations of blank flanges needed to meet the above piping separation requirements. A copy of each cargo handling plan shall be kept on board. Reference to the approved cargo handling plans shall be included in the certificate of approval.

(q) Before loading of these substances and before carriage is resumed a qualified person approved by the competent authority shall certify that the prescribed separation of the piping has been effected; this certificate shall be kept on board. Each connection between a blank flange and a shut-off valve in the piping shall be fitted with a sealed wire to prevent the flange from being disassembled inadvertently.

(r) During the voyage, the cargo shall be covered with nitrogen. An automatic nitrogen make-up system shall be installed to prevent the cargo tank pressure from falling below 7 kPa (0.07 bar) gauge in the event of a cargo temperature fall due to ambient temperature conditions or to some other reason. Sufficient nitrogen shall be available on board to satisfy the demand of automatic pressure control. Nitrogen of commercially pure quality of 99.9 %, by volume, shall be used for padding. A battery of nitrogen cylinders connected to the cargo tanks through a pressure reduction valve satisfies the intention of the expression "automatic" in this context.

(s) The vapour space of the cargo tanks shall be checked before and after each loading operation to ensure that the oxygen content is 2 %, by volume, or less.

(t) Loading flow

The loading flow (L_R) of cargo tank shall not exceed the following value:

$$L_R = 3600 \times U/t \ (m^3/h)$$

In this formula:

U = the free volume (m^3) during loading for the activation of the overflow prevention system;

T = the time (s) required between the activation of the overflow prevention system and the complete stop of the flow of cargo into the cargo tank;

The time is the sum of the partial times needed for successive operations, e.g. reaction time of the service personnel, the time needed to stop the pumps and the time needed to close the shut-off valves;

The loading flow shall also take account of the design pressure of the piping system.

13. If no stabilizer is supplied or if the supply is inadequate, the oxygen content in the vapour phase shall not exceed 0.1 %. Overpressure must be constantly maintained in cargo tanks. This

requirement applies also to voyages on ballast or empty with uncleaned cargo tanks between cargo transport operations.

14. The following substances may not be carried in a type N vessel:

   – substances with self-ignition temperatures ≤ 200 °C;

   – substances with a flash point < 23°C and an explosion range > 15 percentage points;

   – mixtures containing halogenated hydrocarbons;

   – mixtures containing more than 10 % benzene;

   – substances and mixtures carried in a stabilized state.

15. Provision shall be made to ensure that alkaline or acidic substances such as sodium hydroxide solution or sulphuric acid do not contaminate this cargo.

16. If there is a possibility of a dangerous reaction such as polymerisation, decomposition, thermal instability or evolution of gases resulting from local overheating of the cargo in either the cargo tank or associated piping system, this cargo shall be loaded and carried adequately segregated from other substances the temperature of which is sufficiently high to initiate such reaction. Heating coils inside cargo tanks carrying this substance shall be blanked off or secured by equivalent means.

17. The melting point of the cargo shall be shown in the transport documents.

18. *(Reserved)*

19. Provision shall be made to ensure that the cargo does not come into contact with water. The following additional requirements apply:

   Carriage of the cargo is not permitted in cargo tanks adjacent to slop tanks or cargo tanks containing ballast water, slops or any other cargo containing water. Pumps, piping and vent lines connected to such tanks shall be separated from similar equipment of tanks carrying these substances. Pipes from slop tanks or ballast water pipes shall not pass through cargo tanks containing this cargo unless they are encased in a tunnel.

20. The maximum permitted transport temperature given in column (20) shall not be exceeded.

21. *(Reserved)*

22. The relative density of the cargo shall be shown in the transport document.

23. The instrument for measuring the pressure of the vapour phase in the cargo tank shall activate the alarm when the internal pressure reaches 40 kPa (0.4 bar). The water-spray system shall immediately be activated and remain in operation until the internal pressure drops to 30 kPa (0.3 bar).

24. Substances having a flash-point above 60 °C which are handed over for carriage or which are carried heated within a limiting range of 15 K below their flash-point shall be carried under the conditions of substance number 9001.

25. Type 3 cargo tank may be used for the carriage of this substance provided that the construction of the cargo tank has been accepted by a recognized classification society for the maximum permitted transport temperature.

26. Type 2 cargo tank may be used for the carriage of this substance provided that the construction of the cargo tank has been accepted by a recognized classification society for the maximum permitted transport temperature.

27. The requirements of 3.1.2.8.1 are applicable.

28. (a) When UN 2448 SULPHUR, MOLTEN is carried, the forced ventilation of the cargo tanks shall be brought into service at latest when the concentration of hydrogen sulphide reaches 1.0 %, by volume.

    (b) When during the carriage of UN 2448 SULPHUR, MOLTEN, the concentration of hydrogen sulphide exceeds 1.85 %, the boat master shall immediately notify the nearest competent authority.

    When a significant increase in the concentration of hydrogen sulphide in a hold space leads it to be supposed that the sulphur has leaked, the cargo tanks shall be unloaded as rapidly as possible. A new load may only be taken on board once the authority which issued the certificate of approval has carried out a further inspection.

    (c) When UN 2448 SULPHUR, MOLTEN is carried, the concentration of hydrogen sulphide shall be measured in the vapour phase of the cargo tanks and concentrations of sulphur dioxide and hydrogen sulphide in the hold spaces.

    (d) The measurements prescribed in (c) shall be made every eight hours. The results of the measurements shall be recorded in writing.

29. When particulars concerning the vapour pressure or the boiling point are given in column (2), the relevant information shall be added to the proper shipping name in the transport document, e.g.

UN 1224 KETONES, LIQUID, N.O.S.,

110 kPa < vp 50 ≤ 175 kPa or

UN 2929 TOXIC LIQUID, FLAMMABLE, ORGANIC, N.O.S., boiling point ≤ 60°C

30. When these substances are carried, the hold spaces of open type N tank vessels may contain auxiliary equipment.

31. When these substances are carried, the vessel shall be equipped with a rapid blocking valve placed directly on the shore connection.

32. In the case of transport of this substance, the following additional requirements are applicable:

   (a) The outside of the cargo tanks shall be equipped with insulation of low flammability. This insulation shall be strong enough to resist shocks and vibration. Above deck, the insulation shall be protected by a covering.

   The outside temperature of this covering shall not exceed 70 °C.

   (b) The hold spaces containing the cargo tanks shall be provided with ventilation. Connections for forced ventilation shall be fitted.

   (c) The cargo tanks shall be equipped with forced ventilation installations which, in all transport conditions, will reliably keep the concentration of hydrogen sulphide above the liquid phase below 1.85 % by volume.

   The ventilation installations shall be fitted in such a way as to prevent the deposit of the goods to be transported.

   The exhaust line of the ventilation shall be fitted in such a way as not to present a risk to personnel.

   (d) The cargo tank and the hold spaces shall be fitted with outlets and piping to allow gas sampling.

   (e) The outlets of the cargo tanks shall be situated at a height such that for a trim of 2° and a list of 10°, no sulphur can escape. All the outlets shall be situated above the deck in the open air. Each outlet shall be equipped with a permanently fixed closing mechanism.

   One of these mechanisms shall be capable of being opened for slight overpressure within the tank.

   (f) The piping for loading and unloading shall be equipped with adequate insulation. They shall be capable of being heated.

   (g) The heat transfer fluid shall be such that in the event of a leak into a tank, there is no risk of a dangerous reaction with the sulphur.

33. The following provisions are applicable to transport of this substance:

Construction requirements:

   (a) Hydrogen peroxide solutions may be transported only in cargo tanks equipped with deep-well pumps.

(b)   Cargo tanks and their equipment shall be constructed of solid stainless steel of a type appropriate to hydrogen peroxide solutions (for example, 304, 304L, 316, 316L or 316 Ti). None of the non-metallic materials used for the system of cargo tanks shall be attacked by hydrogen peroxide solutions or cause the decomposition of the substance.

(c)   The temperature sensors shall be installed in the cargo tanks directly under the deck and at the bottom. Remote temperature read-outs and monitoring shall be provided for in the wheelhouse.

(d)   Fixed oxygen monitors (or gas-sampling lines) shall be provided in the areas adjacent to the cargo tanks so that leaks in such areas can be detected. Account shall be taken of the increased flammability arising from the increased presence of oxygen. Remote read-outs, continuous monitoring (if the sampling lines are used, intermittent monitoring will suffice) and visible and audible alarms similar to those for the temperature sensors shall also be located in the wheelhouse. The visible and audible alarms shall be activated if the oxygen concentration in these void spaces exceeds 30 % by volume. Two additional oxygen monitors shall also be available.

(e)   The cargo tank venting systems which are equipped with filters shall be fitted with pressure/vacuum relief valves appropriate to closed-circuit ventilation and with an extraction installation should cargo tank pressure rise rapidly as a result of an uncontrolled decomposition (see under m). These air supply and extraction systems shall be so designed that water cannot enter the cargo tanks. In designing the emergency extraction installation account shall be taken of the design pressure and the size of the cargo tanks.

(f)   A fixed water-spray system shall be provided for diluting and washing away any hydrogen peroxide solutions spilled onto the deck. The area covered by the jet of water shall include the shore connections and the deck containing the cargo tanks designated for carrying hydrogen peroxide solutions.

The following minimum requirements shall be complied with:

.1   The substance shall be diluted from the original concentration to a 35 % concentration within five minutes from the spillage on the deck;

.2   The rate and estimated size of the spill shall be determined in the light of the maximum permissible loading or unloading rates, the time required to halt the spillage in the event of tank overfill or a pipe or hose assembly failure, and the time necessary to begin application of dilution water with actuation of the alarm at the cargo control location or in the wheelhouse.

(g) The outlets of the pressure valves shall be situated at least 2 metres above the walkways if they are less than 4 metres from the walkway.

(h) A temperature sensor shall be installed by each pump to make it possible to monitor the temperature of the cargo during unloading and detect any overheating due to defective operation of the pump.

Servicing requirements:

Carrier

(i) Hydrogen peroxide solutions may only be carried in cargo tanks which have been thoroughly cleaned and passivated, in accordance with the procedure described under (j), of all traces of previous cargoes, their vapours or their ballast waters. A certificate stating that the procedure described under (j) has been duly complied with must be carried on board.

Particular care in this respect is essential to ensure the safe carriage of hydrogen peroxide solutions:

.1 When a hydrogen peroxide solution is being carried, no other cargo may be carried simultaneously;

.2 Tanks which have contained hydrogen peroxide solutions may be reused for other cargoes after they have been cleaned by persons or companies approved for this purpose by the competent authority;

.3 In the design of the cargo tanks, efforts must be made to keep to a minimum any internal tank structure, to ensure free draining, no entrapment and ease of visual inspection.

(j) Procedures for inspection, cleaning, passivation and loading for the transport of hydrogen peroxide solutions with a concentration of 8 to 60 per cent in cargo tanks which have previously carried other cargoes.

Before their reuse for the transport of hydrogen peroxide solutions, cargo tanks which have previously carried cargoes other than hydrogen peroxide must be inspected, cleaned and passivated. The procedures described in paragraphs .1 to .7 below for inspection and cleaning apply to stainless steel cargo tanks. The procedure for passivating stainless steel is described in paragraph .8. Failing any other instructions, all the measures apply to cargo tanks and to all their structures which have been in contact with other cargoes.

.1 After unloading of the previous cargo, the cargo tank must be degassed and inspected for any remaining traces, carbon residues and rust.

.2   The cargo tanks and their equipment must be washed with clear filtered water. The water used must be at least of the same quality as drinking water and have a low chlorine content.

.3   Traces of the residues and vapours of the previous cargo must be removed by the steam cleaning of the cargo tanks and their equipment.

.4   The cargo tanks and their equipment must then be rewashed with clear water of the quality specified in paragraph 2 above and dried in filtered, oil-free air.

.5   Samples must be taken of the atmosphere in the cargo tanks and these must be analysed for their content of organic gases and oxygen.

.6   The cargo tank must be reinspected for any traces of the previous cargo, carbon residues or rust or odours of the previous cargo.

.7   If the inspection and the other measures point to the presence of traces of the previous cargo or of its gases, the measures described in paragraphs .2 to .4 above must be repeated.

.8   Stainless steel cargo tanks and their structures which have contained cargoes other than hydrogen peroxide solutions and which have been repaired must, regardless of whether or not they have previously been passivated, be cleaned and passivated in accordance with the following procedure:

.8.1   The new weld seams and other repaired parts must be cleaned and scrubbed with stainless steel brushes, graving tools, sandpaper and polishers. Rough surfaces must be made smooth and a final polishing must be carried out;

.8.2   Fatty and oily residues must be removed with the use of organic solvents or appropriate cleaning products diluted with water. The use of chlorinated products shall be avoided because these might seriously interfere with the passivation procedure;

.8.3   Any residues that have been removed must be eliminated and the tanks must then be washed.

(k)   During the transfer of the hydrogen peroxide solutions, the related piping system must be separated from all other systems. Loading and unloading piping used for the transfer of hydrogen peroxide solutions must be marked as follows:

"For Hydrogen Peroxide
Solution Transfer only"

(l)   If the temperature in the cargo tanks rises above 35 °C, visible and audible alarms shall activate in the wheelhouse.

Master

(m)   If the temperature rise exceeds 4 °C for 2 hours or if the temperature in the cargo tanks exceeds 40 °C, the master must contact the consignor directly, with a view to taking any action that might be necessary.

Filler

(n)   Hydrogen peroxide solutions must be stabilized to prevent decomposition. The manufacturer must provide a stabilization certificate which must be carried on board and must specify:

    .1   The disintegration date of the stabilizer and the duration of its effectiveness;

    .2   Actions to be taken should the product become unstable during the voyage.

(o)   Only those hydrogen peroxide solutions which have a maximum decomposition rate of 1.0 per cent per year at 25 °C may be carried. A certificate from the filler stating that the product meets this standard must be presented to the master and kept on board. An authorized representative of the manufacturer must be on board to monitor the loading operations and to test the stability of the hydrogen peroxide solutions to be transported. He shall certify to the master that the cargo has been loaded in a stable condition.

34. For type N carriage, the flanges and stuffing boxes of the loading and unloading piping must be fitted with a protection device to protect against splashing.

35. Only an indirect system for the cargo refrigerating system is permitted for this substance. Direct or combined systems are not permitted.

36. Merged with remark 35.

37. For this substance, the cargo tank system shall be capable of resisting the vapour pressure of the cargo at higher ambient temperatures whatever the system that has been adopted for treating the boil-off gas.

38. For an initial boiling point above 60 °C and under or equal to 85 °C as determined in accordance with ASTMD 86-01, the applicable conditions of transport are identical to those stipulated for an initial boiling point under or equal to 60 °C.

39. (a)   The joints, outlets, closing devices and other technical equipment shall be of such a sort that there cannot be any leakage of carbon dioxide during normal transport operations (cold, fracturing of materials, freezing of fixtures, run-off outlets etc.).

(b)   The loading temperature (at the loading station) shall be mentioned in the transport document.

(c)   An oxygen meter shall be kept on board, together with instructions on its use which can be read by everyone on board. The oxygen meter shall be used as a testing device when entering holds, pump rooms, areas situated at depth and when work is being carried out on board.

(d)   At the entry of accommodation and in other places where the crew may spend time there shall be a measuring device which lets off an alarm when the oxygen level is too low or when the CO_2 level is too high.

(e)   The loading temperature (established after loading) and the maximum duration of the journey shall be mentioned in the transport document.

40.   *(Deleted)*

41.   n-BUTYLBENZENE is assigned to the entry UN No. 2709 BUTYLBENZENES (n-BUTYLBENZENE).

42.   Loading of refrigerated liquefied gases shall be carried out in such a manner as to ensure that unsatisfactory temperature gradients do not occur in any cargo tank, piping or other ancillary equipment. When determining the holding time (as described in 7.2.4.16.17), it shall be assured that the degree of filling does not exceed 98% in order to prevent the safety valves from opening when the tank is in liquid full condition. When refrigerated liquefied gases are carried using a system according to 9.3.1.24.1 (b) or 9.3.1.24.1 (c), a refrigeration system is not required.

43.   It may be that the mixture has been classified as a floater as a precautionary measure, because some of its components meet the relevant criteria.

3.2.3.2   Table C

(1) UN No. or substance identification No.	(2) Name and description	(3a) Class	(3b) Classification code	(4) Packing group	(5) Dangers	(6) Type of tank vessel	(7) Cargo tank design	(8) Cargo tank type	(9) Cargo tank equipment	(10) Opening pressure of the high-velocity vent valve in kPa	(11) Maximum degree of filling in %	(12) Relative density at 20 °C	(13) Type of sampling device	(14) Pump room below deck permitted	(15) Temperature class	(16) Explosion group	(17) Anti-explosion protection required	(18) Equipment required	(19) Number of cones/blue lights	(20) Additional requirements/Remarks
1005	AMMONIA, ANHYDROUS	2	2TC		2.3+8+2.1+N1	G	1	1	3		91		1	no	T1	II A	yes	PP, EP, EX, TOX, A	2	1; 2; 31
1010	1,2-BUTADIENE, STABILIZED	2	2F		2.1+unst.	G	1	1			91		1	no	T2	II B[4]	yes	PP, EX, EX, A	1	2; 3; 31
1010	1,3-BUTADIENE, STABILIZED	2	2F		2.1+unst.+CMR	G	1	1			91		1	no	T2	II B (II B2[4])	yes	PP, EP, EX, TOX, A	1	2; 3; 31
1010	BUTADIENES STABILIZED or BUTADIENES AND HYDROCARBON MIXTURE, STABILIZED, having a vapour pressure at 70 °C not exceeding 1.1 MPa (11 bar) and a density at 50 °C not lower than 0.525 kg/l (contains less than 0.1% 1.3-butadiene)	2	2F		2.1+unst.	G	1	1			91		1	no	T2	II B[4] (II B2[4])	yes	PP, EX, A	1	
1010	BUTADIENES, STABILIZED or BUTADIENES AND HYDROCARBON MIXTURE, STABILIZED, having a vapour pressure at 70° C not exceeding 1.1 MPa (11 bar) and a density at 50° C not lower than 0.525 kg/l, (with 0.1% or more 1.3-butadiene)	2	2F		2.1+unst.+CMR	G	1	1			91		1	no	T2	II B[4] (II B2[4])	yes	PP, EP, EX, TOX, A	1	2; 3; 31
1011	BUTANE (contains less than 0.1% 1.3-butadiene)	2	2F		2.1	G	1	1			91		1	no	T2	II A	yes	PP, EX, A	1	2; 31
1011	BUTANE (with 0.1% or more 1.3-butadiene)	2	2F		2.1+CMR	G	1	1			91		1	no	T2	II A	yes	PP, EP, EX, TOX, A	1	2 ; 31
1012	1-BUTYLENE	2	2F		2.1	G	1	1			91		1	no	T2	II A	yes	PP, EX, A	1	2; 31
1020	CHLOROPENTAFLUORO-ETHANE (REFRIGERANT GAS R 115)	2	2A		2.2	G	1	1			91		1	no			no	PP	0	31
1030	1,1-DIFLUOROETHANE (REFRIGERANT GAS R 152a)	2	2F		2.1	G	1	1			91		1	no	T1	II A	yes	PP, EX, A	1	2; 31

(1)	(2)	(3a)	(3b)	(4)	(5)	(6)	(7)	(8)	(9)	(10)	(11)	(12)	(13)	(14)	(15)	(16)	(17)	(18)	(19)	(20)
UN No. or substance identification No.	Name and description	Class	Classification code	Packing group	Dangers	Type of tank vessel	Cargo tank design	Cargo tank type	Cargo tank equipment	Opening pressure of the high-velocity vent valve in kPa	Maximum degree of filling in %	Relative density at 20 °C	Type of sampling device	Pump room below deck permitted	Temperature class	Explosion group	Anti-explosion protection required	Equipment required	Number of cones/blue lights	Additional requirements/Remarks
1033	DIMETHYL ETHER	2	2F		2.1	G	1	1			91		1	no	T3	II B (II B2)	yes	PP, EX, A	1	2; 31
1038	ETHYLENE, REFRIGERATED LIQUID	2	3F		2.1	G	1	1	1		95		1	no	T1	II B (II B3)	yes	PP, EX, A	1	2; 31; 42
1040	ETHYLENE OXIDE WITH NITROGEN up to a total pressure of 1 MPa (10 bar) at 50 °C	2	2TF		2.3+2.1	G	1	1			91		1	no	T2	II B (II B3)	yes	PP, EP, EX, TOX, A	2	2; 3; 11; 31; 35
1055	ISOBUTYLENE	2	2F		2.1	G	1	1			91		1	no	T2 [1)]	II A	yes	PP, EX, A	1	2; 31
1063	METHYL CHLORIDE (REFRIGERANT GAS R 40)	2	2F		2.1	G	1	1			91		1	no	T1	II A	yes	PP, EX, A	1	2; 31
1077	PROPYLENE	2	2F		2.1	G	1	1			91		1	no	T1	II A	yes	PP, EX, A	1	2; 31
1083	TRIMETHYLAMINE, ANHYDROUS	2	2F		2.1	G	1	1			91		1	no	T4	II A	yes	PP, EX, A	1	2; 31
1086	VINYL CHLORIDE, STABILIZED	2	2F		2.1+unst.	G	1	1			91		1	no	T2	II A	yes	PP, EX, A	1	2; 3; 13; 31
1088	ACETAL	3	F1	II	3	N	2	2		10	97	0.83	3	yes	T3	II B [4)]	yes	PP, EX, A	1	
1089	ACETALDEHYDE (ethanal)	3	F1	I	3+N3	C	1	1			95	0.78	1	yes	T4	II A	yes	PP, EX, A	1	35
1090	ACETONE	3	F1	II	3	N	2	2		10	97	0.79	3	yes	T1	II A	yes	PP, EX, A	1	
1092	ACROLEINE, STABILIZED	6.1	TF1	I	6.1+3+unst.+N1	C	2	2	3	50	95	0.84	1	no	T3 [2)]	II B (II B3)	yes	PP, EP, EX, TOX, A	2	2; 3; 5; 23
1093	ACRYLONITRILE, STABILIZED	3	FT1	I	3+6.1+unst.+N2+CMR	C	2	2	3	50	95	0.8	1	no	T1	II B (II B2)	yes	PP, EP, EX, TOX, A	2	3; 5; 23
1098	ALLYL ALCOHOL	6.1	TF1	I	6.1+3+N1	C	2	2		40	95	0.85	1	no	T2	II B (II B3)	yes	PP, EP, EX, TOX, A	2	
1100	ALLYL CHLORIDE	3	FT1	I	3+6.1+N1	C	2	2	3	50	95	0.94	1	no	T2	II A	yes	PP, EP, EX, TOX, A	2	23
1105	PENTANOLS (n-PENTANOL)	3	F1	III	3	N	3	2			97	0.81	3	yes	T2	II A	yes	PP, EX, A	0	
1106	AMYLAMINE (n-AMYLAMINE)	3	FC	II	3+8	C	2	2		40	95	0.76	2	yes	T4 [3)]	II A [7)]	yes	PP, EP, EX, A	1	

(1)	(2)	(3a)	(3b)	(4)	(5)	(6)	(7)	(8)	(9)	(10)	(11)	(12)	(13)	(14)	(15)	(16)	(17)	(18)	(19)	(20)
UN No. or substance identification No.	Name and description	Class	Classification code	Packing group	Dangers	Type of tank vessel	Cargo tank design	Cargo tank type	Cargo tank equipment	Opening pressure of the high-velocity vent valve in kPa	Maximum degree of filling in %	Relative density at 20 °C	Type of sampling device	Pump room below deck permitted	Temperature class	Explosion group	Anti-explosion protection required	Equipment required	Number of cones/blue lights	Additional requirements/Remarks
1107	AMYL CHLORIDES (1-CHLOROPENTANE)	3	F1	II	3	C	2	2		40	95	0.88	2	yes	T3	II A	yes	PP, EX, A	1	
1107	AMYL CHLORIDES (1-CHLORO-3-METHYLBUTANE)	3	F1	II	3	C	2	2		45	95	0.89	2	yes	T3	II A	yes	PP, EX, A	1	
1107	AMYL CHLORIDES (2-CHLORO-2-METHYLBUTANE)	3	F1	II	3	C	2	2		50	95	0.87	2	yes	T2	II A	yes	PP, EX, A	1	
1107	AMYL CHLORIDES (1-CHLORO-2,2-DIMETHYL-PROPANE)	3	F1	II	3	C	2	2		50	95	0.87	2	yes	T3 2)	II A	yes	PP, EX, A	1	
1107	AMYL CHLORIDES	3	F1	II	3	C	1	1			95	0.9	1	yes	T3 2)	II A	yes	PP, EX, A	1	27
1108	1-PENTENE (n-AMYLENE)	3	F1	I	3+N3	N	1	1			97	0.64	1	yes	T3	II B 4)	yes	PP, EX, A	1	
1114	BENZENE	3	F1	II	3+N3+CMR	C	2	2	3	50	95	0.88	2	yes	T1	II A	yes	PP, EP, EX, TOX, A	1	6: +10 °C; 17; 23
1120	BUTANOLS (tert- BUTYLALCOHOL)	3	F1	II	3	N	2	2	2	10	97	0.79	3	yes	T1	II A 7)	yes	PP, EX, A	1	7; 17
1120	BUTANOLS (sec-BUTYLALCOHOL)	3	F1	III	3	N	3	2			97	0.81	3	yes	T2	II B 7)	yes	PP, EX, A	0	
1120	BUTANOLS (n- BUTYL ALCOHOL)	3	F1	III	3	N	3	2			97	0.81	3	yes	T2	II B (II B2)	yes	PP, EX, A	0	
1123	BUTYL ACETATES (sec-BUTYLACETATE)	3	F1	II	3	N	2	2		10	97	0.86	3	yes	T2	II A 7)	yes	PP, EX, A	1	
1123	BUTYL ACETATES (n-BUTYL ACETATE)	3	F1	III	3+N3	N	3	2			97	0.86	3	yes	T2	II A	yes	PP, EX, A	0	
1125	n-BUTYLAMINE	3	FC	II	3+8+N3	C	2	2	3	50	95	0.75	2	yes	T2	II A	yes	PP, EP, EX, A	1	23
1127	CHLOROBUTANES (1-CHLOROBUTANE)	3	F1	II	3	C	2	2	3	50	95	0.89	2	yes	T3	II A	yes	PP, EX, A	1	23
1127	CHLOROBUTANES (2-CHLOROBUTANE)	3	F1	II	3	C	2	2	3	50	95	0.87	2	yes	T3	II A	yes	PP, EX, A	1	23

(1)	(2)	(3a)	(3b)	(4)	(5)	(6)	(7)	(8)	(9)	(10)	(11)	(12)	(13)	(14)	(15)	(16)	(17)	(18)	(19)	(20)
1127	CHLOROBUTANES (1-CHLORO-2-METHYLPROPANE)	3	F1	II	3	C	2	2	3	50	95	0.88	2	yes	T3	II A	yes	PP, EX, A	1	23
1127	CHLOROBUTANES (2-CHLORO-2-METHYLPROPANE)	3	F1	II	3	C	2	2	3	50	95	0.84	2	yes	T1	II A	yes	PP, EX, A	1	23
1127	CHLOROBUTANES	3	F1	II	3	C	1	1			95	0.89	1	yes	T4 [3)]	II A	yes	PP, EX, A	1	27
1129	BUTYRALDEHYDE (n-BUTYRALDEHYDE)	3	F1	II	3+N3	C	2	2	3	50	95	0.8	2	yes	T4	II A	yes	PP, EX, A	1	15; 23
1131	CARBON DISULPHIDE	3	FT1	I	3+6.1+N2	C	2	2	3	50	95	1.26	1	no	T6	II C	yes	PP, EP, EX, TOX, A	2	2; 9; 23
1134	CHLOROBENZENE (phenyl chloride)	3	F1	III	3+N2+S	C	2	2	3	30	95	1.11	2	yes	T1	II A [8)]	yes	PP, EX, A	0	
1135	ETHYLENE CHLOROHYDRIN (2-CHLOROETHANOL)	6.1	TF1	I	6.1+3+N3	C	2	2	3	30	95	1.21	1	no	T2	II A [8)]	yes	PP, EP, EX, TOX, A	2	
1143	CROTONALDEHYDE, STABILIZED	6.1	TF1	I	6.1+3+unst.+N1	C	2	2	3	40	95	0.85	1	no	T3	II B (II B2)	yes	PP, EP, EX, TOX, A	2	3; 5; 15
1145	CYCLOHEXANE	3	F1	II	3+N1	C	2	2	3	50	95	0.78	2	yes	T3	II A	yes	PP, EX, A	1	6: +11 °C; 17
1146	CYCLOPENTANE	3	F1	II	3+N2	N	2	3		10	97	0.75	3	yes	T2	II A	yes	PP, EX, A	1	
1150	1,2-DICHLOROETHYLENE (cis-1,2-DICHLOROETHYLENE)	3	F1	II	3+N2	C	2	2	3	50	95	1.28	2	yes	T2 [1)]	II A	yes	PP, EX, A	1	23
1150	1,2-DICHLOROETHYLENE (trans-1,2-DICHLOROETHYLENE)	3	F1	II	3+N2	C	2	2	3	50	95	1.26	2	yes	T2	II A	yes	PP, EX, A	1	23
1153	ETHYLENE GLYCOL DIETHYL ETHER	3	F1	III	3	N	3	2			97	0.84	3	yes	T4	II B (II B2)	yes	PP, EX, A	0	
1154	DIETHYLAMINE	3	FC	II	3+8+N3	C	2	2	3	50	95	0.7	2	yes	T2	II A	yes	PP, EP, EX, A	1	23
1155	DIETHYL ETHER	3	F1	I	3	C	1	1			95	0.71	1	yes	T4	II B (II B1)	yes	PP, EX, A	1	
1157	DIISOBUTYL KETONE	3	F1	III	3+N3+F	N	3	3			97	0.81	3	yes	T2	II B [4)]	yes	PP, EX, A	0	

(1)	(2)	(3a)	(3b)	(4)	(5)	(6)	(7)	(8)	(9)	(10)	(11)	(12)	(13)	(14)	(15)	(16)	(17)	(18)	(19)	(20)
1159	DIISOPROPYL ETHER	3	F1	II	3+N2	C	2	2	3	50	95	0.72	2	yes	T2	II A	yes	PP, EX, A	1	
1160	DIMETHYLAMINE AQUEOUS SOLUTION	3	FC	II	3+8+N3	C	2	2	3	50	95	0.82	2	yes	T2	II A	yes	PP, EP, EX, A	1	23
1163	DIMETHYLHYDRAZINE, UNSYMMETRICAL	6.1	TFC	I	6.1+3+8+N2+CMR	C	2	2	3	50	95	0.78	1	no	T3	II C	yes	PP, EP, EX, TOX, A	2	23
1165	DIOXANE	3	F1	II	3	N	2	2		10	97	1.03	3	yes	T2	II B (II B3)	yes	PP, EX, A	1	6: +14 °C; 17
1167	DIVINYL ETHER, STABILIZED	3	F1	I	3+unst.	C	1	1			95	0.77	1	yes	T2	II B	yes	PP, EX, A	1	2; 3
1170	ETHANOL (ETHYL ALCOHOL) or ETHANOL SOLUTION (ETHYL ALCOHOL SOLUTION), aqueous solution with more than 70 % alcohol by volume	3	F1	II	3	N	2	2		10	97	0,79 - 0,87	3	yes	T2	II B (II B1)	yes	PP, EX, A	1	
1170	ETHANOL SOLUTION (ETHYL ALCOHOL SOLUTION), aqueous solution with more than 24 % and not more than 70 % alcohol by volume	3	F1	III	3	N	3	2			97	0,87 - 0,96	3	yes	T2	II B (II B1 4))	yes	PP, EX, A	0	
1171	ETHYLENE GLYCOL MONOETHYL ETHER	3	F1	III	3+CMR	N	2	3	3	10	97	0.93	3	yes	T3	II B (II B2)	yes	PP, EP, EX, TOX, A	0	
1172	ETHYLENE GLYCOL MONOETHYL ETHER ACETATE	3	F1	III	3+N3+CMR	N	2	3	3	10	97	0.98	3	yes	T2	II A	yes	PP, EP, EX, TOX, A	0	
1173	ETHYL ACETATE	3	F1	II	3	N	2	2		10	97	0.9	3	yes	T1	II A	yes	PP, EX, A	1	
1175	ETHYLBENZENE	3	F1	II	3+N3	N	2	2		10	97	0.87	3	yes	T2	II A	yes	PP, EX, A	1	
1177	2-ETHYLBUTYL ACETATE	3	F1	III	3	N	3	2			97	0.88	3	yes	T3	II A 7)	yes	PP, EX, A	0	
1179	ETHYL BUTYL ETHER (ETHYL tert-BUTYL ETHER)	3	F1	II	3+N3	N	2	2		10	97	0.74	3	yes	T2	II B 4)	yes	PP, EX, A	1	
1184	ETHYLENE DICHLORIDE (1,2-dichloroethane)	3	FT1	II	3+6.1+CMR	C	2	2		50	95	1.25	2	no	T2	II A	yes	PP, EP, EX, TOX, A	2	
1188	ETHYLENE GLYCOL MONOMETHYL ETHER	3	F1	III	3+CMR	N	2	3	3	10	97	0.97	3	yes	T3	II B	yes	PP, EP, EX, TOX, A	0	

Column headers:
(1) UN No. or substance identification No.
(2) Name and description
(3a) Class
(3b) Classification code
(4) Packing group
(5) Dangers
(6) Type of tank vessel
(7) Cargo tank design
(8) Cargo tank type
(9) Cargo tank equipment
(10) Opening pressure of the high-velocity vent valve in kPa
(11) Maximum degree of filling in %
(12) Relative density at 20 °C
(13) Type of sampling device
(14) Pump room below deck permitted
(15) Temperature class
(16) Explosion group
(17) Anti-explosion protection required
(18) Equipment required
(19) Number of cones/blue lights
(20) Additional requirements/Remarks

(1) UN No. or substance identification No.	(2) Name and description	(3a) Class	(3b) Classification code	(4) Packing group	(5) Dangers	(6) Type of tank vessel	(7) Cargo tank design	(8) Cargo tank type	(9) Cargo tank equipment	(10) Opening pressure of the high-velocity vent valve in kPa	(11) Maximum degree of filling in %	(12) Relative density at 20 °C	(13) Type of sampling device	(14) Pump room below deck permitted	(15) Temperature class	(16) Explosion group	(17) Anti-explosion protection required	(18) Equipment required	(19) Number of cones/blue lights	(20) Additional requirements/Remarks
1191	OCTYL ALDEHYDES (2-ETHYLCAPRONALDEHYDE)	3	F1	III	3+N3+F	C	2	2		30	95	0.82	2	yes	T4	II A [7]	yes	PP, EX, A	0	
1191	OCTYL ALDEHYDES (n-OCTALDEHYDE)	3	F1	III	3+N3+F	N	3	3			97	0.82	3	yes	T3	II B [4]	yes	PP, EX, A	0	
1193	ETHYL METHYL KETONE (METHYL ETHYL KETONE)	3	F1	II	3	N	2	2		10	97	0.8	3	yes	T1	II A	yes	PP, EX, A	1	
1198	FORMALDEHYDE SOLUTION, FLAMMABLE	3	FC	III	3+8+N3	N	3	2			97	1.09	3	yes	T2	II B	yes	PP, EP, EX, A	0	34
1199	FURALDEHYDES (a-FURALDEHYDE) or FURFURALDEHYDES (a-FURFURYLALDEHYDE)	6.1	TF1	II	6.1+3	C	2	2		25	95	1.16	2	no	T3 [2]	II B (II B1)	yes	PP, EP, EX, TOX, A	2	15
1202	GAS OIL or DIESEL FUEL (LIGHT) (flash-point not more than 60 °C)	3	F1	III	3+(N1, N2, N3, CMR, F or S)	*	*	*	*	*	*	< 0,85	*	yes		no	*	*	0	*see 3.2.3.3
1202	GAS OIL complying with standard EN 590: 2009 + A1:2010 or DIESEL FUEL or HEATING OIL (LIGHT) with flash-point as specified in EN 590:2009 + A1:2010	3	F1	III	3+N2+F	N	4	3			97	0,82 – 0,85	3	yes		no		PP	0	
1202	GAS OIL or DIESEL FUEL (LIGHT) (flash-point more than 60 °C but not more than 100 °C)	3	F1	III	3+(N1, N2, N3, CMR, F or S)	*	*	*	*	*	*	< 1,1	*	yes		no	*	*	0	*see 3.2.3.3
1203	MOTOR SPIRIT or GASOLINE or PETROL	3	F1	II	3+N2+CMR+F	N	2	3	3	10	97	0,68 – 0,72 [10]	3	yes	T3	II A	yes	PP, EP, EX, TOX, A	1	
1203	MOTOR SPIRIT or GASOLINE or PETROL, WITH MORE THAN 10 % BENZENE BOILING POINT ≤ 60 °C	3	F1	II	3+N2+CMR+F	C	1	1			95		1	yes	T3	II A	yes	PP, EP, EX, TOX, A	1	29

(1) UN No. or substance identification No.	(2) Name and description	(3a) Class	(3b) Classification code	(4) Packing group	(5) Dangers	(6) Type of tank vessel	(7) Cargo tank design	(8) Cargo tank type	(9) Cargo tank equipment	(10) Opening pressure of the high-velocity vent valve in kPa	(11) Maximum degree of filling in %	(12) Relative density at 20 °C	(13) Type of sampling device	(14) Pump room below deck permitted	(15) Temperature class	(16) Explosion group	(17) Anti-explosion protection required	(18) Equipment required	(19) Number of cones/blue lights	(20) Additional requirements/Remarks
1203	MOTOR SPIRIT or GASOLINE or PETROL WITH MORE THAN 10 % BENZENE 60 °C < BOILING POINT ≤ 85 °C	3	F1	II	3+N2+CMR+F	C	2	2	3	50	95		2	yes	T3	II A	yes	PP, EP, EX, TOX, A	1	23; 29
1203	MOTOR SPIRIT or GASOLINE or PETROL WITH MORE THAN 10 % BENZENE 85 °C < BOILING POINT ≤ 115 °C	3	F1	II	3+N2+CMR+F	C	2	2		50	95		2	yes	T3	II A	yes	PP, EP, EX, TOX, A	1	29
1203	MOTOR SPIRIT or GASOLINE or PETROL WITH MORE THAN 10 % BENZENE BOILING POINT > 115 °C	3	F1	II	3+N2+CMR+F	C	2	2		35	95		2	yes	T3	II A	yes	PP, EP, EX, TOX, A	1	29
1206	HEPTANES	3	F1	II	3+N1	C	2	2		50	95	0.68	2	yes	T3	II A	yes	PP, EX, A	1	
1208	HEXANES	3	F1	II	3+N2	N	2	3		50	97	0.66	2	yes	T3	II A	yes	PP, EX, A	1	
1212	ISOBUTANOL or ISOBUTYL ALCOHOL	3	F1	III	3	N	3	2	3		97	0.8	3	yes	T2	II A	yes	PP, EX, A	0	
1213	ISOBUTYLACETATE	3	F1	II	3+N3	N	2	2		10	97	0.87	3	yes	T2	II A[7)]	yes	PP, EX, A	1	
1214	ISOBUTYLAMINE	3	FC	II	3+8+N3	C	2	2	3	50	95	0.73	2	yes	T2	II A[7)]	yes	PP, EP, EX, A	1	23
1216	ISOOCTENES	3	F1	II	3+N2	N	2	3		10	97	0.73	3	yes	T3	II B[4)]	yes	PP, EX, A	1	
1218	ISOPRENE, STABILIZED	3	F1	I	3+unst.+N2+CMR	N	1	1			95	0.68	1	yes	T3	II B (II B2)	yes	PP, EP, EX, TOX, A	1	2; 3; 5; 16
1219	ISOPROPANOL or ISOPROPYL ALCOHOL	3	F1	II	3	N	2	2		10	97	0.78	3	yes	T2	II A	yes	PP, EX, A	1	
1220	ISOPROPYLE ACETATE	3	F1	II	3	N	2	2		10	97	0.88	3	yes	T2	II A[7)]	yes	PP, EX, A	1	
1221	ISOPROPYLAMINE	3	FC	I	3+8+N3	C	1	1			95	0.69	1	yes	T2	II A[7)]	yes	PP, EP, EX, A	1	
1223	KEROSENE	3	F1	III	3+N2+F	N	3	3			97	≤ 0,83	3	yes	T3	II A[7)]	yes	PP, EX, A	0	14
1224	KETONES, LIQUID, N.O.S.	3	F1	II	3+(N1, N2, N3, CMR, F or S)	*	*	*	*	*	*	*	*	yes	T4 [3)]	II B[4)]	yes	*	1	14; 27; 29 *see 3.2.3.3

(1)	(2)	(3a)	(3b)	(4)	(5)	(6)	(7)	(8)	(9)	(10)	(11)	(12)	(13)	(14)	(15)	(16)	(17)	(18)	(19)	(20)
UN No. or substance identification No.	Name and description	Class	Classification code	Packing group	Dangers	Type of tank vessel	Cargo tank design	Cargo tank type	Cargo tank equipment	Opening pressure of the high-velocity vent valve in kPa	Maximum degree of filling in %	Relative density at 20 °C	Type of sampling device	Pump room below deck permitted	Temperature class	Explosion group	Anti-explosion protection required	Equipment required	Number of cones/blue lights	Additional requirements/Remarks
1224	KETONES, LIQUID, N.O.S.	3	F1	III	3+(N1, N2, N3, CMR, F or S)	*	*	*	*	*	*		*	yes	T4³⁾	II B⁴⁾	yes	*	0	14; 27 *see 3.2.3.3
1229	MESITYL OXIDE	3	F1	III	3	N	3	2			97	0.85	3	yes	T2	II B⁴⁾	yes	PP, EX, A	0	
1230	METHANOL	3	FT1	II	3+6.1	N	2	2	3	50	95	0.79	2	yes	T2	II A	yes	PP, EP, EX, TOX, A	2	23
1231	METHYL ACETATE	3	F1	II	3	N	2	2		10	97	0.93	3	yes	T1	II A	yes	PP, EX, A	1	
1235	METHYLAMINE, AQUEOUS SOLUTION	3	FC	II	3+8+N3	C	2	2		50	95		2	yes	T2	II A	yes	PP, EP, EX, A	1	
1243	METHYL FORMATE	3	F1	I	3	C	1	1			95	0.97	1	yes	T2	II A	yes	PP, EX, A	1	
1244	METHYLHYDRAZINE	6.1	TFC	I	6.1+3+8	C	2	2		45	95	0.88	1	no	T4	II C⁻⁵⁾	yes	PP, EP, EX, TOX, A	2	
1245	METHYL ISOBUTYL KETONE	3	F1	II	3	N	2	2	*	10	97	0.8	3	yes	T1	II A	yes	PP, EX, A	1	
1247	METHYL METHACRYLATE MONOMER, STABILIZED	3	F1	II	3+unst.+N3	C	2	2	*	40	95	0.94	1	yes	T2	II A	yes	PP, EX, A	1	3; 5; 16
1262	OCTANES	3	F1	II	3+N1	C	2	2		45	95	0.7	2	yes	T1	II A	yes	PP, EX, A	1	
1264	PARALDEHYDE	3	F1	III	3	N	3	2			97	0.99	3	yes	T2	II A⁷⁾	yes	PP, EX, A	0	6: +16 °C; 17
1265	PENTANES, liquid	3	F1	I	3+N2	*	*	*	*	*	*	*	*	yes	*	II A	yes	PP, EX, A	1	14; * see 3.2.3.3
1265	PENTANES, liquid	3	F1	II	3+N2	*	*	*	*	*	*	*	*	yes	*	II A	yes	PP, EX, A	1	14; * see 3.2.3.3
1265	PENTANES, liquid (2-METHYLBUTANE)	3	F1	I	3+N2	N	1	1		50	97	0.62	1	yes	T2	II A	yes	PP, EX, A	1	
1265	PENTANES, liquid (n-PENTANE)	3	F1	II	3+N2	N	2	3		10	97	0.63	3	yes	T3	II A	yes	PP, EX, A	1	
1265	PENTANES, liquid (n-PENTANE)	3	F1	II	3+N2	N	2	3	3		97	0.63	3	yes	T3	II A	yes	PP, EX, A	1	
1267	PETROLEUM CRUDE OIL	3	F1	I	3+(N1, N2, N3, CMR, F)	*	*	*	*	*	*		*	yes	T4³⁾	II B⁴⁾	yes	*	1	14; *see 3.2.3.3

(1) UN No. or substance identification No.	(2) Name and description	(3a) Class	(3b) Classification code	(4) Packing group	(5) Dangers	(6) Type of tank vessel	(7) Cargo tank design	(8) Cargo tank type	(9) Cargo tank equipment	(10) Opening pressure of the high-velocity vent valve in kPa	(11) Maximum degree of filling in %	(12) Relative density at 20 °C	(13) Type of sampling device	(14) Pump room below deck permitted	(15) Temperature class	(16) Explosion group	(17) Anti-explosion protection required	(18) Equipment required	(19) Number of cones/blue lights	(20) Additional requirements/Remarks
1267	PETROLEUM CRUDE OIL	3	F1	II	3+(N1, N2, N3, CMR, F)	*	*	*	*	*	*		*	yes	$T4^{3)}$	II $B^{4)}$	yes	*	1	14; *see 3.2.3.3
1267	PETROLEUM CRUDE OIL	3	F1	III	3+(N1, N2, N3, CMR, F)	*	*	*	*	*	*		*	yes	$T4^{3)}$	II $B^{4)}$	yes	*	0	14; *see 3.2.3.3
1267	PETROLEUM CRUDE OIL WITH MORE THAN 10% BENZENE INITIAL BOILING POINT ≤ 60 °C	3	F1	I	3+CMR+F+ (N1, N2, N3)	C	1	1			95		1	yes	$T4^{3)}$	II $B^{4)}$	yes	PP, EP, EX, TOX, A	1	29; 43
1267	PETROLEUM CRUDE OIL WITH MORE THAN 10% BENZENE INITIAL BOILING POINT ≤ 60 °C	3	F1	II	3+CMR+F+ (N1, N2, N3)	C	1	1			95		1	yes	$T4^{3)}$	II $B^{4)}$	yes	PP, EP, EX, TOX, A	1	29
1267	PETROLEUM CRUDE OIL WITH MORE THAN 10% BENZENE 60 °C < INITIAL BOILING POINT ≤ 85 °C	3	F1	II	3+CMR+F+ (N1, N2, N3)	C	2	2	3	50	95		2	yes	$T4^{3)}$	II $B^{4)}$	yes	PP, EP, EX, TOX, A	1	23; 29; 38
1267	PETROLEUM CRUDE OIL WITH MORE THAN 10% BENZENE 85 °C < INITIAL BOILING POINT ≤ 115 °C	3	F1	II	3+CMR+F+ (N1, N2, N3)	C	2	2		50	95		2	yes	$T4^{3)}$	II $B^{4)}$	yes	PP, EP, EX, TOX, A	1	29
1267	PETROLEUM CRUDE OIL WITH MORE THAN 10% BENZENE INITIAL BOILING POINT > 115 °C	3	F1	II	3+CMR+F+ (N1, N2, N3)	C	2	2		35	95		2	yes	$T4^{3)}$	II $B^{4)}$	yes	PP, EP, EX, TOX, A	1	29
1267	PETROLEUM CRUDE OIL WITH MORE THAN 10% BENZENE INITIAL BOILING POINT ≤ 60 °C	3	F1	III	3+CMR+F+ (N1, N2, N3)	C	1	1			95		1	yes	$T4^{3)}$	II $B^{4)}$	yes	PP, EP, EX, TOX, A	0	29
1267	PETROLEUM CRUDE OIL WITH MORE THAN 10% BENZENE 60 °C < INITIAL BOILING POINT ≤ 85 °C	3	F1	III	3+CMR+F+ (N1, N2, N3)	C	2	2	3	50	95		2	yes	$T4^{3)}$	II $B^{4)}$	yes	PP, EP, EX, TOX, A	0	23; 29; 38
1267	PETROLEUM CRUDE OIL WITH MORE THAN 10% BENZENE 85 °C < INITIAL BOILING POINT ≤ 115 °C	3	F1	III	3+CMR+F+ (N1, N2, N3)	C	2	2		50	95		2	yes	$T4^{3)}$	II $B^{4)}$	yes	PP, EP, EX, TOX, A	0	29

(1) UN No. or substance identification No.	(2) Name and description	(3a) Class	(3b) Classification code	(4) Packing group	(5) Dangers	(6) Type of tank vessel	(7) Cargo tank design	(8) Cargo tank type	(9) Cargo tank equipment	(10) Opening pressure of the high-velocity vent valve in kPa	(11) Maximum degree of filling in %	(12) Relative density at 20 °C	(13) Type of sampling device	(14) Pump room below deck permitted	(15) Temperature class	(16) Explosion group	(17) Anti-explosion protection required	(18) Equipment required	(19) Number of cones/blue lights	(20) Additional requirements/Remarks
1267	PETROLEUM CRUDE OIL WITH MORE THAN 10% BENZENE INITIAL BOILING POINT > 115 °C	3	F1	III	3+CMR+F+ (N1, N2, N3)	C	2	2		35	95		2	yes	$T4^{3)}$	$II\ B^{4)}$	yes	PP, EP, EX, TOX, A	0	29
1268	PETROLEUM DISTILLATES, N.O.S. or PETROLEUM PRODUCTS, N.O.S.	3	F1	I	3+(N1, N2, N3, CMR, F)	*	*	*	*	*	*		*	yes	$T4^{3)}$	$II\ B^{4)}$	yes	*	1	14; 27 *see 3.2.3.3
1268	PETROLEUM DISTILLATES, N.O.S. or PETROLEUM PRODUCTS, N.O.S.	3	F1	II	3+(N1, N2, N3, CMR, F)	*	*	*	*	*	*		*	yes	$T4^{3)}$	$II\ B^{4)}$	yes	*	1	14; 27 *see 3.2.3.3
1268	PETROLEUM DISTILLATES, N.O.S. or PETROLEUM PRODUCTS, N.O.S.	3	F1	III	3+(N1, N2, N3, CMR, F)	*	*	*	*	*	*		*	yes	$T4^{3)}$	$II\ B^{4)}$	yes	*	0	14; 27 *see 3.2.3.3
1268	PETROLEUM DISTILLATES, N.O.S. or PETROLEUM PRODUCTS, N.O.S. WITH MORE THAN 10% BENZENE INITIAL BOILING POINT ≤ 60 °C	3	F1	I	3+CMR+F+ (N1, N2, N3)	C	1	1			95		1	yes	$T4^{3)}$	$II\ B^{4)}$	yes	PP, EP, EX, TOX, A	1	27; 29; 43
1268	PETROLEUM DISTILLATES, N.O.S. or PETROLEUM PRODUCTS, N.O.S. WITH MORE THAN 10% BENZENE INITIAL BOILING POINT ≤ 60 °C	3	F1	II	3+CMR+F+ (N1, N2, N3)	C	1	1			95		1	yes	$T4^{3)}$	$II\ B^{4)}$	yes	PP, EP, EX, TOX, A	1	27; 29
1268	PETROLEUM DISTILLATES, N.O.S. or PETROLEUM PRODUCTS, N.O.S. WITH MORE THAN 10% BENZENE 60 °C < INITIAL BOILING POINT ≤ 85 °C	3	F1	II	3+CMR+F+ (N1, N2, N3)	C	2	2	3	50	95		2	yes	$T4^{3)}$	$II\ B^{4)}$	yes	PP, EP, EX, TOX, A	1	23; 27; 29; 38
1268	PETROLEUM DISTILLATES, N.O.S. or PETROLEUM PRODUCTS, N.O.S. WITH MORE THAN 10% BENZENE 85 °C < INITIAL BOILING POINT ≤ 115 °C	3	F1	II	3+CMR+F+ (N1, N2, N3)	C	2	2		50	95		2	yes	$T4^{3)}$	$II\ B^{4)}$	yes	PP, EP, EX, TOX, A	1	27; 29

(1)	(2)	(3a)	(3b)	(4)	(5)	(6)	(7)	(8)	(9)	(10)	(11)	(12)	(13)	(14)	(15)	(16)	(17)	(18)	(19)	(20)
UN No. or substance identification No.	Name and description	Class	Classification code	Packing group	Dangers	Type of tank vessel	Cargo tank design	Cargo tank type	Cargo tank equipment	Opening pressure of the high-velocity vent valve in kPa	Maximum degree of filling in %	Relative density at 20 °C	Type of sampling device	Pump room below deck permitted	Temperature class	Explosion group	Anti-explosion protection required	Equipment required	Number of cones/blue lights	Additional requirements/Remarks
1268	PETROLEUM DISTILLATES, N.O.S. or PETROLEUM PRODUCTS, N.O.S. WITH MORE THAN 10% BENZENE INITIAL BOILING POINT > 115 °C	3	F1	II	3+CMR+F+ (N1, N2, N3)	C	2	2		35	95		2	yes	T4[3)]	II B[4)]	yes	PP, EP, EX, TOX, A	1	27; 29
1268	PETROLEUM DISTILLATES, N.O.S. or PETROLEUM PRODUCTS, N.O.S. (NAPHTA) $110\,kPa < vp50 \le 175\,kPa$	3	F1	II	3+N2+ CMR+F	N	2	3	3	50	97	0,735	3	yes	T3	II A	yes	PP, EP, EX, TOX, A	1	14; 29
1268	PETROLEUM DISTILLATES, N.O.S. or PETROLEUM PRODUCTS, N.O.S. (NAPHTA) $110\,kPa < vp50 \le 150\,kPa$	3	F1	II	3+N2+ CMR+F	N	2	3		10	97	0,735	3	yes	T3	II A	yes	PP, EP, EX, TOX, A	1	14; 29
1268	PETROLEUM DISTILLATES, N.O.S. or PETROLEUM PRODUCTS, N.O.S. (NAPHTA) $vp50 \le 110\,kPa$	3	F1	II	3+N2+ CMR+F	N	2	3		10	97	0,735	3	yes	T3	II A	yes	PP, EP, EX, TOX, A	1	14; 29
1268	PETROLEUM DISTILLATES, N.O.S. or PETROLEUM PRODUCTS, N.O.S (BENZENE HEART CUT) $vp50 \le 110\,kPa$	3	F1	II	3+N2+ CMR+F	N	2	3		10	97	0,765	3	yes	T3	II A	yes	PP, EP, EX, TOX, A	1	14; 29
1274	n-PROPANOL or PROPYL ALCOHOL, NORMAL	3	F1	II	3	N	2	2		10	97	0.8	3	yes	T2	II B	yes	PP, EX, A	1	
1274	n-PROPANOL or PROPYL ALCOHOL, NORMAL	3	F1	III	3	N	3	2		10	97	0.8	3	yes	T2	II B	yes	PP, EX, A	0	
1275	PROPIONALDEHYDE	3	F1	II	3+N3	C	2	2	3	50	95	0.81	2	yes	T4	II B	yes	PP, EX, A	1	15; 23
1276	n-PROPYL ACETATE	3	F1	II	3+N3	N	2	2		10	97	0.88	3	yes	T1	II A	yes	PP, EX, A	1	
1277	PROPYLAMINE (1-aminopropane)	3	FC	II	3+8	C	2	2	3	50	95	0.72	2	yes	T2	II A	yes	PP, EP, EX, A	1	23

(1) UN No. or substance identification No.	(2) Name and description	(3a) Class	(3b) Classification code	(4) Packing group	(5) Dangers	(6) Type of tank vessel	(7) Cargo tank design	(8) Cargo tank type	(9) Cargo tank equipment	(10) Opening pressure of the high-velocity vent valve in kPa	(11) Maximum degree of filling in %	(12) Relative density at 20 °C	(13) Type of sampling device	(14) Pump room below deck permitted	(15) Temperature class	(16) Explosion group	(17) Anti-explosion protection required	(18) Equipment required	(19) Number of cones/blue lights	(20) Additional requirements/Remarks
1278	1-CHLOROPROPANE (propyl chloride)	3	F1	II	3	C	2	2	3	50	95	0.89	2	yes	T1	II A	yes	PP, EX, A	1	23
1279	1,2-DICHLOROPROPANE or PROPYL DICHLORIDE	3	F1	II	3+N2	C	2	2		45	95	1.16	2	yes	T1	II A 8)	yes	PP, EX, A	1	
1280	PROPYLENE OXIDE	3	F1	I	3+unst.+N3+CMR	C	1	1			95	0.83	1	yes	T2	II B	yes	PP, EP, EX, TOX, A	1	2; 12; 31; 35
1282	PYRIDINE	3	F1	II	3+N3	N	2	2		10	97	0.98	3	yes	T1	II A 8)	yes	PP, EX, A	1	
1289	SODIUM METHYLATE SOLUTION in alcohol	3	FC	III	3+8	N	3	2			97	0.969	3	yes	T2	II A	yes	PP, EP, EX, A	0	34
1294	TOLUENE	3	F1	II	3+N3	N	2	2		10	97	0.87	3	yes	T1	II A	yes	PP, EX, A	1	
1296	TRIETHYLAMINE	3	FC	II	3+8+N3	C	2	2		50	95	0.73	2	yes	T3	II A 8)	yes	PP, EP, EX, A	1	
1300	TURPENTINE SUBSTITUTE	3	F1	III	3+N2+F	N	3	3			97	0.78	3	yes	T3	II B 4)	yes	PP, EX, A	0	
1301	VINYL ACETATE, STABILIZED	3	F1	II	3+unst.+N3	N	2	2	2	10	97	0.93	2	yes	T2	II A	yes	PP, EX, A	1	3; 5; 16
1307	XYLENES (o- XYLENE)	3	F1	III	3+N2	N	3	3			97	0.88	3	yes	T1	II A	yes	PP, EX, A	0	
1307	XYLENES (m- XYLENE)	3	F1	III	3+N2	N	3	3			97	0.86	3	yes	T1	II A	yes	PP, EX, A	0	
1307	XYLENES (p- XYLENE)	3	F1	III	3+N2	N	3	3	2		97	0.86	3	yes	T1	II A	yes	PP, EX, A	0	
1307	XYLENES (mixture with melting point ≤ 0° C)	3	F1	II	3+N2	N	3	3			97		3	yes	T1	II A	yes	PP, EX, A	1	6: +17 °C; 17
1307	XYLENES (mixture with melting point ≤ 0° C)	3	F1	III	3+N2	N	3	3			97		3	yes	T1	II A	yes	PP, EX, A	0	
1307	XYLENES (mixture with 0° C < melting point < 13° C)	3	F1	III	3+N2	N	3	3	2		97		3	yes	T1	II A	yes	PP, EX, A	0	6: +17 °C; 17
1541	ACETONE CYANOHYDRIN, STABILIZED	6.1	T1	I	6.1+unst.+N1	C	2	2		50	95	0.932	1	no	T1		no	PP, EP, TOX, A	2	3
1545	ALLYL ISOTHIOCYANATE, STABILIZED	6.1	TF1	II	6.1+3+unst.	C	2	2		30	95	1.02	1	no	T4 3)	II B 4)	yes	PP, EP, EX, TOX, A	2	2; 3
1547	ANILINE	6.1	T1	II	6.1+N1	C	2	2		25	95	1.02	2	no			no	PP, EP, TOX, A	2	

(1)	(2)	(3a)	(3b)	(4)	(5)	(6)	(7)	(8)	(9)	(10)	(11)	(12)	(13)	(14)	(15)	(16)	(17)	(18)	(19)	(20)
UN No, or substance identification No.	Name and description	Class	Classification code	Packing group	Dangers	Type of tank vessel	Cargo tank design	Cargo tank type	Cargo tank equipment	Opening pressure of the high-velocity vent valve in kPa	Maximum degree of filling in %	Relative density at 20 °C	Type of sampling device	Pump room below deck permitted	Temperature class	Explosion group	Anti-explosion protection required	Equipment required	Number of cones/blue lights	Additional requirements/Remarks
1578	CHLORONITROBENZENES, SOLID, MOLTEN (p-CHLORONITROBENZENE)	6.1	T2	II	6.1+N2+S	C	2	1	2	25	95	1.37	2	no	T1	II B[4]	yes	PP, EP, EX, TOX, A	2	7; 17; 26
1578	CHLORONITROBENZENES, SOLID, MOLTEN (p-CHLORONITROBENZENE)	6.1	T2	II	6.1+N2+S	C	2	1	4	25	95	1.37	2	no			no	PP, EP, TOX, A	2	7; 17; 20: +112°C; 26
1591	o-DICHLOROBENZENE	6.1	T1	III	6.1+N1+S	C	2	2		25	95	1.32	2	no			no	PP, EP, TOX, A	0	
1593	DICHLOROMETHANE (methyl chloride)	6.1	T1	III	6.1	C	2	2	3	50	95	1.33	2	no			no	PP, EP, TOX, A	0	23
1594	DIETHYL SULPHATE	6.1	T1	II	6.1+N2 +CMR	C	2	2		25	95	1.18	2	no			no	PP, EP, TOX, A	2	
1595	DIMETHYL SULPHATE	6.1	TC1	I	6.1+8+N3+ CMR	C	2	2		25	95	1.33	1	no			no	PP, EP, TOX, A	2	
1604	ETHYLENEDIAMINE	8	CF1	II	8+3+N3	N	3	2			97	0.9	3	yes	T2	II A	yes	PP, EP, EX, A	1	6: +12 °C; 17; 34
1605	ETHYLENE DIBROMIDE	6.1	T1	I	6.1+N2 +CMR	C	2	2		30	95	2.18	1	no			no	PP, EP, TOX, A	2	6: +14 °C; 17
1648	ACETONITRILE (methyl cyanide)	3	F1	II	3	N	2	2		10	97	0.78	3	yes	T1	II A	yes	PP, EX, A	1	
1662	NITROBENZENE	6.1	T1	II	6.1+N2	C	2	2	2	25	95	1.21	2	no	T1	II B (II B1)	yes	PP, EP, EX, TOX, A	2	6: +10°C; 17
1663	NITROPHENOLS	6.1	T2	III	6.1+N3+S	C	2	2	2	25	95		2	no	T1	II B[4]	yes	PP, EP, EX, TOX, A	0	7; 17
1663	NITROPHENOLS	6.1	T2	III	6.1+N3+S	C	2	2	4	25	95		2	no			no	PP, EP, TOX, A	0	7; 17; 20: +65 °C
1664	NITROTOLUENES, LIQUID (o-NITROTOLUENE)	6.1	T1	II	6.1+N2 +CMR+S	C	2	2		25	95	1.16	2	no			no	PP, EP, TOX, A	2	17
1708	TOLUIDINES, LIQUID (o-TOLUIDINE)	6.1	T1	II	6.1+N1+CMR	C	2	2		25	95	1	2	no			no	PP, EP, TOX, A	2	
1708	TOLUIDINES, LIQUID (m-TOLUIDINE)	6.1	T1	II	6.1+N1	C	2	2		25	95	1.03	2	no			no	PP, EP, TOX, A	2	

(1) UN No. or substance identification No.	(2) Name and description	(3a) Class	(3b) Classification code	(4) Packing group	(5) Dangers	(6) Type of tank vessel	(7) Cargo tank design	(8) Cargo tank type	(9) Cargo tank equipment	(10) Opening pressure of the high-velocity vent valve in kPa	(11) Maximum degree of filling in %	(12) Relative density at 20 °C	(13) Type of sampling device	(14) Pump room below deck permitted	(15) Temperature class	(16) Explosion group	(17) Anti-explosion protection required	(18) Equipment required	(19) Number of cones/blue lights	(20) Additional requirements/Remarks
1710	TRICHLOROETHYLENE	6.1	T1	III	6.1+N2 +CMR	C	2	2		50	95	1.46	2	no			no	PP, EP, TOX, A	0	15
1715	ACETIC ANHYDRIDE	8	CF1	II	8+3	N	2	3		10	97	1.08	3	yes	T2	II A	yes	PP, EP, EX, A	1	34
1717	ACETYL CHLORIDE	3	FC	II	3+8	C	2	2	3	50	95	1.1	2	yes	T2	II A [8)]	yes	PP, EP, EX, A	1	23
1718	BUTYL ACID PHOSPHATE	8	C3	III	8+N3	N	4	3			97	0.98	3	yes			no	PP, EP	0	34
1719	CAUSTIC ALKALI LIQUID, N.O.S.	8	C5	II	8+(N1, N2, N3, CMR, F or S)	*	*	*	*	*	*		*	yes			no	*	0	27; 30; 34 *see 3.2.3.3
1719	CAUSTIC ALKALI LIQUID, N.O.S.	8	C5	III	8+(N1, N2, N3, CMR, F or S)	*	*	*	*	*	*		*	yes			no	*	0	27; 30; 34 *see 3.2.3.3
1738	BENZYL CHLORIDE	6.1	TC1	II	6.1+8+N3+ CMR+S	C	2	2		25	95	1.1	2	no	T1	II A [8)]	yes	PP, EP, EX, TOX, A	2	
1742	BORON TRIFLUORIDE ACETIC ACID COMPLEX, LIQUID	8	C3	II	8	N	4	2	2		97	1.35	3	yes			no	PP, EP	0	34
1750	CHLORACETIC ACID SOLUTION	6.1	TC1	II	6.1+8+N1	C	2	2	2	25	95	1.58	2	no	T1	II A	yes	PP, EP, EX, A	2	7; 17
1750	CHLORACETIC ACID SOLUTION	6.1	TC1	II	6.1+8+N1	C	2	1	4	25	95	1.58	2	no			no	PP, EP, TOX, A	2	7; 17; 20: +111°C; 26
1760	CORROSIVE LIQUID, N.O.S.	8	C9	I	8+(N1, N2, N3, CMR, F or S)	*	*	*	*	*	*		*	yes			no	*	0	27; 34 *see 3.2.3.3
1760	CORROSIVE LIQUID, N.O.S.	8	C9	II	8+(N1, N2, N3, CMR, F or S)	*	*	*	*	*	*		*	yes			no	*	0	27; 34 *see 3.2.3.3
1760	CORROSIVE LIQUID, N.O.S.	8	C9	III	8+(N1, N2, N3, CMR, F or S)	*	*	*	*	*	*		*	yes			no	*	0	27; 34 *see 3.2.3.3

(1)	(2)	(3a)	(3b)	(4)	(5)	(6)	(7)	(8)	(9)	(10)	(11)	(12)	(13)	(14)	(15)	(16)	(17)	(18)	(19)	(20)
UN No. or substance identification No.	Name and description	Class	Classification code	Packing group	Dangers	Type of tank vessel	Cargo tank design	Cargo tank type	Cargo tank equipment	Opening pressure of the high-velocity vent valve in kPa	Maximum degree of filling in %	Relative density at 20 °C	Type of sampling device	Pump room below deck permitted	Temperature class	Explosion group	Anti-explosion protection required	Equipment required	Number of cones/blue lights	Additional requirements/Remarks
1760	CORROSIVE LIQUID, N.O.S. (SODIUM MERCAPTOBENZOTHIAZOLE, 50 % AQUEOUS SOLUTION)	8	C9	II	8+N1+F	C	2	2		40	95	1.25	2	yes			no	PP, EP	0	
1760	CORROSIVE LIQUID, N.O.S. (FATTY ALCOHOL, C_{12}-C_{14})	8	C9	III	8+F	N	4	3			97	0.89	3	yes			no	PP, EP	0	34
1760	CORROSIVE LIQUID, N.O.S. (ETHYLENEDIAMINE-TETRAACETIC ACID, TETRASODIUM SALT, 40 % AQUEOUS SOLUTION)	8	C9	III	8+N2	N	4	3			97	1.28	3	yes			no	PP, EP	0	34
1764	DICHLOROACETIC ACID	8	C3	II	8+N1	N	3	3			97	1.56	2	yes	T1	II A	yes	PP, EP, EX, A	0	17
1778	FLUOROSILICIC ACID	8	C1	II	8+N3	N	2	3		10	97		3	yes			no	PP, EP	0	34
1779	FORMIC ACID with more than 85% acid by mass	8	CF1	II	8+3+N3	N	2	3		10	97	1.22	3	yes	T1	II A	yes	PP, EP, EX, A	1	6: +12 °C; 17; 34
1780	FUMARYL CHLORIDE	8	C3	II	8+N3	N	2	3		10	97	1.41	3	yes			no	PP, EP	0	8; 34
1783	HEXAMETHYLENEDIAMINE SOLUTION	8	C7	II	8+N3	N	3	2	2		97		3	yes	T4 [3]	II B[4]	yes	PP, EP, EX, A	0	7; 17; 34
1783	HEXAMETHYLENEDIAMINE SOLUTION	8	C7	III	8+N3	N	3	2	2		97		3	yes	T3	II B[4]	yes	PP, EP, EX, A	0	7; 17; 34
1789	HYDROCHLORIC ACID	8	C1	II	8	N	2	3		10	97		3	yes			no	PP, EP	0	34
1789	HYDROCHLORIC ACID	8	C1	III	8	N	4	3			97		3	yes			no	PP, EP	0	34
1805	PHOSPHORIC ACID, SOLUTION, WITH MORE THAN 80% (VOLUME) ACID	8	C1	III	8	N	4	3	2		95	> 1,6	3	yes			no	PP, EP	0	7; 17; 22; 34
1805	PHOSPHORIC ACID, SOLUTION, WITH 80% (VOLUME) ACID, OR LESS	8	C1	III	8	N	4	3			97	1,00 - 1,6	3	yes			no	PP, EP	0	22; 34
1814	POTASSIUM HYDROXIDE SOLUTION	8	C5	II	8+N3	N	4	2			97		3	yes			no	PP, EP	0	30; 34

(1) UN No. or substance identification No.	(2) Name and description	(3a) Class	(3b) Classification code	(4) Packing group	(5) Dangers	(6) Type of tank vessel	(7) Cargo tank design	(8) Cargo tank type	(9) Cargo tank equipment	(10) Opening pressure of the high-velocity vent valve in kPa	(11) Maximum degree of filling in %	(12) Relative density at 20 °C	(13) Type of sampling device	(14) Pump room below deck permitted	(15) Temperature class	(16) Explosion group	(17) Anti-explosion protection required	(18) Equipment required	(19) Number of cones/blue lights	(20) Additional requirements/Remarks
1814	POTASSIUM HYDROXIDE SOLUTION	8	C5	III	8+N3	N	4	2			97		3	yes			no	PP, EP	0	30; 34
1823	SODIUM HYDROXIDE, SOLID, MOLTEN	8	C6	II	8+N3	N	4	1	4		95	2.13	3	yes			no	PP, EP	0	7; 17; 34
1824	SODIUM HYDROXIDE SOLUTION	8	C5	II	8+N3	N	4	2			97		3	yes			no	PP, EP	0	30; 34
1824	SODIUM HYDROXIDE SOLUTION	8	C5	III	8+N3	N	4	2			97		3	yes			no	PP, EP	0	30; 34
1830	SULPHURIC ACID with more than 51% acid	8	C1	II	8+N3	N	4	3			97	1,4 - 1,84	3	yes			no	PP, EP	0	8; 22; 30; 34
1831	SULPHURIC ACID, FUMING	8	CT1	I	8+6.1	C	2	2		50	95	1.94	1	no			no	PP, EP, TOX, A	2	8
1832	SULPHURIC ACID, SPENT	8	C1	II	8	N	4	3			97		3	yes			no	PP, EP	0	8; 30; 34
1846	CARBON TETRACHLORIDE	6.1	T1	II	6.1+N2+S	C	2	2	3	50	95	1.59	2	no			no	PP, EP, TOX, A	2	23
1848	PROPIONIC ACID with not less than 10% and less than 90% acid by mass	8	C3	III	8+N3	N	3	3			97	0.99	3	yes			no	PP, EP	0	34
1863	FUEL, AVIATION, TURBINE ENGINE	3	F1	I	3+(N1, N2, N3, CMR, F)	*	*	*	*	*	*		*	yes	T4[3]	II B[4]	yes	*	1	14; *see 3.2.3.3
1863	FUEL, AVIATION, TURBINE ENGINE	3	F1	II	3+(N1, N2, N3, CMR, F)	*	*	*	*	*	*		*	yes	T4[3]	II B[4]	yes	*	1	14; *see 3.2.3.3
1863	FUEL, AVIATION, TURBINE ENGINE	3	F1	III	3+(N1, N2, N3, CMR, F)	*	*	*	*	*	*		*	yes	T4[3]	II B[4]	yes	*	0	14; *see 3.2.3.3
1863	FUEL, AVIATION, TURBINE ENGINE WITH MORE THAN 10% BENZENE INITIAL BOILING POINT ≤60 °C	3	F1	I	3+CMR+F+ (N1, N2, N3)	C	1	1			95		1	yes	T4[3]	II B[4]	yes	PP, EP, EX, TOX, A	1	29; 43
1863	FUEL, AVIATION, TURBINE ENGINE WITH MORE THAN 10% BENZENE INITIAL BOILING POINT ≤60 °C	3	F1	II	3+CMR+F+ (N1, N2, N3)	C	1	1			95		1	yes	T4[3]	II B[4]	yes	PP, EP, EX, TOX, A	1	29

(1)	(2)	(3a)	(3b)	(4)	(5)	(6)	(7)	(8)	(9)	(10)	(11)	(12)	(13)	(14)	(15)	(16)	(17)	(18)	(19)	(20)
UN No. or substance identification No.	Name and description	Class	Classification code	Packing group	Dangers	Type of tank vessel	Cargo tank design	Cargo tank type	Cargo tank equipment	Opening pressure of the high-velocity vent valve in kPa	Maximum degree of filling in %	Relative density at 20 °C	Type of sampling device	Pump room below deck permitted	Temperature class	Explosion group	Anti-explosion protection required	Equipment required	Number of cones/blue lights	Additional requirements/Remarks
1863	FUEL, AVIATION, TURBINE ENGINE WITH MORE THAN 10% BENZENE 60 °C < INITIAL BOILING POINT ≤ 85°C	3	F1	III	3+CMR+F+ (N1, N2, N3)	C	2	2	3	50	95		2	yes	T4[3]	II B[4]	yes	PP, EP, EX, TOX, A	0	23; 29; 38
1863	FUEL, AVIATION, TURBINE ENGINE WITH MORE THAN 10% BENZENE 85 °C < INITIAL BOILING POINT ≤ 115 °C	3	F1	III	3+CMR+F+ (N1, N2, N3)	C	2	2		50	95		2	yes	T4[3]	II B[4]	yes	PP, EP, EX, TOX, A	0	29
1863	FUEL, AVIATION, TURBINE ENGINE WITH MORE THAN 10% BENZENE INITIAL BOILING POINT > 115 °C	3	F1	III	3+CMR+F+ (N1, N2, N3)	C	2	2		35	95		2	yes	T4[3]	II B[4]	yes	PP, EP, EX, TOX, A	0	29
1888	CHLOROFORM	6.1	T1	III	6.1+N2+CMR	C	2	2	3	50	95	1.48	2	no			no	PP, EP, TOX, A	0	23
1897	TETRACHLOROETHYLENE	6.1	T1	III	6.1+N2+S	C	2	2		50	95	1.62	2	no			no	PP, EP, TOX, A	0	
1912	METHYL CHLORIDE AND METHYLENE CHLORIDE MIXTURE	2	2F		2.1	G	1	1			91		1	no	T1	II A[8]	yes	PP, EX, A	1	2; 31
1915	CYCLOHEXANONE	3	F1	III	3	N	3	2			97	0.95	3	yes	T2	II A	yes	PP, EX, A	0	
1917	ETHYL ACRYLATE, STABILIZED	3	F1	II	3+unst.+N3	C	2	2		40	95	0.92	1	yes	T2	II B (II B1)	yes	PP, EX, A	1	3; 5
1918	ISOPROPYLBENZENE (cumene)	3	F1	III	3+N2	N	3	3			97	0.86	3	yes	T2	II A[8]	yes	PP, EX, A	0	
1919	METHYL ACRYLATE, STABILIZED	3	F1	II	3+unst.+N3	C	2	2	3	50	95	0.95	1	yes	T2	II B (II B1)	yes	PP, EX, A	1	3; 5; 23
1920	NONANES	3	F1	III	3+N2+F	N	3	3			97	0,70 - 0,75	3	yes	T3	II A	yes	PP, EX, A	0	
1922	PYRROLIDINE	3	FC	II	3+8	C	2	2		50	95	0.86	2	yes	T2	II A[7]	yes	PP, EP, EX, A	1	
1965	HYDROCARBON GAS MIXTURE, LIQUEFIED, N.O.S., (MIXTURE A)	2	2F		2.1	G	1	1			91		1	no	T4[3]	II B[4]	yes	PP, EX, A	1	2; 31

(1) UN No. or substance identification No.	(2) Name and description	(3a) Class	(3b) Classification code	(4) Packing group	(5) Dangers	(6) Type of tank vessel	(7) Cargo tank design	(8) Cargo tank type	(9) Cargo tank equipment	(10) Opening pressure of the high-velocity vent valve in kPa	(11) Maximum degree of filling in %	(12) Relative density at 20 °C	(13) Type of sampling device	(14) Pump room below deck permitted	(15) Temperature class	(16) Explosion group	(17) Anti-explosion protection required	(18) Equipment required	(19) Number of cones/blue lights	(20) Additional requirements/Remarks
1965	HYDROCARBON GAS MIXTURE, LIQUEFIED, N.O.S., (MIXTURE A0)	2	2F		2.1	G	1	1			91		1	no	T4³⁾	II B⁴⁾	yes	PP, EX, A	1	2; 31
1965	HYDROCARBON GAS MIXTURE, LIQUEFIED, N.O.S., (MIXTURE A01)	2	2F		2.1	G	1	1			91		1	no	T4³⁾	II B⁴⁾	yes	PP, EX, A	1	2; 31
1965	HYDROCARBON GAS MIXTURE, LIQUEFIED, N.O.S., (MIXTURE A02)	2	2F		2.1	G	1	1			91		1	no	T4³⁾	II B⁴⁾	yes	PP, EX, A	1	2; 31
1965	HYDROCARBON GAS MIXTURE, LIQUEFIED, N.O.S., (MIXTURE A1)	2	2F		2.1	G	1	1			91		1	no	T4³⁾	II B⁴⁾	yes	PP, EX, A	1	2; 31
1965	HYDROCARBON GAS MIXTURE, LIQUEFIED, N.O.S., (MIXTURE B)	2	2F		2.1	G	1	1			91		1	no	T4³⁾	II B⁴⁾	yes	PP, EX, A	1	2; 31
1965	HYDROCARBON GAS MIXTURE, LIQUEFIED, N.O.S., (MIXTURE B1)	2	2F		2.1	G	1	1			91		1	no	T4³⁾	II B⁴⁾	yes	PP, EX, A	1	2; 31
1965	HYDROCARBON GAS MIXTURE, LIQUEFIED, N.O.S., (MIXTURE B2)	2	2F		2.1	G	1	1			91		1	no	T4³⁾	II B⁴⁾	yes	PP, EX, A	1	2; 31
1965	HYDROCARBON GAS MIXTURE, LIQUEFIED, N.O.S., (MIXTURE C)	2	2F		2.1	G	1	1			91		1	no	T4³⁾	II B⁴⁾	yes	PP, EX, A	1	2; 31
1969	ISOBUTANE (contains less than 0.1% 1.3-butadiene)	2	2F		2.1	G	1	1			91		1	no	T2¹⁾	II A⁷⁾	yes	PP, EX, A	1	2; 31
1969	ISOBUTANE (with 0.1% or more 1.3-butadiene)	2	2F		2.1+CMR	G	1	1	1		91		1	no	T2¹⁾	II A	yes	PP, EP, EX, TOX, A	1	2; 31
1972	METHANE, REFRIGERATED LIQUID or NATURAL GAS, REFRIGERATED LIQUID, with high methane content	2	3F		2.1	G	1	1			95			no	T1	IIA	yes	PP, EX, A	1	2; 31; 42
1978	PROPANE	2	2F		2.1	G	1	1			91		1	no	T1	II A	yes	PP, EX, A	1	2; 31

(1)	(2)	(3a)	(3b)	(4)	(5)	(6)	(7)	(8)	(9)	(10)	(11)	(12)	(13)	(14)	(15)	(16)	(17)	(18)	(19)	(20)
UN No. or substance identification No.	Name and description	Class	Classification code	Packing group	Dangers	Type of tank vessel	Cargo tank design	Cargo tank type	Cargo tank equipment	Opening pressure of the high-velocity vent valve in kPa	Maximum degree of filling in %	Relative density at 20 °C	Type of sampling device	Pump room below deck permitted	Temperature class	Explosion group	Anti-explosion protection required	Equipment required	Number of cones/blue lights	Additional requirements/Remarks
1986	ALCOHOLS, FLAMMABLE, TOXIC, N.O.S.	3	FT1	I	3+6.1+(N1, N2, N3, CMR, F or S)	C	1	1	*	*	95		1	no	T4[3]	II B[4]	yes	PP, EP, EX, TOX, A	2	27; 29; *see 3.2.3.3
1986	ALCOHOLS, FLAMMABLE, TOXIC, N.O.S.	3	FT1	I	3+6.1+(N1, N2, N3, CMR, F or S)	C	2	2	*	*	95		1	no	T4[3]	II B[4]	yes	PP, EP, EX, TOX, A	2	27; 29; *see 3.2.3.3
1986	ALCOHOLS, FLAMMABLE, TOXIC, N.O.S.	3	FT1	II	3+6.1+(N1, N2, N3, CMR, F or S)	C	2	2	*	*	95		2	no	T4[3]	II B[4]	yes	PP, EP, EX, TOX, A	2	27; 29; *see 3.2.3.3
1986	ALCOHOLS, FLAMMABLE, TOXIC, N.O.S.	3	FT1	III	3+6.1+(N1, N2, N3, CMR, F or S)	C	2	2	*	*	95		2	no	T4[3]	II B[4]	yes	PP, EP, EX, TOX, A	0	27; 29; *see 3.2.3.3
1987	ALCOHOLS, N.O.S. (tert-BUTANOL 90 % (MASS)/METHANOL 10 % (MASS) MIXTURE)	3	F1	II	3	N	2	2		10	97		3	yes	T1	II A	yes	PP, EX, A	1	
1987	ALCOHOLS, N.O.S.	3	F1	II	3+(N1, N2, N3, CMR, F or S)	*	*	*	*	*	*		*	yes	T4[3]	II B[4]	yes	*	1	14; 27; 29; *see 3.2.3.3
1987	ALCOHOLS, N.O.S.	3	F1	III	3+(N1, N2, N3, CMR, F or S)	*	*	*	*	*	*		*	yes	T4[3]	II B[4]	yes	*	0	14; 27; *see 3.2.3.3
1987	ALCOHOLS, N.O.S. (CYCLOHEXANOL)	3	F1	III	3+N3+F	N	3	3	2	*	95	0.95	3	yes	T3	II A	yes	PP, EX, A	0	7; 17
1987	ALCOHOLS, N.O.S. (CYCLOHEXANOL)	3	F1	III	3+N3+F	N	3	3	4	*	95	0.95	3	yes			no	PP	0	7; 17; 20: +46 °C
1989	ALDEHYDES, N.O.S.	3	F1	II	3+(N1, N2, N3, CMR, F or S)	*	*	*	*	*	*		*	yes	T4[3]	II B[4]	yes	*	1	14; 27; 29; *see 3.2.3.3
1989	ALDEHYDES, N.O.S.	3	F1	III	3+(N1, N2, N3, CMR, F or S)	*	*	*	*	*	*		*	yes	T4[3]	II B[4]	yes	*	0	14; 27; *see 3.2.3.3

(1) UN No. or substance identification No.	(2) Name and description	(3a) Class	(3b) Classification code	(4) Packing group	(5) Dangers	(6) Type of tank vessel	(7) Cargo tank design	(8) Cargo tank type	(9) Cargo tank equipment	(10) Opening pressure of the high-velocity vent valve in kPa	(11) Maximum degree of filling in %	(12) Relative density at 20 °C	(13) Type of sampling device	(14) Pump room below deck permitted	(15) Temperature class	(16) Explosion group	(17) Anti-explosion protection required	(18) Equipment required	(19) Number of cones/blue lights	(20) Additional requirements/Remarks
1991	CHLOROPRENE, STABILIZED	3	FT1	I	3+6.1+unst.+CMR	C	2	2	3	50	95	0.96	1	no	T2	II B[4]	yes	PP, EP, EX, TOX, A	2	3; 5; 23
1992	FLAMMABLE LIQUID, TOXIC, N.O.S.	3	FT1	I	3+6.1+(N1, N2, N3, CMR, F or S)	C	1	1	*	*	95		1	no	T4[3]	II B[4]	yes	PP, EP, EX, TOX, A	2	27; 29 *see 3.2.3.3
1992	FLAMMABLE LIQUID, TOXIC, N.O.S	3	FT1	I	3+6.1+(N1, N2, N3, CMR, F or S)	C	2	2	*	*	95		1	no	T4[3]	II B[4]	yes	PP, EP, EX, TOX, A	2	27; 29 *see 3.2.3.3
1992	FLAMMABLE LIQUID, TOXIC, N.O.S	3	FT1	II	3+6.1+(N1, N2, N3, CMR, F or S)	C	2	2	*	*	95		2	no	T4[3]	II B[4]	yes	PP, EP, EX, TOX, A	2	27; 29 *see 3.2.3.3
1992	FLAMMABLE LIQUID, TOXIC, N.O.S	3	FT1	III	3+6.1+(N1, N2, N3, CMR, F or S)	C	2	2	*	*	95		2	no	T4[3]	II B[4]	yes	PP, EP, EX, TOX, A	0	27; 29 *see 3.2.3.3
1993	FLAMMABLE LIQUID, N.O.S.	3	F1	I	3+(N1, N2, N3, CMR, F)	*	*	*	*	*	*		*	yes	T4[3]	II B[4]	yes	*	1	14; *see 3.2.3.3
1993	FLAMMABLE LIQUID, N.O.S.	3	F1	II	3+(N1, N2, N3, CMR, F)	*	*	*	*	*	*		*	yes	T4[3]	II B[4]	yes	*	1	14; *see 3.2.3.3
1993	FLAMMABLE LIQUID, N.O.S.	3	F1	III	3+(N1, N2, N3, CMR, F)	*	*	*	*	*	*		*	yes	T4[3]	II B[4]	yes	*	0	14; *see 3.2.3.3
1993	FLAMMABLE LIQUID, N.O.S. WITH MORE THAN 10% BENZENE INITIAL BOILING POINT ≤ 60 °C	3	F1	I	3+(N1, N2, N3, CMR, F)	C	1	1			95		1	yes	T4[3]	II B[4]	yes	PP, EP, EX, TOX, A	1	29
1993	FLAMMABLE LIQUID, N.O.S. WITH MORE THAN 10% BENZENE INITIAL BOILING POINT ≤ 60 °C	3	F1	II	3+(N1, N2, N3, CMR, F)	C	1	1			95		1	yes	T4[3]	II B[4]	yes	PP, EP, EX, TOX, A	1	29
1993	FLAMMABLE LIQUID, N.O.S. WITH MORE THAN 10% BENZENE 60 °C < INITIAL BOILING POINT ≤ 85 °C	3	F1	II	3+(N1, N2, N3, CMR, F)	C	2	2	3	50	95		2	yes	T4[3]	II B[4]	yes	PP, EP, EX, TOX, A	1	23; 29; 38

(1) UN No. or substance identification No.	(2) Name and description	(3a) Class	(3b) Classification code	(4) Packing group	(5) Dangers	(6) Type of tank vessel	(7) Cargo tank design	(8) Cargo tank type	(9) Cargo tank equipment	(10) Opening pressure of the high-velocity vent valve in kPa	(11) Maximum degree of filling in %	(12) Relative density at 20 °C	(13) Type of sampling device	(14) Pump room below deck permitted	(15) Temperature class	(16) Explosion group	(17) Anti-explosion protection required	(18) Equipment required	(19) Number of cones/blue lights	(20) Additional requirements/Remarks
1993	FLAMMABLE LIQUID, N.O.S. WITH MORE THAN 10% BENZENE 85 °C < INITIAL BOILING POINT ≤ 115 °C	3	F1	II	3+(N1, N2, N3, CMR, F)	C	2	2		50	95		2	yes	T4$^{3)}$	II B$^{4)}$	yes	PP, EP, EX, TOX, A	1	29
1993	FLAMMABLE LIQUID, N.O.S. WITH MORE THAN 10% BENZENE INITIAL BOILING POINT > 115 °C	3	F1	II	3+(N1, N2, N3, CMR, F)	C	2	2		35	95		2	yes	T4$^{3)}$	II B$^{4)}$	yes	PP, EP, EX, TOX, A	1	29
1993	FLAMMABLE LIQUID, N.O.S. WITH MORE THAN 10% BENZENE INITIAL BOILING POINT ≤ 60 °C	3	F1	III	3+(N1, N2, N3, CMR, F)	C	1	1	3		95		1	yes	T4$^{3)}$	II B$^{4)}$	yes	PP, EP, EX, TOX, A	0	29
1993	FLAMMABLE LIQUID, N.O.S. WITH MORE THAN 10% BENZENE 60 °C < INITIAL BOILING POINT ≤ 85 °C	3	F1	III	3+(N1, N2, N3, CMR, F)	C	2	2		50	95		2	yes	T4$^{3)}$	II B$^{4)}$	yes	PP, EP, EX, TOX, A	0	23; 29; 38
1993	FLAMMABLE LIQUID, N.O.S. WITH MORE THAN 10% BENZENE 85 °C < INITIAL BOILING POINT ≤ 115 °C	3	F1	III	3+(N1, N2, N3, CMR, F)	C	2	2		50	95		2	yes	T4$^{3)}$	II B$^{4)}$	yes	PP, EP, EX, TOX, A	0	29
1993	FLAMMABLE LIQUID, N.O.S. WITH MORE THAN 10% BENZENE INITIAL BOILING POINT > 115 °C	3	F1	III	3+(N1, N2, N3, CMR, F)	C	2	2		35	95		2	yes	T4$^{3)}$	II B$^{4)}$	yes	PP, EP, EX, TOX, A	0	29
1993	FLAMMABLE LIQUID, N.O.S. (CYCLOHEXANONE/ CYCLOHEXANOL MIXTURE)	3	F1	III	3+F	N	3	3			97	0,95	3	yes	T3	II A	yes	PP, EX, A	0	
1999	TARS, LIQUID, including road oils, and cutback bitumens	3	F1	III	3+S	N	4	3	2		97		3	yes	T3	II A$^{7)}$	yes	PP, EX, A	0	

(1)	(2)	(3a)	(3b)	(4)	(5)	(6)	(7)	(8)	(9)	(10)	(11)	(12)	(13)	(14)	(15)	(16)	(17)	(18)	(19)	(20)
UN No. or substance identification No.	Name and description	Class	Classification code	Packing group	Dangers	Type of tank vessel	Cargo tank design	Cargo tank type	Cargo tank equipment	Opening pressure of the high-velocity vent valve in kPa	Maximum degree of filling in %	Relative density at 20 °C	Type of sampling device	Pump room below deck permitted	Temperature class	Explosion group	Anti-explosion protection required	Equipment required	Number of cones/blue lights	Additional requirements/Remarks
2014	HYDROGEN PEROXIDE, AQUEOUS SOLUTION with not less than 20 % but not more than 60 % hydrogen peroxide (stabilized as necessary)	5.1	OC1	II	5.1+8+unst.	C	2	2		35	95	1.2	2	yes			no	PP, EP	0	3; 33
2021	CHLOROPHENOLS, LIQUID (2-CHLOROPHENOL)	6.1	T1	III	6.1+N2	C	2	2		25	95	1.23	2	no	T1	II A[7]	yes	PP, EP, EX, TOX, A	0	6: +10 °C; 17
2022	CRESYLIC ACID	6.1	TC1	II	6.1+8+3+S	C	2	2		25	95	1.03	2	no	T1	II A[7]	yes	PP, EP, EX, TOX, A	2	6: +16 °C; 17
2023	EPICHLORHYDRINE	6.1	TF1	II	6.1+3+N3	C	2	2		35	95	1.18	2	no	T2	II B (II B3)	yes	PP, EP, EX, TOX, A	2	5
2031	NITRIC ACID, other than red fuming, with more than 70 % acid	8	CO1	I	8+5.1+N3	N	2	3		10	97	1,41-1,48	3	yes			no	PP, EP	0	34
2031	NITRIC ACID, other than red fuming with at least 65 % but not more than 70 % acid	8	CO1	II	8+5.1+N3	N	2	3		10	97	1,39-1,41	3	yes			no	PP, EP	0	34
2031	NITRIC ACID, other than red fuming, with less than 65 % acid	8	CO1	II	8+N3	N	2	3		10	97	1,02-1,39	3	yes			no	PP, EP	0	34
2032	NITRIC ACID, RED FUMING	8	COT	I	8+5.1+6.1+N3	C	2	2		50	95	1,48-1.51	1	no			no	PP, EP	0	34
2045	ISOBUTYRALDEHYDE (ISOBUTYL ALDEHYDE)	3	F1	II	3+N3	C	2	2	3	50	95	0.79	2	yes	T4	II A[7]	yes	PP, EP, TOX, A	2	15; 23
2046	CYMENES	3	F1	III	3+N2+F	N	3	3			97	0.88	3	yes	T2	II A[7]	yes	PP, EX, A	1	
2047	DICHLOROPROPENES (2,3-DICHLOROPROP-1-ENE)	3	F1	II	3+N2+CMR	C	2	2		45	95	1.2	2	yes	T1	II A[7]	yes	PP, EX, A	0	
2047	DICHLOROPROPENES (MIXTURES of 2,3-DICHLOROPROP-1-ENE and 1,3-DICHLOROPROPENE)	3	F1	II	3+N1+CMR	C	2	2		45	95	1.23	2	yes	T2[1]	II A[7]	yes	PP, EP, EX, TOX, A	1	

(1) UN No. or substance identification No.	(2) Name and description	(3a) Class	(3b) Classification code	(4) Packing group	(5) Dangers	(6) Type of tank vessel	(7) Cargo tank design	(8) Cargo tank type	(9) Cargo tank equipment	(10) Opening pressure of the high-velocity vent valve in kPa	(11) Maximum degree of filling in %	(12) Relative density at 20 °C	(13) Type of sampling device	(14) Pump room below deck permitted	(15) Temperature class	(16) Explosion group	(17) Anti-explosion protection required	(18) Equipment required	(19) Number of cones/blue lights	(20) Additional requirements/Remarks
2047	DICHLOROPROPENES (MIXTURES of 2,3-DICHLOROPROP-1-ENE and 1,3-DICHLOROPROPENE)	3	F1	III	3+N1+CMR	C	2	2		45	95	1.23	2	yes	T2[1]	II A[7]	yes	PP, EP, EX, TOX, A	0	
2047	DICHLOROPROPENES (1,3-DICHLOROPROPENE)	3	F1	III	3+N1+CMR	C	2	2		40	95	1.23	2	yes	T2[1]	II A[7]	yes	PP, EP, EX, TOX, A	0	
2048	DICYCLOPENTADIENE	3	F1	III	3+N2+F	N	3	3	2		95	0.94	3	yes	T1	II B[4]	yes	PP, EX, A	0	7; 17
2050	DIISOBUTYLENE, ISOMERIC COMPOUNDS	3	F1	II	3+N2+F	N	2	3		10	97	0.72	3	yes	T3[2]	II A[7]	yes	PP, EX, A	1	
2051	2-DIMETHYLAMINO ETHANOL	8	CF1	II	8+3+N3	N	3	2			97	0.89	3	yes	T3	II A[7]	yes	PP, EP, EX, A	1	34
2053	METHYL ISOBUTYL CARBINOL	3	F1	III	3	N	3	2			97	0.81	3	yes	T2	II B[4]	yes	PP, EX, A	0	
2054	MORPHOLINE	8	CF1	I	8+3+N3	N	3	2			97	1	3	yes	T3	II A	yes	PP, EP, EX, A	1	34
2055	STYRENE MONOMER, STABILIZED	3	F1	III	3+unst.+N3	N	3	2			97	0.91	3	yes	T1	II A	yes	PP, EX, A	0	3; 5; 16
2056	TETRAHYDROFURAN	3	F1	II	3	N	2	2		10	97	0.89	3	yes	T3	II B (II B1)	yes	PP, EX, A	1	
2057	TRIPROPYLENE	3	F1	II	3+N3	N	2	3		10	97	0.744	3	yes	T3	II B[4]	yes	PP, EX, A	1	
2057	TRIPROPYLENE	3	F1	III	3+N3	N	3	3			97	0.73	3	yes	T3	II B[4]	yes	PP, EX, A	0	
2078	TOLUENE DIISOCYANATE (and isomeric mixtures) (2,4-TOLUENE DIISOCYANATE)	6.1	T1	II	6.1+N2+S	C	2	2	2	25	95	1.22	2	no	T1	II B[4]	yes	PP, EP, EX, TOX, A	2	2; 7; 8; 17
2078	TOLUENE DIISOCYANATE (and isomeric mixtures) (2,4-TOLUENE DIISOCYANATE)	6.1	T1	II	6.1+N2+S	C	2	1	4	25	95	1.22	2	no			no	PP, EP, TOX, A	2	2; 7; 8; 17; 20: +112°C; 26
2079	DIETHYLENETRIAMINE	8	C7	II	8+N3	N	4	2	1		97	0.96	3	yes			no	PP, EP	0	34
2187	CARBON DIOXIDE, REFRIGERATED LIQUID	2	3A		2.2	G	1	1			95		1	yes			no	PP	0	31,39
2205	ADIPONITRILE	6.1	T1	III	6.1	C	2	2		25	95	0.96	2	no	T4	II B[4]	yes	PP, EP, EX, TOX, A	0	6: 6°C; 17

(1)	(2)	(3a)	(3b)	(4)	(5)	(6)	(7)	(8)	(9)	(10)	(11)	(12)	(13)	(14)	(15)	(16)	(17)	(18)	(19)	(20)
UN No. or substance identification No.	Name and description	Class	Classification code	Packing group	Dangers	Type of tank vessel	Cargo tank design	Cargo tank type	Cargo tank equipment	Opening pressure of the high-velocity vent valve in kPa	Maximum degree of filling in %	Relative density at 20 °C	Type of sampling device	Pump room below deck permitted	Temperature class	Explosion group	Anti-explosion protection required	Equipment required	Number of cones/blue lights	Additional requirements/Remarks
2206	ISOCYANATES, TOXIC, N.O.S. (4-CHLOROPHENYL ISOCYANATE)	6.1	T1	II	6.1+S	C	2	2	4	25	95	1.25	2	no			no	PP, EP, TOX, A	2	7; 17
2209	FORMALDEHYDE SOLUTION with not less than 25 % formaldehyde	8	C9	III	8+N3	N	4	2			97	1.09	3	yes			no	PP, EP	0	15; 34
2215	MALEIC ANHYDRIDE, MOLTEN	8	C3	III	8+N3	N	3	3	2		95	0.93	3	yes	T2	II B$^{4)}$	yes	PP, EP, EX, A	0	7; 17; 25; 34
2215	MALEIC ANHYDRIDE, MOLTEN	8	C3	III	8+N3	N	3	1	4		95	0.93	3	yes			no	PP, EP	0	7; 17; 20: +88 °C; 25; 34
2218	ACRYLIC ACID, STABILIZED	8	CF1	II	8+3+unst.+N1	C	2	2	4	30	95	1.05	1	yes	T2	II B (II B1)	yes	PP, EP, EX, A	1	3; 4; 5; 17
2227	n-BUTYL METHACRYLATE, STABILIZED	3	F1	III	3+unst.+N3+F	C	2	2		25	95	0.9	1	yes	T3	II A	yes	PP, EX, A	0	3; 5
2238	CHLOROTOLUENES (m-CHLOROTOLUENE)	3	F1	III	3+N2+S	C	2	2	2	30	95	1.08	2	yes	T1	II A$^{7)}$	yes	PP, EX, A	0	
2238	CHLOROTOLUENES (o-CHLOROTOLUENE)	3	F1	III	3+N2+S	C	2	2	2	30	95	1.08	2	yes	T1	II A$^{7)}$	yes	PP, EX, A	0	
2238	CHLOROTOLUENES (p-CHLOROTOLUENE)	3	F1	III	3+N2+S	C	2	2	2	30	95	1.07	2	yes	T1	II A$^{7)}$	yes	PP, EX, A	0	6: +11 °C; 17
2241	CYCLOHEPTANE	3	F1	II	3+N2	N	2	3		10	97	0.81	3	yes	T4 $^{3)}$	II A$^{7)}$	yes	PP, EX, A	1	
2247	n-DECANE	3	F1	III	3+F	C	2	2		30	95	0.73	2	yes	T4	II A	yes	PP, EX, A	0	
2248	DI-n-BUTYLAMINE	8	CF1	II	8+3+N3	N	3	2				0.76	3	yes	T3	II A$^{7)}$	yes	PP, EP, EX, A	1	34
2259	TRIETHYLENETETRAMINE	8	C7	II	8+N2	N	3	3			97	0.98	3	yes	T2	II B$^{4)}$	yes	PP, EP, EX, A	0	6: 16°C; 17; 34
2263	DIMETHYLCYCLOHEXANES (cis-1,4- DIMETHYL-CYCLOHEXANE)	3	F1	II	3	C	2	2		35	95	0.78	2	yes	T4 $^{3)}$	II A$^{7)}$	yes	PP, EX, A	1	
2263	DIMETHYLCYCLOHEXANES (trans-1,4- DIMETHYL-CYCLOHEXANE)	3	F1	II	3	C	2	2		35	95	0.76	2	yes	T4 $^{3)}$	II A$^{7)}$	yes	PP, EX, A	1	

(1)	(2)	(3a)	(3b)	(4)	(5)	(6)	(7)	(8)	(9)	(10)	(11)	(12)	(13)	(14)	(15)	(16)	(17)	(18)	(19)	(20)
UN No. or substance identification No.	Name and description	Class	Classification code	Packing group	Dangers	Type of tank vessel	Cargo tank design	Cargo tank type	Cargo tank equipment	Opening pressure of the high-velocity vent valve in kPa	Maximum degree of filling in %	Relative density at 20 °C	Type of sampling device	Pump room below deck permitted	Temperature class	Explosion group	Anti-explosion protection required	Equipment required	Number of cones/blue lights	Additional requirements/Remarks
2264	N,N-DIMETHYL-CYCLOHEXYLAMINE	8	CF1	II	8+3+N2	N	3	3			97	0.85	3	yes	T3	II B[4]	yes	PP, EP, EX, A	1	34
2265	N,N-DIMETHYLFORMAMIDE	3	F1	III	3+CMR	N	2	3	3		97	0.95	3	yes	T2	II A	yes	PP, EP, EX, TOX, A	0	
2266	DIMETHYL-N-PROPYLAMINE	3	FC	II	3+8	C	2	2	3	10	95	0.72	2	yes	T4	II A[7]	yes	PP, EP, EX, A	1	23
2276	2-ETHYLHEXYLAMINE	3	FC	III	3+8+N3	N	3	2	3	50	97	0.79	3	yes	T3	II A[7]	yes	PP, EP, EX, A	0	34
2278	n-HEPTENE	3	F1	II	3+N3	N	2	2		10	97	0.7	3	yes	T3	II B[4] (II B1)	yes	PP, EX, A	1	
2280	HEXAMETHYLENEDIAMINE, SOLID, MOLTEN	8	C8	III	8+N3	N	3	3	2		95	0.83	3	yes	T3	II B[4]	yes	PP, EP, EX, A	0	7; 17; 34
2280	HEXAMETHYLENEDIAMINE, SOLID, MOLTEN	8	C8	III	8+N3	N	3	3	4		95	0.83	3	yes			no	PP, EP	0	7; 17; 20: +66 °C; 34
2282	HEXANOLS	3	F1	III	3+N3	N	3	2			97	0.83	3	yes	T3	II A	yes	PP, EX, A	0	
2286	PENTAMETHYLHEPTANE	3	F1	III	3+F	N	3	3		50	97	0.75	3	yes	T2	II A[7]	yes	PP, EX, A	0	
2288	ISOHEXENES	3	F1	II	3+unst.+N3	C	2	2			95	0.735	2	yes	T2	II B[4]	yes	PP, EX, A	1	3; 23
2289	ISOPHORONEDIAMINE	8	C7	III	8+N2	N	3	3	3		97	0.92	3	yes	T2	II A[7]	yes	PP, EP, EX, A	0	6: 14°C; 17; 34
2302	5-METHYLHEXAN-2-ONE	3	F1	III	3	N	3	2			97	0.81	3	yes	T1	II A	yes	PP, EX, A	0	
2303	ISOPROPENYLBENZENE	3	F1	III	3+N2+F	N	3	3			97	0.91	3	yes	T2	II B (II B1)	yes	PP, EX, A	0	
2309	OCTADIENE (1,7-OCTADIENE)	3	F1	II	3+N2	N	2	3		10	97	0.75	3	yes	T3	II B[4]	yes	PP, EX, A	1	
2311	PHENETIDINES	6.1	T1	III	6.1	C	2	2		25	95	1.07	2	no			no	PP, EP, TOX, A	0	6: +7 °C; 17
2312	PHENOL, MOLTEN	6.1	T1	II	6.1+N3+S	C	2	2	4	25	95	1.07	2	no	T1	II A[8]	yes	PP, EP, EX, TOX, A	2	7; 17
2312	PHENOL, MOLTEN	6.1	T1	II	6.1+N3+S	C	2	2	4	25	95	1.07	2	no			no	PP, EP, TOX, A	2	7; 17; 20: +67 °C
2320	TETRAETHYLENEPENTAMINE	8	C7	III	8+N2	N	4	3			97	1	3	yes			no	PP, EP	0	34

(1)	(2)	(3a)	(3b)	(4)	(5)	(6)	(7)	(8)	(9)	(10)	(11)	(12)	(13)	(14)	(15)	(16)	(17)	(18)	(19)	(20)
UN No. or substance identification No.	Name and description	Class	Classification code	Packing group	Dangers	Type of tank vessel	Cargo tank design	Cargo tank type	Cargo tank equipment	Opening pressure of the high-velocity vent valve in kPa	Maximum degree of filling in %	Relative density at 20 °C	Type of sampling device	Pump room below deck permitted	Temperature class	Explosion group	Anti-explosion protection required	Equipment required	Number of cones/blue lights	Additional requirements/Remarks
2321	TRICHLOROBENZENES, LIQUID (1,2,4-TRICHLOROBENZENE)	6.1	T1	III	6.1+N1+S	C	2	2	2	25	95	1.45	2	no		II A[7]	yes	PP, EP, EX, TOX, A	0	7; 17
2321	TRICHLOROBENZENES, LIQUID (1,2,4-TRICHLOROBENZENE)	6.1	T1	III	6.1+N1+S	C	2	1	4	25	95	1.45	2	no			no	PP, EP, TOX, A	0	7; 17; 20: +95 °C; 26
2323	TRIETHYL PHOSPHITE	3	F1	III	3	N	3	2			97	0.8	3	yes	T3	II B[4]	yes	PP, EX, A	0	
2324	TRIISOBUTYLENE	3	F1	III	3+N1+F	C	2	2		35	95	0.76	2	yes	T2	II B[4]	yes	PP, EX, A	0	
2325	1,3,5-TRIMETHYLBENZENE	3	F1	III	3+N1	C	2	2		35	95	0.87	2	yes	T1	II A[7]	yes	PP, EX, A	0	
2333	ALLYL ACETATE	3	FT1	II	3+6.1	C	2	2		40	95	0.93	2	no	T2	II A[7]	yes	PP, EP, EX, TOX, A	2	
2348	BUTYL ACRYLATES, STABILIZED (n- BUTYL ACRYLATE, STABILIZED)	3	F1	III	3+unst.+N3	C	2	2		30	95	0.9	1	yes	T3	II B (II B1)	yes	PP, EX, A	0	3; 5
2350	BUTYL METHYL ETHER	3	F1	II	3	N	2	2		10	97	0.74	3	yes	T4[3]	II B[4]	yes	PP, EX, A	1	
2356	2-CHLOROPROPANE	3	F1	I	3	C	2	2	3	50	95	0.86	2	yes	T1	II A	yes	PP, EX, A	1	23
2357	CYCLOHEXYLAMINE	8	CF1	II	8+3+N3	N	3	2			97	0.86	3	yes	T3	II B[4]	yes	PP, EP, EX, A	1	34
2362	1,1-DICHLOROETHANE	3	F1	II	3+N2	C	2	2	3	50	95	1.17	2	yes	T2	II A	yes	PP, EX, A	1	23
2370	1-HEXENE	3	F1	II	3+N3	N	2	2		10	97	0.67	3	yes	T3	II B[4]	yes	PP, EX, A	1	
2381	DIMÉTHYL DISULPHIDE	3	FT1	II	3+6.1	C	2	2		40	95	1.063	2	yes	T2	IIB	yes	PP, EP, EX, TOX, A	2	
2382	DIMETHYLHYDRAZINE, SYMMETRICAL	6.1	TF1	I	6.1+3+CMR	C	2	2		50	95	0.83	1	no	T4[3]	II C[5]	yes	PP, EP, EX, TOX, A	2	
2383	DIPROPYLAMINE	3	FC	II	3+8+N3	C	2	2		35	95	0.74	2	yes	T3	II A	yes	PP, EP, EX, A	1	
2397	3-METHYLBUTAN-2-ONE	3	F1	II	3	N	2	2		10	97	0.81	3	yes	T1	II A[7]	yes	PP, EX, A	1	
2398	METHYL tert-BUTYL ETHER	3	F1	II	3	N	2	2		10	97	0.74	3	yes	T1	II A	yes	PP, EX, A	1	
2404	PROPIONITRILE	3	FT1	II	3+6.1	C	2	2		45	95	0.78	2	no	T1[9]	II A[7]	yes	PP, EP, EX, TOX, A	2	
2414	THIOPHENE	3	F1	II	3+N3+S	N	2	3		10	97	1.06	3	yes	T2	II A	yes	PP, EX, A	1	

(1)	(2)	(3a)	(3b)	(4)	(5)	(6)	(7)	(8)	(9)	(10)	(11)	(12)	(13)	(14)	(15)	(16)	(17)	(18)	(19)	(20)
UN No. or substance identification No.	Name and description	Class	Classification code	Packing group	Dangers	Type of tank vessel	Cargo tank design	Cargo tank type	Cargo tank equipment	Opening pressure of the high-velocity vent valve in kPa	Maximum degree of filling in %	Relative density at 20 °C	Type of sampling device	Pump room below deck permitted	Temperature class	Explosion group	Anti-explosion protection required	Equipment required	Number of cones/blue lights	Additional requirements/Remarks
2430	ALKYLPHENOLS, SOLID, N.O.S. (NONYLPHENOL, ISOMERIC MIXTURE, MOLTEN)	8	C4	II	8+N1+F	N	3	1	2		95	0.95	2	yes	T2	II A[7]	yes	PP, EP, EX, A	0	7; 17
2430	ALKYLPHENOLS, SOLID, N.O.S. (NONYLPHENOL, ISOMERIC MIXTURE, MOLTEN)	8	C4	II	8+N1+F	N	3	2	4		95	0.95	2	yes			no	PP, EP	0	7; 17; 20: +125 °C
2432	N,N-DIETHYLANILINE	6.1	T1	III	6.1+N2	C	2	2		25	95	0.93	2	no			no	PP, EP, TOX, A	0	
2448	SULPHUR, MOLTEN	4.1	F3	III	4.1+S	N	4	1	4		95	2.07	3	yes			no	PP, EP, TOX*, A	0	* Toximeter for H2S; 7; 20: +150°C; 28; 32
2458	HEXADIENES	3	F1	II	3+N3	N	2	2		10	97	0.72	3	yes	T4[3]	II A[7]	yes	PP, EX, A	1	
2477	METHYL ISOTHIOCYANATE	6.1	TF1	I	6.1+3+N1	C	2	2	2	35	95	1,07[11]	1	no	T4[3]	II B[4]	yes	PP, EP, EX, TOX, A	2	7; 17
2485	n-BUTYL ISOCYANATE	6.1	TF1	I	6.1+3	C	2	2		35	95	0.89	1	no	T2	II B[4]	yes	PP, EP, EX, TOX, A	2	
2486	ISOBUTYL ISOCYANATE	6.1	TF1	I	6.1+3	C	2	2		40	95		1	no	T4[3]	II B[4]	yes	PP, EP, EX, TOX, A	2	
2487	PHENYL ISOCYANATE	6.1	TF1	I	6.1+3	C	2	2		25	95	1.1	1	no	T1	II A	yes	PP, EP, EX, TOX, A	2	
2490	DICHLOROISOPROPYL ETHER	6.1	T1	II	6.1	C	2	2		25	95	1.11	2	no			no	PP, EP, TOX, A	2	
2491	ETHANOLAMINE or ETHANOLAMINE SOLUTION	8	C7	III	8+N3	N	3	2			97	1.02	3	yes	T2	II A[7]	yes	PP, EP, EX, A	0	6: 14°C; 17; 34
2493	HEXAMETHYLENEIMINE	3	FC	II	3+8+N3	N	3	2			97	0.88	3	yes	T3[2]	II A	yes	PP, EP, EX, A	1	34
2496	PROPIONIC ANHYDRIDE	8	C3	III	8+N3	N	4	3			97	1.02	3	yes			no	PP, EP	0	34
2518	1,5,9-CYCLODODECATRIENE	6.1	T1	III	6.1+F	C	2	2		25	95	0.9	2	no			no	PP, EP, TOX, A	0	
2527	ISOBUTYL ACRYLATE, STABILIZED	3	F1	III	3+unst.	C	2	2		30	95	0.89	1	yes	T2	II B[9]	yes	PP, EX, A	0	3; 5

(1) UN No. or substance identification No.	(2) Name and description	(3a) Class	(3b) Classification code	(4) Packing group	(5) Dangers	(6) Type of tank vessel	(7) Cargo tank design	(8) Cargo tank type	(9) Cargo tank equipment	(10) Opening pressure of the high-velocity vent valve in kPa	(11) Maximum degree of filling in %	(12) Relative density at 20 °C	(13) Type of sampling device	(14) Pump room below deck permitted	(15) Temperature class	(16) Explosion group	(17) Anti-explosion protection required	(18) Equipment required	(19) Number of cones/blue lights	(20) Additional requirements/Remarks
2528	ISOBUTYL ISOBUTYRATE	3	F1	III	3+N3	N	3	2			97	0.86	3	yes	T2	II A	yes	PP, EX, A	0	
2531	METHACRYLIC ACID, STABILIZED	8	C3	II	8+unst.+N3	C	2	2	4	25	95	1.02	1	yes	T2	II B[4]	yes	PP, EP, EX, A	0	3; 4; 5; 7; 17
2564	TRICHLOROACETIC ACID SOLUTION	8	C3	II	8+N1	C	2	2	2	25	95	1,62[11]	2	yes	T1	II A[7]	yes	PP, EP, EX, A	0	7; 17; 22
2564	TRICHLOROACETIC ACID SOLUTION	8	C3	III	8+N1	C	2	2	2	25	95	1,62[11]	2	yes			no	PP, EP	0	22
2574	TRICRESYL PHOSPHATE with more than 3% ortho isomer	6.1	T1	II	6.1+N1+S	C	2	2	2	25	95	1.18	2	no			no	PP, EP, TOX, A	2	
2579	PIPERAZINE, MOLTEN	8	C8	III	8+N2	N	3	3			95	0.9	3	yes			no	PP, EP	0	7; 17; 34
2582	FERRIC CHLORIDE SOLUTION	8	C1	III	8	N	4	3	2		97	1.45	3	yes			no	PP, EP	0	22; 30; 34
2586	ALKYLSULPHONIC ACIDS, LIQUID or ARYLSULPHONIC ACIDS, LIQUID with not more than 5% free sulphuric acid	8	C3	III	8	N	4	3			97		3	yes				PP, EP	0	34
2608	NITROPROPANES	3	F1	III	3	N	3	2			97	1	3	yes	T2	II B[7] (II B2)	yes	PP, EX, A	0	
2615	ETHYL PROPYL ETHER	3	F1	II	3	N	2	2		10	97	0.73	3	yes	T4[3]	II A[7]	yes	PP, EX, A	1	3; 5
2618	VINYLTOLUENES, STABILIZED	3	F1	III	3+unst.+N2+F	C	2	2	2	25	95	0.92	1	yes	T1	II B[4]	yes	PP, EX, A	0	7; 17
2651	4,4'-DIAMINO-DIPHENYLMETHANE	6.1	T2	III	6.1+N2+ CMR+S	C	2	2	2	25	95	1	2	no			no	PP, EP, TOX, A	0	7; 17
2672	AMMONIA SOLUTION, relative density between 0.880 and 0.957 at 15°C in water, with more than 10% but not more than 35% ammonia (more than 25% but not more than 35% ammonia)	8	C5	III	8+N1	C	2	2	1	50	95	0,88[10] – 0,96[10]	2	yes			no	PP, EP	0	
2672	AMMONIA SOLUTION, relative density between 0.880 and 0.957 at 15°C in water, with more than 10% but not more than 35% ammonia (not more than 25% ammonia)	8	C5	III	8+N3	N	2	2	2	10	95	0,88[10] – 0,96[10]	2	yes			no	PP, EP	0	34

(1)	(2)	(3a)	(3b)	(4)	(5)	(6)	(7)	(8)	(9)	(10)	(11)	(12)	(13)	(14)	(15)	(16)	(17)	(18)	(19)	(20)
2683	AMMONIUM SULPHIDE SOLUTION	8	CFT	II	8+3+6.1	C	2	2		50	95		2	no	T4 3)	II B 4)	yes	PP, EP, EX, TOX, A	2	15; 16
2693	BISULPHITES, AQUEOUS SOLUTION, N.O.S.	8	C1	III	8	N	4	3			97		3	yes			no	PP, EP	0	27; 34
2709	BUTYLBENZENES	3	F1	III	3+N1+F	N	2	3		35	97	0.87	2	yes	T2	II A 7)	yes	PP, EX, A	0	41
2709	BUTYLBENZENES (n-BUTYLBENZENE)	3	F1	III	3+N1+F	N	3	3			97	0.87	2	yes	T2	II A	yes	PP, EX, A	0	41
2733	AMINES, FLAMMABLE, CORROSIVE, N.O.S. or POLYAMINES, FLAMMABLE, CORROSIVE, N.O.S. (2-AMINOBUTANE)	3	FC	II	3+8+N1	C	2	2	3	50	95	0.72	2	yes	T4 3)	II A 7)	yes	PP, EP, EX, A	1	23
2735	AMINES, LIQUID, CORROSIVE, N.O.S. or POLYAMINES, LIQUID, CORROSIVE, N.O.S.	8	C7	I	8+(N1, N2, N3, CMR, F or S)	*	*	*	*	*	*		*	yes			no	*	0	27; 34 *see 3.2.3.3
2735	AMINES, LIQUID, CORROSIVE, N.O.S. or POLYAMINES, LIQUID, CORROSIVE, N.O.S.	8	C7	II	8+(N1, N2, N3, CMR, F or S)	*	*	*	*	*	*		*	yes			no	*	0	27; 34 *see 3.2.3.3
2735	AMINES, LIQUID, CORROSIVE, N.O.S. or POLYAMINES, LIQUID, CORROSIVE, N.O.S.	8	C7	III	8+(N1, N2, N3, CMR, F or S)	*	*	*	*	*	*		*	yes			no	*	0	27; 34 *see 3.2.3.3
2754	N-ETHYLTOLUIDINES (N-ETHYL-o-TOLUIDINE)	6.1	T1	II	6.1+F	C	2	2		25	95	0.94	2	no			no	PP, EP, TOX, A	2	
2754	N-ETHYLTOLUIDINES (N-ETHYL-m-TOLUIDINE)	6.1	T1	II	6.1+F	C	2	2		25	95	0.94	2	no			no	PP, EP, TOX, A	2	
2754	N-ETHYLTOLUIDINES (N-ETHYL-o-TOLUIDINE and N-ETHYL-m-TOLUIDINE MIXTURES)	6.1	T1	II	6.1+F	C	2	2		25	95	0.94	2	no			no	PP, EP, TOX, A	2	
2754	N-ETHYLTOLUIDINES (N-ETHYL-p-TOLUIDINE)	6.1	T1	II	6.1+F	C	2	2	2	25	95	0.94	2	no			no	PP, EP, TOX, A	2	7; 17

(1)	(2)	(3a)	(3b)	(4)	(5)	(6)	(7)	(8)	(9)	(10)	(11)	(12)	(13)	(14)	(15)	(16)	(17)	(18)	(19)	(20)
UN No. or substance identification No.	Name and description	Class	Classification code	Packing group	Dangers	Type of tank vessel	Cargo tank design	Cargo tank type	Cargo tank equipment	Opening pressure of the high-velocity vent valve in kPa	Maximum degree of filling in %	Relative density at 20 °C	Type of sampling device	Pump room below deck permitted	Temperature class	Explosion group	Anti-explosion protection required	Equipment required	Number of cones/blue lights	Additional requirements/Remarks
2785	4-THIAPENTANAL (3-MÉTHYLMERCAPTO-PROPIONALDÉHYDE)	6.1	T1	III	6.1	C	2	2		25	95	1.04	2	no			no	PP, EP, TOX, A	0	
2789	ACETIC ACID, GLACIAL or ACETIC ACID SOLUTION, more than 80 % acid, by mass	8	CF1	II	8+3	N	2	3	2	10	95	1,05 with 100% acid	3	yes	T1	II A$^{7)}$	yes	PP, EP, EX, A	1	7; 17; 34
2790	ACETIC ACID SOLUTION, not less than 50 % but not more than 80 % acid, by mass	8	C3	II	8	N	2	3		10	97		3	yes			no	PP, EP	0	34
2790	ACETIC ACID SOLUTION, more than 10 % and less than 50 % acid, by mass	8	C3	III	8	N	2	3		10	97		3	yes			no	PP, EP	0	34
2796	BATTERY FLUID, ACID	8	C1	II	8+N3	N	4	3			97	1,00 - 1,84	3	yes			no	PP, EP	0	8; 22; 30; 34
2796	SULPHURIC ACID with not more than 51 % acid	8	C1	II	8+N3	N	4	3			97	1,00 - 1,41	3	yes			no	PP, EP	0	8; 22; 30; 34
2797	BATTERY FLUID, ALKALI	8	C5	II	8+N3	N	4	3			97	1,00 - 2,13	3	yes			no	PP, EP	0	22; 30; 34
2810	TOXIC LIQUID, ORGANIC, N.O.S.	6.1	T1	I	6.1+(N1, N2, N3, CMR, F or S)	C	2	2	*	*	95		1	no			no	PP, EP, TOX, A	2	27; 29 *see 3.2.3.3
2810	TOXIC LIQUID, ORGANIC, N.O.S.	6.1	T1	II	6.1+(N1, N2, N3, CMR, F or S)	C	2	2	*	*	95		2	no			no	PP, EP, TOX, A	2	27; 29 *see 3.2.3.3
2810	TOXIC LIQUID, ORGANIC, N.O.S.	6.1	T1	III	6.1+(N1, N2, N3, CMR, F or S)	C	2	2	*	*	95		2	no			no	PP, EP, TOX, A	0	27; 29 *see 3.2.3.3
2811	TOXIC SOLID, ORGANIC, N.O.S. (1,2,3-TRICHLOROBENZENE, MOLTEN)	6.1	T2	III	6.1+S	C	2	2	2	25	95		2	no	T4 $^{3)}$	II A$^{7)}$	yes	PP, EP, EX, TOX, A	0	7; 17; 22

(1) UN No. or substance identification No.	(2) Name and description	(3a) Class	(3b) Classification code	(4) Packing group	(5) Dangers	(6) Type of tank vessel	(7) Cargo tank design	(8) Cargo tank type	(9) Cargo tank equipment	(10) Opening pressure of the high-velocity vent valve in kPa	(11) Maximum degree of filling in %	(12) Relative density at 20 °C	(13) Type of sampling device	(14) Pump room below deck permitted	(15) Temperature class	(16) Explosion group	(17) Anti-explosion protection required	(18) Equipment required	(19) Number of cones/blue lights	(20) Additional requirements/Remarks
2811	TOXIC SOLID, ORGANIC, N.O.S. (1,2,3-TRICHLOROBENZENE, MOLTEN)	6.1	T2	III	6.1+S	C	2	1	4	25	95		2	no			no	PP, EP, TOX, A	0	7; 17; 20: +92 °C; 22; 26
2811	TOXIC SOLID, ORGANIC, N.O.S. (1,3,5-TRICHLOROBENZENE, MOLTEN)	6.1	T2	III	6.1+S	C	2	2	2	25	95		2	no	T4 [3]	II A [7]	yes	PP, EP, EX, TOX, A	0	7; 17; 22
2811	TOXIC SOLID, ORGANIC, N.O.S. (1,3,5-TRICHLOROBENZENE, MOLTEN)	6.1	T2	III	6.1+S	C	2	1	4	25	95		2	no			no	PP, EP, TOX, A	0	7; 17; 20: +92 °C; 22; 26
2815	N-AMINOETHYL PIPERAZINE	8	C7	III	8+N2	N	4	3			97	0.98	3	yes			no	PP, EP	0	34
2820	BUTYRIC ACID	8	C3	III	8+N3	N	2	3		10	97	0.96	3	yes			no	PP, EP	0	34
2829	CAPROIC ACID	8	C3	III	8+N3	N	4	3			97	0.92	3	yes			no	PP, EP	0	34
2831	1,1,1-TRICHLOROETHANE	6.1	T1	III	6.1+N2	C	2	2	3	50	95	1.34	2	no			no	PP, EP, TOX, A	0	23
2850	PROPYLENE TETRAMER	3	F1	III	3+N1+F	N	4	3			97	0.76	2	yes			no	PP	0	
2874	FURFURYL ALCOHOL	6.1	T1	III	6.1+N3	C	2	2		25	95	1.13	2	no			no	PP, EP, TOX, A	0	
2904	PHENOLATES, LIQUID	8	C9	III	8	N	4	2			97	1,13-1,18	3	yes			no	PP, EP	0	34
2920	CORROSIVE LIQUID, FLAMMABLE, N.O.S. (2-PROPANOL AND DODECYLDIMETHYL-AMMONIUM CHLORIDE, AQUEOUS SOLUTION)	8	CF1	II	8+3+F	N	3	3			97	0.95	3	yes	T3	II A	yes	PP, EP, EX, A	1	34;
2920	CORROSIVE LIQUID, FLAMMABLE, N.O.S. (AQUEOUS SOLUTION OF HEXADECYLTRIMETHYL-AMMONIUM CHLORIDE (50 %) AND ETHANOL (35 %))	8	CF1	II	8+3+F	N	2	3		10	95	0.9	3	yes	T2	II B	yes	PP, EP, EX, A	1	6: +7 °C; 17; 34;

(1) UN No. or substance identification No.	(2) Name and description	(3a) Class	(3b) Classification code	(4) Packing group	(5) Dangers	(6) Type of tank vessel	(7) Cargo tank design	(8) Cargo tank type	(9) Cargo tank equipment	(10) Opening pressure of the high-velocity vent valve in kPa	(11) Maximum degree of filling in %	(12) Relative density at 20 °C	(13) Type of sampling device	(14) Pump room below deck permitted	(15) Temperature class	(16) Explosion group	(17) Anti-explosion protection required	(18) Equipment required	(19) Number of cones/blue lights	(20) Additional requirements/Remarks
2922	CORROSIVE LIQUID, TOXIC, N.O.S.	8	CT1	I	8+6.1+(N1, N2, N3, CMR, F or S)	C	2	2	*	*	95		1	no			no	PP, EP, TOX, A	2	27; 29 *see 3.2.3.3
2922	CORROSIVE LIQUID, TOXIC, N.O.S.	8	CT1	II	8+6.1+(N1, N2, N3, CMR, F or S)	C	2	2	*	*	95		2	no			no	PP, EP, TOX, A	2	27; 29 *see 3.2.3.3
2922	CORROSIVE LIQUID, TOXIC, N.O.S.	8	CT1	III	8+6.1+(N1, N2, N3, CMR, F or S)	C	2	2	*	*	95		2	no			no	PP, EP, TOX, A	0	27; 29 *see 3.2.3.3
2924	FLAMMABLE LIQUID, CORROSIVE, N.O.S.	3	FC	I	3+8+(N1, N2, N3, CMR, F or S)	C	1	1	*	*	95		1	yes	T4 3)	II B 4)	yes	*	1	27; 29 *see 3.2.3.3
2924	FLAMMABLE LIQUID, CORROSIVE, N.O.S.	3	FC	I	3+8+(N1, N2, N3, CMR, F or S)	C	2	2	*	*	95		1	yes	T4 3)	II B 4)	yes	*	1	27; 29 *see 3.2.3.3
2924	FLAMMABLE LIQUID, CORROSIVE, N.O.S.	3	FC	II	3+8+(N1, N2, N3, CMR, F or S)	C	2	2	*	*	95		2	yes	T4 3)	II B 4)	yes	*	1	27; 29 *see 3.2.3.3
2924	FLAMMABLE LIQUID, CORROSIVE, N.O.S.	3	FC	III	3+8+(N1, N2, N3, CMR, F or S)	*	*	*	*	*	*		*	yes	T4 3)	II B 4)	yes	*	0	27; 34 *see 3.2.3.3
2924	FLAMMABLE LIQUID, CORROSIVE, N.O.S. (AQUEOUS SOLUTION OF DIALKYL-(C_8-C_{18})-DIMETHYLAMMONIUM CHLORIDE AND 2-PROPANOL)	3	FC	II	3+8+F	C	2	2	*	50	95	0.88	2	yes	T2	II A	yes	PP, EP, EX, A	1	
2927	TOXIC LIQUID, CORROSIVE, ORGANIC, N.O.S.	6.1	TC1	I	6.1+8+(N1, N2, N3, CMR, F or S)	C	2	2	*	*	95		1	no			no	PP, EP, TOX, A	2	27; 29 *see 3.2.3.3
2927	TOXIC LIQUID, CORROSIVE, ORGANIC, N.O.S.	6.1	TC1	II	6.1+8+(N1, N2, N3, CMR, F or S)	C	2	2	*	*	95		2	no			no	PP, EP, TOX, A	2	27; 29 *see 3.2.3.3

(1)	(2)	(3a)	(3b)	(4)	(5)	(6)	(7)	(8)	(9)	(10)	(11)	(12)	(13)	(14)	(15)	(16)	(17)	(18)	(19)	(20)
	Name and description	Class	Classification code	Packing group	Dangers	Type of tank vessel	Cargo tank design	Cargo tank type	Cargo tank equipment	Opening pressure of the high-velocity vent valve in kPa	Maximum degree of filling in %	Relative density at 20 °C	Type of sampling device	Pump room below deck permitted	Temperature class	Explosion group	Anti-explosion protection required	Equipment required	Number of cones/blue lights	Additional requirements/Remarks
UN No. or substance identification No.																				
2929	TOXIC LIQUID, FLAMMABLE, ORGANIC, N.O.S.	6.1	TF1	I	6.1+3+ (N1, N2, N3, CMR, F or S)	C	2	2	*	*	95		1	no	T4³⁾	II B⁴⁾	yes	PP, EP, EX, TOX, A	2	27; 29 *see 3.2.3.3
2929	TOXIC LIQUID, FLAMMABLE, ORGANIC, N.O.S.	6.1	TF1	II	6.1+3+ (N1, N2, N3, CMR, F or S)	C	2	2	*	*	95		2	no	T4³⁾	II B⁴⁾	yes	PP, EP, EX, TOX, A	2	27; 29 *see 3.2.3.3
2935	ETHYL-2-CHLORO-PROPIONATE	3	F1	III	3	C	2	2		30	95	1.08	2	yes	T4³⁾	II A	yes	PP, EX, A	0	
2947	ISOPROPYL CHLOROACETATE	3	F1	III	3	C	2	2		30	95	1.09	2	yes	T4³⁾	II A	yes	PP, EX, A	0	
2966	THIOGLYCOL	6.1	T1	II	6.1	C	2	2		25	95	1.12	2	no			no	PP, EP, TOX, A	2	
2983	ETHYLENE OXIDE AND PROPYLENE OXIDE MIXTURE, with not more than 30% ethylene oxide	3	FT1	I	3+6.1+unst.	C	1	1	3		95	0.85	1	no	T2	II B	yes	PP, EP, EX, TOX, A	2	2; 3; 12; 31; 35
2984	HYDROGEN PEROXIDE AQUEOUS SOLUTION with not less than 8%, but less than 20% hydrogen peroxide (stabilized as necessary)	5.1	O1	III	5.1+unst.	C	2	2		35	95	1.06	2	yes			no	PP	0	3; 33
3077	ENVIRONMENTALLY HAZARDOUS SUBSTANCE, SOLID, N.O.S., MOLTEN, (ALKYLAMINE (C_{12} to C_{18}))	9	M7	III	9+F	N	4	3	2		95	0.79	3	yes			no	PP	0	7; 17
3079	METHACRYLONITRILE, STABILIZED	6.1	TF1	I	6.1+3+unst.+ N3	C	2	2	*	45	95	0.8	1	no	T1	II B⁴⁾	yes	PP, EP, EX, TOX, A	2	3; 5
3082	ENVIRONMENTALLY HAZARDOUS SUBSTANCE, LIQUID, N.O.S.	9	M6	III	9+(N1, N2, CMR, F or S)	*	*	*	*	*	*		*	yes			no	*	0	22; 27 *see 3.2.3.3
3082	ENVIRONMENTALLY HAZARDOUS SUBSTANCE, LIQUID, N.O.S. (BILGE WATER)	9	M6	III	9+N2+F	N	4	3			97		3	yes			no	PP	0	

(1)	(2)	(3a)	(3b)	(4)	(5)	(6)	(7)	(8)	(9)	(10)	(11)	(12)	(13)	(14)	(15)	(16)	(17)	(18)	(19)	(20)
UN No. or substance identification No.	Name and description	Class	Classification code	Packing group	Dangers	Type of tank vessel	Cargo tank design	Cargo tank type	Cargo tank equipment	Opening pressure of the high-velocity vent valve in kPa	Maximum degree of filling in %	Relative density at 20 °C	Type of sampling device	Pump room below deck permitted	Temperature class	Explosion group	Anti-explosion protection required	Equipment required	Number of cones/blue lights	Additional requirements/Remarks
3082	ENVIRONMENTALLY HAZARDOUS SUBSTANCE; LIQUID, N.O.S. (HEAVY HEATING OIL)	9	M6	III	9+CMR (N1, N2, F or S)	N	2	3		10	97		3	yes			no	PP	0	
3092	1-METHOXY-2-PROPANOL	3	F1	III	3	N	3	2			97	0.92	3	yes	T3	II B (II B1)	yes	PP, EX, A	0	
3145	ALKYLPHENOLS, LIQUID, N.O.S. (including C_2-C_{12} homologues)	8	C3	II	8+N3	N	4	3			97	0.95	3	yes			no	PP, EP	0	34
3145	ALKYLPHENOLS, LIQUID, N.O.S. (including C_2-C_{12} homologues)	8	C3	III	8+N3	N	4	3			97	0.95	3	yes			no	PP, EP	0	34
3175	SOLIDS CONTAINING FLAMMABLE LIQUID, N.O.S., MOLTEN, having a flash-point up to 60 °C (2- PROPANOL AND DIALKYL-(C_{12} to C_{18})-DIMETHYLAMMONIUM CHLORIDE)	4.1	F1	II	4.1	N	3	3	4	*	95	0.86	3	yes	T2	II A[7]	yes	PP, EX, A	1	7; 17
3256	ELEVATED TEMPERATURE LIQUID, FLAMMABLE, N.O.S. with flash-point above 60 °C, at or above its flash-point	3	F2	III	3+(N1, N2, N3, CMR, F or S)	*	*	*	*		95		*	yes	T4[3]	II B[4]	yes	*	0	7; 27 *see 3.2.3.3
3256	ELEVATED TEMPERATURE LIQUID, FLAMMABLE, N.O.S. with flash-point above 60 °C, at or above its flash-point (CARBON BLACK REEDSTOCK) (PYROLYSIS OIL)	3	F2	III	3+F	N	3	3	2		95		3	yes	T1	II B	yes	PP, EX, A	0	7
3256	ELEVATED TEMPERATURE LIQUID, FLAMMABLE, N.O.S. with flash-point above 60 °C, at or above its flash-point (PYROLYSIS OIL A)	3	F2	III	3+F	N	3	3	2		95		3	yes	T1	II B	yes	PP, EX, A	0	7

(1) UN No. or substance identification No.	(2) Name and description	(3a) Class	(3b) Classification code	(4) Packing group	(5) Dangers	(6) Type of tank vessel	(7) Cargo tank design	(8) Cargo tank type	(9) Cargo tank equipment	(10) Opening pressure of the high-velocity vent valve in kPa	(11) Maximum degree of filling in %	(12) Relative density at 20 °C	(13) Type of sampling device	(14) Pump room below deck permitted	(15) Temperature class	(16) Explosion group	(17) Anti-explosion protection required	(18) Equipment required	(19) Number of cones/blue lights	(20) Additional requirements/Remarks
3256	ELEVATED TEMPERATURE LIQUID, FLAMMABLE, N.O.S. with flash-point above 60 °C, at or above its flash-point (RESIDUAL OIL)	3	F2	III	3+F	N	3	3	2		95		3	yes	T 1	II B	yes	PP, EX, A	0	7
3256	ELEVATED TEMPERATURE LIQUID, FLAMMABLE, N.O.S. with flash-point above 60 °C, at or above its flash-point (MIXTURE OF CRUDE NAPHTHALINE)	3	F2	III	3+F	N	3	3	2		95		3	yes	T 1	II B	yes	PP, EX, A	0	7
3256	ELEVATED TEMPERATURE LIQUID, FLAMMABLE, N.O.S. with flash-point above 60 °C, at or above its flash-point (CREOSOTE OIL)	3	F2	III	3+N1+F	C	2	2	2	10	95		2	yes	T 2	II B	yes	PP, EX, A	0	7
3256	ELEVATED TEMPERATURE LIQUID, FLAMMABLE, N.O.S. with flash-point above 60 °C, at or above its flash-point (Low QI Pitch)	3	F2	III	3+N2+CMR+S	N	3	1	4		95	1,1-1,3	3	yes	T2	II B	yes	PP, EP, EX, TOX, A	0	7
3257	ELEVATED TEMPERATURE LIQUID, N.O.S. at or above 100 °C and below its flash-point (including molten metals, molten salts, etc.)	9	M9	III	9+(N1, N2, N3, CMR, F or S)	*	*	*	*	*	95		*	yes			no	*	0	7; 20:+115 °C; 22; 24; 25; 27 *see 3.2.3.3
3257	ELEVATED TEMPERATURE LIQUID, N.O.S. at or above 100 °C and below its flash-point (including molten metals, molten salts, etc.)	9	M9	III	9+(N1, N2, N3, CMR, F or S)	*	*	*	*	*	95		*	yes			no	*	0	7; 20:+225 °C; 22; 24; 27 *see 3.2.3.3
3257	ELEVATED TEMPERATURE LIQUID, N.O.S. at or above 100 °C and below its flash-point (including molten metals, molten salts, etc.)	9	M9	III	9+(N1, N2, N3, CMR, F or S)	*	*	*	*	*	95		*	yes			no	*	0	7; 20:+250°C; 22; 24; 27 *see 3.2 3.3
3259	AMINES, SOLID, CORROSIVE, N.O.S. (MONOALKYL-(C_{12} to C_{18})-AMINE ACETATE, MOLTEN)	8	C8	III	8	N	4	3	2		95	0.87	3	yes			no	PP, EP	0	7; 17; 34

(1)	(2)	(3a)	(3b)	(4)	(5)	(6)	(7)	(8)	(9)	(10)	(11)	(12)	(13)	(14)	(15)	(16)	(17)	(18)	(19)	(20)
UN No. or substance identification No.	Name and description	Class	Classification code	Packing group	Dangers	Type of tank vessel	Cargo tank design	Cargo tank type	Cargo tank equipment	Opening pressure of the high-velocity vent valve in kPa	Maximum degree of filling in %	Relative density at 20 °C	Type of sampling device	Pump room below deck permitted	Temperature class	Explosion group	Anti-explosion protection required	Equipment required	Number of cones/blue lights	Additional requirements/Remarks
3264	CORROSIVE LIQUID, ACIDIC, INORGANIC, N.O.S.	8	C1	I	8+(N1, N2, N3, CMR, F or S)	*	*	*	*	*	*		*	yes			no	*	0	27; 34 *see 3.2.3.3
3264	CORROSIVE LIQUID, ACIDIC, INORGANIC, N.O.S.	8	C1	II	8+(N1, N2, N3, CMR, F or S)	*	*	*	*	*	*		*	yes			no	*	0	27; 34 *see 3.2.3.3
3264	CORROSIVE LIQUID, ACIDIC, INORGANIC, N.O.S.	8	C1	III	8+(N1, N2, N3, CMR, F or S)	*	*	*	*	*	*		*	yes			no	*	0	27; 34 *see 3.2.3.3
3264	CORROSIVE LIQUID, ACIDIC, INORGANIC, N.O.S. (AQUEOUS SOLUTION OF PHOSPHORIC ACID AND NITRIC ACID)	8	C1	I	8	N	2	3		10	97		3	yes			no	PP, EP	0	34
3264	CORROSIVE LIQUID, ACIDIC, INORGANIC, N.O.S. (AQUEOUS SOLUTION OF PHOSPHORIC ACID AND NITRIC ACID)	8	C1	II	8	N	4	3			97		3	yes			no	PP, EP	0	34
3264	CORROSIVE LIQUID, ACIDIC, INORGANIC, N.O.S. (AQUEOUS SOLUTION OF PHOSPHORIC ACID AND NITRIC ACID)	8	C1	III	8	N	4	3			97		3	yes			no	PP, EP	0	34
3265	CORROSIVE LIQUID, ACIDIC, ORGANIC, N.O.S.	8	C3	I	8+(N1, N2, N3, CMR, F or S)	*	*	*	*	*	*		*	yes			no	*	0	27; 34 *see 3.2.3.3
3265	CORROSIVE LIQUID, ACIDIC, ORGANIC, N.O.S.	8	C3	II	8+(N1, N2, N3, CMR, F or S)	*	*	*	*	*	*		*	yes			no	*	0	27; 34 *see 3.2.3.3
3265	CORROSIVE LIQUID, ACIDIC, ORGANIC, N.O.S.	8	C3	III	8+(N1, N2, N3, CMR, F or S)	*	*	*	*	*	*		*	yes			no	*	0	27; 34 *see 3.2.3.3
3266	CORROSIVE LIQUID, BASIC, INORGANIC, N.O.S.	8	C5	I	8+(N1, N2, N3, CMR, F or S)	*	*	*	*	*	*		*	yes			no	*	0	27; 34 *see 3.2.3.3

(1)	(2)	(3a)	(3b)	(4)	(5)	(6)	(7)	(8)	(9)	(10)	(11)	(12)	(13)	(14)	(15)	(16)	(17)	(18)	(19)	(20)
3266	CORROSIVE LIQUID, BASIC, INORGANIC, N.O.S.	8	C5	II	8+(N1, N2, N3, CMR, F or S)	*	*	*	*	*	*		*	yes			no	*	0	27; 34 *see 3.2.3.3
3266	CORROSIVE LIQUID, BASIC, INORGANIC, N.O.S.	8	C5	III	8+(N1, N2, N3, CMR, F or S)	*	*	*	*	*	*		*	yes			no	*	0	27; 34 *see 3.2.3.3
3267	CORROSIVE LIQUID, BASIC, ORGANIC, N.O.S.	8	C7	I	8+(N1, N2, N3, CMR, F or S)	*	*	*	*	*	*		*	yes			no	*	0	27; 34 *see 3.2.3.3
3267	CORROSIVE LIQUID, BASIC, ORGANIC, N.O.S.	8	C7	II	8+(N1, N2, N3, CMR, F or S)	*	*	*	*	*	*		*	yes			no	*	0	27; 34 *see 3.2.3.3
3267	CORROSIVE LIQUID, BASIC, ORGANIC, N.O.S.	8	C7	III	8+(N1, N2, N3, CMR, F or S)	*	*	*	*	*	*		*	yes			no	*	0	27; 34 *see 3.2.3.3
3271	ETHERS, N.O.S.	3	F1	II	3+(N1, N2, N3, CMR, F or S)	*	*	*	*	*	*		*	yes	T4 3)	II B 4)	yes	*	1	14, 27; 29 *see 3.2.3.3
3271	ETHERS, N.O.S. (tert-AMYL-METHYL ETHER)	3	F1	II	3+N1	C	2	2	3	50	95	0.77	2	yes	T2	II B 4)	yes	PP, EX, A	1	
3271	ETHERS, N.O.S.	3	F1	III	3+(N1, N2, N3, CMR, F or S)	*	*	*	*	*	*		*	yes	T4 3)	II B 4)	yes	*	0	14, 27 *see 3.2.3.3
3272	ESTERS, N.O.S.	3	F1	II	3+(N1, N2, N3, CMR, F or S)	*	*	*	*	*	*		*	yes	T2	II B 4)	yes	*	1	14, 27; 29 *see 3.2.3.3
3272	ESTERS, N.O.S.	3	F1	III	3+(N1, N2, N3, CMR, F or S)	*	*	*	*	*	*		*	yes	T4 3)	II B 4)	yes	*	0	14, 27 *see 3.2.3.3
3276	NITRILES, TOXIC, LIQUID, N.O.S. (2-METHYLGLUTARONITRILE)	6.1	T1	II	6.1	C	2	2		10	95	0.95	2	no			no	PP, EP, TOX, A	2	

(1) UN No. or substance identification No.	(2) Name and description	(3a) Class	(3b) Classification code	(4) Packing group	(5) Dangers	(6) Type of tank vessel	(7) Cargo tank design	(8) Cargo tank type	(9) Cargo tank equipment	(10) Opening pressure of the high-velocity vent valve in kPa	(11) Maximum degree of filling in %	(12) Relative density at 20 °C	(13) Type of sampling device	(14) Pump room below deck permitted	(15) Temperature class	(16) Explosion group	(17) Anti-explosion protection required	(18) Equipment required	(19) Number of cones/blue lights	(20) Additional requirements/Remarks
3286	FLAMMABLE LIQUID, TOXIC, CORROSIVE, N.O.S.	3	FTC	I	3+6.1+8+(N1, N2, N3, CMR, F or S)	C	1	1	*	*	95		1	no	T4[3]	II B[4]	yes	PP, EP, EX, TOX, A	2	27; 29 *see 3.2.3.3
3286	FLAMMABLE LIQUID, TOXIC, CORROSIVE, N.O.S.	3	FTC	I	3+6.1+8+(N1, N2, N3, CMR, F or S)	C	2	2	*	*	95		1	no	T4[3]	II B[4]	yes	PP, EP, EX, TOX, A	2	27; 29 *see 3.2.3.3
3286	FLAMMABLE LIQUID, TOXIC, CORROSIVE, N.O.S.	3	FTC	II	3+6.1+8+(N1, N2, N3, CMR, F or S)	C	2	2	*	*	95		2	no	T4[3]	II B[4]	yes	PP, EP, EX, TOX, A	2	27; 29 *see 3.2.3.3
3287	TOXIC LIQUID, INORGANIC, N.O.S.	6.1	T4	I	6.1+(N1, N2, N3, CMR, F or S)	C	2	2	*	*	95		1	no			no	PP, EP, TOX, A	2	27; 29 *see 3.2.3.3
3287	TOXIC LIQUID, INORGANIC, N.O.S.	6.1	T4	II	6.1+(N1, N2, N3, CMR, F or S)	C	2	2	*	*	95		2	no			no	PP, EP, TOX, A	2	27; 29 *see 3.2.3.3
3287	TOXIC LIQUID, INORGANIC, N.O.S.	6.1	T4	III	6.1+(N1, N2, N3, CMR, F or S)	C	2	2	*	*	95		2	no			no	PP, EP, TOX, A	0	27; 29 *see 3.2.3.3
3287	TOXIC LIQUID, INORGANIC, N.O.S. (SODIUM DICHROMATE SOLUTION)	6.1	T4	III	6.1+CMR	C	2	2	*	30	95	1.68	2	no			no	PP, EP, TOX, A	0	
3289	TOXIC LIQUID, CORROSIVE, INORGANIC, N.O.S. BOILING POINT > 115 °C	6.1	TC3	I	6.1+8+(N1, N2, N3, CMR, F or S)	C	2	2	*	*	95		1	no			no	PP, EP, TOX, A	2	27; 29 *see 3.2.3.3
3289	TOXIC LIQUID, CORROSIVE, INORGANIC, N.O.S. BOILING POINT > 115 °C	6.1	TC3	II	6.1+8+(N1, N2, N3, CMR, F or S)	C	2	2	*	*	95		2	no			no	PP, EP, TOX, A	2	27; 29 *see 3.2.3.3
3295	HYDROCARBONS, LIQUID, N.O.S.	3	F1	I	3+(N1, N2, N3, CMR, F)	*	*	*	*	*	*		*	yes	T4[3]	II B[4]	yes	*	1	14; *see 3.2.3.3
3295	HYDROCARBONS, LIQUID, N.O.S.	3	F1	II	3+(N1, N2, N3, CMR, F)	*	*	*	*	*	*		*	yes	T4[3]	II B[4]	yes	*	1	14; *see 3.2.3.3
3295	HYDROCARBONS, LIQUID, N.O.S.	3	F1	III	3+(N1, N2, N3, CMR, F)	*	*	*	*	*	*		*	yes	T4[3]	II B[4]	yes	*	0	14; *see 3.2.3.3

(1)	(2)	(3a)	(3b)	(4)	(5)	(6)	(7)	(8)	(9)	(10)	(11)	(12)	(13)	(14)	(15)	(16)	(17)	(18)	(19)	(20)
UN No. or substance identification No.	Name and description	Class	Classification code	Packing group	Dangers	Type of tank vessel	Cargo tank design	Cargo tank type	Cargo tank equipment	Opening pressure of the high-velocity vent valve in kPa	Maximum degree of filling in %	Relative density at 20 °C	Type of sampling device	Pump room below deck permitted	Temperature class	Explosion group	Anti-explosion protection required	Equipment required	Number of cones/blue lights	Additional requirements/Remarks
3295	HYDROCARBONS, LIQUID, N.O.S. WITH MORE THAN 10% BENZENE INITIAL BOILING POINT ≤ 60 °C	3	F1	I	3+CMR+ (N1, N2, N3)	C	1	1			95		1	yes	T4³⁾	II B⁴⁾	yes	PP, EP, EX, TOX, A	1	29
3295	HYDROCARBONS, LIQUID, N.O.S. WITH MORE THAN 10% BENZENE INITIAL BOILING POINT ≤ 60 °C	3	F1	II	3+CMR+ (N1, N2, N3)	C	1	1			95		1	yes	T4³⁾	II B⁴⁾	yes	PP, EP, EX, TOX, A	1	29
3295	HYDROCARBONS, LIQUID, N.O.S. WITH MORE THAN 10% BENZENE 60 °C < INITIAL BOILING POINT ≤ 85 °C	3	F1	II	3+CMR+ (N1, N2, N3)	C	2	2	3	50	95		2	yes	T4³⁾	II B⁴⁾	yes	PP, EP, EX, TOX, A	1	23; 29; 38
3295	HYDROCARBONS, LIQUID, N.O.S. WITH MORE THAN 10% BENZENE 85 °C < INITIAL BOILING POINT ≤ 115 °C	3	F1	II	3+CMR+ (N1, N2, N3)	C	2	2		50	95		2	yes	T4³⁾	II B⁴⁾	yes	PP, EP, EX, TOX, A	1	29
3295	HYDROCARBONS, LIQUID, N.O.S. WITH MORE THAN 10% BENZENE INITIAL BOILING POINT > 115°C	3	F1	II	3+CMR+ (N1, N2, N3)	C	2	2		35	95		2	yes	T4³⁾	II B⁴⁾	yes	PP, EP, EX, TOX, A	1	29
3295	HYDROCARBONS, LIQUID, N.O.S. WITH MORE THAN 10% BENZENE INITIAL BOILING POINT ≤ 60 °C	3	F1	III	3+CMR+ (N1, N2, N3)	C	1	1			95		1	yes	T4³⁾	II B⁴⁾	yes	PP, EP, EX, TOX, A	0	29
3295	HYDROCARBONS, LIQUID, N.O.S. WITH MORE THAN 10% BENZENE 60 °C < INITIAL BOILING POINT ≤ 85 °C	3	F1	III	3+CMR+ (N1, N2, N3)	C	2	2	3	50	95		2	yes	T4³⁾	II B⁴⁾	yes	PP, EP, EX, TOX, A	0	23; 29; 38
3295	HYDROCARBONS, LIQUID, N.O.S. WITH MORE THAN 10% BENZENE 85 °C < INITIAL BOILING POINT ≤ 115 °C	3	F1	III	3+CMR+ (N1, N2, N3)	C	2	2		50	95		2	yes	T4³⁾	II B⁴⁾	yes	PP, EP, EX, TOX, A	0	29

(1) UN No. or substance identification No.	(2) Name and description	(3a) Class	(3b) Classification code	(4) Packing group	(5) Dangers	(6) Type of tank vessel	(7) Cargo tank design	(8) Cargo tank type	(9) Cargo tank equipment	(10) Opening pressure of the high-velocity vent valve in kPa	(11) Maximum degree of filling in %	(12) Relative density at 20 °C	(13) Type of sampling device	(14) Pump room below deck permitted	(15) Temperature class	(16) Explosion group	(17) Anti-explosion protection required	(18) Equipment required	(19) Number of cones/blue lights	(20) Additional requirements/Remarks
3295	HYDROCARBONS, LIQUID, N.O.S. WITH MORE THAN 10% BENZENE INITIAL BOILING POINT > 115 °C	3	F1	III	3+CMR+(N1, N2, N3)	C	2	2		35	95		2	yes	T4[3]	II B[4]	yes	PP, EP, EX, TOX, A	0	29
3295	HYDROCARBONS, LIQUID, N.O.S. CONTAINING ISOPRENE AND PENTADIENE, STABILIZED	3	F1	I	3+inst.+N2+CMR	C	2	2	3	50	95	0,678	1	yes	T4[3]	II B[4]	yes	PP, EX, A	1	3; 27
3295	HYDROCARBONS, LIQUID, N.O.S. (1-OCTEN)	3	F1	II	3+N2+F	N	2	3		10	97	0,71	3	yes	T3	II B[4]	yes	PP, EP, EX, TOX, A	1	14
3295	HYDROCARBONS, LIQUID, N.O.S. (POLYCYCLIC AROMATIC HYDOCARBONS MIXTURE)	3	F1	III	3+CMR+F	N	2	3	3	10	97	1,08	3	yes	T1	II A	yes	PP, EP, EX, TOX, A	0	14
3412	FORMIC ACID with not less than 10% but not more than 85% acid by mass	8	C3	II	8+N3	N	2	3		10	97	1,22	3	yes	T1	II A	yes	PP, EP, EX, A	0	6: +12 °C; 17; 34
3412	FORMIC ACID with not less than 5% but less than 10% acid by mass	8	C3	III	8	N	2	3		10	97	1,22	3	yes	T1	II A	yes	PP, EP, EX, A	0	6: +12 °C; 17; 34
3426	ACRYLAMIDE, SOLUTION	6.1	T1	III	6.1	C	2	2		30	95	1,03	2	no	T1		no	PP, EP, TOX, A	0	3; 5; 16
3429	CHLOROTOLUIDINES, LIQUID	6.1	T1	III	6.1+S	C	2	2		25	95	1,15	2	no	T1	II A[7]	yes	PP, EP, EX, TOX, A	0	6: +6 °C; 17;
3446	NITROTOLUENES, SOLID, MOLTEN (p-NITROTOLUENE)	6.1	T2	II	6.1+N2+S	C	2	2	2	25	95	1,16	2	no	T2	II B[4]	yes	PP, EP, EX, TOX, A	2	7; 17
3446	NITROTOLUENES, SOLID, MOLTEN (p-NITROTOLUENE)	6.1	T2	II	6.1+N2+S	C	2	1	4	25	95	1,16	2	no			no	PP, EP, TOX, A	2	7; 17; 20: +88 °C; 26
3451	TOLUIDINES, SOLID, MOLTEN (p-TOLUIDINE)	6.1	T2	II	6.1+N1	C	2	2	2	25	95	1,05	2	no	T1	II A[8]	yes	PP, EP, EX, TOX, A	2	7; 17
3451	TOLUIDINES, SOLID, MOLTEN (p-TOLUIDINE)	6.1	T2	II	6.1+N1	C	2	2	4	25	95	1,05	2	no			no	PP, EP, TOX, A	2	7; 17; 20: +60 °C
3455	CRESOLS, SOLID, MOLTEN	6.1	TC2	II	6.1+8+N3	C	2	2	2	25	95	1,03 - 1,05	2	no	T1	II A[8]	yes	PP, EP, EX, TOX, A	2	7; 17

(1) UN No.	(2) Name and description	(3a) Class	(3b) Classification code	(4) Packing group	(5) Dangers	(6) Type of tank vessel	(7) Cargo tank design	(8) Cargo tank type	(9) Cargo tank equipment	(10) Opening pressure of the high-velocity vent valve in kPa	(11) Maximum degree of filling in %	(12) Relative density at 20 °C	(13) Type of sampling device	(14) Pump room below deck permitted	(15) Temperature class	(16) Explosion group	(17) Anti-explosion protection required	(18) Equipment required	(19) Number of cones/blue lights	(20) Additional requirements/Remarks
3455	CRESOLS, SOLID, MOLTEN	6.1	TC2	II	6.1+8+N3	C	2	2	4	25	95	1,03 - 1,05	2	no			no	PP, EP, TOX, A	2	7; 17; 20: +66 °C
3463	PROPIONIC ACID with not less than 90% acid by mass	8	CF1	II	8+3+N3	N	3	3			97	0.99	3	yes	T1	II A[7]	yes	PP, EP, EX, A	1	34
3475	ETHANOL AND GASOLINE MIXTURE or ETHANOL AND MOTOR SPIRIT MIXTURE or ETHANOL AND PETROL MIXTURE, with more than 10% but not more than 90% ethanol	3	F1	II	3+N2+CMR+F	N	2	3	3	10	97	0.69 − 0.78[10]	3	yes	T3	II A	yes	PP, EP, EX, TOX, A	1	
3475	ETHANOL AND GASOLINE MIXTURE or ETHANOL AND MOTOR SPIRIT MIXTURE or ETHANOL AND PETROL MIXTURE, with more than 90% ethanol	3	F1	II	3+N2+CMR+F	N	2	3	3	10	97	0.78 − 0.79[10]	3	yes	T2	II B	yes	PP, EP, EX, TOX, A	1	
3494	PETROLEUM SOUR CRUDE OIL, FLAMMABLE, TOXIC	3	FT1	I	3+6.1+(N1, N2, N3, CMR, F)	C	*	*	*	*	95		1	no	T4[3]	II B[4]	yes	PP, EP, EX, TOX, A	2	14; 27; *see 3.2.3.3
3494	PETROLEUM SOUR CRUDE OIL, FLAMMABLE, TOXIC	3	FT1	II	3+6.1+(N1, N2, N3, CMR, F)	C	*	*	*	*	95		2	no	T4[3]	II B[4]	yes	PP, EP, EX, TOX, A	2	14; 27; *see 3.2.3.3
3494	PETROLEUM SOUR CRUDE OIL, FLAMMABLE, TOXIC	3	FT1	III	3+6.1+(N1, N2, N3, CMR, F)	C	*	*	*	*	95		2	no	T4[3]	II B[4]	yes	PP, EP, EX, TOX, A	0	14; 27; *see 3.2.3.3
9000	AMMONIA, ANHYDROUS, DEEPLY REFRIGERATED	2	3TC		2.1+2.3+8+N1	G	1	1	1; 3		95		1	no	T1	II A	yes	PP, EP, EX, TOX, A	2	1; 2; 31

UN No. or substance identification No. (1)	Name and description (2)	Class (3a)	Classification code (3b)	Packing group (4)	Dangers (5)	Type of tank vessel (6)	Cargo tank design (7)	Cargo tank type (8)	Cargo tank equipment (9)	Opening pressure of the high-velocity vent valve in kPa (10)	Maximum degree of filling in % (11)	Relative density at 20 °C (12)	Type of sampling device (13)	Pump room below deck permitted (14)	Temperature class (15)	Explosion group (16)	Anti-explosion protection required (17)	Equipment required (18)	Number of cones/blue lights (19)	Additional requirements/Remarks (20)
9001	SUBSTANCES WITH A FLASH-POINT ABOVE 60 °C handed over for carriage or carried at a TEMPERATURE WITHIN A RANGE OF 15K BELOW THEIR FLASH-POINT OR SUBSTANCES WITH A FLASH-POINT > 60 °C, HEATED TO LESS THAN 15 K FROM THE FLASH-POINT	3	F4		3+(N1, N2, N3, CMR, F or S)	*	*	*	*	*	*		*	yes	T4 [3]	II B [4]	yes	*	0	27 *see 3.2.3.3
9002	SUBSTANCES HAVING A SELF-IGNITION TEMPERATURE ≤ 200 °C, N.O.S.	3	F5		3+(N1, N2, N3, CMR, F or S)	C	1	1	*	*	95		1	yes	T4	II B [4]	yes	*	0	*see 3.2.3.3
9003	SUBSTANCES WITH A FLASH-POINT ABOVE 60 °C BUT NOT MORE THAN 100 °C or SUBSTANCES WHERE 60° C < flash-point ≤ 100° C, which are not affected to another class	9			9+(N1, N2, N3, CMR, F or S)	*	*	*	*	*	*		*	yes			no	*	0	27 *see 3.2.3.3
9003	SUBSTANCES WITH A FLASH-POINT ABOVE 60 °C BUT NOT MORE THAN 100 °C or SUBSTANCES WHERE 60° C < flash-point ≤ 100° C, which are not affected to another class (ETHYLENE GLYCOL MONOBUTYL ETHER)	9			9+N3+F	N	4	3			97	0.9	3	yes			no	PP	0	
9003	SUBSTANCES WITH A FLASH-POINT ABOVE 60 °C BUT NOT MORE THAN 100 °C or SUBSTANCES WHERE 60° C < flash-point ≤ 100° C, which are not affected to another class (2-ETHYLHEXYLACRYLATE)	9			9+N3+F	N	4	3			97	0.89	3	yes			no	PP	0	3; 5; 16;
9004	DIPHENYLMETHANE-4,4'-DIISOCYANATE	9			S	N	2	3	4	10	95	1,21 [11]	3	yes			no	PP	0	7; 8; 17; 19

(1)	(2)	(3a)	(3b)	(4)	(5)	(6)	(7)	(8)	(9)	(10)	(11)	(12)	(13)	(14)	(15)	(16)	(17)	(18)	(19)	(20)
UN No. or substance identification No.	Name and description	Class	Classification code	Packing group	Dangers	Type of tank vessel	Cargo tank design	Cargo tank type	Cargo tank equipment	Opening pressure of the high-velocity vent valve in kPa	Maximum degree of filling in %	Relative density at 20 °C	Type of sampling device	Pump room below deck permitted	Temperature class	Explosion group	Anti-explosion protection required	Equipment required	Number of cones/blue lights	Additional requirements/Remarks
9005	ENVIRONMENTALLY HAZARDOUS SUBSTANCE, SOLID, N.O.S, MOLTEN	9			9+(N2, N3, CMR, F or S)	*	*	*	*	*	95		*	yes			no	*	0	*see 3.2.3.3
9006	ENVIRONMENTALLY HAZARDOUS SUBSTANCE, LIQUID, N.O.S.	9			9+(N2, N3, CMR, F or S)	*	*	*	*	*	97		*	yes			no	*	0	*see 3.2.3.3

Footnotes related to the list of substances

1) The ignition temperature has not been determined in accordance with a standardized determination procedure; therefore, provisional assignment has been made to temperature class T2 which is considered safe.

2) The ignition temperature has not been determined in accordance with a standardized determination procedure; therefore, provisional assignment has been made to temperature class T3 which is considered safe.

3) The ignition temperature has not been determined in accordance with a standardized determination procedure; therefore, provisional assignment has been made to temperature class T4 which is considered safe.

4) The maximum experimental safe gap (MESG) has not been measured in accordance with a standardized determination procedure; therefore, assignment has been made to explosion group II B which is considered safe.

5) The maximum experimental safe gap (MESG) has not been measured in accordance with a standardized determination procedure; therefore, assignment has been made to explosion group II C which is considered safe.

6) (*Deleted*)

7) The maximum experimental safe gap (MESG) has not been measured in accordance with a standardized determination procedure; therefore, assignment has been made to the explosion group which is considered safe.

8) The maximum experimental safe gap (MESG) has not been measured in accordance with a standardized determination procedure; therefore, assignment has been made to the explosion group in compliance with IEC 60079-20-1.

9) Assignment in accordance with IMO IBC Code.

10) Relative density at 15 °C.

11) Relative density at 25 °C.

12) (*Deleted*)

13) (*Deleted*)

3.2.3.3 *Flowchart, schemes and criteria for determining applicable special requirements (columns (6) to (20) of Table C)*

Flowchart for classification of liquids of Classes 3, 6.1, 8 and 9 for carriage in tanks in inland navigation

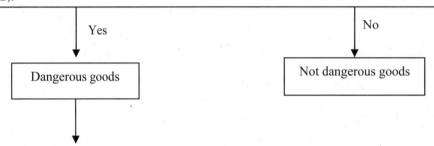

- Flash-point ≤ 100 °C,
- Flash-point > 60 °C and heated to T ≤ 15 K from flash-point,
- Toxic substances (see 2.2.61),
- Corrosive substances (see 2.2.8),
- Elevated temperature liquids at or above 100 °C (UN No. 3257), or
- Substances characterized by acute or chronic aquatic toxicity LC/EC$_{50}$ ≤ 100 mg/l (criteria according to 2.2.9.1.10.2).

Yes → **Dangerous goods**

No → **Not dangerous goods**

- Flash-point < 23 °C and explosivity range > 15% at 20°C,
- Flash-point < 23 °C and corrosive (see 2.2.8),
- Auto-ignition temperature ≤ 200 °C,
- Toxic substances (see 2.2.61),
- Halogenated hydrocarbons,
- Benzene and mixtures containing more than 10% benzene,
- Substances that may only be transported while stabilized, or
- Substances characterized by acute or chronic 1 aquatic toxicity (N1: criteria according to 2.2.9.1.10.2) and vapour pressure at 50 °C ≥ 1 kPa.

No

Yes → **Vessel of type C (continued under A)**

- Flash-point < 23 °C and chronic 2 or 3 aquatic toxicity (N2: criteria according to 2.2.9.1.10.2),
- Flash-point < 23 °C and floating on water surface (floater) or sinking to bottom of water (sinker) (criteria according to 2.2.9.1.10.5),
- Corrosive substances (packing group I or II) with vapour pressure at 50 °C > 12.5 kPa,
- Corrosive substances that react dangerously with water,
- Acute or chronic toxicity 1 (N1: criteria according to 2.2.9.1.10.2) and vapour pressure at 50°C < 1 kPa, or
- Substances with long-term effects on health - CMR (criteria: Categories 1A or 1B of chapters 3.5, 3.6 and 3.7 of GHS).

No

Yes → **Vessel of type N: closed cargo tank walls must be distinct from vessel hull (continued under B)**

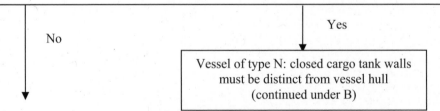

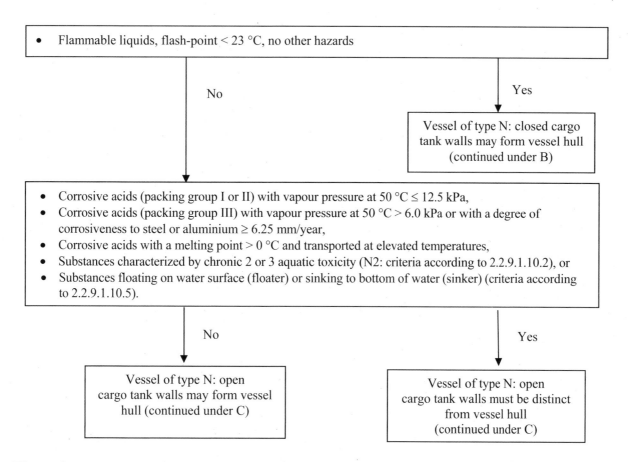

Elevated temperature substances

Irrespective of the above classifications, for substances that must be transported at elevated temperatures, the type of cargo tank shall be determined on the basis of the transport temperature, using the following table:

Maximum transport temperature T in °C	Type N	Type C
T ≤ 80	Integral cargo tank	Integral cargo tank
80 < T ≤ 115	Independent cargo tank, remark 25	Independent cargo tank, remark 26
T > 115	Independent cargo tank	Independent cargo tank

Remark 25 = remark No. 25 in column (20) of the list of substances contained in Chapter 3.2, Table C.

Remark 26 = remark No. 26 in column (20) of the list of substances contained in Chapter 3.2, Table C.

Scheme A: Criteria for cargo tank equipment in vessels of type C

Cargo tank equipment	Cargo tank internal pressure at liquid temperature of 30 °C and gaseous phase temperature of 37.8 °C > 50 kPa	Cargo tank internal pressure at liquid temperature of 30 °C and gaseous phase temperature of 37.8 °C > 50 kPa	Cargo tank internal pressure unknown, owing to absence of certain data
With refrigeration (No. 1 in column (9))	Refrigerated		
Pressure tank (400 kPa)	Non-refrigerated	Cargo tank internal pressure at 50 °C > 50 kPa without water spraying	Boiling point ≤ 60°C
High-velocity vent valve opening pressure: 50 kPa, with water-spraying system (No. 3 in column (9))		Cargo tank internal pressure at 50 °C > 50 kPa with water spraying	60 °C < boiling point ≤ 85°C
High-velocity vent valve opening pressure as calculated, but at least 10 kPa		Cargo tank internal pressure at 50 °C ≤ 50 kPa	
High-velocity vent valve opening pressure: 50 kPa			85 °C < boiling point ≤ 115°C
High-velocity vent valve opening pressure: 35 kPa			Boiling point > 115°C

Scheme B: Criteria for equipment of vessels of type N with closed cargo tanks

Cargo tank equipment	Class 3, flash-point < 23°C		Corrosive substances	CMR substances
Pressure tank (400 kPa)	175 kPa $\leq$ P$_{d\,50}$ < 300 kPa without refrigeration			
High-velocity vent valve opening pressure: 50 kPa	175 kPa $\leq$ P$_{d\,50}$ < 300 kPa, with refrigeration (No. 1 in column (9))	110 kPa $\leq$ P$_{d\,50}$ < 175 kPa without water spraying		
High-velocity vent valve opening pressure: 10 kPa	110 kPa $\leq$ P$_{d\,50}$ < 150 kPa with water spraying (No. 3 in column (9))	P$_{d\,50}$ < 110 kPa	Packing group I or II with P$_{d\,50}$ > 12.5 kPa or reacting dangerously with water	High-velocity vent valve opening pressure: 10 kPa; with water spraying when vapour pressure > 10 kPa (calculation of the vapour pressure according to the formula for column 10, except that v_a = 0.03)

Scheme C: Criteria for equipment of vessels of type N with open cargo tanks

Cargo tank equipment	Classes 3 and 9	Flammable substances	Corrosive substances
With flame-arrester	23°C $\leq$ flash-point $\leq$ 60°C	Flash-point > 60 °C carried while heated to $\leq$ 15 K below flash-point or Flash-point > 60 °C, at or above their flash-point	Acids, transported while heated or flammable substances
Without flame-arrester	60 °C < flash-point $\leq$ 100 °C or elevated temperature substances of Class 9		Non-flammable substances

Column (9): Cargo tank equipment for substances transported in a molten state

-   **Possibility of heating the cargo (number 2 in column (9))**

    A possibility of heating the cargo shall be required on board:

    -   When the melting point of the substance to be transported is + 15 °C or greater,

        or

    -   When the melting point of the substance to be transported is greater than 0 °C but less than + 15 °C and the outside temperature is no more than 4 K above the melting point. In column (20), reference shall be made to remark 6 with the temperature derived as follows: melting point + 4 K

-   **Heating system on board (number 4 in column (9))**

    A cargo heating system shall be required on board:

    -   For substances that must not be allowed to solidify owing to the possibility of dangerous reactions on reheating, and

    -   For substances that must be maintained at a guaranteed temperature not less than 15 K below their flash-point

Column (10): Determination of opening pressure of high-velocity vent valve in kPa

For vessels of type C, the opening pressure of the high-velocity vent valve shall be determined on the basis of the internal pressure of the tanks, rounded up to the nearest 5 kPa

To calculate the internal pressure, the following formula shall be used:

$$P_{max} = P_{Ob\,max} + \frac{k \cdot v_a \left(P_0 - P_{Da}\right)}{v_a - \alpha \cdot \delta_t + \alpha \cdot \delta_t \cdot v_a} - P_0$$

$$k = \frac{T_{D\,max}}{T_a}$$

In this formula:

P_{max}	:	Maximum internal pressure in kPa
P_{Obmax}	:	Absolute vapour pressure at maximum liquid surface temperature in kPa
P_{Da}	:	Absolute vapour pressure at filling temperature in kPa
P_0	:	Atmospheric pressure in kPa
v_a	:	Free relative volume at filling temperature compared with cargo tank volume
α	:	Cubic expansion coefficient in K^{-1}
δ_t	:	Average temperature increase of the liquid due to heating in K
T_{Dmax}	:	Maximum gaseous phase temperature in K
T_a	:	Filling temperature in K
k	:	Temperature correction factor
t_{Ob}	:	Maximum liquid surface temperature in °C

In the formula, the following basic data are used:

P_{Obmax}	:	At 50 °C and 30 °C
P_{Da}	:	At 15 °C
P_0	:	101.3 kPa
v_a	:	5% = 0.05
δ_t	:	5 K
T_{Dmax}	:	323 K and 310.8 K
T_a	:	288 K
t_{Ob}	:	50 °C and 30 °C

Column (11): Determination of maximum degree of filling of cargo tanks

If, in accordance with the provisions under A above:

−   Type G is required: 91%; however, in the case of deeply refrigerated substances: 95%

−   Type C is required: 95%

−   Type N is required: 97%; however, in the case of substances in a molten state and of flammable liquids with $175 \text{ kPa} \leq P_{v50} < 300 \text{ kPa}$: 95%

Column (12): Relative density of substance at 20 °C

These data are provided for information only.

Column (13): Determination of type of sampling device

1 = *closed*:	− Substances to be transported in pressure cargo tanks
	− Substances with T in column (3b) and assigned to packing group I
	− Stabilized substances to be transported under inert gas
2 = *partly closed*:	− All other substances for which type C is required
3 = *open:*	− All other substances

Column (14): Determination of whether a pump-room is permitted below deck

| No | −   All substances with T in column (3b) with the exception of substances of Class 2 |
| Yes | −   All other substances |

Column (15): Determination of temperature class

Flammable substances shall be assigned to a temperature class on the basis of their auto-ignition point:

Temperature class	Auto-ignition temperature T of flammable liquids and gases in °C
T1	T > 450
T2	300 < T ≤ 450
T3	200 < T ≤ 300
T4	135 < T ≤ 200
T5	100 < T ≤ 135
T6	85 < T ≤ 100

When anti-explosion protection is required and the auto-ignition temperature is not known, reference shall be made to temperature class T4, considered safe.

Column (16): Determination of explosion group

Flammable substances shall be assigned to an explosion group on the basis of their maximum experimental safe gaps. The maximum experimental safe gaps shall be determined in accordance with standard IEC 60079-20-1.

The different explosion groups are as follows:

Explosion group	Maximum experimental safe gap in mm
II A	> 0.9
II B	≥ 0.5 to ≤ 0.9
II C	< 0.5

When anti-explosion protection is required and the relevant data are not provided, reference shall be made to explosion group II B, considered safe.

Column (17): Determination of whether anti-explosion protection is required for electrical equipment and systems

Yes
- Substances with a flash-point ≤ 60 °C
- Substances that must be transported while heated to a temperature less than 15 K from their flash-point
- Flammable gases

No
- All other substances

Column (18): Determination of whether personal protective equipment, escape devices, portable flammable gas detectors, portable toximeters or ambient-air-dependent breathing apparatus is required

- PP: For all substances of Classes 1 to 9;
- EP: For all substances
    - of Class 2 with letter T or letter C in the classification code indicated in column (3b),
    - of Class 3 with letter T or letter C in the classification code indicated in column (3b),
    - of Class 4.1,
    - of Class 6.1, and
    - of Class 8,
    - CMR substances of Category 1A or 1B according to chapters 3.5, 3.6 and 3.7 of GHS;

- EX: For all substances for which anti-explosion protection is required;

- TOX: For all substances of Class 6.1,

    For all substances of other classes with T in column (3b),

    For CMR substances of Category 1A or 1B according to chapters 3.5, 3.6 and 3.7 of GHS;

- A: For all substances for which EX or TOX is required

Column (19): Determination of the number of cones or blue lights

For all substances of Class 2 with letter F in the classification code indicated in column (3b):	1 cone/light
For all substances of Classes 3 to 9 with letter F in the classification code indicated in column (3b) and assigned to packing group I or II:	1 cone/light
For all substances of Class 2 with letter T in the classification code indicated in column (3b)	2 cones/lights
For all substances of Classes 3 to 9 with letter T in the classification code indicated in column (3b) and assigned to packing group I or II:	2 cones/lights

Column (20): Determination of additional requirements and remarks

Remark 1: Reference shall be made in column (20) to remark 1 for transport of UN No. 1005 AMMONIA, ANHYDROUS.

Remark 2: Reference shall be made in column (20) to remark 2 for stabilized substances that react with oxygen and for gases for which danger 2.1 is mentioned in column 5.

Remark 3: Reference shall be made in column (20) to remark 3 for substances that must be stabilized.

Remark 4: Reference shall be made in column (20) to remark 4 for substances that must not be allowed to solidify owing to the possibility of dangerous reactions on reheating.

Remark 5: Reference shall be made in column (20) to remark 5 for substances liable to polymerization.

Remark 6: Reference shall be made in column (20) to remark 6 for substances liable to crystallization and for substances for which a heating system or possibility of heating is required and the vapour pressure of which at 20 °C is greater than 0.1 kPa.

Remark 7: Reference shall be made in column (20) to remark 7 for substances with a melting point of + 15 °C or greater.

Remark 8: Reference shall be made in column (20) to remark 8 for substances that react dangerously with water.

Remark 9: Reference shall be made in column (20) to remark 9 for transport of UN No. 1131 CARBON DISULPHIDE.

Remark 10: *No longer used.*

Remark 11: Reference shall be made in column (20) to remark 11 for transport of UN No. 1040 ETHYLENE OXIDE WITH NITROGEN.

Remark 12: Reference shall be made in column (20) to remark 12 for transport of UN No. 1280 PROPYLENE OXIDE and UN No. 2983 ETHYLENE OXIDE AND PROPYLENE OXIDE MIXTURE.

Remark 13: Reference shall be made in column (20) to remark 13 for transport of UN No. 1086 VINYL CHLORIDE, STABILIZED.

Remark 14: Reference shall be made in column (20) to remark 14 for mixtures or N.O.S. entries which are not clearly defined and for which type N is stipulated under the classification criteria.

Remark 15: Reference shall be made in column (20) to remark 15 for substances that react dangerously with alkalis or acids such as sodium hydroxide or sulphuric acid.

Remark 16: Reference shall be made in column (20) to remark 16 for substances that may react dangerously to local overheating.

Remark 17: Reference shall be made in column (20) to remark 17 when reference is made to remark 6 or 7.

Remark 18: *No longer used.*

Remark 19: Reference shall be made in column (20) to remark 19 for substances that must under no circumstances come into contact with water.

Remark 20: Reference shall be made in column (20) to remark 20 for substances the transport temperature of which must not exceed a maximum temperature in combination with the cargo tank materials. Reference shall be made to this maximum permitted temperature immediately after the number 20.

Remark 21: *No longer used.*

Remark 22: Reference shall be made in column (20) to remark 22 for substances for which a range of values or no value of the density is indicated in column (12).

Remark 23: Reference shall be made in column (20) to remark 23 for substances the internal pressure of which at 30 °C is less than 50 kPa and which are transported with water spraying.

Remark 24: Reference shall be made in column (20) to remark 24 for transport of UN No. 3257 ELEVATED TEMPERATURE LIQUID, N.O.S.

Remark 25: Reference shall be made in column (20) to remark 25 for substances that must be transported while heated in a type 3 cargo tank.

Remark 26: Reference shall be made in column (20) to remark 26 for substances that must be transported while heated in a type 2 cargo tank.

Remark 27: Reference shall be made in column (20) to remark 27 for substances for which the reference N.O.S. or a generic reference is made in column (2).

Remark 28: Reference shall be made in column (20) to remark 28 for transport of UN No. 2448 SULPHUR, MOLTEN.

Remark 29: Reference shall be made in column (20) to remark 29 for substances for which the vapour pressure or boiling point is indicated in column (2).

Remark 30: Reference shall be made in column (20) to remark 30 for transport of UN Nos. 1719, 1794, 1814, 1819, 1824, 1829, 1830, 1832, 1833, 1906, 2240, 2308, 2583, 2584, 2677, 2679, 2681, 2796, 2797, 2837 and 3320 under the entries for which open type N is required.

Remark 31: Reference shall be made in column (20) to remark 31 for transport of substances of Class 2 and UN Nos. 1280 PROPYLENE OXIDE and 2983 ETHYLENE OXIDE AND PROPYLENE OXIDE MIXTURE of Class 3.

Remark 32: Reference shall be made in column (20) to remark 32 for transport of UN No. 2448 SULPHUR, MOLTEN of Class 4.1.

Remark 33: Reference shall be made in column (20) to remark 33 for transport of UN Nos. 2014 and 2984 HYDROGEN PEROXIDE, AQUEOUS SOLUTION of Class 5.1.

Remark 34: Reference shall be made in column (20) to remark 34 for transport of substances for which hazard 8 is mentioned in column (5) and type N in column (6).

Remark 35: Reference shall be made in column (20) to remark 35 for substances for which complete refrigeration may cause dangerous reactions in the event of compression. This is also applicable if the refrigeration is partly done by compression.

Remark 36: *No longer used.*

Remark 37: Reference shall be made in column (20) to remark 37 for substances for which the cargo storage system must be capable of resisting the full vapour pressure of the cargo at the upper limits of the ambient design temperatures, whatever the system adopted for the boil-off gas.

Remark 38: Reference shall be made in column (20) to remark 38 for mixtures with an initial boiling point above 60 °C or under or equal to 85 °C in accordance with ASTMD 86-01.

Remark 39: Reference shall be made in column (20) to remark 39 for the carriage of UN No. 2187 CARBON DIOXIDE, REFRIGERATED LIQUID of Class 2.

Remark 40: *No longer used.*

Remark 41: Reference shall be made in column (20) to remark 41 for UN No. 2709 BUTYLBENZENES (n-BUTYLBENZENE).

Remark 42: Reference shall be made in column (20) to remark 42 for UN No. 1038 ETHYLENE, REFRIGERATED LIQUID and for UN No. 1972 METHANE REFRIGERATED LIQUID or NATURAL GAS, REFRIGERATED LIQUID, with high methane content.

Remark 43: Reference shall be made in column (20) to remark 43 for all packing group I entries with letter F (flammable) in the classification code indicated in column (3b), and with letter F (floater) in column (5), Dangers.

3.2.4 Modalities for the application of section 1.5.2 on special authorizations concerning transport in tank vessels

3.2.4.1 Model special authorization under section 1.5.2

Special authorization
under 1.5.2 of ADN

Under 1.5.2 of ADN, the transport in tank vessels of the substance specified in the annex to this special authorization shall be authorized in the conditions referred to therein.

Before transporting the substance, the carrier shall be required to have it added to the list referred to in 1.16.1.2.5 of ADN by a recognized classification society.

This special authorization shall be valid ....
                            (places and/or routes of validity)

It shall be valid for two years from the date of signature, unless it is repealed at an earlier date.

Issuing State: ....

Competent authority: ....

Date: ....

Signature: ....

3.2.4.2 Application form for special authorizations under section 1.5.2

For applications for special authorizations, please answer the following questions and points.[*] Data are used for administrative purposes only and are treated confidentially.

Applicant

....
(Name)   (Company)

....
(   ) ....

....
(Address)

Summary of the application

Authorization for transport in tank vessels of .... as a substance of Class
....

Annexes
(with brief description)

Application made:

At: ....

Date: ....

_____

[*] *For questions not relevant to the subject of the application, write "not applicable".*

Signature: ....
                    (of the person responsible for the data)

1.   General data on the dangerous substance

1.1   Is it a pure substance ☐, a mixture ☐, a solution ☐ ?

1.2   Technical name (if possible ADN nomenclature or possibly the IBC Code).

1.3   Synonym.

1.4   Trade name.

1.5   Structure formula and, for mixtures, composition and/or concentration.

1.6   Hazard class and, where applicable classification code, packing group.

1.7   UN No. or substance identification number (if known).

2.   Physico-chemical properties

2.1   State during transport (e.g. gas, liquid, molten, ...).

2.2   Relative density of liquid at 20 ° C or at the transport temperature if the substance is to be heated or refrigerated during transport.

2.3   Transport temperature (for substances heated or refrigerated during transport).

2.4   Melting point or range ...... ° C.

2.5   Boiling point or range ...... ° C.

2.6   Vapour pressure at 15 ° C ...., 20 ° C ...., 30 ° C ...., 37.8 ° C ...., 50 ° C ....,
         (for liquefied gases, vapour pressure at 70 ° C ....), (for permanent gases, filling pressure at
         15 ° C ....).

2.7   Cubic expansion coefficient ..... K^{-1}

2.8   Solubility in water at 20 ° C
         Saturation concentration ...... mg/l

         or

         Miscibility with water at 15 ° C

         ☐ Complete ☐ partial   ☐ none
         (If possible, in the case of solutions and mixtures, indicate concentration)

2.9   Colour.

2.10   Odour.

2.11   Viscosity ..... mm^2/s.

2.12   Flow time (ISO 2431-1996) ....s.

2.13   Solvent separation test ....

2.14   pH of the substance or aqueous solution (indicate concentration).

2.15   Other information.

3.   Technical safety properties

3.1   Auto-ignition temperature in accordance with IEC 60079-20-1:2010, EN 14522:2005, DIN 51 794:2003 in °C; where applicable, indicate the temperature class in accordance with IEC 60079-20-1:2010.

3.2   Flash-point

    For flash-points up to 175 ° C

    Closed-cup test methods - non-equilibrium procedure

    Abel method: EN ISO 13736:2008

    Abel-Pensky method: DIN 51755–1:1974 or NF M T60-103:1968

    Pensky-Martens method: EN ISO 2719:2012

    Luchaire apparatus: French standard NF T60-103:1968

    Tag method: ASTM D56-05(2010)

    Closed-cup test methods – equilibrium procedure

    Rapid equilibrium procedure: EN ISO 3679:2004; ASTM D3278-96 (2011)

    Closed-cup equilibrium procedure: EN ISO 1523:2002+AC1:2006; ASTM D3941-90 (2007)

    For flash-points above 175 ° C

    In addition to the above-mentioned methods, the following open-cup test method may be applied:

    Cleveland method: EN ISO 2592:2002; ASTM D92-12.

3.3   Explosion limits:

    Determination of upper and lower explosion limits in accordance with EN 1839:2012.

3.4   Maximum safe gap in accordance with IEC 60079-20-1:2010 in mm.

3.5   Is the substance stabilized during transport? If so, provide data on the stabilizer:

    ....

3.6   Decomposition products in the event of combustion on contact with air or under the influence of an external fire:

3.7   Is the substance fire intensifying?

3.8   Abrasion (corrosion) ..... mm/year.

3.9   Does the substance react with water or moist air by releasing flammable or toxic gases? Yes/No. Gases released: ....

3.10   Does the substance react dangerously in any other way?

3.11   Does the substance react dangerously when reheated?
        Yes/no

4.   Physiological hazards

4.1 LD_{50} and/or LC_{50} value. Necrosis value (where applicable, other toxicity criteria in accordance with 2.2.61.1 of ADN).

        CMR properties according to Categories 1A and 1B of chapters 3.5, 3.6 and 3.7 of GHS.

4.2   Does decomposition or reaction produce substances posing physiological hazards? (Indicate which substances where known)

4.3   Environmental properties (see 2.4.2.1 of ADN)

        Acute toxicity:

        LC_{50} 96 hr for fish .... mg/l

        EC_{50} 48 hr for crustacea .... mg/l

        E_rC_{50} 72 hr for algae .... mg/l

        Chronic toxicity:

        NOEC .... mg/l

        BCF .... mg/l or log K_{ow} ....

        Easily biodegradable .... yes/no

5.   Data on hazard potential

5.1   What specific damage is to be expected if the hazard characteristics produce their effect?

        ☐   Combustion

        ☐   Injury

        ☐   Corrosion

        ☐   Intoxication in the event of dermal absorption

        ☐   Intoxication in the event of absorption by inhalation

        ☐   Mechanical damage

        ☐   Destruction

        ☐   Fire

        ☐   Abrasion (corrosion to metals)

        ☐   Environmental pollution

6. **Data on the transport equipment**

6.1 Are particular loading requirements envisaged/necessary (what are they)?

7. **Transport of dangerous substances in tanks**

7.1 With which materials is the substance to be carried compatible?

8. **Technical safety requirements**

8.1 Taking into account the current state of science and technology, what safety measures are necessary in the light of the hazards posed by the substance or liable to arise in the course of the transport process as a whole?

8.2 Additional safety measures

Use of stationary or mobile techniques to measure flammable gases and flammable liquid vapours.

Use of stationary or mobile techniques (toximeters) to measure concentrations of toxic substances.

3.2.4.3 Criteria for assignment of substances

A. **Columns (6), (7) and (8): Determination of the type of tank vessel**

1. **Gases** (criteria according to 2.2.2 of ADN)

- Without refrigeration: type G   pressure
- With refrigeration: type G   refrigerated

2. **Halogenated hydrocarbons**

Substances that may only be transported in a stabilized state

Toxic substances (see 2.2.61.1 of ADN)

Flammable (flash-point < 23 °C) and corrosive substances (see 2.2.8 of ADN)

Substances with an auto-ignition temperature ≤ 200 °C

Substances with a flash-point < 23 °C and an explosivity range > 15 % at 20 °C

Benzene and mixtures of non-toxic and non-corrosive substances containing more than 10% benzene

Environmentally hazardous substances, aquatic toxicity category Acute 1 or Chronic 1 (group N1 in accordance with 2.2.9.1.10.2 of ADN) and vapour pressure at 50 °C ≥ 1 kPa

- Cargo tank internal pressure > 50 kPa at the following temperatures: liquid 30 °C, gaseous phase 37.8 °C

  - Without refrigeration: type C   pressure (400 kPa)

  - With refrigeration: type C   refrigerated

- Cargo tank internal pressure ≤ 50 kPa at the following temperatures: liquid 30 °C, gaseous phase 37.8 °C but with cargo tank internal pressure > 50 kPa at 50 °C

  - Without water spraying: type C   pressure (400 kPa)

  - With water spraying: type C   with high-velocity vent valve opening pressure of 50 kPa

- Cargo tank internal pressure ≤ 50 kPa at the following temperatures: liquid 30 °C, gaseous phase 37.8 °C with cargo tank internal pressure ≤ 50 kPa at 50 °C

         type C   with high-velocity vent valve opening pressure as calculated, but at least 10 kPa

2.1 **Mixtures for which type C is required in accordance with the criteria referred to in 2 above but for which certain data are lacking**

In cases where the internal pressurization of the tank cannot be calculated owing to a lack of data, the following criteria may be used

–	Initial boiling point ≤ 60 °C	type C	(400 kPa)
–	60 °C < initial boiling point ≤ 85 °C	type C	with high-velocity vent valve opening pressure of 50 kPa and with water spraying
–	85 °C < initial boiling point ≤ 115 °C	type C	with high-velocity vent valve opening pressure of 50 kPa
–	115 °C < initial boiling point	type C	with high-velocity vent valve opening pressure of 35 kPa

3. **Substances which are flammable only** (see 2.2.3 of ADN)

– Flash-point < 23 °C
with 175 kPa ≤ P_v 50 < 300 kPa

•	Without refrigeration:	closed type N	pressure (400 kPa)
•	With refrigeration:	closed type N	refrigerated with high-velocity vent valve opening pressure of 50 kPa

– Flash-point < 23 °C   closed type N   with eductor opening
with 150 kPa ≤ P_v 50 < 175 kPa: pressure of 50 kPa

– Flash-point < 23 °C
with 110 kPa ≤ P_v 50 < 150 kPa

•	Without water spraying:	closed type N	with high-velocity vent valve opening pressure of 50 kPa
•	With water spraying:	closed type N	with high-velocity vent valve opening pressure of 10 kPa

– Flash-point < 23 °C   closed type N   with high-velocity
with P_v 50 < 110 kPa: vent valve opening
                                   pressure of 10 kPa

– Flash-point ≥ 23 °C but ≤ 60 °C: open type N   with flame-arrester

– Substances with a flash-point > 60 °C   open type N   with flame-arrester
heated to less than 15 K from the flash-
point, N.O.S. (...):

– Substances with a flash-point > 60 °C   open type N   with flame-arrester
heated to or above the flash-point,
N.O.S. (...):

4. Corrosive substances (see 2.2.8 of ADN)

- **Corrosive substances liable to produce corrosive vapours**

  - Substances assigned to packing group I or II in the list of substances and having a vapour pressure[1] greater than 12.5 kPa (125 mbar) at 50 °C or

    closed type N — cargo tank walls must be distinct from vessel hull; high-velocity vent valve/safety valve opening pressure of 10 kPa

  - Substances liable to react dangerously with water (for example acid chlorides)

  - Substances containing gases in solution

- **Corrosive acids:**

  - Substances assigned to packing group I or II in the list of substances and having a vapour pressure[1] of 12.5 kPa (125 mbar) or less at 50 °C or

    open type N — cargo tank walls must be distinct from vessel hull

  - Substances assigned to packing group III in the list of substances and having a vapour pressure[1] of > 6.0 kPa (60 mbar) at 50 °C or

    open type N — cargo tank walls must be distinct from vessel hull

  - Substances assigned to packing group III in the list of substances because of their degree of corrosiveness to steel or aluminium or

    open type N — cargo tank walls must be distinct from vessel hull

  - Substances with a melting point greater than 0 °C and transported at elevated temperatures

    open type N — cargo tank walls must be distinct from vessel hull

  - Flammable substances — open type N — with flame-arresters

  - Elevated temperature substances — open type N — with flame-arresters

  - Non-flammable substances — open type N — without flame-arresters

- **All other corrosive substances:**

  - Flammable substance — open type N — with flame-arresters

  - Non-flammable substances — open type N — without flame-arresters

---

[1] *If the data are available, the sum of the partial pressures of the dangerous substances may be used in place of the vapour pressure.*

5. **Environmentally hazardous substances (see 2.2.9.1 of ADN)**

• Aquatic toxicity Acute 1 or Chronic 1 (group N1 in accordance with 2.2.9.1.10.2) and vapour pressure below 1 kPa at 50 °C	closed type N	cargo tank walls must be distinct from vessel hull
• Chronic 2 and 3 (group N2 in accordance with 2.2.9.1.10.2)	open type N	cargo tank walls must be distinct from vessel hull
• Acute 2 and 3 (group N3 in accordance with 2.2.9.1.10.2)	open type N	_____

6. **Substances of Class 9, UN No. 3257**   open type N   independent cargo tanks

7. **Substances of Class 9, Identification No. 9003**

Flash-point > 60 °C and ≤ 100 °C: open type N   _____

8. **Substances that must be transported at elevated temperatures**

For substances that must be transported at elevated temperatures, the type of cargo tank shall be determined on the basis of the transport temperature, using the following table:

Maximum transport temperature T in °C	Type N	Type C
T ≤ 80	2	2
80 < T ≤ 115	1 + remark 25	1 + remark 26
T > 115	1	1

1 = cargo tank type: independent tank

2 = cargo tank type: integral tank

Remark 25 = remark No. 25 in column (20) of the list of substances contained in Chapter 3.2, Table C.

Remark 26 = remark No. 26 in column (20) of the list of substances contained in Chapter 3.2, Table C.

9. **Substances with long-term effects on health - CMR substances (Categories 1A and 1B in accordance with the criteria of chapters 3.5, 3.6 and 3.7 of GHS[*]), provided that they are already assigned to Classes 2 to 9 by virtue of other criteria**

C   carcinogenic

M   mutagenic

---

[*]   *Since there is no official international list of CMR substances of Categories 1A and 1B, pending the availability of such a list, the list of CMR substances of Categories 1 and 2 in Directives 67/548/EEC and 88/379/EEC of the Council of the European Union, as amended, shall apply.*

R toxic to reproduction

| | closed type N | cargo tank walls must be distinct from vessel hull; high-velocity vent valve opening pressure of at least 10 kPa, with water-spray system, if the internal pressurization of the tank is more than 10 kPa (calculation of the vapour pressure according to the formula for column 10, except that $v_a = 0.03$) |

10.   **Substances that float on the water surface ('floaters') or sink to the bottom of the water ('sinkers') (criteria according to 2.2.9.1.10.5) provided that they are already assigned to Classes 3 to 9 and that type N is required on that basis**

| | open type N | cargo tank walls must be distinct from vessel hull |

B.   **Column (9): Determination of cargo tank equipment**

(1)   Refrigeration system

Determined in accordance with A.

(2)   Possibility of heating the cargo

A possibility of heating the cargo shall be required:

−   When the melting point of the substance to be transported is + 15 °C or greater, or

−   When the melting point of the substance to be transported is greater than 0 °C but less than + 15 °C and the outside temperature is no more than 4 K above the melting point. In column (20), reference shall be made to remark 6 with the temperature derived as follows: melting point + 4 K.

(3)   Water-spray system

Determined in accordance with A.

(4)   Cargo heating system on board

−   For substances that must not be allowed to solidify owing to the possibility of dangerous reactions on reheating, and

−   For substances that must be maintained at a guaranteed temperature of not less than 15 K below their flash-point.

C.   **Column (10): Determination of opening pressure of high-velocity vent valve in kPa**

For vessels of type C, the opening pressure of the high-velocity vent valve shall be determined on the basis of the internal pressure of the tanks, rounded up to the nearest 5 kPa.

To calculate the internal pressure, the following formula shall be used:

$$P_{max} = P_{Ob\,max} + \frac{k \cdot v_a \left(P_0 - P_{Da} \right)}{v_a - \alpha \cdot \delta_t + \alpha \cdot \delta_t \cdot v_a} - P_o$$

$$k = \frac{T_{D\,max}}{T_a}$$

In this formula:

P_{max} : Maximum internal pressure in kPa

P_{Obmax} : Absolute vapour pressure at maximum liquid surface temperature in kPa

P_{Da} : Absolute vapour pressure at filling temperature in kPa

P_0 : Atmospheric pressure in kPa

v_a : Free relative volume at filling temperature compared with cargo tank volume

α : Cubic expansion coefficient in K^{-1}

δ_t : Average temperature increase of the liquid due to heating in K

T_{Dmax} : Maximum gaseous phase temperature in K

T_a : Filling temperature in K

k : Temperature correction factor

t_{Ob} : Maximum liquid surface temperature in °C

In the formula, the following basic data are used:

P_{Obmax} : At 50 °C and 30 °C

P_{Da} : At 15 °C

P_0 : 101.3 kPa

v_a : 5% = 0.05

δ_t : 5 K

T_{Dmax} : 323 K and 310.8 K

T_a : 288 K

t_{Ob} : 50 °C and 30 °C

D.   **Column (11): Determination of maximum degree of filling of cargo tanks**

If, in accordance with the provisions under A above:

-     Type G is required: 91%   however, in the case of deeply refrigerated substances: 95%

-     Type C is required: 95%

-     Type N is required: 97%   however, in the case of substances in a molten state and of flammable liquids with $175 \text{ kPa} \leq P_{v50} < 300 \text{ kPa}$: 95%.

E. **Column (13): Determination of type of sampling device**

1 = *closed*:	-	Substances to be transported in pressure cargo tanks
	-	Substances with T in column (3b) and assigned to packing group I
	-	Stabilized substances to be transported under inert gas.
2 = *partly closed*:	-	All other substances for which type C is required
3 = *open*:	-	All other substances

F. **Column (14): Determination of whether a pump-room is permitted below deck**

No	-	All substances with letter T in the classification code indicated in column (3b) with the exception of substances of Class 2.
Yes	-	All other substances

G. **Column (15): Determination of temperature class**

Flammable substances shall be assigned to a temperature class on the basis of their auto-ignition point:

Temperature class	Auto-ignition temperature T of flammable liquids and gases in °C
T1	$T > 450$
T2	$300 < T \leq 450$
T3	$200 < T \leq 300$
T4	$135 < T \leq 200$
T5	$100 < T \leq 135$
T6	$85 < T \leq 100$

When anti-explosion protection is required and the auto-ignition temperature is not known, reference shall be made to temperature class T4, considered safe.

H. **Column (16): Determination of explosion group**

Flammable substances shall be assigned to an explosion group on the basis of their maximum experimental safe gaps. The maximum experimental safe gaps shall be determined in accordance with standard IEC 60079-20-1.

The different explosion groups are as follows:

Explosion group	Maximum experimental safe gap in mm
II A	> 0.9
II B	≥ 0.5 to ≤ 0.9
II C	< 0.5

When anti-explosion protection is required and the relevant data are not provided, reference shall be made to explosion group II B, considered safe.

I.   **Column (17): Determination of whether anti-explosion protection is required for electrical equipment and systems**

Yes
- Substances with a flash-point $\leq 60\ °C$.
- Substances that must be transported while heated to a temperature less than 15 K from their flash-point.
- Flammable gases

No
- All other substances

J.   **Column (18): Determination of whether personal protective equipment, escape devices, portable flammable gas detectors, portable toximeters or ambient-air-dependent breathing apparatus is required**

- PP: For all substances of Classes 1 to 9;
- EP: For all substances
  - of Class 2 with letter T or letter C in the classification code indicated in column (3b);
  - of Class 3 with letter T or letter C in the classification code indicated in column (3b);
  - of Class 4.1;
  - of Class 6.1;
  - of Class 8; and

  for CMR substances of Category 1A or 1B according to chapters 3.5, 3.6 and 3.7 of GHS;[*]
- EX: For all substances for which anti-explosion protection is required;
- TOX: For all substances of Class 6.1;

  For all substances of other classes with T in column (3b);

  For CMR substances of Category 1A or 1B according to chapters 3.5, 3.6 and 3.7 of GHS;[*]
- A: For all substances for which EX or TOX is required.

K.   **Column (19): Determination of the number of cones or blue lights**

For all substances of Class 2 with letter F in the classification code indicated in column (3b):	1 cone/light
For all substances of Classes 3 to 9 with letter F in the classification code indicated in column (3b) and assigned to packing group I or II:	1 cone/light
For all substances of Class 2 with letter T in the classification code indicated in column (3b):	2 cones/lights
For all substances of Classes 3 to 9 with letter T in the classification code indicated in column (3b) and assigned to packing group I or II:	2 cones/lights

L.   **Column (20): Determination of additional requirements and remarks**

Remark 1:   Reference shall be made in column (20) to remark 1 for transport of UN No. 1005 AMMONIA, ANHYDROUS.

Remark 2:   Reference shall be made in column (20) to remark 2 for stabilized substances that react with oxygen and for gases for which danger 2.1 is mentioned in column (5).

---

[*]   *Since there is no official international list of CMR substances of Categories 1A and 1B, pending the availability of such a list, the list of CMR substances of Categories 1 and 2 in Directives 67/548/EEC and 88/379/EEC of the Council of the European Union, as amended, shall apply.*

Remark 3: Reference shall be made in column (20) to remark 3 for substances that must be stabilized.

Remark 4: Reference shall be made in column (20) to remark 4 for substances that must not be allowed to solidify owing to the possibility of dangerous reactions on reheating.

Remark 5: Reference shall be made in column (20) to remark 5 for substances liable to polymerization.

Remark 6: Reference shall be made in column (20) to remark 6 for substances liable to crystallization and for substances for which a heating system or possibility of heating is required and the vapour pressure of which at 20 ºC is greater than 0.1 kPa.

Remark 7: Reference shall be made in column (20) to remark 7 for substances with a melting point of + 15 ºC or greater.

Remark 8: Reference shall be made in column (20) to remark 8 for substances that react dangerously with water.

Remark 9: Reference shall be made in column (20) to remark 9 for transport of UN No. 1131 CARBON DISULPHIDE.

Remark 10: *No longer used.*

Remark 11: Reference shall be made in column (20) to remark 11 for transport of UN No. 1040 ETHYLENE OXIDE WITH NITROGEN.

Remark 12: Reference shall be made in column (20) to remark 12 for transport of UN No. 1280 PROPYLENE OXIDE and UN No. 2983 ETHYLENE OXIDE AND PROPYLENE OXIDE MIXTURE.

Remark 13: Reference shall be made in column (20) to remark 13 for transport of UN No. 1086 VINYL CHLORIDE, STABILIZED.

Remark 14: Reference shall be made in column (20) to remark 14 for mixtures or N.O.S. entries which are not clearly defined and for which type N is stipulated under the classification criteria.

Remark 15: Reference shall be made in column (20) to remark 15 for substances that react dangerously with alkalis or acids such as sodium hydroxide or sulphuric acid.

Remark 16: Reference shall be made in column (20) to remark 16 for substances that may react dangerously to local overheating.

Remark 17: Reference shall be made in column (20) to remark 17 when reference is made to remark 6 or 7.

Remark 18: *No longer used.*

Remark 19: Reference shall be made in column (20) to remark 19 for substances that must under no circumstances come into contact with water.

Remark 20: Reference shall be made in column (20) to remark 20 for substances the transport temperature of which must not exceed a maximum temperature in combination with the cargo tank materials. Reference shall be made to this maximum permitted temperature immediately after the number 20.

Remark 21: *No longer used.*

Remark 22: Reference shall be made in column (20) to remark 22 for substances for which a range of values or no value of the density is indicated in column (12).

Remark 23: Reference shall be made in column (20) to remark 23 for substances the internal pressure of which at 30 ºC is less than 50 kPa and which are transported with water spraying.

Remark 24: Reference shall be made in column (20) to remark 24 for transport of UN No. 3257 ELEVATED TEMPERATURE LIQUID, N.O.S.

Remark 25: Reference shall be made in column (20) to remark 25 for substances that must be transported while heated in a type 3 cargo tank.

Remark 26: Reference shall be made in column (20) to remark 26 for substances that must be transported while heated in a type 2 cargo tank.

Remark 27: Reference shall be made in column (20) to remark 27 for substances for which the reference N.O.S. or a generic reference is made in column (2).

Remark 28: Reference shall be made in column (20) to remark 28 for transport of UN No. 2448 SULPHUR, MOLTEN.

Remark 29: Reference shall be made in column (20) to remark 29 for substances for which the vapour pressure or boiling point is indicated in column (2).

Remark 30: Reference shall be made in column (20) to remark 30 for transport of UN Nos. 1719, 1794, 1814, 1819, 1824, 1829, 1830, 1832, 1833, 1906, 2240, 2308, 2583, 2584, 2677, 2679, 2681, 2796, 2797, 2837 and 3320 under the entries for which open type N is required.

Remark 31: Reference shall be made in column (20) to remark 31 for transport of substances of Class 2 and UN Nos. 1280 PROPYLENE OXIDE and 2983 ETHYLENE OXIDE AND PROPYLENE OXIDE MIXTURE of Class 3.

Remark 32: Reference shall be made in column (20) to remark 32 for transport of UN No. 2448 SULPHUR, MOLTEN of Class 4.1.

Remark 33: Reference shall be made in column (20) to remark 33 for transport of UN Nos. 2014 and 2984 HYDROGEN PEROXIDE, AQUEOUS SOLUTION of Class 5.1.

Remark 34: Reference shall be made in column (20) to remark 34 for transport of substances for which hazard 8 is mentioned in column (5) and type N in column (6).

Remark 35: Reference shall be made in column (20) to remark 35 for substances for which complete refrigeration may cause dangerous reactions in the event of compression. This is also applicable if the refrigeration is partly done by compression.

Remark 36: *No longer used.*

Remark 37: Reference shall be made in column (20) to remark 37 for substances for which the cargo storage system must be capable of resisting the full vapour pressure of the cargo at the upper limits of the ambient design temperatures, whatever the system adopted for the boil-off gas.

Remark 38: Reference shall be made in column (20) to remark 38 for mixtures with an initial boiling point above 60 °C or under or equal to 85 °C in accordance with ASTMD 86-01.

Remark 39: Reference shall be made in column (20) to remark 39 for the carriage of UN No. 2187 CARBON DIOXIDE, REFRIGERATED LIQUID of Class 2.

Remark 40: *No longer used.*

Remark 41: Reference shall be made in column (20) to remark 41 for UN No. 2709 BUTYLBENZENES (n-BUTYLBENZENE).

Remark 42: Reference shall be made in column (20) to remark 42 for UN No. 1038 ETHYLENE, REFRIGERATED LIQUID and for UN No. 1972 METHANE REFRIGERATED LIQUID or NATURAL GAS, REFRIGERATED LIQUID, with high methane content.

Remark 43: Reference shall be made in column (20) to remark 43 for all packing group I entries with letter F (flammable) in the classification code indicated in column (3b), and with letter F (floater) in column (5), Dangers.

PART 4

Provisions concerning the use of packagings, tanks and bulk cargo transport units

CHAPTER 4.1

GENERAL PROVISIONS

4.1.1   Packagings and tanks shall be used in accordance with the requirements of one of the international Regulations, bearing in mind the indications given in the list of substances of these international Regulations, namely:

– For packagings (including IBCs and large packagings): columns (9a) and (9b) of Chapter 3.2, Table A of RID or ADR, or the list of substances in Chapter 3.2 of the IMDG Code or the ICAO Technical Instructions;

– For portable tanks: columns (10) and (11) of Chapter 3.2, Table A of RID or ADR or the list of substances in the IMDG Code;

– For RID or ADR tanks: columns (12) and (13) of Chapter 3.2, Table A of RID or ADR.

4.1.2   The requirements to be implemented are as follows:

– For packagings (including IBCs and large packagings): Chapter 4.1 of RID, ADR, the IMDG Code or the ICAO Technical Instructions;

– For portable tanks: Chapter 4.2 of RID, ADR or the IMDG Code;

– For RID or ADR tanks: Chapter 4.3 of RID or ADR, and, where applicable, sections 4.2.5 or 4.2.6 of the IMDG Code;

– For fibre-reinforced plastics tanks: Chapter 4.4 of ADR;

– For vacuum-operated waste tanks: Chapter 4.5 of ADR.

– For mobile explosive manufacturing units (MEMUs): Chapter 4.7 of ADR.

4.1.3   For carriage in bulk of solids in vehicles, wagons, containers or bulk containers, the following requirements of the international Regulations shall be complied with:

– Chapter 4.3 of the IMDG Code; or

– Chapter 7.3 of ADR, taking account of indications in columns (10) or (17) of Table A of Chapter 3.2 of ADR, except that sheeted vehicles and containers are not allowed;

– Chapter 7.3 of RID, taking account of indications in columns (10) or (17) of Table A of Chapter 3.2 of RID, except that sheeted wagons and containers are not allowed.

4.1.4   Only packagings and tanks which meet the requirements of Part 6 of ADR or RID may be used.

PART 5

Consignment procedures

CHAPTER 5.1

GENERAL PROVISIONS

5.1.1　　　**Application and general provisions**

This Part sets forth the provisions for dangerous goods consignments relative to marking, labelling, and documentation, and, where appropriate, authorisation of consignments and advance notifications.

5.1.2　　　**Use of overpacks**

5.1.2.1　　　(a)　　Unless marks and labels required in Chapter 5.2, except 5.2.1.3 to 5.2.1.6, 5.2.1.7.2 to 5.2.1.7.8 and 5.2.1.10, representative of all dangerous goods in the overpack are visible, the overpack shall be:

(i)　　marked with the word "OVERPACK". The lettering of the "OVERPACK" mark shall be at least 12 mm high. The mark shall be in an official language of the country of origin and also, if that language is not English, French or German, in English, French or German, unless agreements, if any, concluded between the countries concerned in the transport operation provide otherwise; and

(ii)　　labelled and marked with the UN number and other marks, as required for packages in Chapter 5.2 except 5.2.1.3 to 5.2.1.6, 5.2.1.7.2 to 5.2.1.7.8 and 5.2.1.10, for each item of dangerous goods contained in the overpack. Each applicable mark or label only needs to be applied once.

Labelling of overpacks containing radioactive material shall be in accordance with 5.2.2.1.11.

(b)　　Orientation arrows illustrated in 5.2.1.10 shall be displayed on two opposite sides of overpacks containing packages which shall be marked in accordance with 5.2.1.10.1, unless the marks remains visible.

5.1.2.2　　　Each package of dangerous goods contained in an overpack shall comply with all applicable provisions of ADN. The intended function of each package shall not be impaired by the overpack.

5.1.2.3　　　Each package bearing package orientation marks as prescribed in 5.2.1.10 and which is overpacked or placed in a large packaging shall be oriented in accordance with such marks.

5.1.2.4　　　The prohibitions on mixed loading also apply to these overpacks.

5.1.3　　　**Empty uncleaned packagings (including IBCs and large packagings), tanks, MEMUs, vehicles, wagons and containers for carriage in bulk**

5.1.3.1　　　Empty uncleaned packagings (including IBCs and large packagings), tanks (including tank-vehicles, tank-wagons, battery-vehicles, battery-wagons, demountable tanks, portable tanks, tank-containers, MEGCs, MEMUs), vehicles, wagons and containers for carriage in bulk having contained dangerous goods of the different classes other than Class 7, shall be marked and labelled as if they were full.

NOTE: For documentation, see Chapter 5.4.

| 5.1.3.2 | Containers, tanks, IBCs, as well as other packagings and overpacks, used for the carriage of radioactive material shall not be used for the storage or carriage of other goods unless decontaminated below the level of 0.4 Bq/cm^2 for beta and gamma emitters and low toxicity alpha emitters and 0.04 Bq/cm^2 for all other alpha emitters. |

5.1.4 Mixed packing

When two or more dangerous goods are packed within the same outer packaging, the package shall be labelled and marked as required for each substance or article. If the same label is required for different goods, it only needs to be applied once.

5.1.5 General provisions for Class 7

5.1.5.1 *Approval of shipments and notification*

5.1.5.1.1 *General*

In addition to the approval of package designs described in Chapter 6.4 of ADR, multilateral shipment approval is also required in certain circumstances (5.1.5.1.2 and 5.1.5.1.3). In some circumstances it is also necessary to notify competent authorities of a shipment (5.1.5.1.4).

5.1.5.1.2 *Shipment approvals*

Multilateral approval shall be required for:

(a) the shipment of Type B(M) packages not conforming with the requirements of 6.4.7.5 of ADR or designed to allow controlled intermittent venting;

(b) the shipment of Type B(M) packages containing radioactive material with an activity greater than 3000 A_1 or 3000 A_2, as appropriate, or 1000 TBq, whichever is the lower;

(c) the shipment of packages containing fissile materials if the sum of the criticality safety indexes of the packages in a single vessel, vehicle, wagon or container exceeds 50;

(d) radiation protection programmes for shipments by special use vessels in accordance with 7.1.4.14.7.3.7;

except that a competent authority may authorise carriage into or through its country without shipment approval, by a specific provision in its design approval (see 5.1.5.2.1).

5.1.5.1.3 *Shipment approval by special arrangement*

Provisions may be approved by a competent authority under which a consignment, which does not satisfy all of the applicable requirements of ADN may be carried under special arrangement (see 1.7.4).

5.1.5.1.4 *Notifications*

Notification to competent authorities is required as follows:

(a) Before the first shipment of any package requiring competent authority approval, the consignor shall ensure that copies of each applicable competent authority certificate applying to that package design have been submitted to the competent authority of the country of origin of the shipment and to the competent authority of each country through or into which the consignment is to be carried. The consignor is not required to await an acknowledgement from the competent authority, nor is the competent authority required to make such acknowledgement of receipt of the certificate;

(b)   For each of the following types of shipments:

   (i)   Type C packages containing radioactive material with an activity greater than 3000 A_1 or 3000 A_2, as appropriate, or 1000 TBq, whichever is the lower;

   (ii)   Type B(U) packages containing radioactive material with an activity greater than 3000 A_1 or 3000 A_2, as appropriate, or 1000 TBq, whichever is the lower;

   (iii)   Type B(M) packages;

   (iv)   Shipment under special arrangement.

   The consignor shall notify the competent authority of the country of origin of the shipment and the competent authority of each country through or into which the consignment is to be carried. This notification shall be in the hands of each competent authority prior to the commencement of the shipment, and preferably at least 7 days in advance;

(c)   The consignor is not required to send a separate notification if the required information has been included in the application for approval of shipment (see 6.4.23.2 of ADR);

(d)   The consignment notification shall include:

   (i)   sufficient information to enable the identification of the package or packages including all applicable certificate numbers and identification marks;

   (ii)   information on the date of shipment, the expected date of arrival and proposed routeing;

   (iii)   the name(s) of the radioactive material(s) or nuclide(s);

   (iv)   descriptions of the physical and chemical forms of the radioactive material, or whether it is special form radioactive material or low dispersible radioactive material; and

   (v)   the maximum activity of the radioactive contents during carriage expressed in becquerels (Bq) with an appropriate SI prefix symbol (see 1.2.2.1). For fissile material, the mass of fissile material (or of each fissile nuclide for mixtures when appropriate) in grams (g), or multiples thereof, may be used in place of activity.

5.1.5.2   *Certificates issued by the competent authority*

5.1.5.2.1   Certificates issued by the competent authority are required for the following:

(a)   Designs for:

   (i)   special form radioactive material;

   (ii)   low dispersible radioactive material;

   (iii)   fissile material excepted under 2.2.7.2.3.5 (f);

   (iv)   packages containing 0.1 kg or more of uranium hexafluoride;

   (v)   packages containing fissile material unless excepted by 2.2.7.2.3.5 of the present Regulations or 6.4.11.2 or 6.4.11.3 of ADR;

(vi)   Type B(U) packages and Type B(M) packages;

(vii)   Type C packages;

(b)   Special arrangements;

(c)   Certain shipments (see 5.1.5.1.2);

(d)   Determination of the basic radionuclide values referred to in 2.2.7.2.2.1 for individual radionuclides which are not listed in Table 2.2.7.2.2.1 (see 2.2.7.2.2.2 (a));

(e)   Alternative activity limits for an exempt consignment of instruments or articles (see 2.2.7.2.2.2 (b)).

The certificates shall confirm that the applicable requirements are met, and for design approvals shall attribute to the design an identification mark.

The certificates of approval for the package design and the shipment may be combined into a single certificate.

Certificates and applications for these certificates shall be in accordance with the requirements in 6.4.23 of ADR.

5.1.5.2.2   The consignor shall be in possession of a copy of each applicable certificate.

5.1.5.2.3   For package designs where it is not required that a competent authority issue a certificate of approval, the consignor shall, on request, make available for inspection by the competent authority, documentary evidence of the compliance of the package design with all the applicable requirements.

5.1.5.3   *Determination of transport index (TI) and criticality safety index (CSI)*

5.1.5.3.1   The transport index (TI) for a package, overpack or container, or for unpackaged LSA-I or SCO-I, shall be the number derived in accordance with the following procedure:

(a)   Determine the maximum radiation level in units of millisieverts per hour (mSv/h) at a distance of 1 m from the external surfaces of the package, overpack, container, or unpackaged LSA-I and SCO-I. The value determined shall be multiplied by 100 and the resulting number is the transport index. For uranium and thorium ores and their concentrates, the maximum radiation level at any point 1 m from the external surface of the load may be taken as:

0.4 mSv/h   for ores and physical concentrates of uranium and thorium;

0.3 mSv/h   for chemical concentrates of thorium;

0.02 mSv/h   for chemical concentrates of uranium, other than uranium hexafluoride;

(b)   For tanks, containers and unpackaged LSA-I and SCO-I, the value determined in step (a) above shall be multiplied by the appropriate factor from Table 5.1.5.3.1;

(c)   The value obtained in steps (a) and (b) above shall be rounded up to the first decimal place (e.g. 1.13 becomes 1.2), except that a value of 0.05 or less may be considered as zero.

Table 5.1.5.3.1: **Multiplication factors for tanks, containers and unpackaged LSA-I and SCO-I**

Size of load [a]	Multiplication factor
size of load ≤ 1 m^2	1
1 m^2 < size of load ≤ 5 m^2	2
5 m^2 < size of load ≤ 20 m^2	3
20 m^2 < size of load	10

[a]   *Largest cross-sectional area of the load being measured.*

5.1.5.3.2   The transport index for each overpack, vessel or cargo transport unit shall be determined as either the sum of the TIs of all the packages contained, or by direct measurement of radiation level, except in the case of non-rigid overpacks for which the transport index shall be determined only as the sum of the TIs of all the packages.

5.1.5.3.3   The criticality safety index for each overpack or container shall be determined as the sum of the CSIs of all the packages contained. The same procedure shall be followed for determining the total sum of the CSIs in a consignment or aboard a vessel or cargo transport unit.

5.1.5.3.4   Packages, overpacks and containers shall be assigned to either category I-WHITE, II-YELLOW or III-YELLOW in accordance with the conditions specified in Table 5.1.5.3.4 and with the following requirements:

(a)   For a package, overpack or container, both the transport index and the surface radiation level conditions shall be taken into account in determining which is the appropriate category. Where the transport index satisfies the condition for one category but the surface radiation level satisfies the condition for a different category, the package, overpack or container shall be assigned to the higher category. For this purpose, category I-WHITE shall be regarded as the lowest category;

(b)   The transport index shall be determined following the procedures specified in 5.1.5.3.1 and 5.1.5.3.2;

(c)   If the surface radiation level is greater than 2 mSv/h, the package or overpack shall be carried under exclusive use and under the provisions of 7.1.4.14.7.1.3 and 7.1.4.14.7.3.5 (a) as appropriate;

(d)   A package carried under a special arrangement shall be assigned to category III-YELLOW except under the provisions of 5.1.5.3.5;

(e)   An overpack or container which contains packages carried under special arrangement shall be assigned to category III-YELLOW except under the provisions of 5.1.5.3.5.

Table 5.1.5.3.4: Categories of packages, overpacks and containers

Conditions		Category
Transport index	**Maximum radiation level at any point on external surface**	
0[a]	Not more than 0.005 mSv/h	I-WHITE
More than 0 but not more than 1[a]	More than 0.005 mSv/h but not more than 0.5 mSv/h	II-YELLOW
More than 1 but not more than 10	More than 0.5 mSv/h but not more than 2 mSv/h	III-YELLOW
More than 10	More than 2 mSv/h but not more than 10 mSv/h	III-YELLOW [b]

[a] *If the measured TI is not greater than 0.05, the value quoted may be zero in accordance with 5.1.5.3.1(c).*

[b] *Shall also be carried under exclusive use except for containers (see Table D in 7.1.4.14.7.3.3).*

5.1.5.3.5   In all cases of international carriage of packages requiring competent authority approval of design or shipment, for which different approval types apply in the different countries concerned by the shipment, the categorization shall be in accordance with the certificate of the country of origin of design.

5.1.5.4   ***Specific provisions for excepted packages of radioactive material of Class 7***

5.1.5.4.1   Excepted packages of radioactive material of Class 7 shall be legibly and durably marked on the outside of the packaging with:

(a)   The UN number preceded by the letters "UN";

(b)   An identification of either the consignor or consignee, or both; and

(c)   The permissible gross mass if this exceeds 50 kg.

5.1.5.4.2   The documentation requirements of Chapter 5.4 do not apply to excepted packages of radioactive material of Class 7, except that:

(a)   The UN number preceded by the letters "UN" and the name and address of the consignor and the consignee and, if relevant, the identification mark for each competent authority certificate of approval (see 5.4.1.2.5.1 (g)) shall be shown on a transport document such as a bill of lading, air waybill or CMR, CIM or CMNI consignment note;

(b)   If relevant, the requirements of 5.4.1.2.5.1 (g), 5.4.1.2.5.3 and 5.4.1.2.5.4 shall apply;

(c)   The requirements of 5.4.2 and 5.4.4 shall apply.

5.1.5.4.3   The requirements of 5.2.1.7.8 and 5.2.2.1.11.5 shall apply if relevant.

5.1.5.5 *Summary of approval and prior notification requirements*

NOTE 1: Before first shipment of any package requiring competent authority approval of the design, the consignor shall ensure that a copy of the approval certificate for that design has been submitted to the competent authority of each country en route (see 5.1.5.1.4 (a)).

NOTE 2: Notification required if contents exceed $3 \times 10^3 \, A_1$, or $3 \times 10^3 \, A_2$, or 1000 TBq (see 5.1.5.1.4 (b)).

NOTE 3: Multilateral approval of shipment required if contents exceed $3 \times 10^3 \, A_1$, or $3 \times 10^3 \, A_2$, or 1000 TBq, or if controlled intermittent venting is allowed (see 5.1.5.1).

NOTE 4: See approval and prior notification provisions for the applicable package for carrying this material.

Subject	UN Number	Competent authority approval required		Consignor required to notify the competent authorities of the country of origin and of the countries en route[a] before each shipment	Reference
		Country of origin	Countries en route[a]		
Calculation of unlisted A_1 and A_2 values	-	Yes	Yes	No	2.2.7.2.2.2 (a), 5.1.5.2.1 (d)
Excepted packages - package design - shipment	2908, 2909, 2910, 2911	 No No	 No No	 No No	-
LSA material[b] and SCO[b] Industrial packages types 1, 2 or 3, non fissile and fissile excepted - package design - shipment	2912, 2913, 3321, 3322	 No No	 No No	 No No	-
Type A packages,[b] non fissile and fissile excepted - package design - shipment	2915, 3332	 No No	 No No	 No No	-
Type B(U) packages,[b] non fissile and fissile excepted - package design - shipment	2916	 Yes No	 No No	 See Note 1 See Note 2	5.1.5.1.4 (b), 5.1.5.2.1 (a) 6.4.22.2 (ADR)
Type B(M) packages,[b] non fissile and fissile excepted - package design - shipment	2917	 Yes See Note 3	 Yes See Note 3	 No Yes	5.1.5.1.4 (b), 5.1.5.2.1 (a), 5.1.5.1.2. 6.4.22.3 (ADR)

Subject	UN Number	Competent authority approval required		Consignor required to notify the competent authorities of the country of origin and of the countries en route[a] before each shipment	Reference
		Country of origin	Countries en route[a]		
Type C packages,[b] non fissile and fissile excepted - package design - shipment	3323	 Yes No	 No No	 See Note 1 See Note 2	5.1.5.1.4 (b), 5.1.5.2.1 (a) 6.4.22.2 of ADR
Packages for fissile material - package design - shipment - sum of criticality safety indexes not more than 50 - sum of criticality safety indexes greater than 50	2977, 3324, 3325, 3326, 3327, 3328, 3329, 3330, 3331, 3333	Yes[c] No[d] Yes	Yes[c] No[d] Yes	No See Note 2 See Note 2	5.1.5.2.1 (a), 5.1.5.1.2, 6.4.22.4 (ADR)
Special form radioactive material - design - shipment	- See Note 4	 Yes See Note 4	 No See Note 4	 No See Note 4	1.6.6.4 (ADR), 5.1.5.2.1 (a) 6.4.22.5 (ADR)
Low dispersable radioactive material - design - shipment	- See Note 4	 Yes See Note 4	 No See Note 4	 No See Note 4	5.1.5.2.1 (a), 6.4.22.5 (ADR)
Packages containing 0.1 kg or more of uranium hexafluoride - design - shipment	- See Note 4	 Yes See Note 4	 No See Note 4	 No See Note 4	5.1.5.2.1 (a), 6.4.22.1 (ADR)
Special arrangement - shipment	2919, 3331	Yes	Yes	Yes	1.7.4.2, 5.1.5.2.1 (b), 5.1.5.1.4 (b)

Subject	UN Number	Competent authority approval required		Consignor required to notify the competent authorities of the country of origin and of the countries en route[a] before each shipment	Reference
		Country of origin	Countries en route[a]		
Approved packages designs subjected to transitional measures	-	See 1.6.6	See 1.6.6	See Note 1	1.6.6.2, (ADR), 5.1.5.1.4 (b), 5.1.5.2.1 (a), 5.1.5.1.2. 6.4.22.9 (ADR)
Alternative activity limits for an exempt consignment of instruments or articles	-	Yes	Yes	No	5.1.5.2.1(e), 6.4.22.7 (ADR)
Fissile material excepted in accordance with 2.2.7.2.3.5 (f)	-	Yes	Yes	No	5.1.5.2.1 (a) (iii), 6.4.22.6 (ADR)

[a]   *Countries from, through or into which the consignment is carried.*

[b]   *If the radioactive contents are fissile material which is not excepted from the provisions for packages containing fissile material, then the provisions for fissile material packages apply (see 6.4.11 of ADR).*

[c]   *Designs of packages for fissile material may also require approval in respect of one of the other items in the table.*

[d]   *Shipments may, however, require approval in respect of one of the other items in the table.*

CHAPTER 5.2

MARKING AND LABELLING

5.2.1 **Marking of packages**

NOTE: For marks related to the construction, testing and approval of packagings, large packagings, pressure receptacles and IBCs, see Part 6 of ADR.

5.2.1.1   Unless provided otherwise in ADN, the UN number corresponding to the dangerous goods contained, preceded by the letters "UN" shall be clearly and durably marked on each package. The UN number and the letters "UN" shall be at least 12 mm high, except for packages of 30 litres capacity or less or of 30 kg maximum net mass and for cylinders of 60 litres water capacity or less, when they shall be at least 6 mm in height and except for packages of 5 litres or 5 kg or less when they shall be of an appropriate size. In the case of unpackaged articles the mark shall be displayed on the article, on its cradle or on its handling, storage or launching device.

5.2.1.2   All package marks required by this Chapter:

    (a)   shall be readily visible and legible;

    (b)   shall be able to withstand open weather exposure without a substantial reduction in effectiveness.

5.2.1.3   Salvage packagings and salvage pressure receptacles shall additionally be marked with the word "**SALVAGE**". The lettering of the "SALVAGE" mark shall be at least 12 mm high.

5.2.1.4   Intermediate bulk containers of more than 450 litres capacity and large packagings shall be marked on two opposite sides.

5.2.1.5   *Additional provisions for goods of Class 1*

For goods of Class 1, packages shall, in addition, bear the proper shipping name as determined in accordance with 3.1.2. The mark, which shall be clearly legible and indelible, shall be in an official language of the country of origin and also, if that language is not English, French or German, in English, French or German unless any agreements concluded between the countries concerned in the transport operation provide otherwise.

5.2.1.6   *Additional provisions for goods of Class 2*

Refillable receptacles shall bear the following particulars in clearly legible and durable characters:

    (a)   the UN number and the proper shipping name of the gas or mixture of gases, as determined in accordance with 3.1.2.

        In the case of gases classified under an N.O.S. entry, only the technical name[1] of the gas has to be indicated in addition to the UN number.

---

[1]   *Instead of the proper shipping name or, if applicable, of the proper shipping name of the n.o.s. entry followed by the technical name, the use of the following names is permitted:*

- *for UN No. 1078 refrigerant gas, n.o.s: mixture F1, mixture F2, mixture F3;*
- *for UN No. 1060 methylacetylene and propadiene mixtures, stabilized: mixture P1, mixture P2;*
- *for UN No. 1965 hydrocarbon gas mixture, liquefied, n.o.s: mixture A or butane, mixture A01 or butane, mixture A02 or butane, mixture A0 or butane, mixture A1, mixture B1, mixture B2, mixture B, mixture C or propane.*
- *for UN No. 1010 Butadienes, stabilized: 1,2-Butadiene, stabilized, 1,3-Butadiene, stabilized .*

In the case of mixtures, not more than the two constituents which most predominantly contribute to the hazards have to be indicated;

(b) for compressed gases filled by mass and for liquefied gases, either the maximum filling mass and the tare of the receptacle with fittings and accessories as fitted at the time of filling, or the gross mass;

(c) the date (year) of the next periodic inspection.

These particulars can either be engraved or indicated on a durable information disk or label attached on the receptacle or indicated by an adherent and clearly visible mark such as by printing or by any equivalent process.

NOTE 1: See also 6.2.2.7 of ADR.

NOTE 2: For non refillable receptacles, see 6.2.2.8 of ADR.

5.2.1.7 *Special marking provisions for radioactive material*

5.2.1.7.1 Each package shall be legibly and durably marked on the outside of the packaging with an identification of either the consignor or consignee, or both. Each overpack shall be legibly and durably marked on the outside of the overpack with an identification of either the consignor or consignee, or both unless these marks of all packages within the overpack are clearly visible.

5.2.1.7.2 For each package, other than excepted packages, the UN number preceded by the letters "UN" and the proper shipping name shall be legibly and durably marked on the outside of the packaging. The marking of excepted packages shall be as required by 5.1.5.4.1.

5.2.1.7.3 Each package of gross mass exceeding 50 kg shall have its permissible gross mass legibly and durably marked on the outside of the packaging.

5.2.1.7.4 Each package which conforms to:

(a) a Type IP-1 package, a Type IP-2 package or a Type IP-3 package design shall be legibly and durably marked on the outside of the packaging with "TYPE IP-1", "TYPE IP-2" or "TYPE IP-3" as appropriate;

(b) a Type A package design shall be legibly and durably marked on the outside of the packaging with "TYPE A";

(c) a Type IP-2 package, a Type IP-3 package or a Type A package design shall be legibly and durably marked on the outside of the packaging with the distinguishing sign used on vehicles in international road traffic[2] of the country of origin of design and either the name of the manufacturer or other identification of the packaging specified by the competent authority of the country of origin of design.

5.2.1.7.5 Each package which conforms to a design approved under one or more of paragraphs 5.1.5.2.1 of these Regulations, 1.6.6.2.1, 6.4.22.1 to 6.4.22.4 and 6.4.23.4 to 6.4.23.7 of ADR shall be legibly and durably marked on the outside of the package with the following information:

(a) the identification mark allocated to that design by the competent authority;

---

[2] *Distinguishing sign of the State of registration used on motor vehicles and trailers in international road traffic, e.g. in accordance with the Geneva Convention on Road Traffic of 1949 or the Vienna Convention on Road Traffic of 1968.*

(b)   a serial number to uniquely identify each packaging which conforms to that design;

(c)   "Type B(U)", "Type B(M)" or "Type C", in the case of a Type B(U), Type B(M) or Type C package design.

5.2.1.7.6   Each package which conforms to a Type B(U), Type B(M) or Type C package design shall have the outside of the outermost receptacle which is resistant to the effects of fire and water plainly marked by embossing, stamping or other means resistant to the effects of fire and water with the trefoil symbol shown in the figure below.

Basic trefoil symbol with proportions based on a central circle of radius X. The minimum allowable size of X shall be 4 mm.

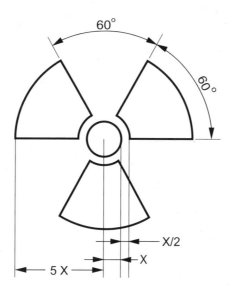

5.2.1.7.7   Where LSA-I or SCO-I material is contained in receptacles or wrapping materials and is carried under exclusive use as permitted by 4.1.9.2.4 of ADR, the outer surface of these receptacles or wrapping materials may bear the mark "RADIOACTIVE LSA-I" or "RADIOACTIVE SCO-I", as appropriate.

5.2.1.7.8   In all cases of international carriage of packages requiring competent authority approval of design or shipment, for which different approval types apply in the different countries concerned by the shipment, marking shall be in accordance with the certificate of the country of origin of the design.

5.2.1.8   ***Special marking provisions for environmentally hazardous substances***

5.2.1.8.1   Packages containing environmentally hazardous substances meeting the criteria of 2.2.9.1.10 shall be durably marked with the environmentally hazardous substance mark shown in 5.2.1.8.3 with the exception of single packagings and combination packagings where such single packagings or inner packagings of such combination packagings have:

-     a quantity of 5 *l* or less for liquids; or

-     a net mass of 5 kg or less for solids.

5.2.1.8.2   The environmentally hazardous substance mark shall be located adjacent to the marks required by 5.2.1.1. The requirements of 5.2.1.2 and 5.2.1.4 shall be met.

5.2.1.8.3   The environmentally hazardous substance mark shall be as shown in Figure 5.2.1.8.3.

Figure 5.2.1.8.3

Environmentally hazardous substance mark

The mark shall be in the form of a square set at an angle of 45° (diamond-shaped). The symbol (fish and tree) shall be black on white or suitable contrasting background. The minimum dimensions shall be 100 mm x 100 mm and the minimum width of the line forming the diamond shall be 2 mm. If the size of the package so requires, the dimensions/line thickness may be reduced, provided the mark remains clearly visible. Where dimensions are not specified, all features shall be in approximate proportion to those shown.

NOTE: The labelling provisions of 5.2.2 apply in addition to any requirement for packages to bear the environmentally hazardous substance mark.

5.2.1.9   *Lithium battery mark*

5.2.1.9.1   Packages containing lithium cells or batteries prepared in accordance with special provision 188 shall be marked as shown in Figure 5.2.1.9.2.

5.2.1.9.2   The mark shall indicate the UN number preceded by the letters "UN", i.e. 'UN 3090' for lithium metal cells or batteries or 'UN 3480' for lithium ion cells or batteries. Where the lithium cells or batteries are contained in, or packed with, equipment, the UN number preceded by the letters "UN", i.e. 'UN 3091' or 'UN 3481' as appropriate shall be indicated. Where a package contains lithium cells or batteries assigned to different UN numbers, all applicable UN numbers shall be indicated on one or more marks.

Figure 5.2.1.9.2

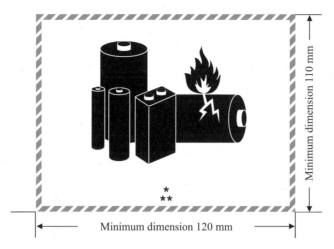

Lithium battery mark

*   Place for UN number(s)

**   Place for telephone number for additional information

The mark shall be in the form of a rectangle with hatched edging. The dimensions shall be a minimum of 120 mm wide x 110 mm high and the minimum width of the hatching shall be 5 mm. The symbol (group of batteries, one damaged and emitting flame, above the UN number for lithium ion or lithium metal batteries or cells) shall be black on white. The hatching shall be red. If the size of the package so requires, the dimensions/line thickness may be reduced to not less than 105 mm wide x 74 mm high. Where dimensions are not specified, all features shall be in approximate proportion to those shown.

5.2.1.10   *Orientation arrows*

5.2.1.10.1   Except as provided in 5.2.1.10.2:

-       *combination packagings having inner packagings containing liquids;*

-       single packagings fitted with vents; and

-       cryogenic receptacles intended for the carriage of refrigerated liquefied gases,

shall be legibly marked with package orientation arrows which are similar to the illustration shown below or with those meeting the specifications of ISO 780:1997. The orientation arrows shall appear on two opposite vertical sides of the package with the arrows pointing in the correct upright direction. They shall be rectangular and of a size that is clearly visible commensurate with the size of the package. Depicting a rectangular border around the arrows is optional.

Figure 5.2.1.10.1.1   **Figure 5.2.1.10.1.2**

   or

Two black or red arrows on white or suitable contrasting background.

The rectangular border is optional.

All features shall be in approximate proportion to those shown.

5.2.1.10.2   Orientation arrows are not required on:

(a)   Outer packagings containing pressure receptacles except cryogenic receptacles;

(b)   Outer packagings containing dangerous goods in inner packagings each containing not more than 120 ml, with sufficient absorbent material between the inner and outer packagings to completely absorb the liquid contents;

(c)   Outer packagings containing Class 6.2 infectious substances in primary receptacles each containing not more than 50 ml;

(d)   Type IP-2, type IP-3, type A, type B(U), type B(M) or type C packages containing Class 7 radioactive material;

(e)   Outer packagings containing articles which are leak-tight in all orientations (e.g. alcohol or mercury in thermometers, aerosols, etc.); or

(f)   Outer packagings containing dangerous goods in hermetically sealed inner packagings each containing not more than 500 ml.

5.2.1.10.3   Arrows for purposes other than indicating proper package orientation shall not be displayed on a package marked in accordance with this sub-section.

5.2.2   Labelling of packages

5.2.2.1   *Labelling provisions*

5.2.2.1.1   For each article or substance listed in Table A of Chapter 3.2, the labels shown in Column (5) shall be affixed unless otherwise provided for by a special provision in Column (6).

5.2.2.1.2   Indelible danger mark corresponding exactly to the prescribed models may be used instead of labels.

5.2.2.1.3 to 5.2.2.1.5   (*Reserved*)

5.2.2.1.6   Except as provided in 5.2.2.2.1.2, each label shall:

(a)   be affixed to the same surface of the package, if the dimensions of the package allow; for packages of Class 1 and 7, near the mark indicating the proper shipping name;

(b)   be so placed on the package that it is not covered or obscured by any part or attachment to the packaging or any other label or mark; and

(c)   be displayed next to each other when more than one label is required.

Where a package is of such an irregular shape or small size that a label cannot be satisfactorily affixed, the label may be attached to the package by a securely affixed tag or other suitable means.

5.2.2.1.7   Intermediate bulk containers of more than 450 litres capacity and large packages shall be labelled on two opposite sides.

5.2.2.1.8   (*Reserved*)

5.2.2.1.9   *Special provisions for the labelling of self-reactive substances and organic peroxides*

(a)   the label conforming to model No. 4.1 also implies that the product may be flammable and hence no label conforming to model No. 3 is required. In addition, a label conforming to model No. 1 shall be applied for self-reactive substances Type B, unless the competent authority has permitted this label to be dispensed with for a specific packaging because test data have proven that the self-reactive substance in such a packaging does not exhibit explosive behaviour.

(b)   the label conforming to model No. 5.2 also implies that the product may be flammable and hence no label conforming to model No. 3 is required. In addition, the following labels shall be applied:

(i)   a label conforming to model No. 1 for organic peroxides type B, unless the competent authority has permitted this label to be dispensed with for a specific packaging because test data have proven that the organic peroxide in such a packaging does not exhibit explosive behaviour;

(ii)   a label conforming to model No. 8 is required when Packing Group I or II criteria of Class 8 are met.

For self-reactive substances and organic peroxides mentioned by name, the labels to be affixed are indicated in the list found in 2.2.41.4 and 2.2.52.4 respectively.

5.2.2.1.10   *Special provisions for the labelling of infectious substances packages*

In addition to the label conforming to model No. 6.2, infectious substances packages shall bear any other label required by the nature of the contents.

5.2.2.1.11   *Special provisions for the labelling of radioactive material*

5.2.2.1.11.1   Except when enlarged labels are used in accordance with 5.3.1.1.3, each package, overpack and container containing radioactive material shall bear the labels conforming to the applicable models Nos. 7A, 7B or 7C, according to the appropriate category. Labels shall be affixed to two opposite sides on the outside of the package or overpack or on the outside of all four sides of a container or tank. In addition, each package, overpack and container containing fissile material, other than fissile material excepted under the provisions of 2.2.7.2.3.5 shall bear labels conforming to model No. 7E; such labels, where applicable, shall be affixed adjacent to the labels conforming to the applicable model Nos. 7A, 7B or 7C. Labels shall not cover the marks specified in 5.2.1. Any labels which do not relate to the contents shall be removed or covered.

5.2.2.1.11.2   Each label conforming to the applicable model No. 7A, 7B or 7C shall be completed with the following information:

(a)   *Contents*:

(i)   except for LSA-I material, the name(s) of the radionuclide(s) as taken from Table 2.2.7.2.2.1, using the symbols prescribed therein. For mixtures of radionuclides, the most restrictive nuclides shall be listed to the extent the space on the line permits. The group of LSA or SCO shall be shown following the name(s) of the radionuclide(s). The terms "LSA-II","LSA-III", "SCO-I" and "SCO-II" shall be used for this purpose;

(ii)   for LSA-I material, only the term "LSA-I" is necessary; the name of the radionuclide is not necessary;

(b)   *Activity*: The maximum activity of the radioactive contents during carriage expressed in becquerels (Bq) with the appropriate SI prefix symbol (see 1.2.2.1). For fissile material, the total mass of fissile nuclides in units of grams (g), or multiples thereof, may be used in place of activity;

(c)   For overpacks and containers the "contents" and "activity" entries on the label shall bear the information required in (a) and (b) above, respectively, totalled together for the entire contents of the overpack or container except that on labels for overpacks or containers containing mixed loads of packages containing different radionuclides, such entries may read "See Transport Documents";

(d)   *Transport index (TI)*: The number determined in accordance with 5.1.5.3.1 and 5.1.5.3.2 (no transport index entry is required for category I-WHITE).

5.2.2.1.11.3   Each label conforming to the model No. 7E shall be completed with the criticality safety index (CSI) as stated in the certificate of approval applicable in the countries through or into which the consignment is carried and issued by the competent authority or as specified in 6.4.11.2 or 6.4.11.3 of ADR.

5.2.2.1.11.4   For overpacks and containers, the label conforming to model No. 7E shall bear the sum of the criticality safety indexes of all the packages contained therein.

5.2.2.1.11.5   In all cases of international carriage of packages requiring competent authority approval of design or shipment, for which different approval types apply in the different countries concerned by the shipment, labelling shall be in accordance with the certificate of the country of origin of design.

5.2.2.2   Provisions for labels

5.2.2.2.1   Labels shall satisfy the provisions below and conform, in terms of colour, symbols and general format, to the models shown in 5.2.2.2.2. Corresponding models required for other modes of transport, with minor variations which do not affect the obvious meaning of the label, are also acceptable.

NOTE: Where appropriate, labels in 5.2.2.2.2 are shown with a dotted outer boundary as provided for in 5.2.2.2.1.1. This is not required when the label is applied on a background of contrasting colour.

5.2.2.2.1.1   Labels shall be configured as shown in Figure 5.2.2.2.1.1.

Figure 5.2.2.2.1.1

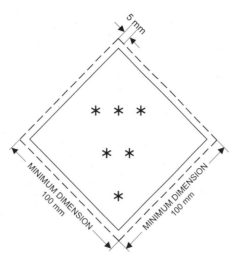

Class/division label

*   *The class or for Classes 4.1, 4.2 and 4.3, the figure "4" or for Classes 6.1 and 6.2, the figure "6", shall be shown in the bottom corner.*

**   *Additional text/numbers/symbol/letters shall (if mandatory) or may (if optional) be shown in this bottom half.*

***   *The class symbol or, for divisions 1.4, 1.5 and 1.6, the division number and for Model No 7E the word "FISSILE" shall be shown in this top half.*

5.2.2.2.1.1.1   Labels shall be displayed on a background of contrasting colour, or shall have either a dotted or solid outer boundary line.

5.2.2.2.1.1.2   The label shall be in the form of a square set at an angle of 45° (diamond-shaped). The minimum dimensions shall be 100 mm x 100 mm and the minimum width of the line inside the edge forming the diamond shall be 2 mm. The line inside the edge shall be parallel and 5 mm from the outside of that line to the edge of the label. The line inside the edge on the upper half of the label shall be the same colour as the symbol and the line inside the edge on the lower half of the label shall be the same colour as the class or division number in the bottom corner. Where dimensions are not specified, all features shall be in approximate proportion to those shown.

5.2.2.2.1.1.3   If the size of the package so requires the dimensions may be reduced, provided the symbols and other elements of the label remain clearly visible. The line inside the edge shall remain 5 mm to the edge of the label. The minimum width of the line inside the edge shall remain 2 mm. Dimensions for cylinders shall comply with 5.2.2.2.1.2.

5.2.2.2.1.2   Gas cylinders for Class 2 may, on account of their shape, orientation and securing mechanisms for carriage, bear labels representative of those specified in this section and the environmentally hazardous substance mark when appropriate, which have been reduced in size, according to the dimensions outlined in ISO 7225:2005, *"Gas cylinders - Precautionary labels"*, for display on the non-cylindrical part (shoulder) of such cylinders.

> **NOTE:**   *When the diameter of the cylinder is too small to permit the display of the reduced size labels on the non-cylindrical upper part of the cylinder, the reduced sized labels may be displayed on the cylindrical part.*

Notwithstanding the provisions of 5.2.2.1.6, labels and the environmentally hazardous substance mark (see 5.2.1.8.3) may overlap to the extent provided for by ISO 7225:2005. However, in all cases, the primary risk label and the figures appearing on any label shall remain fully visible and the symbols recognizable.

Empty uncleaned pressure receptacles for gases of Class 2 may be carried with obsolete or damaged labels for the purposes of refilling or inspection as appropriate and the application of a new label in conformity with current regulations or for the disposal of the pressure receptacle.

5.2.2.2.1.3   With the exception of labels for Divisions 1.4, 1.5 and 1.6 of Class 1, the upper half of the label shall contain the pictorial symbol and the lower half shall contain:

(a)   For Classes 1, 2, 3, 5.1, 5.2, 7, 8 and 9, the class number;

(b)   For Classes 4.1, 4.2 and 4.3, the figure "4";

(c)   For Classes 6.1 and 6.2, the figure "6".

However for label model No. 9A, the upper half of the label shall only contain the seven vertical stripes of the symbol and the lower half shall contain the group of batteries of the symbol and the class number.

Except for label model No. 9A, the label may include text such as the UN number or words describing the hazard (e.g. "flammable") in accordance with 5.2.2.2.1.5 provided the text does not obscure or detract from the other required label elements.

5.2.2.2.1.4   In addition, except for Divisions 1.4, 1.5 and 1.6, labels for Class 1 shall show in the lower half, above the class number, the division number and the compatibility group letter for the substance or article. Labels for Divisions 1.4, 1.5 and 1.6 shall show in the upper half the division number, and in the lower half the class number and the compatibility group letter.

5.2.2.2.1.5   On labels other than those for material of Class 7, the optional insertion of any text (other than the class number) in the space below the symbol shall be confined to particulars indicating the nature of the risk and precautions to be taken in handling.

5.2.2.2.1.6   The symbols, text and numbers shall be clearly legible and indelible and shall be shown in black on all labels except for:

(a)   the Class 8 label, where the text (if any) and class number shall appear in white;

(b)   labels with entirely green, red or blue backgrounds where they may be shown in white;

(c)   the Class 5.2 label, where the symbol may be shown in white; and

(d)   labels conforming to model No. 2.1 displayed on cylinders and gas cartridges for gases of UN Nos. 1011, 1075, 1965 and 1978, where they may be shown in the background colour of the receptacle if adequate contrast is provided.

5.2.2.2.1.7   All labels shall be able to withstand open weather exposure without a substantial reduction in effectiveness.

5.2.2.2.2   Specimen labels

CLASS 1 HAZARD
Explosive substances or articles

(No.1)
Divisions 1.1, 1.2 and 1.3
Symbol (exploding bomb): black; Background: orange; Figure '1' in bottom corner

(No. 1.4)
Division 1.4

(No. 1.5)
Division 1.5

(No. 1.6)
Division 1.6

Background: orange; Figures: black; Numerals shall be about 30 mm in height
and be about 5 mm thick (for a label measuring 100 mm x 100 mm); Figure '1' in bottom corner

** Place for division - to be left blank if explosive is the subsidiary risk
* Place for compatibility group - to be left blank if explosive is the subsidiary risk

CLASS 2 HAZARD
Gases

(No.2.1)
Flammable gases
Symbol (flame): black or white;
(except as provided for in 5.2.2.2.1.6 d))
Background: red; Figure '2' in bottom corner

(No.2.2)
Non flammable, non-toxic gases
Symbol (gas cylinder): black or white;
Background: green; Figure '2' in bottom corner

CLASS 3 HAZARD
Flammable liquids

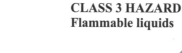

(No 2.3)
Toxic gases
Symbol (skull and crossbones): black;
Background: white; Figure '2' in bottom corner

(No. 3)
Symbol (flame): black or white;
Background: red; Figure '3' in bottom corner

CLASS 4.1 HAZARD
Flammable solids, self-reactive substances, polymerizing substances and solid desensitized explosives

(No. 4.1)
Symbol (flame): black;
Background: white with seven vertical red stripes;
Figure '4' in bottom corner

CLASS 4.2 HAZARD
Substances liable to spontaneous combustion

(No. 4.2)
Symbol (flame): black;
Background: upper half white, lower half red;
Figure '4' in bottom corner

CLASS 4.3 HAZARD
Substances which, in contact with water, emit flammable gases

(No. 4.3)
Symbol (flame): black or white;
Background: blue;
Figure '4' in bottom corner

CLASS 5.1 HAZARD
Oxidizing substances

(No. 5.1)
Symbol (flame over circle): black;
Background: yellow;
Figure '5.1' in bottom corner

CLASS 5.2 HAZARD
Organic peroxides

(No. 5.2)
Symbol (flame): black or white;
Background: upper half red; lower half yellow;
Figure '5.2' in bottom corner

CLASS 6.1 HAZARD
Toxic substances

(No. 6.1)
Symbol (skull and crossbones): black;
Background: white; Figure '6' in bottom corner

CLASS 6.2 HAZARD
Infectious substances

(No. 6.2)
The lower half of the label may bear the inscriptions: 'INFECTIOUS SUBSTANCE'
and 'In the case of damage or leakage immediately notify Public Health Authority';
Symbol (three crescents superimposed on a circle) and inscriptions: black;
Background: white; Figure '6' in bottom corner

CLASS 7 HAZARD
Radioactive material

(No. 7A)
Category I - White
Symbol (trefoil): black;
Background: white;
Text (mandatory): black in lower half of label:
'RADIOACTIVE'
'CONTENTS ....'
'ACTIVITY ......'
One red bar shall
follow the word 'RADIOACTIVE';
Figure '7' in bottom corner.

(No. 7B)
Category II - Yellow

(No. 7C)
Category III - Yellow

Symbol (trefoil): black;
Background: upper half yellow with white border, lower half white;
Text (mandatory): black in lower half of label:
'RADIOACTIVE'
'CONTENTS ....'
'ACTIVITY ......'
In a black outlined box: 'TRANSPORT INDEX';

Two red vertical bars shall
follow the word 'RADIOACTIVE';

Three red vertical bars shall
follow the word 'RADIOACTIVE';
Figure '7' in bottom corner.

(No. 7E)
Class 7 fissile material
Background: white;
Text (mandatory): black in upper half of label: 'FISSILE';
In a black outlined box in the lower half of the label:
'CRITICALITY SAFETY INDEX'
Figure '7' in bottom corner.

CLASS 8 HAZARD
Corrosive substances

CLASS 9 HAZARD
Miscellaneous dangerous substances and articles

(No. 8)
Symbol (liquids, spilling from two glass vessels
and attacking a hand and a metal): black;
Background: upper half white;
lower half black with white border;
Figure '8' in bottom corner

(No. 9)
Symbol (seven vertical stripes
in upper half): black;
Background: white;
Figure '9' underlined
in bottom corner

(No. 9A)
Symbol (seven vertical stripes
in upper half; battery group,
one broken and emitting flame
in lower half): black;
Background: white;
Figure '9' underlined
in bottom corner

CHAPTER 5.3

PLACARDING AND MARKING OF CONTAINERS, MEGCs, MEMUs, TANK-CONTAINERS, PORTABLE TANKS, VEHICLES AND WAGONS

NOTE: For marking and placarding of containers, MEGCs, tank-containers and portable tanks for carriage in a transport chain including a maritime journey, see also 1.1.4.2.1. If the provisions of 1.1.4.2.1 (c) are applied, only 5.3.1.3 and 5.3.2.1.1 of this Chapter are applicable.

5.3.1 **Placarding**

5.3.1.1 *General provisions*

5.3.1.1.1   As and when required in this section, placards shall be affixed to the exterior surface of containers, MEGCs, MEMUs, tank-containers, portable tanks, vehicles and wagons. Placards shall correspond to the labels required in Column (5) and, where appropriate, Column (6) of Table A of Chapter 3.2 for the dangerous goods contained in the container, MEGC, MEMU, tank-container, portable tank, vehicle or wagon and shall conform to the specifications given in 5.3.1.7. Placards shall be displayed on a background of contrasting colour, or shall have either a dotted or solid outer boundary line.

5.3.1.1.2   For Class 1, compatibility groups shall not be indicated on placards if the vehicle or wagon or container or special compartments of MEMUs are carrying substances or articles belonging to two or more compatibility groups. Vehicles or wagons or containers or special compartments of MEMUs carrying substances or articles of different divisions shall bear only placards conforming to the model of the most dangerous division in the order:

1.1 (most dangerous), 1.5, 1.2, 1.3, 1.6, 1.4 (least dangerous).

When 1.5D substances are carried with substances or articles of Division 1.2, the vehicle, wagon or container shall be placarded as Division 1.1.

Placards are not required for the carriage of explosives of Division 1.4, Compatibility Group S.

5.3.1.1.3   For Class 7, the primary risk placard shall conform to model No. 7D as specified in 5.3.1.7.2. This placard is not required for vehicles, wagons or containers carrying excepted packages and for small containers.

Where both Class 7 labels and placards would be required to be affixed to vehicles, wagons, containers, MEGCs, tank-containers or portable tanks, an enlarged label corresponding to the required label of model No. 7A, 7B or 7C may be displayed instead of placard No. 7D to serve both purposes. In that case, the dimensions shall be not less than 250 mm by 250 mm.

5.3.1.1.4   For Class 9 the placard shall correspond to the label model No. 9 as in 5.2.2.2.2; label model No. 9A shall not be used for placarding purposes.

5.3.1.1.5   Containers, MEGCs, MEMUs, tank-containers, portable tanks, vehicles or wagons containing goods of more than one class need not bear a subsidiary risk placard if the hazard represented by that placard is already indicated by a primary or subsidiary risk placard.

5.3.1.1.6   Placards which do not relate to the dangerous goods being carried, or residues thereof, shall be removed or covered.

5.3.1.1.7   When the placarding is affixed to folding panels, they shall be designed and secured so that they cannot unfold or come loose from the holder during carriage (especially as a result of impacts or unintentional actions).

5.3.1.2   *Placarding of containers, MEGCs, tank-containers and portable tanks*

NOTE: This subsection does not apply to swap-bodies, except tank swap bodies carried on vehicles bearing the orange markings stipulated in 5.3.2.

The placards shall be affixed to both sides and at each end of the container, MEGC, tank-container or portable tank.

When the tank-container or portable tank has multiple compartments and carries two or more dangerous goods, the appropriate placards shall be displayed along each side at the position of the relevant compartments and one placard of each model shown on each side at both ends. If all compartments have to bear the same placards, these placards need to be displayed only once along each side and at both ends of the tank container or portable tank.

5.3.1.3   *Placarding of vehicles and wagons carrying containers, MEGCs, tank-containers or portable tanks*

NOTE: This subsection does not apply to swap-bodies, except tank swap bodies carried on vehicles bearing the orange markings stipulated in 5.3.2.

If the placards affixed to the containers, MEGCs, tank-containers or portable tanks are not visible from outside the carrying vehicles or wagons, the same placards shall also be affixed to both sides and at the rear of the vehicle or to both sides of the wagon. Otherwise, no placard need be affixed on the carrying vehicle or wagon.

5.3.1.4   *Placarding of vehicles for carriage in bulk, wagons for carriage in bulk, tank-vehicles, tank-wagons, battery vehicles, battery-wagons, MEMUs, vehicles with demountable tanks and wagons with demountable tanks*

5.3.1.4.1   Placards shall be affixed to both sides and at the rear of the vehicle, or, for wagons, to both sides.

When the tank-vehicle, tank-wagon, the demountable tank carried on the vehicle or the demountable tank carried on the wagon has multiple compartments and carries two or more dangerous goods, the appropriate placards shall be displayed along each side at the position of the relevant compartments and (vehicles only) one placard of each model shown on each side at the rear of the vehicle. If all compartments have to bear the same placards, these placards need be displayed only once along each side and (vehicles only) at the rear of the vehicle.

Where more than one placard is required for the same compartment, these placards shall be displayed adjacent to each other.

NOTE: When a tank semi-trailer is separated from its tractor to be loaded on board a ship or a vessel, placards shall also be displayed at the front of the semi-trailer.

5.3.1.4.2   MEMUs with tanks and bulk containers shall be placarded in accordance with 5.3.1.4.1 for the substances contained therein. For tanks with a capacity of less than 1 000 litres placards may be replaced by labels conforming to 5.2.2.2.

5.3.1.4.3   For MEMUs carrying packages containing substances or articles of Class 1 (other than of Division 1.4, Compatibility group S), placards shall be affixed to both sides and at the rear of the MEMU.

Special compartments for explosives shall be placarded in accordance with the provisions of 5.3.1.1.2. The last sentence of 5.3.1.1.2 does not apply.

5.3.1.5   *Placarding of vehicles and wagons carrying packages only*

> *NOTE: This sub-section applies also to vehicles or wagons carrying swap-bodies loaded with packages.*

5.3.1.5.1   For vehicles carrying packages containing substances or articles of Class 1 (other than of Division 1.4, Compatibility Group S), placards shall be affixed to both sides and at the rear of the vehicle.

5.3.1.5.2   For vehicles carrying radioactive material of Class 7 in packagings or IBCs (other than excepted packages), placards shall be affixed to both sides and at the rear of the vehicle.

> *NOTE: If a vehicle carrying packages containing dangerous goods of classes other than Classes 1 and 7 is loaded on board a vessel for an ADN journey preceding a voyage by sea, placards shall be affixed to both sides and at the rear of the vehicle. Such placards may remain affixed to a vehicle for an ADN journey following a sea voyage.*

5.3.1.5.3   For wagons carrying packages, placards corresponding to the goods carried shall be affixed to both sides.

5.3.1.6   *Placarding of empty tank-vehicles, tank-wagons, vehicles with demountable tanks, wagons with demountable tanks, battery-vehicles, battery-wagons, MEGCs, MEMUs, tank-containers, portable tanks and empty vehicles, wagons and containers for carriage in bulk*

5.3.1.6.1   Empty tank-vehicles, tank-wagons, vehicles with demountable tanks, wagons with demountable tanks, battery-vehicles, battery-wagons, MEGCs, MEMUs, tank-containers and portable tanks uncleaned and not degassed, and empty vehicles, wagons and containers for carriage in bulk, uncleaned, shall continue to display the placards required for the previous load.

5.3.1.7   *Specifications for placards*

5.3.1.7.1   Except as provided in 5.3.1.7.2 for the Class 7 placard, and in 5.3.6.2 for the environmentally hazardous substance mark, a placard shall be configured as shown in Figure 5.3.1.7.1.

Figure 5.3.1.7.1

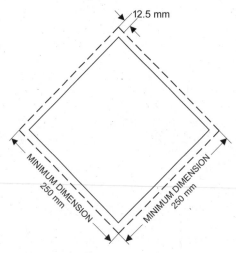

Placard (except for Class 7)

The placard shall be in the form of a square set at an angle of 45° (diamond-shaped). The minimum dimensions shall be 250 mm x 250 mm (to the edge of the placard). The line inside the edge shall be parallel and 12.5 mm from the outside of that line to the edge of the placard. The symbol and line inside the edge shall correspond in colour to the label for the class or division of the dangerous goods in question. The class or division symbol/numeral shall be positioned and sized in proportion to those prescribed in 5.2.2.2 for the corresponding class or division of the dangerous goods in question. The placard shall display the number of the class or division (and for goods in Class 1, the compatibility group letter) of the dangerous goods in question in the manner prescribed in 5.2.2.2 for the corresponding label, in digits not less than 25 mm high. Where dimensions are not specified, all features shall be in approximate proportion to those shown.

5.3.1.7.2   The Class 7 placard shall be not less than 250 mm by 250 mm with a black line running 5 mm inside the edge and parallel with it and is otherwise as shown below (Model No. 7D). The number "7" shall not be less than 25 mm high. The background colour of the upper half of the placard shall be yellow and of the lower half white, the colour of the trefoil and the printing shall be black. The use of the word "RADIOACTIVE" in the bottom half is optional to allow the use of this placard to display the appropriate UN number for the consignment.

Placard for radioactive material of Class 7

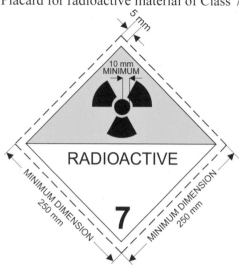

(No.7D)

Symbol (trefoil): black; Background: upper half yellow with white border, lower half white; The lower half shall show the word "RADIOACTIVE" or alternatively the appropriate UN Number and the figure "7" in the bottom corner.

5.3.1.7.3   For tanks with a capacity of not more than 3 m³ and for small containers, placards may be replaced by labels conforming to 5.2.2.2. If these labels are not visible from outside the carrying vehicle or wagon, placards according to 5.3.1.7.1 shall also be affixed to both sides of the wagon or to both sides and at the rear of the vehicle.

5.3.1.7.4   For Classes 1 and 7, if the size and construction of the vehicle are such that the available surface area is insufficient to affix the prescribed placards, their dimensions may be reduced to 100 mm on each side. The dimensions of the placards to be affixed to wagons may be reduced to 150 mm by 150 mm. In this case, the upper dimensions prescribed for the trefoil, lines, figures and letters do not apply.

5.3.2   Orange-coloured plate marking

5.3.2.1   *General orange-coloured plate marking provisions*

5.3.2.1.1   Transport units carrying dangerous goods shall display two rectangular orange-coloured plates conforming to 5.3.2.2.1, set in a vertical plane. They shall be affixed one at the front and the other at the rear of the transport unit, both perpendicular to the longitudinal axis of the transport unit. They shall be clearly visible.

If a trailer containing dangerous goods is detached from its motor vehicle during carriage of dangerous goods, an orange-coloured plate shall remain affixed to the rear of the trailer. When tanks are marked in accordance with 5.3.2.1.3, this plate shall correspond to the most hazardous substance carried in the tank.

5.3.2.1.2   When a hazard identification number is indicated in Column (20) of Table A of Chapter 3.2 of ADR, tank-vehicles, battery vehicles or transport units having one or more tanks carrying dangerous goods shall in addition display on the sides of each tank, each tank compartment or each element of battery-vehicles, clearly visible and parallel to the longitudinal axis of the vehicle, orange-coloured plates identical with those prescribed in 5.3.2.1.1. These orange-coloured plates shall bear the hazard identification number and the UN number prescribed respectively in Columns (20) and (1) of Table A of Chapter 3.2 of ADR for each of the substances carried in the tank, in a compartment of the tank or in an element of a battery-vehicle.

The provisions of this paragraph are also applicable to tank-wagons, battery-wagons and wagons with demountable tanks. In the latter case the hazard identification number to be used is that indicated in column (20) of table A of Chapter 3.2 of RID. For MEMUs these requirements shall only apply to tanks with a capacity of 1 000 litres or more and bulk containers.

5.3.2.1.3   For tank-vehicles or transport units having one or more tanks carrying substances with UN Nos. 1202, 1203 or 1223, or aviation fuel classed under UN Nos. 1268 or 1863, but no other dangerous substance, the orange-coloured plates prescribed in 5.3.2.1.2 need not be affixed if the plates affixed to the front and rear in accordance with 5.3.2.1.1 bear the hazard identification number and the UN number prescribed for the most hazardous substance carried, i.e. the substance with the lowest flashpoint.

5.3.2.1.4   When a hazard identification number is indicated in Column (20) of Table A of Chapter 3.2 of ADR, transport units and containers carrying unpackaged solids or articles or packaged radioactive material with a single UN number required to be carried under exclusive use and no other dangerous goods shall in addition display on the sides of each transport unit or container, clearly visible and parallel to the longitudinal axis of the vehicle, orange-coloured plates identical with those prescribed in 5.3.2.1.1. These orange-coloured plates shall bear the hazard identification number and the UN number prescribed respectively in Columns (20) and (1) of Table A of Chapter 3.2 of ADR for each of the substances carried in bulk in the transport unit or in the container or for the packaged radioactive material when required to be carried under exclusive use in the transport unit or in the container.

The provisions of this paragraph are also applicable to wagons for carriage in bulk and full wagon loads comprising packages containing only one substance. In the latter case the hazard identification number to be used is that indicated in Column (20) of Table A of Chapter 3.2 of RID.

5.3.2.1.5   If the orange-coloured plates prescribed in 5.3.2.1.2 and 5.3.2.1.4 affixed to the containers, tank-containers, MEGCs or portable tanks are not clearly visible from outside the carrying vehicle or wagon, the same plates shall also be affixed to both sides of the vehicle or wagon.

NOTE: This paragraph need not be applied to the marking with orange coloured plates of closed and sheeted wagons or vehicles, carrying tanks with a maximum capacity of 3 000 litres.

5.3.2.1.6   For transport units carrying only one dangerous substance and no non-dangerous substance, the orange-coloured plates prescribed in 5.3.2.1.2, 5.3.2.1.4 and 5.3.2.1.5 shall not be necessary provided that those displayed at the front and rear in accordance with 5.3.2.1.1 bear the hazard identification number and the UN number for that substance prescribed respectively in Columns (20) and (1) of Table A of Chapter 3.2 of ADR.

5.3.2.1.7   The requirements of 5.3.2.1.1 to 5.3.2.1.5 are also applicable to empty fixed or demountable tanks, battery-vehicles, tank-containers, portable tanks, MEGCs, tank-wagons, battery-wagons and wagons with demountable tanks, uncleaned, not degassed or not decontaminated, MEMUs, uncleaned as well as to empty vehicles, wagons and containers for carriage in bulk, uncleaned or not decontaminated.

5.3.2.1.8   Any orange-coloured plates which does not relate to dangerous goods carried, or residues thereof, shall be removed or covered. If plates are covered, the covering shall be total and remain effective after 15 minutes' engulfment in fire.

5.3.2.2   *Specifications for the orange-coloured plates*

5.3.2.2.1   The orange-coloured plates shall be reflectorized and shall be of 40 cm base and of 30 cm high; they shall have a black border of 15 mm wide. The material used shall be weather-resistant and ensure durable marking. The plate shall not become detached from its mount in the event of a 15 minutes' engulfment in fire. It shall remain affixed irrespective of the orientation of the vehicle or wagon. The orange-coloured plates may be separated in their middle with a black horizontal line of 15 mm thickness.

If the size and construction of the vehicle are such that the available surface area is insufficient to affix these orange-coloured plates, their dimensions may be reduced to a minimum of 300 mm for the base, 120 mm for the height and 10 mm for the black border. In this case, a different set of dimensions within the specified range may be used for the two orange-coloured plates specified in 5.3.2.1.1.

When reduced dimensions of orange-coloured plates are used for a packaged radioactive material carried under exclusive use, only the UN number is required and the size of the digits stipulated in 5.3.2.2.2 may be reduced to 65 mm in height and 10 mm in stroke thickness.

A non-reflectorized colour is permitted for wagons.

For containers carrying dangerous solid substances in bulk and for tank-containers, MEGCs and portable tanks, the plates prescribed in 5.3.2.1.2, 5.3.2.1.4 and 5.3.2.1.5 may be replaced by a self-adhesive sheet, by paint or by any other equivalent process.

This alternative marking shall conform to the specifications set in this sub-section except for the provisions concerning resistance to fire mentioned in 5.3.2.2.1 and 5.3.2.2.2.

NOTE: The colour of the orange plates in conditions of normal use should have chromaticity coordinates lying within the area on the chromaticity diagram formed by joining the following coordinates:

Chromaticity coordinates of points at the corners of the area on the chromaticity diagram				
x	0.52	0.52	0.578	0.618
y	0.38	0.40	0.422	0.38

Luminance factor of reflectorized colour: $\beta > 0.12$.
Luminance factor of non-reflectorized colour (wagons): $\beta \geq 0.22$
Reference centre E, standard illuminant C, normal incidence 45°, viewed at 0°.
Coefficient of reflex luminous intensity at an angle of illumination of 5°, viewed at 0.2°: not less than 20 candelas per lux per m^2 (not required for wagons).

5.3.2.2.2   The hazard identification number and the UN number shall consist of black digits 100 mm high and of 15 mm stroke thickness. The hazard identification number shall be inscribed in the upper part of the plate and the UN number in the lower part; they shall be separated by a horizontal black line, 15 mm in stroke width, extending from side to side of the plate at mid-height (see 5.3.2.2.3). The hazard identification number and the UN number shall be indelible and shall remain legible after 15 minutes engulfment in fire. Interchangeable numbers and letters on plates presenting the hazard identification number and the UN number shall remain in place during carriage and irrespective of the orientation of the wagon or vehicle.

5.3.2.2.3   *Example of orange-coloured plate with hazard identification number and UN number*

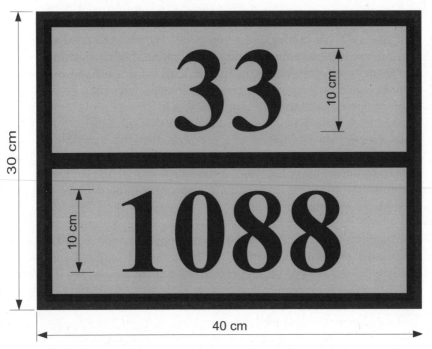

Hazard identification number (2 or 3 figures preceded where appropriate by the letter X, see 5.3.2.3)

UN number (4 figures)

Background orange.
Border, horizontal line and figures black, 15 mm thickness.

5.3.2.2.4   The permitted tolerances for dimensions specified in this sub-section are ± 10%.

5.3.2.2.5   When the orange-coloured plate is affixed to folding panels, they shall be designed and secured so that they cannot unfold or come loose from the holder during carriage (especially as a result of impacts or unintentional actions).

5.3.2.3   *Meaning of hazard identification numbers*

5.3.2.3.1   The hazard identification number consists of two or three figures. In general, the figures indicate the following hazards:

2	Emission of gas due to pressure or to chemical reaction
3	Flammability of liquids (vapours) and gases or self-heating liquid
4	Flammability of solids or self-heating solid
5	Oxidizing (fire-intensifying) effect
6	Toxicity or risk of infection
7	Radioactivity
8	Corrosivity
9	Risk of spontaneous violent reaction

NOTE: *The risk of spontaneous violent reaction within the meaning of figure 9 includes the possibility following from the nature of a substance of a risk of explosion, disintegration and polymerization reaction following the release of considerable heat or flammable and/or toxic gases.*

Doubling of a figure indicates an intensification of that particular hazard.

Where the hazard associated with a substance can be adequately indicated by a single figure, this is followed by zero.

The following combinations of figures, however, have a special meaning: 22, 323, 333, 362, 382, 423, 44, 446, 462, 482, 539, 606, 623, 642, 823, 842, 90 and 99 (see 5.3.2.3.2 below).

If a hazard identification number is prefixed by the letter "X", this indicates that the substance will react dangerously with water. For such substances, water may only be used by approval of experts.

For substances of Class 1, the classification code in accordance with Column (3b) of Table A of Chapter 3.2, shall be used as the hazard identification number. The classification code consists of:

- the division number in accordance with 2.2.1.1.5; and

- the compatibility group letter in accordance with 2.2.1.1.6.

5.3.2.3.2   The hazard identification numbers listed in Column (20) of Table A of Chapter 3.2 of ADR or RID have the following meanings:

20	asphyxiant gas or gas with no subsidiary risk
22	refrigerated liquefied gas, asphyxiant
223	refrigerated liquefied gas, flammable
225	refrigerated liquefied gas, oxidizing (fire-intensifying)
23	flammable gas
238	gas, flammable corrosive
239	flammable gas, which can spontaneously lead to violent reaction
25	oxidizing (fire-intensifying) gas
26	toxic gas
263	toxic gas, flammable
265	toxic gas, oxidizing (fire-intensifying)
268	toxic gas, corrosive
28	gas, corrosive
285	gas, corrosive, oxidizing

30	flammable liquid (flashpoint between 23 °C and 60 °C, inclusive) or flammable liquid or solid in the molten state with a flashpoint above 60 °C, heated to a temperature equal to or above its flashpoint, or self-heating liquid
323	flammable liquid which reacts with water, emitting flammable gases
X323	flammable liquid which reacts dangerously with water, emitting flammable gases[1]
33	highly flammable liquid (flashpoint below 23 °C)
333	pyrophoric liquid
X333	pyrophoric liquid which reacts dangerously with water[1]
336	highly flammable liquid, toxic
338	highly flammable liquid, corrosive
X338	highly flammable liquid, corrosive, which reacts dangerously with water[1]
339	highly flammable liquid which can spontaneously lead to violent reaction
36	flammable liquid (flashpoint between 23 °C and 60 °C, inclusive), slightly toxic, or self-heating liquid, toxic
362	flammable liquid, toxic, which reacts with water, emitting flammable gases
X362	flammable liquid, toxic, which reacts dangerously with water, emitting flammable gases[1]
368	flammable liquid, toxic, corrosive
38	flammable liquid (flashpoint between 23 °C and 60 °C, inclusive), slightly corrosive or self-heating liquid, corrosive
382	flammable liquid, corrosive, which reacts with water, emitting flammable gases
X382	flammable liquid, corrosive, which reacts dangerously with water, emitting flammable gases[1]

---

[1]   Water not to be used except by approval of experts.

39	flammable liquid, which can spontaneously lead to violent reaction
40	flammable solid, or self-reactive substance, or self-heating substance, or polymerizing substance
423	solid which reacts with water, emitting flammable gases, or flammable solid which reacts with water, emitting flammable gases or self-heating solid which reacts with water, emitting flammable gases
X423	solid which reacts dangerously with water, emitting flammable gases, or flammable solid which reacts dangerously with water, emitting flammable gases, or self-heating solid which reacts dangerously with water, emitting flammable gases [1]
43	spontaneously flammable (pyrophoric) solid
X432	spontaneously flammable (pyrophoric) solid which reacts dangerously with water, emitting flammable gases[1]
44	flammable solid, in the molten state at an elevated temperature
446	flammable solid, toxic, in the molten state, at an elevated temperature
46	flammable or self-heating solid, toxic
462	toxic solid which reacts with water, emitting flammable gases
X462	solid which reacts dangerously with water, emitting toxic gases[1]
48	flammable or self-heating solid, corrosive
482	corrosive solid which reacts with water, emitting flammable gases
X482	solid which reacts dangerously with water, emitting corrosive gases[1]
50	oxidizing (fire-intensifying) substance
539	flammable organic peroxide
55	strongly oxidizing (fire-intensifying) substance
556	strongly oxidizing (fire-intensifying) substance, toxic
558	strongly oxidizing (fire-intensifying) substance, corrosive
559	strongly oxidizing (fire-intensifying) substance, which can spontaneously lead to violent reaction
56	oxidizing substance (fire-intensifying), toxic
568	oxidizing substance (fire-intensifying), toxic, corrosive
58	oxidizing substance (fire-intensifying), corrosive
59	oxidizing substance (fire-intensifying), which can spontaneously lead to violent reaction
60	toxic or slightly toxic substance
606	infectious substance
623	toxic liquid, which reacts with water, emitting flammable gases
63	toxic substance, flammable (flashpoint between 23 °C and 60 °C, inclusive)
638	toxic substance, flammable (flashpoint between 23 °C and 60 °C, inclusive), corrosive
639	toxic substance, flammable (flashpoint not above 60 °C) which can spontaneously lead to violent reaction
64	toxic solid, flammable or self-heating
642	toxic solid, which reacts with water, emitting flammable gases
65	toxic substance, oxidizing (fire-intensifying)
66	highly toxic substance
663	highly toxic substance, flammable (flashpoint not above 60 °C)
664	highly toxic solid, flammable or self-heating
665	highly toxic substance, oxidizing (fire-intensifying)
668	highly toxic substance, corrosive
X668	highly toxic substance, corrosive, which reacts dangerously with water[1]
669	highly toxic substance which can spontaneously lead to violent reaction

---

[1]   *Water not to be used except by approval of experts.*

68	toxic substance, corrosive
687	toxic substance, corrosive, radioactive
69	toxic or slightly toxic substance, which can spontaneously lead to violent reaction
70	radioactive material
768	radioactive material, toxic, corrosive
78	radioactive material, corrosive
80	corrosive or slightly corrosive substance
X80	corrosive or slightly corrosive substance, which reacts dangerously with water[1]
823	corrosive liquid which reacts with water, emitting flammable gases
83	corrosive or slightly corrosive substance, flammable (flashpoint between 23 °C and 60 °C, inclusive)
X83	corrosive or slightly corrosive substance, flammable (flashpoint between 23 °C and 60 °C, inclusive), which reacts dangerously with water[1]
839	corrosive or slightly corrosive substance, flammable (flashpoint between 23 °C and 60 °C inclusive) which can spontaneously lead to violent reaction
X839	corrosive or slightly corrosive substance, flammable (flashpoint between 23 °C and 60 °C inclusive), which can spontaneously lead to violent reaction and which reacts dangerously with water[1]
84	corrosive solid, flammable or self-heating
842	corrosive solid which reacts with water, emitting flammable gases
85	corrosive or slightly corrosive substance, oxidizing (fire-intensifying)
856	corrosive or slightly corrosive substance, oxidizing (fire-intensifying) and toxic
86	corrosive or slightly corrosive substance, toxic
88	highly corrosive substance
X88	highly corrosive substance, which reacts dangerously with water[1]
883	highly corrosive substance, flammable (flashpoint between 23 °C and 60 °C inclusive)
884	highly corrosive solid, flammable or self-heating
885	highly corrosive substance, oxidizing (fire-intensifying)
886	highly corrosive substance, toxic
X886	highly corrosive substance, toxic, which reacts dangerously with water[1]
89	corrosive or slightly corrosive substance, which can spontaneously lead to violent reaction
90	environmentally hazardous substance; miscellaneous dangerous substances
99	miscellaneous dangerous substance carried at an elevated temperature.

5.3.3 Mark for elevated temperature substances

Tank-vehicles, tank-wagons, tank-containers, portable tanks, special vehicles, special wagons or special containers or specially equipped vehicles, specially equipped wagons or specially equipped containers containing a substance that is carried or handed over for carriage in a liquid state at or above 100 °C or in a solid state at or above 240 °C shall bear on both sides for wagons, on both sides and at the rear for vehicles, and on both sides and at each end for containers, tank-containers and portable tanks, the mark shown in Figure 5.3.3.

---

[1]   *Water not to be used except by approval of experts.*

Figure 5.3.3

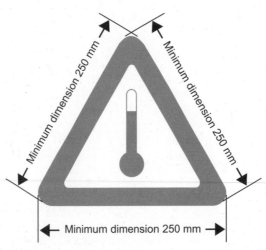

Mark for carriage at elevated temperature

The mark shall be an equilateral triangle. The colour of the mark shall be red. The minimum dimension of the sides shall be 250 mm. For tank-containers or portable tanks with a capacity of not more than 3 000 litres and with an available surface area insufficient to affix the prescribed marks, the minimum dimensions of the sides may be reduced to 100 mm. Where dimensions are not specified, all features shall be in approximate proportion to those shown.

5.3.4   Marking for carriage in a transport chain including maritime transport

5.3.4.1   For carriage in a transport chain including maritime transport, containers, portable tanks and MEGCs are not required to carry the orange-coloured plate marking according to section 5.3.2 if they carry the marking prescribed in section 5.3.2 of the IMDG Code, where:

(a)   The proper shipping name of the contents is durably marked on at least two sides:

–   of portable tanks and MEGCs;

–   of containers for carriage in bulk;

–   of containers containing dangerous goods in packages constituting only one substance for which the IMDG Code does not require a placard or the marine pollutant mark;

(b)   The UN number for the goods is displayed in black digits not less than 65 mm high:

–   either on a white background in the lower half of the placards affixed to the cargo transport unit;

–   or on an orange rectangular panel not less than 120 mm high and 300 mm wide, with a 10 mm black border, to be placed immediately adjacent to the placard or the marine pollutant marks of the IMDG Code, or, if no placard or marine pollutant mark is prescribed, adjacent to the proper shipping name.

**Example of marking for a portable tank carrying acetal,
class 3, UN No 1088, according to the IMDG Code**

FIRST VARIANT

black flame on
red background

SECOND VARIANT

black flame on
red background

orange background
border and digits in black

5.3.4.2   If portable tanks, MEGCs or containers marked in accordance with 5.3.4.1 are carried on board a vessel loaded on vehicles, only paragraph 5.3.2.1.1 applies to the carrying vehicle.

5.3.4.3   In addition to the placards, orange-coloured plate marking and marks prescribed or permitted by ADN, cargo transport units may carry additional marks, placards and other markings prescribed where appropriate by the IMDG Code, for example, the marine pollutant mark or the "LIMITED QUANTITIES" mark.

5.3.5   *(Reserved)*

5.3.6   **Environmentally hazardous substance mark**

5.3.6.1   When a placard is required to be displayed in accordance with the provisions of section 5.3.1, containers, MEGCs, tank-containers, portable tanks, vehicles and wagons containing environmentally hazardous substances meeting the criteria of 2.2.9.1.10 shall be marked with the environmentally hazardous substance mark shown in 5.2.1.8.3.

5.3.6.2   The environmentally hazardous substance mark for containers, MEGCs, tank-containers, portable tanks, wagons and vehicles shall be as described in 5.2.1.8.3 and Figure 5.2.1.8.3, except that the minimum dimensions shall be 250 mm x 250 mm. For tank-containers or portable tanks with a capacity of not more than 3 000 litres and with an available surface area insufficient to affix the prescribed marks, the minimum dimensions may be reduced to 100 mm x 100 mm. The other provisions of section 5.3.1 concerning placards shall apply mutatis mutandis to the mark.

CHAPTER 5.4

DOCUMENTATION

5.4.0   **General**

5.4.0.1   Unless otherwise specified, any carriage of goods governed by ADN shall be accompanied by the documentation prescribed in this Chapter, as appropriate.

NOTE: For the list of documentation to be carried on board vessels, see 8.1.2.

5.4.0.2   The use of electronic data processing (EDP) or electronic data interchange (EDI) techniques as an aid to or instead of paper documentation is permitted, provided that the procedures used for the capture, storage and processing of electronics data meet the legal requirements as regards the evidential value and availability of data during carriage in a manner at least equivalent to that of paper documentation.

5.4.0.3   When the dangerous goods transport information is given to the carrier by EDP or EDI techniques, the consignor shall be able to give the information to the carrier as a paper document, with the information in the sequence required by this Chapter.

5.4.1   **Dangerous goods transport document and related information**

5.4.1.1   *General information required in the transport document*

5.4.1.1.1   *General information required in the transport document for carriage in bulk or in packages*

The transport document(s) shall contain the following information for each dangerous substance, material or article offered for carriage:

(a)   the UN number, preceded by the letters "UN" or substance identification number;

(b)   the proper shipping name supplemented, when applicable (see 3.1.2.8.1) with the technical name in brackets (see 3.1.2.8.1.1), as determined in accordance with 3.1.2.

(c)   –   For substances and articles of Class 1: the classification code given in Column (3b) of Table A of Chapter 3.2.

When, in Column (5) of Table A of Chapter 3.2, label model numbers are given other than 1, 1.4, 1.5 and 1.6, these label model numbers, in brackets, shall follow the classification code;

–   For radioactive material of Class 7: the Class number: "7";

NOTE: For radioactive material with a subsidiary risk, see also special provision 172 in Chapter 3.3.

–   For lithium batteries of UN numbers 3090, 3091, 3480 and 3481: the Class number "9";

–   For other substances and articles: the label model numbers given in Column (5) of Table A of Chapter 3.2 or applicable according to a special provision referred to in Column (6). When more than one label model number is given, the numbers following the first one shall be given in brackets. For substances and articles for which no label model is given in Column (5) of Table A in Chapter 3.2, their class according to Column (3a) shall be given instead;

(d)   where assigned, the packing group for the substance which may be preceded by the letters "PG" (e.g. "PG II"), or the initials corresponding to the words "Packing Group" in the languages used according to 5.4.1.4.1;

NOTE: For radioactive material of Class 7 with subsidiary risks, see special provision 172 (d) in Chapter 3.3.

(e)   the number and a description of the packages when applicable. UN packaging codes may only be used to supplement the description of the kind of package (e.g. one box (4G));

NOTE: The number, type and capacity of each inner packaging within the outer packaging of a combination packaging is not required to be indicated.

(f)   the total quantity of each item of dangerous goods bearing a different UN number, proper shipping name (as a volume or as a gross mass, or as a net mass as appropriate);

NOTE: For dangerous goods in machinery and or equipment specified in these Regulations, the quantity indicated shall be the total quantity of dangerous goods contained therein in kilograms or litres as appropriate.

(g)   the name and address of the consignor;

(h)   the name and address of the consignee(s);

(i)   a declaration as required by the terms of any special agreement.

The location and order in which the elements of information required appear in the transport document is left optional, except that (a), (b), (c) and (d) shall be shown in the order listed above (i.e. (a), (b), (c), (d)) with no information interspersed, except as provided in ADN.

Examples of such permitted dangerous goods descriptions are:

"UN 1098 ALLYL ALCOHOL, 6.1 (3), I" or
"UN1098, ALLYL ALCOHOL, 6.1 (3), PG I"

The information required on a transport document shall be legible.

Although upper case is used in Chapter 3.1 and in Table A of Chapter 3.2 to indicate the elements which shall be part of the proper shipping name, and although upper and lower case are used in this Chapter to indicate the information required in the transport document, the use of upper or of lower case for entering the information in the transport document is left optional.

5.4.1.1.2   *General information required in the transport document for carriage in tank vessels*

The transport document(s) shall contain the following information for each dangerous substance or article offered for carriage:

(a)   the UN number preceded by the letters "UN" or the substance identification number;

(b)   the proper shipping name given in Column (2) of Table C of Chapter 3.2, supplemented, when applicable, by the technical name in parenthesis;

(c)   the data contained in column (5) of Table C of Chapter 3.2. When more than one number is given, the numbers following the first one shall be given in brackets. For substances not mentioned by name in Table C (assigned to a generic entry or a N.O.S.

entry and for which the flowchart in 3.2.3.3 is applicable) only the actual dangerous properties of the substance shall be mentioned;

(d)   where assigned, the packing group for the substance, which may be preceded by the letters 'PG' (e.g. 'PG II'), or the initials corresponding to the words 'Packing Group' in the languages used in accordance with 5.4.1.4.1;

(e)   the mass in tonnes;

(f)   the name and address of the consignor;

(g)   the name and address of the consignee(s).

The location and order in which the elements of information required appear in the transport document is left optional, except that (a), (b), (c) and (d) shall be shown in the order listed above (i.e. (a), (b), (c), (d)) with no information interspersed, except as provided in ADN.

Examples of such permitted dangerous goods descriptions are:

"**UN 1203 MOTOR SPIRIT, 3 (N2, CMR, F), II**"; or
"**UN 1203 MOTOR SPIRIT, 3 (N2, CMR, F), PG II**"

The information required on a transport document shall be legible.

Although upper case is used in Chapter 3.1 and in Table C of Chapter 3.2 to indicate the elements which shall be part of the proper shipping name, and although upper and lower case are used in this Chapter to indicate the information required in the transport document, the use of upper or of lower case for entering the information in the transport document is left optional.

5.4.1.1.3   *Special provisions for wastes*

If waste containing dangerous goods (other than radioactive wastes) is being carried, the proper shipping name shall be preceded by the word "**WASTE**", unless this term is part of the proper shipping name, e.g.:

"**UN 1230 WASTE METHANOL, 3 (6.1), II**", or
"**UN 1230 WASTE METHANOL, 3 (6.1), PG II,**", or
"**UN 1993 WASTE FLAMMABLE LIQUID, N.O.S. (toluene and ethyl alcohol), 3, II,**", or
"**UN 1993 WASTE FLAMMABLE LIQUID, N.O.S. (toluene and ethyl alcohol), 3, PG II**"

If the provision for waste as set out in 2.1.3.5.5 is applied, the following shall be added to the dangerous goods description required in 5.4.1.1.1 (a) to (d) and (k):

"**WASTE IN ACCORDANCE WITH 2.1.3.5.5**" (e.g. "**UN 3264, CORROSIVE LIQUID, ACIDIC, INORGANIC, N.O.S., 8, II, WASTE IN ACCORDANCE WITH 2.1.3.5.5**").

The technical name, as prescribed in Chapter 3.3, special provision 274, need not be added.

5.4.1.1.4   *(Deleted)*

| 5.4.1.1.5 | *Special provisions for salvage packagings and salvage pressure receptacles* |

When dangerous goods are carried in a salvage packaging or salvage pressure receptacle, the words **"SALVAGE PACKAGING"** or **"SALVAGE PRESSURE RECEPTACLE"** shall be added after the description of the goods in the transport document.

| 5.4.1.1.6 | *Special provision for empty means of containment and for empty cargo tanks of tank vessels* |

5.4.1.1.6.1 For empty means of containment, uncleaned, which contain the residue of dangerous goods of classes other than Class 7, the words "EMPTY, UNCLEANED" or "RESIDUE, LAST CONTAINED" shall be indicated before or after the dangerous goods description specified in 5.4.1.1.1 (a) to (d).. Moreover, 5.4.1.1.1 (f) does not apply.

5.4.1.1.6.2 The special provision of 5.4.1.1.6.1 may be replaced with the provisions of 5.4.1.1.6.2.1, 5.4.1.1.6.2.2 or 5.4.1.1.6.2.3, as appropriate.

5.4.1.1.6.2.1 For empty packagings, uncleaned, which contain the residue of dangerous goods of classes other than Class 7, including empty uncleaned receptacles for gases with a capacity of not more than 1000 litres, the particulars according to 5.4.1.1.1 (a), (b), (c), (d), (e) and (f) are replaced with "EMPTY PACKAGING", "EMPTY RECEPTACLE", "EMPTY IBC" or "EMPTY LARGE PACKAGING", as appropriate, followed by the information of the goods last loaded, as described in 5.4.1.1.1 (c).

Example:

"EMPTY PACKAGING, 6.1 (3)".

In addition, in such a case

(a) if the dangerous goods last loaded are goods of Class 2, the information prescribed in 5.4.1.1.1 (c) may be replaced by the number of the class "2".

(b) if the dangerous goods last loaded are goods of Classes 3, 4.1, 4.2, 4.3, 5.1, 5.2, 6.1, 8 or 9, the information of the goods last loaded, as described in 5.4.1.1.1 (c) may be replaced by the words "WITH RESIDUES OF [...]" followed by the class(es) and subsidiary risk(s) corresponding to the different residues, in the class numbering order.

Example: Empty packagings, uncleaned, having contained goods of Class 3 carried together with empty packagings, uncleaned, having contained goods of Class 8 with a Class 6.1 subsidiary risk may be referred to in the transport document as:

"EMPTY PACKAGINGS, WITH RESIDUES OF 3, 6.1, 8".

5.4.1.1.6.2.2 For empty means of containment other than packagings, uncleaned, which contain the residue of dangerous goods of classes other than Class 7 and for empty uncleaned receptacles for gases with a capacity of more than 1000 litres, the particulars according to 5.4.1.1.1 (a) to (d) are preceded by "EMPTY TANK-WAGON", "EMPTY TANK-VEHICLE", "EMPTY DEMOUNTABLE TANK", "EMPTY TANK-CONTAINER", "EMPTY PORTABLE TANK", "EMPTY BATTERY-WAGON", "EMPTY BATTERY-VEHICLE", "EMPTY MEGC", "EMPTY MEMU", "EMPTY WAGON", "EMPTY VEHICLE", "EMPTY CONTAINER" or "EMPTY RECEPTACLE", as appropriate, followed by the words "LAST LOAD:". Moreover, paragraph 5.4.1.1.1 (f) does not apply.

See example as follows:

"EMPTY TANK-CONTAINER, LAST LOAD: UN 1098 ALLYL ALCOHOL, 6.1 (3), I" or
"EMPTY TANK-CONTAINER, LAST LOAD: UN 1098 ALLYL ALCOHOL, 6.1 (3), PG I".

5.4.1.1.6.2.3   When empty means of containment, uncleaned, which contain the residue of dangerous goods of classes other than Class 7, are returned to the consignor, the transport documents prepared for the full-capacity carriage of these goods may also be used. In such cases, the indication of the quantity is to be eliminated (by effacing it, striking it out or any other means) and replaced by the words "EMPTY, UNCLEANED RETURN".

5.4.1.1.6.3   (a)   If empty tanks, battery-vehicles, battery wagons and MEGCs, uncleaned, are carried to the nearest place where cleaning or repair can be carried out in accordance with the provisions of 4.3.2.4.3 of ADR or RID, the following additional entry shall be made in the transport document: **"Carriage in accordance with 4.3.2.4.3 of ADR (or RID)"**.

(b)   If empty vehicles, wagons and containers, uncleaned, are carried to the nearest place where cleaning or repair can be carried out in accordance with the provisions of 7.5.8.1 of ADR or RID, the following additional entry shall be made in the transport document: **"Carriage in accordance with 7.5.8.1 of ADR (or RID)"**.

5.4.1.1.6.4   For the carriage of tank wagons, fixed tanks (tank vehicles), wagons with removable tanks, vehicles with demountable tanks, battery-wagons, battery-vehicles, tank-containers and MEGCs under the conditions of 4.3.2.4.4 of ADR or RID, the following entry shall be included in the transport document: "Carriage in accordance with 4.3.2.4.4 of ADR (or RID)" as appropriate.

5.4.1.1.6.5   For tank vessels with empty cargo tanks or cargo tanks that have been discharged, the master is deemed to be the consignor for the purpose of the transport documents required. In this case, the following particulars shall be entered on the transport document for each empty cargo tank or cargo tank that has been discharged:

(a)   the number of the cargo tank;

(b)   the UN number preceded by the letters "UN" or the substance identification number;

(c)   the proper shipping name of the last substance carried, the class and, if applicable, the packing group in accordance with 5.4.1.1.2.

5.4.1.1.7   *Special provisions for carriage in a transport chain including maritime, road, rail or air carriage*

For carriage in accordance with 1.1.4.2.1, a statement shall be included in the transport document, as follows: **"Carriage in accordance with 1.1.4.2.1"**.

5.4.1.1.8 and 5.4.1.1.9   (*Reserved*)

5.4.1.1.10   (*Deleted*)

5.4.1.1.11   *Special provisions for the carriage of IBCs, tanks, battery-vehicles, portable tanks and MEGCs after the date of expiry of the last periodic test or inspection*

For carriage in accordance with 4.1.2.2 (b), 4.3.2.3.7 (b), 6.7.2.19.6 (b), 6.7.3.15.6 (b) or 6.7.4.14.6 (b) of ADR (or RID), a statement to this effect shall be included in the transport document, as follows:

"CARRIAGE IN ACCORDANCE WITH 4.1.2.2 (b) of ADR (or RID)",

"CARRIAGE IN ACCORDANCE WITH 4.3.2.3.7 (b) of ADR (or RID)",

"CARRIAGE IN ACCORDANCE WITH 6.7.2.19.6 (b) of ADR (or RID)",

"CARRIAGE IN ACCORDANCE WITH 6.7.3.15.6 (b) of ADR (or RID)"; or

"CARRIAGE IN ACCORDANCE WITH 6.7.4.14.6 (b) of ADR (or RID)" as appropriate.

5.4.1.1.12 and 5.4.1.1.13   (*Reserved*)

5.4.1.1.14   *Special provisions for the carriage of substances carried under elevated temperature*

If the proper shipping name of a substance which is carried or offered for carriage in a liquid state at a temperature equal to or exceeding 100 °C, or in a solid state at a temperature equal to or exceeding 240 °C, does not convey the elevated temperature condition (for example, by using the term **"MOLTEN"** or **"ELEVATED TEMPERATURE"** as part of the proper shipping name), the word **"HOT"** shall immediately precede the proper shipping name.

5.4.1.1.15   *Special provisions for the carriage of substances stabilized by temperature control*

If the word "STABILIZED" is part of the proper shipping name (see also 3.1.2.6), when stabilization is by means of temperature control, the control and emergency temperatures (see 2.2.41.1.17) shall be indicated in the transport document, as follows:

"Control temperature: ... °C   Emergency temperature: ... °C".

5.4.1.1.16   *Information required in accordance with special provision 640 in Chapter 3.3*

Where it is required by special provision 640 of Chapter 3.3, the transport document shall bear the inscription "**Special provision 640X**" where "X" is the capital letter appearing after the pertinent reference to special provision 640 in Column (6) of Table A of Chapter 3.2.

5.4.1.1.17   *Special provisions for the carriage of solids in bulk containers conforming to 6.11.4 of ADR*

When solid substances are carried in bulk containers conforming to 6.11.4 of ADR, the following statement shall be shown on the transport document (see NOTE at the beginning of 6.11.4 of ADR):

"Bulk container BK(x)[1] approved by the competent authority of..."

5.4.1.1.18   *Special provisions for carriage of environmentally hazardous substances (aquatic environment)*

When a substance belonging to one of classes 1 to 9 meets the classification criteria of 2.2.9.1.10, the transport document shall bear the additional inscription "ENVIRONMENTALLY HAZARDOUS" or "MARINE POLLUTANT/ ENVIRONMENTALLY HAZARDOUS". This additional requirement does not apply to UN Nos. 3077 and 3082 or for the exceptions listed in 5.2.1.8.1.

The inscription "MARINE POLLUTANT" (according to 5.4.1.4.3 of the IMDG Code) is acceptable for carriage in a transport chain including maritime carriage.

---

[1]   (x) shall be replaced with "1" or "2" as appropriate.

5.4.1.1.19   *Special provisions for carriage of packagings, discarded, empty, uncleaned (UN No. 3509)*

For packagings, discarded, empty, uncleaned, the proper shipping name specified in 5.4.1.1.1 (b) shall be complemented with the words "(WITH RESIDUES OF [...])" followed by the class(es) and subsidiary risk(s) corresponding to the residues, in the class numbering order. Moreover, 5.4.1.1.1 (f) does not apply.

Example: Packagings, discarded, empty, uncleaned having contained goods of Class 4.1 packed together with packagings, discarded, empty, uncleaned having contained goods of Class 3 with a Class 6.1 subsidiary risk should be referred to in the transport document as:

"UN 3509 PACKAGINGS, DISCARDED, EMPTY, UNCLEANED (WITH RESIDUES OF 3, 4.1, 6.1), 9".

5.4.1.1.20   *Special provisions for the carriage of substances classified in accordance with 2.1.2.8*

For carriage in accordance with 2.1.2.8, a statement shall be included in the transport document, as follows "Classified in accordance with 2.1.2.8".

5.4.1.1.21   *Special provisions for the carriage of UN Nos. 3528, 3529 and 3530*

For carriage of UN Nos. 3528, 3529 and 3530, the transport document, when required according to special provision 363 of Chapter 3.3, shall contain the following additional statement "Transport in accordance with special provision 363".

5.4.1.1.22   *Special provisions for carriage in oil separator vessels and supply vessels*

5.4.1.1.2 and 5.4.1.1.6.5 are not applicable to oil separator vessels or supply vessels.

5.4.1.2   *Additional or special information required for certain classes*

5.4.1.2.1   *Special provisions for Class 1*

(a)   The transport document shall indicate, in addition to the requirements in 5.4.1.1.1 (f):

–   the total net mass, in kg, of explosive contents[2] for each substance or article identified by its UN number;

–   the total net mass, in kg, of explosive contents[2] for all substances and articles covered by the transport document.

(b)   For mixed packing of two different goods, the description of the goods in the transport document shall include the UN numbers and names printed in capitals in Columns (1) and (2) of Table A of Chapter 3.2 of both substances or articles. If more than two different goods are contained in the same package in conformity with the mixed packing provisions given in 4.1.10 of ADR special provisions MP1, MP2 and MP20 to MP24, the transport document shall indicate under the description of the goods the UN numbers of all the substances and articles contained in the package, in the form, **"Goods of UN Nos. ..."**.

(c)   For the carriage of substances and articles assigned to an n.o.s. entry or the entry "0190 SAMPLES, EXPLOSIVE" or packed conforming to packing instruction P101 of 4.1.4.1 of ADR, a copy of the competent authority approval with the conditions of carriage shall be attached to the transport document. It shall be in an official language of the forwarding country and also, if that language is not English, French or German,

---

[2]   *For articles, "explosive contents" means the explosive substance contained in the article.*

in English, French or German unless agreements, if any, concluded between the countries concerned in the transport operation provide otherwise.

(d) If packages containing substances and articles of compatibility groups B and D are loaded together in the same vehicle or wagon in accordance with the requirements of 7.5.2.2 or ADR or RID, the approval certificate of the protective compartment or containment system in accordance with 7.5.2.2, note ^a under the table of ADR or RID, shall be attached to the transport document. It shall be in an official language of the forwarding country and also, if that language is not English, French or German, in English, French or German unless agreements, if any, concluded between the countries concerned in the transport operation provide otherwise.

(e) When explosive substances or articles are carried in packagings conforming to packing instruction P101 of ADR, the transport document shall bear the inscription **"Packaging approved by the competent authority of ..."** (see 4.1.4.1, packing instruction P101).

(f) (*Reserved*)

(g) When fireworks of UN Nos. 0333, 0334, 0335, 0336 and 0337 are carried, the transport document shall bear the inscription:

"Classification of fireworks by the competent authority of XX with the firework reference XX/YYZZZZ".

The classification approval certificate need not be carried with the consignment, but shall be made available by the consignor to the carrier or the competent authorities for control purposes. The classification approval certificate or a copy of it shall be in an official language of the forwarding country, and also, if that language is not German, English or French, in German, English or French.

NOTE 1: The commercial or technical name of the goods may be entered additionally to the proper shipping name in the transport document.

NOTE 2: The classification reference(s) shall consist of the ADN Contracting Party in which the classification code according to special provision 645 of 3.3.1 was approved, indicated by the distinguishing sign used on vehicles in international road traffic (XX)[3], the competent authority identification (YY) and a unique serial reference (ZZZZ). Examples of such classification references are:

GB/HSE123456

D/BAM1234.

5.4.1.2.2   *Additional provisions for Class 2*

(a) For the carriage of mixtures (see 2.2.2.1.1) in tanks (demountable tanks, fixed tanks, tank-wagons, portable tanks, tank-containers or elements of battery-vehicles or battery-wagons or of MEGCs), the composition of the mixture as a percentage of the volume or as a percentage of the mass shall be given. Constituents below 1% need not be indicated (see also 3.1.2.8.1.2). The composition of the mixture need not be given when the technical names authorized by special provisions 581, 582 or 583 are used to supplement the proper shipping name;

---

[3]   *Distinguishing sign of the State of registration used on motor vehicles and trailers in international road traffic, e.g. in accordance with the Geneva Convention on Road Traffic of 1949 or the Vienna Convention on Road Traffic of 1968.*

(b)   For the carriage of cylinders, tubes, pressure drums, cryogenic receptacles and bundles of cylinders under the conditions of 4.1.6.10 of ADR, the following entry shall be included in the transport document: **"Carriage in accordance with 4.1.6.10 of ADR"**.

(c)   *(Reserved)*

(d)   In the case of tank-containers carrying refrigerated liquefied gases the consignor shall enter in the transport document the date (or time) by which the actual holding time will be exceeded.

"End of holding time: .... (DD/MM/YYYY)".

5.4.1.2.3   *Additional provisions for self-reactive substances and polymerizing substances of Class 4.1 and organic peroxides of Class 5.2*

5.4.1.2.3.1   For self-reactive substances of Class 4.1 and for organic peroxides of Class 5.2 that require temperature control during carriage (for self-reactive substances or polymerizing substances see 2.2.41.1.17; for organic peroxides, see 2.2.52.1.15 to 2.2.52.1.17; for polymerizing substance see 2.2.41.1.21), the control and emergency temperatures shall be indicated in the transport document, as follows:

"Control temperature: ... °C   Emergency temperature: ... °C".

5.4.1.2.3.2   When for certain self-reactive substances of Class 4.1 and certain organic peroxides of Class 5.2 the competent authority has permitted the label conforming to model No. 1 to be dispensed with for a specific packaging (see 5.2.2.1.9), a statement to this effect shall be included in the transport document, as follows: **"The label conforming to model No. 1 is not required"**.

5.4.1.2.3.3   When organic peroxides and self-reactive substances are carried under conditions where approval is required (for organic peroxides see 2.2.52.1.8, 4.1.7.2.2 and special provision TA2 of 6.8.4 of ADR; for self-reactive substances see 2.2.41.1.13 and 4.1.7.2.2 of ADR, a statement to this effect shall be included in the transport document, e.g. **"Carriage in accordance with 2.2.52.1.8"**. It shall be in an official language of the forwarding country and also, if that language is not English, French or German, in English, French or German unless agreements, if any, concluded between the countries concerned in the transport operation provide otherwise.

A copy of the approval of the competent authority with the conditions of carriage shall be attached to the transport document.

5.4.1.2.3.4   When a sample of an organic peroxide (see 2.2.52.1.9) or a self-reactive substance (see 2.2.41.1.15) is carried, a statement to this effect shall be included in the transport document, e.g. **"Carriage in accordance with 2.2.52.1.9"**.

5.4.1.2.3.5   When self-reactive substances type G (see Manual of Tests and Criteria, Part II, paragraph 20.4.2 (g)) are carried, the following statement may be given in the transport document: **"Not a self-reactive substance of Class 4.1"**.

When organic peroxides type G (see Manual of Tests and Criteria, Part II, paragraph 20.4.3 (g)) are carried, the following statement may be given in the transport document: **"Not a substance of Class 5.2"**.

5.4.1.2.4   *Additional provisions for Class 6.2*

In addition to the information concerning the consignee (see 5.4.1.1.1 (h)), the name and telephone number of a responsible person shall be indicated.

5.4.1.2.5   *Additional provisions for Class 7*

5.4.1.2.5.1   The following information shall be inserted in the transport document for each consignment of Class 7 material, as applicable, in the order given and immediately after the information required under 5.4.1.1.1 (a) to (c):

(a)   The name or symbol of each radionuclide or, for mixtures of radionuclides, an appropriate general description or a list of the most restrictive nuclides;

(b)   A description of the physical and chemical form of the material, or a notation that the material is special form radioactive material or low dispersible radioactive material. A generic chemical description is acceptable for chemical form. For radioactive material with a subsidiary risk, see sub-paragraph (c) of special provision 172 of Chapter 3.3;

(c)   The maximum activity of the radioactive contents during carriage expressed in becquerels (Bq) with an appropriate SI prefix symbol (see 1.2.2.1). For fissile material, the mass of fissile material (or mass of each fissile nuclide for mixtures when appropriate) in grams (g), or appropriate multiples thereof, may be used in place of activity;

(d)   The category of the package, i.e. I-WHITE, II-YELLOW, III-YELLOW;

(e)   The transport index (categories II-YELLOW and III-YELLOW only);

(f)   For fissile material:

   (i)   Shipped under one exception of 2.2.7.2.3.5 (a) to (f), reference to that paragraph;

   (ii)   Shipped under 2.2.7.2.3.5 (c) to (e), the total mass of fissile nuclides;

   (iii) Contained in a package for which one of 6.4.11.2 (a) to (c) or 6.4.11.3 of ADR is applied, reference to that paragraph;

   (iv)   The criticality safety index, where applicable;

(g)   The identification mark for each competent authority certificate of approval (special form radioactive material, low dispersible radioactive material, fissile material excepted under 2.2.7.2.3.5 (f), special arrangement, package design, or shipment) applicable to the consignment;

(h)   For consignments of more than one package, the information required in 5.4.1.1.1 and in (a) to (g) above shall be given for each package. For packages in an overpack, container, or conveyance, a detailed statement of the contents of each package within the overpack, container, or conveyance and, where appropriate, of each overpack, container, or conveyance shall be included. If packages are to be removed from the overpack, container, or conveyance at a point of intermediate unloading, appropriate transport documents shall be made available;

(i)   Where a consignment is required to be shipped under exclusive use, the statement **"EXCLUSIVE USE SHIPMENT"**; and

(j)   For LSA-II and LSA-III substances, SCO-I and SCO-II, the total activity of the consignment as a multiple of A_2. For radioactive material for which the A_2 value is unlimited, the multiple of A_2 shall be zero.

5.4.1.2.5.2   The consignor shall provide in the transport documents a statement regarding actions, if any, that are required to be taken by the carrier. The statement shall be in the languages deemed necessary by the carrier or the authorities concerned, and shall include at least the following information:

(a)   Supplementary requirements for loading, stowage, carriage, handling and unloading of the package, overpack or container including any special stowage provisions for the safe dissipation of heat (see 7.1.4.14.7.3.2), or a statement that no such requirements are necessary;

(b)   Restrictions on the mode of carriage or vehicle or wagon and any necessary routeing instructions;

(c)   Emergency arrangements appropriate to the consignment.

5.4.1.2.5.3   In all cases of international carriage of packages requiring competent authority approval of design or shipment, for which different approval types apply in the different countries concerned by the shipment, the UN number and proper shipping name required in 5.4.1.1.1 shall be in accordance with the certificate of the country of origin of design.

5.4.1.2.5.4   The applicable competent authority certificates need not necessarily accompany the consignment. The consignor shall make them available to the carrier(s) before loading and unloading.

5.4.1.3   *(Reserved)*

5.4.1.4   *Format and language*

5.4.1.4.1   The document containing the information in 5.4.1.1 and 5.4.1.2 may be that already required by other regulations in force for carriage by another mode of carriage. In case of multiple consignees, the name and address of the consignees and the quantities delivered enabling the nature and quantities carried to be evaluated at any time, may be entered in other documents which are to be used or in any other documents made mandatory according to other specific regulations and which shall be on board.

The particulars to be entered in the document shall be drafted in an official language of the forwarding country, and also, if that language is not English, French or German, in English, French or German, unless agreements concluded between the countries concerned in the transport operation, provide otherwise.

5.4.1.4.2   If by reason of the size of the load, a consignment cannot be loaded in its entirety on a single transport unit, at least as many separate documents, or copies of the single document, shall be made out as transport units loaded. Furthermore, in all cases, separate transport documents shall be made out for consignments or parts of consignments which may not be loaded together on the same vehicle by reason of the prohibitions set forth in 7.5.2 of ADR.

The information relative to the hazards of the goods to be carried (as indicated in 5.4.1.1) may be incorporated in, or combined with, an existing transport or cargo handling document. The layout of the information in the document (or the order of transmission of the corresponding data by electronic data processing (EDP) or electronic data interchange (EDI) techniques) shall be as provided in 5.4.1.1.1 or 5.4.1.1.2 as relevant.

When an existing transport document or cargo handling document cannot be used for the purposes of dangerous goods documentation for multimodal transport, the use of documents corresponding to the example shown in 5.4.5 is considered advisable.[4]

5.4.1.5   *Non-dangerous goods*

When goods mentioned by name in Table A of Chapter 3.2, are not subject to ADN because they are considered as non-dangerous according to Part 2, the consignor may enter in the transport document a statement to that effect, e.g.: "**Not goods of Class ...**"

NOTE: This provision may be used in particular when the consignor considers that, due to the chemical nature of the goods (e.g. solutions and mixtures) carried or to the fact that such goods are deemed dangerous for other regulatory purposes the consignment might be subject to control during the journey.

5.4.2   **Container, vehicle or wagon packing certificate**

If the carriage of dangerous goods in a container precedes a voyage by sea, a container/vehicle packing certificate conforming to section 5.4.2 of the IMDG Code[5] shall be provided with the transport document.[6]

The functions of the transport document required under 5.4.1 and of the container/vehicle packing certificate as provided above may be incorporated into a single document; if not, these documents shall be attached one to the other. If these functions are incorporated into a

---

[4]   *If used, the relevant recommendations of the UNECE United Nations Centre for Trade Facilitation and Electronic Business (UN/CEFACT) may be consulted, in particular Recommendation No. 1 (United Nations Layout Key for Trade Documents) (ECE/TRADE/137, edition 81.3), UN Layout Key for Trade Documents - Guidelines for Applications (ECE/TRADE/270, edition 2002), Recommendation No. 11 (Documentary Aspects of the International Transport of Dangerous Goods) (ECE/TRADE/204, edition 96.1 – currently under revision) and Recommendation No. 22 (Layout Key for Standard Consignment Instructions) (ECE/TRADE/168, edition 1989). Refer also to the UN/CEFACT Summary of Trade Facilitation Recommendations (ECE/TRADE/346, edition 2006) and the United Nations Trade Data Elements Directory (UNTDED) (ECE/TRADE/362, edition 2005).*

[5]   *Guidelines for use in practice and in training for loading goods in transport units have also been drawn up by the International Maritime Organization (IMO), the International Labour Organization (ILO) and the United Nations Economic Commission for Europe (UNECE) and have been published by IMO ("IMO/ILO/UNECE Code of Practice for Packing of Cargo Transport Units (CTU Code)").*

[6]   *Section 5.4.2 of the IMDG Code (Amendment 38-16) requires the following:*

"5.4.2   Container/vehicle packing certificate

5.4.2.1   When dangerous goods are packed or loaded into any container or vehicle, those responsible for packing the container or vehicle shall provide a "container/vehicle packing certificate" specifying the container/vehicle identification number(s) and certifying that the operation has been carried out in accordance with the following conditions:

.1   The container/vehicle was clean, dry and apparently fit to receive the goods;

.2   Packages, which need to be segregated in accordance with applicable segregation requirements, have not been packed together onto or in the container/vehicle (unless approved by the competent authority concerned in accordance with 7.3.4.1 (of the IMDG Code));

.3   All packages have been externally inspected for damage, and only sound packages have been loaded;

.4   Drums have been stowed in an upright position, unless otherwise authorised by the competent authority, and all goods have been properly loaded, and, where necessary, adequately braced with securing material to suit the mode(s) of transport for the intended journey;

.5   Goods loaded in bulk have been evenly distributed within the container/vehicle;

.6   For consignments including goods of class 1, other than division 1.4, the container/vehicle is structurally serviceable in accordance with 7.1.2 (of the IMDG Code);

.7   The container/vehicle and packages are properly marked, labelled, and placarded, as appropriate;

Cont'd on the next page

single document, the inclusion in the transport document of a statement that the loading of the container or vehicle has been carried out in accordance with the applicable modal regulations together with the identification of the person responsible for the container/vehicle packing certificate shall be sufficient.

NOTE: The container/vehicle packing certificate is not required for portable tanks, tank-containers and MEGCs.

If the carriage of dangerous goods in a vehicle precedes a voyage by sea, a "container/vehicle packing certificate" conforming to section 5.4.2 of the IMDG Code[5,6] may be provided with the transport document.

5.4.3   Instructions in writing

5.4.3.1   As an aid during an accident emergency situation that may occur or arise during carriage, instructions in writing in the form specified in 5.4.3.4 shall be carried in the wheelhouse and shall be readily available.

5.4.3.2   These instructions shall be provided by the carrier to the master in the language(s) that the master and the expert can read and understand before loading. The master shall ensure that each member of the crew concerned understands and is capable of carrying out the instructions properly.

5.4.3.3   Before loading, the members of the crew shall inform themselves of the dangerous goods to be loaded and consult the instructions in writing for details on actions to be taken in the event of an accident or emergency.

5.4.3.4   The instructions in writing shall correspond to the following four-page model as regards its form and contents.

5.4.3.5   Contracting Parties shall provide the UNECE secretariat with the official translation of the instructions in writing in their national language(s), in accordance with this section. The UNECE secretariat shall make the national versions of the instructions in writing that it has received available to all Contracting Parties.

Footnote 6 (cont'd)

.8   *When substances presenting a risk of asphyxiation are used for cooling or conditioning purposes (such as dry ice (UN 1845) or nitrogen, refrigerated liquid (UN 1977) or argon, refrigerated liquid (UN 1951)), the container/vehicle is externally marked in accordance with 5.5.3.6 (of the IMDG Code); and*

.9   *A dangerous goods transport document, as indicated in 5.4.1 (of the IMDG Code) has been received for each dangerous goods consignment loaded in the container/vehicle.*

NOTE: The container/vehicle packing certificate is not required for portable tanks.

5.4.2.2   The information required in the dangerous goods transport document and the container/vehicle packing certificate may be incorporated into a single document; if not, these documents shall be attached one to the other. If the information is incorporated into a single document, the document shall a signed declaration such as "It is declared that the packing of the goods into the container/vehicle has been carried out in accordance with the applicable provisions". This declaration shall be dated and the person signing this declaration shall be identified on the document. Facsimile signatures are acceptable where applicable laws and regulations recognize the legal validity of facsimile signatures.

5.4.2.3   If the container/vehicle packing certificate is presented to the carrier by means of EDP or EDI transmission techniques, the signature(s) may be electronic signature(s) or may be replaced by the name(s) (in capitals) of the person authorized to sign.

5.4.2.4   When the container/vehicle packing certificate is given to a carrier by EDP or EDI techniques and subsequently the dangerous goods are transferred to a carrier that requires a paper container/vehicle packing certificate, the carrier shall ensure that the paper document indicates "Original received electronically" and the name of the signatory shall be shown in capital letters."

INSTRUCTIONS IN WRITING ACCORDING TO ADN
Actions in the event of an accident or incident

In the event of an accident or incident that may occur during carriage, the members of the crew shall take the following actions where safe and practicable to do so:

– Inform all other persons on board about the emergency and keep them away as much as possible from the danger zone. Alert other vessels in the vicinity;

– Avoid sources of ignition, in particular, do not smoke, use electronic cigarettes or similar devices or switch on or off any electrical equipment that is not the "certified safe" type and is not designed for use in emergency response;

– Inform the appropriate body, giving as much information about the accident or incident and substances involved as possible;

– Keep the transport documents and the loading plan readily available for responders on arrival;

– Do not walk into or touch spilled substances and avoid inhalation of fumes, smoke, dusts and vapours by staying up wind;

– Where appropriate and safe to do so, tackle small/initial fires;

– Where appropriate and safe to do so, use on-board equipment to prevent leakages into the aquatic environment and contain spillages;

– Where necessary and safe to do so, secure the ship against drifting;

– Where appropriate, move away from the vicinity of the accident or incident, advise other persons to move away and follow the advice of the appropriate body;

– Remove any contaminated clothing and used contaminated protective equipment, dispose of it safely and wash the body by appropriate means;

– Observe the additional guidance assigned to the hazards of all concerned goods in the following table. For carriage in packages or in bulk, the hazards correspond to the number of the danger label model; for carriage in tank vessels to the data in accordance with 5.4.1.1.2 (c).

Additional guidance to members of the crew on the hazard characteristics of dangerous goods by class and on actions to be taken subject to prevailing circumstances		
Danger labels and placards, description of the hazards	**Hazard characteristics**	**Additional guidance**
(1)	**(2)**	**(3)**
Explosive substances and articles 1 1.5 1.6	May have a range of properties and effects such as mass detonation; projection of fragments; intense fire/heat flux; formation of bright light, loud noise or smoke. Sensitive to shocks and/or impacts and/or heat.	Take cover but stay away from windows. Steer the vessel as far away as possible from infrastructure and inhabited areas.
Explosive substances and articles 1.4	Slight risk of explosion and fire.	Take cover.
Flammable gases 2.1	Risk of fire. Risk of explosion. May be under pressure. Risk of asphyxiation. May cause burns and/or frostbite. Containments may explode when heated.	Take cover. Keep out of low areas.
Non-flammable, non-toxic gases 2.2	Risk of asphyxiation. May be under pressure. May cause frostbite. Containments may explode when heated.	Take cover. Keep out of low areas.
Toxic gases 2.3	Risk of intoxication. May be under pressure. May cause burns and/or frostbite. Containments may explode when heated.	Use emergency escape mask. Take cover. Keep out of low areas.
Flammable liquids 3	Risk of fire. Risk of explosion. Containments may explode when heated.	Take cover. Keep out of low areas.
Flammable solids, self-reactive substances, polymerizing substances and solid desensitized explosives 4.1	Risk of fire. Flammable or combustible, may be ignited by heat, sparks or flames. May contain self-reactive substances that are liable to exothermic decomposition in the case of heat supply, contact with other substances (such as acids, heavy-metal compounds or amines), friction or shock. This may result in the evolution of harmful and flammable gases or vapours or self-ignition. Containments may explode when heated. Risk of explosion of desensitized explosives after loss of desensitizer.	
Substances liable to spontaneous combustion 4.2	Risk of fire by spontaneous combustion if packages are damaged or contents spilled. May react vigorously with water.	
Substances which, in contact with water, emit flammable gases 4.3	Risk of fire and explosion in contact with water.	Spilled substances should be kept dry by covering the spillages.

Danger labels and placards, description of the hazards	Hazard characteristics	Additional guidance
(1)	(2)	(3)
Oxidizing substances 5.1	Risk of vigorous reaction, ignition and explosion in contact with combustible or flammable substances.	Avoid mixing with flammable or combustible substances (e.g. sawdust).
Organic peroxides 5.2	Risk of exothermic decomposition at elevated temperatures, contact with other substances (such as acids, heavy-metal compounds or amines), friction or shock. This may result in the evolution of harmful and flammable gases or vapours or self-ignition.	Avoid mixing with flammable or combustible substances (e.g. sawdust).
Toxic substances 6.1	Risk of intoxication by inhalation, skin contact or ingestion. Risk to the aquatic environment.	Use emergency escape mask.
Infectious substances 6.2	Risk of infection. May cause serious disease in humans or animals Risk to the aquatic environment.	
Radioactive material 7A 7B 7C 7D	Risk of intake and external radiation.	Limit time of exposure.
Fissile material 7E	Risk of nuclear chain reaction.	
Corrosive substances 8	Risk of burns by corrosion. May react vigorously with each other, with water and with other substances. Spilled substance may evolve corrosive vapours. Risk to the aquatic environment.	
Miscellaneous dangerous substances and articles 9 9A	Risk of burns. Risk of fire. Risk of explosion. Risk to the aquatic environment	

NOTE: 1. *For dangerous goods with multiple risks and for mixed loads, each applicable entry shall be observed.*

2. *Additional guidance shown in column (3) of the table may be adapted to reflect the classes of dangerous goods to be carried and their means of transport.*

3. *Risks see also entries in the transport document as well as Chapter 3.2, Table C, Column (5).*

Additional guidance to members of the crew on the hazard characteristics of dangerous goods, indicated by marks, and on actions to be taken subject to prevailing circumstances		
Mark	Hazard characteristics	Additional guidance
(1)	(2)	(3)
Environmentally hazardous substances	Risk to the aquatic environment.	
Elevated temperature substances	Risk of burns by heat.	Avoid contact with hot parts of the transport unit and the spilled substance.

Equipment for personal and general protection to carry out general actions and hazard specific emergency actions to be carried on board the vessel in accordance with section 8.1.5 of ADN

The equipment required by Chapter 3.2, Table A, Column (9) and Table C, Column (18) shall be carried on board the vessel for all hazards listed in the transport document.

5.4.4 Retention of dangerous goods transport information

5.4.4.1 The consignor and the carrier shall retain a copy of the dangerous goods transport document and additional information and documentation as specified in ADN, for a minimum period of three months.

5.4.4.2 When the documents are kept electronically or in a computer system, the consignor and the carrier shall be able to reproduce them in a printed form.

5.4.5. Example of a multimodal dangerous goods form

Example of a form which may be used as a combined dangerous goods declaration and container packing certificate for multimodal carriage of dangerous goods.

MULTIMODAL DANGEROUS GOODS FORM

1. Shipper/Consignor/Sender	2. Transport document number
	3. Page 1 of Pages	**4.** Shipper's reference
		5. Freight Forwarder's reference
6. Consignee	7. Carrier (to be completed by the carrier)

SHIPPER'S DECLARATION

I hereby declare that the contents of this consignment are fully and accurately described below by the proper shipping name, and are classified, packaged, marked and labelled/placarded and are in all respects in proper condition for transport according to the applicable international and national governmental regulations.

| 8. This shipment is within the limitations prescribed for: *(Delete non-applicable)* | 9. Additional handling information |
| PASSENGER AND   CARGO AIRCRAFT ONLY
CARGO AIRCRAFT | |

| 10. Vessel/ flight No. and date | 11. Port/place of loading |
| 12. Port/place of discharge | 13. Destination |

14. Shipping marks	* Number and kind of packages; description of goods	Gross mass (kg)	Net mass	Cube (m³)

15. Container identification No./ vehicle registration No.	16. Seal number (s)	17. Container/vehicle size & type	18. Tare (kg)	19. Total gross mass (including tare) (kg)

CONTAINER/VEHICLE PACKING CERTIFICATE	**21.RECEIVING ORGANIZATION RECEIPT**	
I hereby declare that the goods described above have been packed/loaded into the container/vehicle identified above in accordance with the applicable provisions ** **MUST BE COMPLETED AND SIGNED FOR ALL CONTAINER/VEHICLE LOADS BY PERSON RESPONSIBLE FOR PACKING/LOADING**	Received the above number of packages/containers/trailers in apparent good order and condition unless stated hereon: RECEIVING ORGANIZATION REMARKS:	
20. Name of company	Haulier's name	22. Name of company (OF SHIPPER PREPARING THIS NOTE)
Name/Status of declarant	Vehicle reg. No.	Name/Status of declarant
Place and date	Signature and date	Place and date
Signature of declarant	DRIVER'S SIGNATURE	Signature of declarant

** See 5.4.2.

Left margin: and any other element of information required under applicable national and international regulations

Right margin: BLACK HATCHINGS BLACK HATCHINGS BLACK HATCHINGS BLACK HATCHINGS BLACK HATCHINGS BLACK HATCHINGS BLACK HATCHINGS BLACK HATCHINGS BLACK HATCHINGS BLACK HATCHINGS BLACK HATCHINGS

MULTIMODAL DANGEROUS GOODS FORM
Continuation Sheet

1. Shipper/Consignor/Sender	2. Transport document number	
	3. Page 1 of pages	4. Shipper's reference
		5. Freight Forwarder's reference

14. Shipping marks	* Number and kind of packages; description of goods	Gross mass (kg)	Net mass	Cube (m³)

* FOR DANGEROUS GOODS: you must specify: proper shipping name, hazard class, UN No., packing group (where assigned) and any other element of information required under applicable national and international regulations

CHAPTER 5.5

SPECIAL PROVISIONS

5.5.1   *(Deleted)*

5.5.2   **Special provisions applicable to fumigated cargo transport units (UN 3359)**

5.5.2.1   *General*

5.5.2.1.1   Fumigated cargo transport units (UN 3359) containing no other dangerous goods are not subject to any provisions of ADN other than those of this section.

5.5.2.1.2   When the fumigated cargo transport unit is loaded with dangerous goods in addition to the fumigant, any provision of ADN relevant to these goods (including placarding, marking and documentation) applies in addition to the provisions of this section.

5.5.2.1.3   Only cargo transport units that can be closed in such a way that the escape of gas is reduced to a minimum shall be used for the carriage of cargo under fumigation.

5.5.2.2   *Training*

Persons engaged in the handling of fumigated cargo transport units shall be trained commensurate with their responsibilities.

5.5.2.3   *Marking and placarding*

5.5.2.3.1   A fumigated cargo transport unit shall be marked with a warning mark, as specified in 5.5.2.3.2, affixed at each access point in a location where it will be easily seen by persons opening or entering the cargo transport unit. This mark shall remain on the cargo transport unit until the following provisions are met:

(a)   The fumigated cargo transport unit has been ventilated to remove harmful concentrations of fumigant gas; and

(b)   The fumigated goods or materials have been unloaded.

5.5.2.3.2   The fumigation warning mark shall be as shown in Figure 5.5.2.3.2.

Figure 5.5.2.3.2

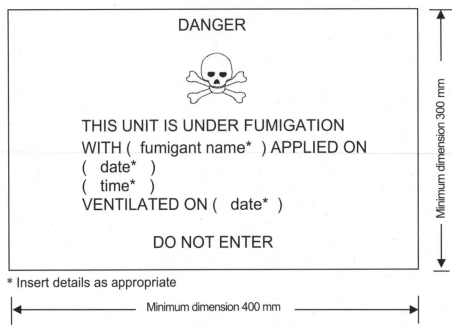

Fumigation warning mark

The mark shall be a rectangle. The minimum dimensions shall be 400 mm wide x 300 mm high and the minimum width of the outer line shall be 2 mm. The mark shall be in black print on a white background with lettering not less than 25 mm high. Where dimensions are not specified, all features shall be in approximate proportion to those shown.

5.5.2.3.3   If the fumigated cargo transport unit has been completely ventilated either by opening the doors of the unit or by mechanical ventilation after fumigation, the date of ventilation shall be marked on the fumigation warning mark.

5.5.2.3.4   When the fumigated cargo transport unit has been ventilated and unloaded, the fumigation warning mark shall be removed.

5.5.2.3.5   Placards conforming to model No. 9 (see 5.2.2.2.2) shall not be affixed to a fumigated cargo transport unit except as required for other Class 9 substances or articles packed therein.

5.5.2.4   *Documentation*

5.5.2.4.1   Documents associated with the carriage of cargo transport units that have been fumigated and have not been completely ventilated before carriage shall include the following information:

–   "UN 3359, fumigated cargo transport unit, 9", or "UN 3359, fumigated cargo transport unit, class 9";

–   The date and time of fumigation; and

–   The type and amount of the fumigant used.

These particulars shall be drafted in an official language of the forwarding country and also, if the language is not English, French or German, in English, French or German, unless agreements, if any, concluded between the countries concerned in the transport operation provide otherwise.

5.5.2.4.2	The documents may be in any form, provided they contain the information required in 5.5.2.4.1. This information shall be easy to identify, legible and durable.

5.5.2.4.3 Instructions for disposal of any residual fumigant including fumigation devices (if used) shall be provided.

5.5.2.4.4 A document is not required when the fumigated cargo transport unit has been completely ventilated and the date of ventilation has been marked on the warning mark (see 5.5.2.3.3 and 5.5.2.3.4).

5.5.3 Special provisions applicable to packages and vehicles and containers containing substances presenting a risk of asphyxiation when used for cooling or conditioning purposes (such as dry ice (UN 1845) or nitrogen, refrigerated liquid (UN 1977) or argon, refrigerated liquid (UN 1951))

5.5.3.1 *Scope*

5.5.3.1.1 This section is not applicable to substances which may be used for cooling or conditioning purposes when carried as a consignment of dangerous goods, except for the carriage of dry ice (UN No. 1845). When they are carried as a consignment, these substances shall be carried under the relevant entry of Table A of Chapter 3.2 in accordance with the associated conditions of carriage

For UN No. 1845, the conditions of carriage specified in this section, except 5.5.3.3.1, apply for all kinds of carriage, as a coolant, conditioner, or as a consignment. For the carriage of UN No. 1845, no other provisions of ADN apply.

5.5.3.1.2 This section is not applicable to gases in cooling cycles.

5.5.3.1.3 Dangerous goods used for cooling or conditioning tanks or MEGCs during carriage are not subject to this section.

5.5.3.1.4 Vehicles, wagons and containers containing substances used for cooling or conditioning purposes include vehicles, wagons and containers containing substances used for cooling or conditioning purposes inside packages as well as vehicles, wagons and containers with unpackaged substances used for cooling or conditioning purposes.

5.5.3.1.5 Sub-sections 5.5.3.6 and 5.5.3.7 only apply when there is an actual risk of asphyxiation in the vehicle, wagon or container. It is for the participants concerned to assess this risk, taking into consideration the hazards presented by the substances being used for cooling or conditioning, the amount of substance to be carried, the duration of the journey, the types of containment to be used and the gas concentration limits given in the note to 5.5.3.3.3.

5.5.3.2 *General*

5.5.3.2.1 Vehicles, wagons and containers containing substances used for cooling or conditioning purposes (other than fumigation) during carriage are not subject to any provisions of ADN other than those of this section.

5.5.3.2.2 When dangerous goods are loaded in vehicles, wagons or containers containing substances used for cooling or conditioning purposes any provisions of ADN relevant to these dangerous goods apply in addition to the provisions of this section.

5.5.3.2.3 *(Reserved)*

5.5.3.2.4 Persons engaged in the handling or carriage of vehicles, wagons and containers containing substances used for cooling or conditioning purposes shall be trained commensurate with their responsibilities.

5.5.3.3   *Packages containing a coolant or conditioner*

5.5.3.3.1   Packaged dangerous goods requiring cooling or conditioning assigned to packing instructions P203, P620, P650, P800, P901 or P904 of 4.1.4.1 of ADR shall meet the appropriate requirements of that packing instruction.

5.5.3.3.2   For packaged dangerous goods requiring cooling or conditioning assigned to other packing instructions, the packages shall be capable of withstanding very low temperatures and shall not be affected or significantly weakened by the coolant or conditioner. Packages shall be designed and constructed to permit the release of gas to prevent a build-up of pressure that could rupture the packaging. The dangerous goods shall be packed in such a way as to prevent movement after the dissipation of any coolant or conditioner.

5.5.3.3.3   Packages containing a coolant or conditioner shall be carried in well ventilated vehicles, wagons and containers. Marking according to 5.5.3.6 is not required in this case.

Ventilation is not required, and marking according to 5.5.3.6 is required, if:

–   the load compartment is insulated, refrigerated or mechanically refrigerated equipment, for example as defined in the Agreement on the International Carriage of Perishable Foodstuffs and on the Special Equipment to be Used for such Carriage (ATP) and separated from the driver's cab;

–   for vehicles, gas exchange between the load compartment and the driver's cab is prevented.

NOTE: In this context "well ventilated" means there is an atmosphere where the carbon dioxide concentration is below 0.5% by volume and the oxygen concentration is above 19.5% by volume.

5.5.3.4   *Marking of packages containing a coolant or conditioner*

5.5.3.4.1   Packages containing dangerous goods used for cooling or conditioning shall be marked with the name indicated in Column (2) of Table A of Chapter 3.2 of these dangerous goods followed by the words "AS COOLANT" or "AS CONDITIONER" as appropriate in an official language of the country of origin and also, if that language is not English, French or German, in English, French or German, unless agreements concluded between the countries concerned in the transport operation provide otherwise.

5.5.3.4.2   The marks shall be durable, legible and placed in such a location and of such a size relative to the package as to be readily visible.

5.5.3.5   *Vehicles, wagons and containers containing unpackaged dry ice*

5.5.3.5.1   If dry ice in unpackaged form is used, it shall not come into direct contact with the metal structure of a vehicle, wagon or container to avoid embrittlement of the metal. Measures shall be taken to provide adequate insulation between the dry ice and the vehicle, wagon or container by providing a minimum of 30 mm separation (e.g. by using suitable low heat conducting materials such as timber planks, pallets etc).

5.5.3.5.2   Where dry ice is placed around packages, measures shall be taken to ensure that packages remain in the original position during carriage after the dry ice has dissipated.

5.5.3.6   *Marking of vehicles, wagons and containers*

5.5.3.6.1   Vehicles, wagons and containers containing dangerous goods used for cooling or conditioning purposes that are not well ventilated shall be marked with a warning mark, as specified in 5.5.3.6.2, affixed at each access point in a location where it will be easily seen by persons opening or entering the vehicle, wagon or container. This mark shall remain on the vehicle, wagon or container until the following provisions are met:

(a)   The vehicle, wagon or container has been well ventilated to remove harmful concentrations of coolant or conditioner; and

(b)   The cooled or conditioned goods have been unloaded.

As long as the vehicle, wagon or container is marked, the necessary precautions have to be taken before entering it. The necessity of ventilating through the cargo doors or other means (e.g. forced ventilation) has to be evaluated and included in training of the involved persons.

5.5.3.6.2   The warning mark shall be as shown in Figure 5.5.3.6.2.

Figure 5.5.3.6.2

Coolant/conditioning warning mark for vehicles, wagons and containers

*   *Insert the name indicated in column (2) of Table A of Chapter 3.2 of the coolant/conditioner. The lettering shall be in capitals, all be on one line and shall be at least 25 mm high. If the length of the proper shipping name is too long to fit in the space provided, the lettering may be reduced to the maximum size possible to fit. For example: "CARBON DIOXIDE, SOLID".*

**   *Insert "AS COOLANT" or "AS CONDITIONER" as appropriate. The lettering shall be in capitals, all be on one line and be at least 25 mm high.*

The mark shall be a rectangle. The minimum dimensions shall be 150 mm wide x 250 mm high. The word "WARNING" shall be in red or white and be at least 25 mm high. Where dimensions are not specified, all features shall be in approximate proportion to those shown.

The word "WARNING" and the words "AS COOLANT" or "AS CONDITIONER", as appropriate, shall be in an official language of the country of origin and also, if that language is not English, French or German, in English, French or German, unless agreements concluded between the countries concerned in the transport operation provide otherwise.

5.5.3.7   *Documentation*

5.5.3.7.1   Documents (such as a bill of lading, cargo manifest or CMR/CIM/CMNI consignment note) associated with the carriage of vehicles, wagons or containers containing or having contained substances used for cooling or conditioning purposes and that have not been completely ventilated before carriage shall include the following information:

(a)   The UN number preceded by the letters "UN"; and

(b)   The name indicated in Column (2) of Table A of Chapter 3.2 followed by the words "AS COOLANT" or "AS CONDITIONER" as appropriate in an official language of the country of origin and also, if that language is not English, French or German, in English, French or German, unless agreements, if any, concluded between the countries concerned in the transport operation provide otherwise.

For example: UN 1845, CARBON DIOXIDE, SOLID, AS COOLANT.

5.5.3.7.2   The transport document may be in any form, provided it contains the information required in 5.5.3.7.1. This information shall be easy to identify, legible and durable.

PART 6

Requirements for the construction and testing of packagings (including IBCs and large packagings), tanks and bulk cargo transport units

CHAPTER 6.1

GENERAL REQUIREMENTS

6.1.1   Packagings (including IBCs and large packagings) and tanks shall meet the following

requirements of ADR in respect of construction and testing:

Chapter 6.1:   Requirements for the construction and testing of packagings;

Chapter 6.2:   Requirements for the construction and testing of pressure receptacles, aerosol dispensers, small receptacles containing gas (gas cartridges) and fuel cell cartridges containing liquefied flammable gas;

Chapter 6.3:   Requirements for the construction and testing of packagings for Class 6.2 infectious substances of category A;

Chapter 6.4:   Requirements for the construction, testing and approval of packages and material of Class 7;

Chapter 6.5   Requirements for the construction and testing of intermediate bulk containers (IBCs);

Chapter 6.6   Requirements for the construction and testing of large packagings;

Chapter 6.7   Requirements for the design, construction, inspection and testing of portable tanks and UN multiple-element gas containers (MEGCs);

Chapter 6.8   Requirements for the construction, equipment, type approval, inspections and tests, and marking of fixed tanks (tank-vehicles), demountable tanks and tank-containers and tank swap bodies, with shell made of metallic materials and battery-vehicles and multiple element gas containers (MEGCs);

Chapter 6.9   Requirements for the design, construction, equipment, type approval, testing and marking of fibre-reinforced plastics (FRP) fixed tanks (tank-vehicles), demountable tanks, tank-containers and tank swap bodies;

Chapter 6.10   Requirements for the construction, equipment, type approval, inspection and marking of vacuum-operated waste tanks;

Chapter 6.11   Requirements for the design, construction, inspection and testing of bulk containers;

Chapter 6.12   Requirements for the construction, equipment, type approval, inspections and tests, and marking of tanks, bulk containers and special compartments for explosives of mobile explosive manufacturing units (MEMUs).

6.1.2   Portable tanks may also meet the requirements of Chapter 6.7 or, if appropriate, Chapter 6.9 of the IMDG Code.

6.1.3   Tank-vehicles may also meet the requirements of Chapter 6.8 of the IMDG Code.

6.1.4   Tank wagons, with fixed or demountable tanks and battery-wagons shall meet the requirements of Chapter 6.8 of the RID.

6.1.5   Bodies of vehicles for bulk carriage shall, if necessary, meet the requirements of Chapter 6.11 or of Chapter 9.5 of ADR.

6.1.6   When the provisions of 7.3.1.1 (a) of RID or ADR are applied, the bulk containers shall meet the requirements of Chapter 6.11 of RID or ADR.

PART 7

Requirements concerning loading, carriage, unloading and handling of cargo

CHAPTER 7.1

DRY CARGO VESSELS

7.1.0   **General requirements**

7.1.0.1   The provisions of 7.1.0 to 7.1.6 are applicable to dry cargo vessels.

7.1.0.2 to 7.1.0.99   *(Reserved)*

7.1.1   **Mode of carriage of goods**

7.1.1.1 to 7.1.1.9   *(Reserved)*

7.1.1.10   *Carriage of packages*

Unless otherwise specified, the masses given for packages shall be the gross masses. When packages are carried in containers or vehicles, the mass of the container or vehicle shall not be included in the gross mass of such packages.

7.1.1.11   *Carriage in bulk*

Carriage of dangerous goods in bulk shall be prohibited except where this mode of carriage is explicitly authorized in column (8) of Table A of Chapter 3.2. The code "B" shall then appear in this column.

7.1.1.12   *Ventilation*

The ventilation of holds is required only if it is prescribed in 7.1.4.12 or by an additional requirement "VE …" in column (10) of Table A of Chapter 3.2.

7.1.1.13   *Measures to be taken prior to loading*

Additional measures to be taken prior to loading are required only if prescribed in 7.1.4.13 or by an additional requirement "LO …" in column (11) of Table A of Chapter 3.2.

7.1.1.14   *Handling and stowage of cargo*

During the handling and stowage of cargo additional measures are required only if prescribed in 7.1.4.14 or by an additional requirement "HA …" in column (11) of Table A of Chapter 3.2.

7.1.1.15   *(Reserved)*

7.1.1.16   *Measures to be taken during loading, carriage, unloading and handling of cargo*

The additional measures to be taken during loading, carriage, unloading and handling of cargo are required only if prescribed in 7.1.4.16 or by an additional requirement "IN …" in column (11) of Table A of Chapter 3.2.

7.1.1.17   *(Reserved)*

7.1.1.18 *Carriage in containers, in bulk containers, in intermediate bulk containers (IBCs) and in large packagings, in MEGCs, in portable tanks and in tank-containers*

The carriage of containers, bulk containers, IBCs, large packagings, MEGCs, portable tanks and tank-containers shall be in accordance with the provisions applicable to the carriage of packages.

7.1.1.19 *Vehicles and wagons*

The carriage of vehicles and wagons shall be in accordance with the provisions applicable to the carriage of packages.

7.1.1.20 *(Reserved)*

7.1.1.21 *Carriage in cargo tanks*

The carriage of dangerous goods in cargo tanks in dry-cargo vessels is prohibited.

7.1.1.22 to 7.1.1.99 *(Reserved)*

7.1.2 **Requirements applicable to vessels**

7.1.2.0 *Permitted vessels*

7.1.2.0.1 Dangerous goods may be carried in quantities not exceeding those indicated in 7.1.4.1.1, or, if applicable, in 7.1.4.1.2:

– In dry cargo vessels conforming to the applicable construction requirements of 9.1.0.0 to 9.1.0.79; or

– In seagoing vessels conforming to the applicable construction requirements of 9.1.0.0 to 9.1.0.79, or otherwise to the requirements of 9.2.0 to 9.2.0.79.

7.1.2.0.2 Dangerous goods of classes 2, 3, 4.1, 4.2, 4.3, 5.1, 5.2, 6.1, 7, 8 or 9, with the exception of those for which a No. 1 model label is required in column (5) of table A of Chapter 3.2, may be carried in quantities greater than those indicated in 7.1.4.1.1 and 7.1.4.1.2:

– In double-hull dry cargo vessels conforming to the applicable construction requirements of 9.1.0.80 to 9.1.0.95; or

– In double-hull seagoing vessels conforming to the applicable construction requirements of 9.1.0.80 to 9.1.0.95, or otherwise to the requirements of 9.2.0 to 9.2.0.95.

7.1.2.1 to 7.1.2.4 *(Reserved)*

7.1.2.5 *Instructions for the use of devices and installations*

Where specific safety rules have to be complied with when using any device or installation, instructions for the use of the particular device or installation shall be readily available for consultation at appropriate places on board in the language normally spoken on board and also if that language is not English, French or German, in English, French or German unless agreements concluded between the countries concerned in the transport operation provide otherwise.

7.1.2.6 to 7.1.2.18 *(Reserved)*

7.1.2.19   *Pushed convoys and side-by-side formations*

7.1.2.19.1   Where at least one vessel of a convoy or side-by-side formation is required to be in possession of a certificate of approval for the carriage of dangerous goods, all vessels of such convoy or side-by-side formation shall be provided with an appropriate certificate of approval.

Vessels not carrying dangerous goods shall comply with the requirements of the following paragraphs:

1.16.1.1, 1.16.1.2, 1.16.1.3, 7.1.2.5, 8.1.5, 8.1.6.1, 8.1.6.3, 8.1.7, 9.1.0.0, 9.1.0.12.3, 9.1.0.17.2, 9.1.0.17.3, 9.1.0.31, 9.1.0.32, 9.1.0.34, 9.1.0.41, 9.1.0.52.2, 9.1.0.52.3, 9.1.0.56, 9.1.0.71 and 9.1.0.74.

7.1.2.19.2   For the purposes of the application of the provisions of this Chapter with the exception of 7.1.4.1.1 and 7.1.4.1.2, the entire pushed convoy or the side-by-side formation shall be deemed to be a single vessel.

7.1.2.20 to 7.1.2.99   *(Reserved)*

7.1.3   **General service requirements**

7.1.3.1   *Access to holds, double-hull spaces and double bottoms; inspections*

7.1.3.1.1   Access to the holds is not permitted except for the purpose of loading or unloading and carrying out inspections or cleaning work.

7.1.3.1.2   Access to the double-hull spaces and the double bottoms is not permitted while the vessel is under way.

7.1.3.1.3   If the concentration of gases or the oxygen content of the air in holds, double-wall spaces or double bottoms has to be measured before entry the results of these measurements shall be recorded in writing. The measurement may only be effected by persons equipped with suitable breathing apparatus for the substance carried.

Entry into the spaces is not permitted for the purpose of measuring.

7.1.3.1.4   In case of suspected damage to packages, the gas concentration in holds containing dangerous goods of Classes 2, 3, 5.2, 6.1 and 8 for which EX and/or TOX appears in column (9) of Table A of Chapter 3.2, shall be measured before any person enters these holds.

7.1.3.1.5   The gas concentration in holds and in adjacent holds containing dangerous goods carried in bulk or without packaging for which EX and/or TOX appears in column (9) of Table A of Chapter 3.2, shall be measured before any person enters these holds.

7.1.3.1.6   Entry into holds where damage is suspected to packages in which dangerous goods of Classes 2, 3, 5.2, 6.1 and 8 are carried as well as entry into double-hull spaces and double bottoms is not permitted except where:

–   there is no lack of oxygen and no measurable amount of dangerous substances in a dangerous concentration; or

–   the person entering the space wears a self-contained breathing apparatus and other necessary protective and rescue equipment and is secured by a line. Entry into these spaces is only permitted if this operation is supervised by a second person for whom the same equipment is readily at hand. Another two persons capable of giving assistance in an emergency shall be on the vessel within calling distance.

7.1.3.1.7   Entry into holds where dangerous goods are carried in bulk or without packaging as well as entry into double-hull spaces and double bottoms is not permitted except where:

–   there is no lack of oxygen and no measurable amount of dangerous substances in a dangerous concentration; or

–   the person entering the space wears a self-contained breathing apparatus and other necessary protective and rescue equipment and is secured by a line. Entry into these spaces is only permitted if this operation is supervised by a second person for whom the same equipment is readily at hand. Another two persons capable of giving assistance in an emergency shall be on the vessel within calling distance.

7.1.3.2 to 7.1.3.14   *(Reserved)*

7.1.3.15   ***Expert on board the vessel***

When dangerous goods are carried, the responsible master shall at the same time be an expert according to 8.2.1.2.

NOTE: Which master of the vessel's crew is the responsible master shall be determined and documented on board by the carrier. If there is no such determination, the requirement applies to every master.

By derogation from this, for the loading and unloading of dangerous goods in a barge, it is sufficient that the person who is responsible for loading and unloading and for ballasting of the barge has the expertise required according to 8.2.1.2.

7.1.3.16 to 7.1.3.19   *(Reserved)*

7.1.3.20   ***Water ballast***

Double-hull spaces and double bottoms may be used for water ballast.

7.1.3.21   *(Reserved)*

7.1.3.22   ***Opening of holds***

7.1.3.22.1   Dangerous goods shall be protected against the influences of weather and against spray water except during loading and unloading or during inspection.

This provision does not apply when dangerous goods are loaded in sprayproof containers, IBCs, or large packagings, or in MEGCs, portable tanks, tank-containers, vehicles or wagons which are closed or sheeted.

7.1.3.22.2   Where dangerous goods are carried in bulk, the holds shall be covered with hatch covers.

7.1.3.23 to 7.1.3.30   *(Reserved)*

7.1.3.31   ***Engines***

The use of engines running on fuels having a flash-point below 55° C (e.g. petrol engines) is prohibited.

This requirement does not apply to the petrol-operated outboard motors of lifeboats.

7.1.3.32   *Oil fuel tanks*

Double bottoms with a height of at least 0.6 m may be used as oil fuel tanks provided that they have been constructed in accordance with Chapters 9.1 or 9.2.

7.1.3.33 to 7.4.3.40   *(Reserved)*

7.1.3.41   *Fire and naked light*

7.1.3.41.1   The use of fire or naked light is prohibited.

This provision does not apply to the accommodation and the wheelhouse.

7.1.3.41.2   Heating, cooking and refrigerating appliances shall not be fuelled with liquid fuels, liquid gas or solid fuels.

Cooking and refrigerating appliances may only be used in the accommodation and in the wheelhouse.

7.1.3.41.3   Heating appliances or boilers fuelled with liquid fuels having a flash-point above 55° C which are installed in the engine room or in another suitable space may, however, be used.

7.1.3.42   *Heating of holds*

The heating of holds or the operation of a heating system in the holds is prohibited.

7.1.3.43   *(Reserved)*

7.1.3.44   *Cleaning operations*

The use of liquids having a flash-point below 55° C for cleaning purposes is prohibited.

7.1.3.45 to 7.1.3.50   *(Reserved)*

7.1.3.51   *Electrical installations*

7.1.3.51.1   The electrical installations shall be properly maintained.

7.1.3.51.2   The use of movable electric cables is prohibited in the protected area. This provision does not apply to:

– intrinsically safe electric circuits;

– electric cables for connecting signal lights or gangway lighting, provided the socket is permanently fitted to the vessel close to the signal mast or gangway;

– electric cables for connecting containers;

– electric cables for electrically operated hatch cover gantries;

– electric cables for connecting submerged pumps;

– electric cables for connecting hold ventilators.

7.1.3.51.3   The sockets for connecting the signal lights and gangway lighting and for connecting containers, submerged pumps, hatch cover gantries, or hold fans shall not be live except when the signal lights or the gangway lighting are switched on or when the containers or the submerged pumps or the hatch cover gantries or hold fans are in operation. In the protected area, connecting or disconnecting shall not be possible except when the sockets are not live.

7.1.3.51.4   The electrical installations in the holds shall be kept switched off and protected against unintentional connection.

This provision does not apply to permanently installed cables passing through the holds, to movable cables connecting containers, stowed according to 7.1.4.4.4, and to apparatus of a certified safe type.

7.1.3.52 to 7.1.3.69   *(Reserved)*

7.1.3.70   *Aerials, lightning conductors, wire cables and masts*

7.1.3.70.1   No part of an aerial for electronic apparatus, no lightning conductor and no wire cable shall be situated above the holds.

7.1.3.70.2   No part of aerials for radiotelephones shall be located within 2.00 m from substances or articles of Class 1.

7.1.3.71 to 7.1.3.99   *(Reserved)*

7.1.4   Additional requirements concerning loading, carriage, unloading and other handling of the cargo

7.1.4.1   *Limitation of the quantities carried*

7.1.4.1.1   Subject to 7.1.4.1.3, the following gross masses shall not be exceeded on any vessel. For pushed convoys and side-by-side formations this gross mass applies to each unit of the convoy or formation.

Class 1	
All substances and articles of Division 1.1 of compatibility group A	90 kg[1]
All substances and articles of Division 1.1 of compatibility groups B, C, D. E, F, G, J or L	15,000 kg[2]
All substances and articles of Division 1.2 of compatibility groups B, C, D, E, F, G, H, J or L	50,000 kg
All substances and articles of Division 1.3 of compatibility groups C, G, H, J or L	300,000 kg[3]
All substances and articles of Division 1.4 of compatibility groups B, C, D, E, F, G or S	1,100,000 kg
All substances of Division 1.5 of compatibility group D	15,000 kg[2]
All articles of Division 1.6 of compatibility group N	300,000 kg[3]
Empty packagings, uncleaned	1,100,000 kg

Note:

   [1]   *In not less than three batches of a maximum of 30 kg each, distance between batches not less than 10.00 m.*

   [2]   *In not less than three batches of a maximum of 5 000 kg each, distance between batches not less than 10.00 m.*

   [3]   *Not more than 100,000 kg per hold. A wooden partition is permitted for subdividing a hold.*

Class 2

All goods for which label No. 2.1 is required in column (5) of Table A of Chapter 3.2: total	300 000 kg
All goods for which label No. 2.3 is required in column (5) of Table A of Chapter 3.2: total	120 000 kg
Other goods	No limitation

Class 3

All goods for which label No. 6.1 is required in column (5) of Table A of Chapter 3.2: total	120 000 kg
Other goods: total	300 000 kg

Class 4.1

UN Nos. 3221, 3222, 3231 and 3232, total	15 000 kg
All goods of packing group I; all goods of packing group II for which label No. 6.1 is required in column (5) of Table A of Chapter 3.2; self-reactive substances of types C, D, E and F (UN Nos. 3223 to 3230 and 3233 to 3240); other substances of classification code SR1 or SR2 (UN Nos. 2956, 3241, 3242 and 3251); and desensitized explosive substances of packing group II (UN Nos. 2907, 3319 and 3344): total	120 000 kg
Other goods	No limitation

Class 4.2

All goods of packing groups I or II for which label No. 6.1 is required in column (5) of Table A of Chapter 3.2: total	300 000 kg
Other goods	No limitation

Class 4.3

All goods of packing groups I or II for which label No. 3, 4.1 or 6.1 is required in column (5) of Table A of Chapter 3.2: total	300 000 kg
Other goods	No limitation

Class 5.1

All goods of packing groups I or II for which label No. 6.1 is required in column (5) of Table A of Chapter 3.2: total	300 000 kg
Other goods	No limitation

Class 5.2

UN Nos. 3101, 3102, 3111 and 3112: total	15 000 kg
Other goods: total	120 000 kg

Class 6.1

All goods of packing group I: total	120 000 kg
All goods of packing group II: total	300 000 kg
All goods carried in bulk	0 kg
Other goods	No limitation

Class 7

UN Nos. 2912, 2913, 2915, 2916, 2917, 2919, 2977, 2978 and 3321 to 3333	0 kg
Other goods	No limitation

Class 8

All goods of packing group I; goods of packing group II for which
label No. 3 or 6.1 is required in column (5) of Table A in Chapter 3.2:
total 　　　　　　　　　　　　　　　　　　　　　　　　　　　300 000 kg

Other goods 　　　　　　　　　　　　　　　　　　　　　　　No limitation

Class 9

All goods of packing group II: total 　　　　　　　　　　　　　300 000 kg

UN No. 3077, for goods carried in bulk and classified as hazardous to
the aquatic environment, categories Acute 1 or Chronic 1, in 　　　0 kg
accordance with 2.4.3:

Other goods 　　　　　　　　　　　　　　　　　　　　　　　No limitation

7.1.4.1.2	Subject to 7.1.4.1.3, the maximum quantity of dangerous goods permitted on board a vessel or on board each unit of a pushed convoy or side-by-side formation is 1,100,000 kg.
7.1.4.1.3	The limitations of 7.1.4.1.1 and 7.1.4.1.2 shall not apply in the case of transport of dangerous goods of classes 2, 3, 4.1, 4.2, 4.3, 5.1, 5.2, 6.1, 7, 8 and 9, except of those for which a label of Model No 1 is required in column (5) of Table A of Chapter 3.2, on board double-hull vessels complying with the additional requirements of 9.1.0.88 to 9.1.0.95 or 9.2.0.88 to 9.2.0.95.
7.1.4.1.4	Where substances and articles of different divisions of Class 1 are loaded in a single vessel in conformity with the provisions for prohibition of mixed loading of 7.1.4.3.3 or 7.1.4.3.4, the entire load shall not exceed the smallest maximum net mass given in 7.1.4.1.1 above for the goods of the most dangerous division loaded, the order of precedence being 1.1, 1.5, 1.2, 1.3, 1.6, 1.4.
7.1.4.1.5	Where the total net mass of the explosive substances carried and of explosive substances contained in articles carried is not known, the gross mass of the cargo shall apply to the mass mentioned in the table in 7.1.4.1.1 above.
7.1.4.1.6	For activity limits, transport index (TI) limits and criticality safety indices (CSI) in the case of the carriage of radioactive material, see 7.1.4.14.7.

7.1.4.2 **_Prohibition of mixed loading (bulk)_**

Vessels carrying substances of Class 5.1 in bulk shall not carry any other goods.

7.1.4.3 **_Prohibition of mixed loading (packages in holds)_**

7.1.4.3.1	Goods of different classes shall be separated by a minimum horizontal distance of 3.00 m. They shall not be stowed one on top of the other.
7.1.4.3.2	Irrespective of the quantity, dangerous goods for which marking with two blue cones or two blue lights is prescribed in column (12) of Table A of Chapter 3.2 shall not be stowed in the same hold together with flammable goods for which marking with one blue cone or one blue light is prescribed in column (12) of Table A of Chapter 3.2.
7.1.4.3.3	Packages containing substances or articles of Class 1 and packages containing substances of Classes 4.1 or 5.2 for which marking with three blue cones or three blue lights is prescribed in column (12) of Table A of Chapter 3.2 shall be separated by a distance of not less than 12 m from goods of all other classes.

7.1.4.3.4 Substances and articles of Class 1 shall not be stowed in the same hold, except as indicated in the following table:

Compatibility group	A	B	C	D	E	F	G	H	J	L	N	S
A	X	-	-	-	-	-	-	-	-	-	-	-
B	-	X	-	1/	-	-	-	-	-	-	-	X
C	-	-	X	X	X	-	X	-	-	-	2/, 3/	X
D	-	1/	X	X	X	-	X	-	-	-	2/, 3/	X
E	-	-	X	X	X	-	X	-	-	-	2/, 3/	X
F	-	-	-	-	-	X	-	-	-	-	-	X
G	-	-	X	X	X	-	X	-	-	-	-	X
H	-	-	-	-	-	-	-	X	-	-	-	X
J	-	-	-	-	-	-	-	-	X	-	-	X
L	-	-	-	-	-	-	-	-	-	4/	-	-
N	-	-	2/, 3/	2/, 3/	2/, 3/	-	-	-	-	-	2/	X
S	-	X	X	X	X	X	X	X	X	-	X	X

"X" indicates that explosive substances or articles of corresponding compatibility groups in accordance with Part 2 of these Regulations may be stowed in the same hold.

1/ Packages containing articles assigned to compatibility group B or substances or articles assigned to compatibility group D may be loaded together in the same hold provided that they are carried in containers or vehicles or wagons with complete metal walls.

2/ Different categories of articles of Division 1.6, compatibility group N, may be carried together as articles of Division 1.6, compatibility group N, only when it is proven by testing or analogy that there is no additional risk of sympathetic detonation between the articles. Otherwise they should be treated as hazard Division 1.1.

3/ When articles of compatibility group N are carried with substances or articles of compatibility groups C, D or E, the articles of compatibility group N should be considered as having the characteristics of compatibility group D.

4/ Packages with substances or articles of compatibility group L may be stowed in the same hold with packages containing the same type of substances or articles of the same compatibility group.

7.1.4.3.5 For the carriage of material Class 7 (UN Nos. 2916, 2917, 3323, 3328, 3329 and 3330) in Type B(U) or Type B(M) or Type C packages, the controls, restrictions or provisions specified in the competent authority approval certificate shall be complied with.

7.1.4.3.6 For the carriage of material of Class 7 under special arrangement (UN Nos. 2919 and 3331), the special provisions specified by the competent authority shall be met. In particular, mixed loading shall not be permitted unless specifically authorized by the competent authority.

7.1.4.4 *Prohibition of mixed loading (containers, vehicles, wagons)*

7.1.4.4.1 7.1.4.3 shall not apply to packages stowed in containers, vehicles or wagons in accordance with international regulations.

7.1.4.4.2 7.1.4.3 shall not apply to:

– closed containers with complete metal walls;

– closed vehicles and closed wagons with complete metal walls;

– tank-containers, portable tanks and MEGCs;

– tank-vehicles and tank-wagons.

7.1.4.4.3 For containers other than those referred to in paragraph 7.1.4.4.1 and 7.1.4.4.2 above the separation distance required by 7.1.4.3.1 may be reduced to 2.40 m (width of container).

7.1.4.4.4   The electrical equipment fitted to the outside of a closed container may be connected with removable electrical cables in accordance with the provisions of 9.1.0.56 and be put into operation provided that:

(a)   Such electrical equipment is of a certified safe type; or

(b)   Such electrical equipment is not of a certified safe type but is separated sufficiently from other containers containing substances of:

- Class 2 for which a label No. 2.1 is required in column (5) of Table A of Chapter 3.2;

- Class 3, packing group I or II;

- Class 4.3;

- Class 6.1; packing group I or II, with an additional hazard of Class 4.3;

- Class 8, packing group I, with an additional hazard of Class 3; and

- Class 8, packing group I or II, with an additional hazard of Class 4.3.

This condition is deemed to be met if no container containing the above-mentioned substances is stowed within an area of cylindrical form with a radius of 2.4 m around the electrical equipment and an unlimited vertical extension.

This condition does not apply if containers with electrical equipment which is not of a certified safe type and containers containing the above-mentioned substances are stowed in separate holds.

Examples of stowage and segregation of containers

Legend

R   Container (e.g. reefer) with electrical equipment which is not of a certified safe type.

Z   Electrical equipment which is not of a certified safe type.

X   Container not allowed when containing dangerous substances for which sufficient separation is required.

Top view
1. On deck

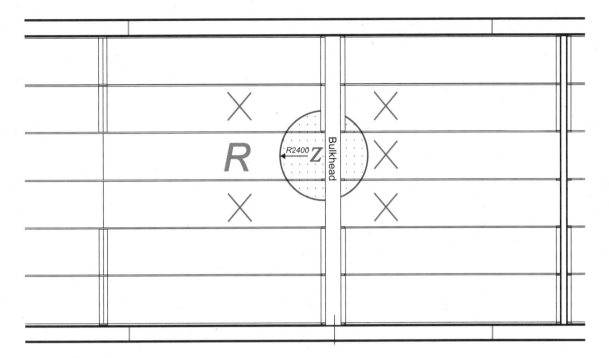

Top view
2. In the hold

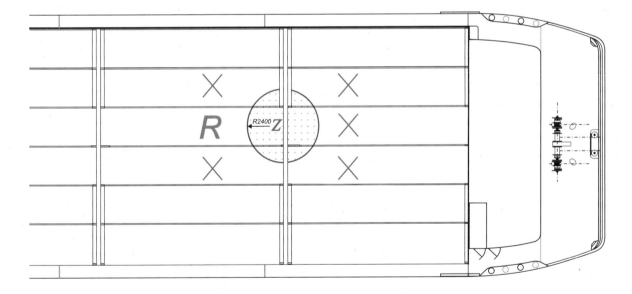

Top view
2. In the hold

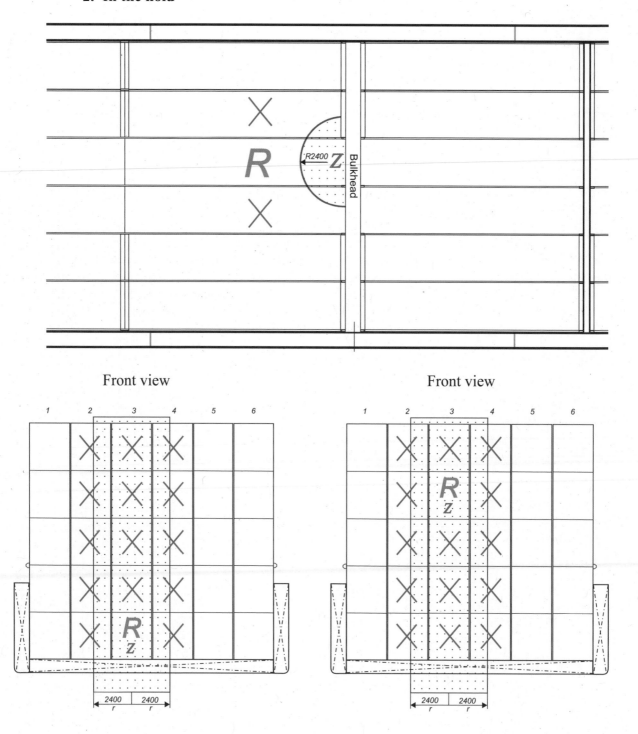

Front view

Front view

7.1.4.4.5   The electrical equipment fitted to an open container may not be connected with removable electrical cables in accordance with the provisions of 9.1.0.56 nor be put into operation unless it is of a certified safe type or the container is placed in a hold which does not contain containers with substances referred to in 7.1.4.4.4 (b).

7.1.4.5   *Prohibition of mixed loading (seagoing vessels; inland navigation vessels carrying containers)*

For seagoing vessels and inland waterway vessels, where the latter only carry containers, the prohibition of mixed loading shall be deemed to have been met if the stowage and segregation requirements of the IMDG Code have been complied with.

7.1.4.6   *(Reserved)*

7.1.4.7   *Places of loading and unloading*

7.1.4.7.1   The dangerous goods shall be loaded or unloaded only at the places designated or approved for this purpose by the competent authority. In those places the means of evacuation mentioned in subsection 7.1.4.77 should be made available. Otherwise trans-shipment is permitted only with the authorization of the competent authority.

7.1.4.7.2   When substances or articles of Class 1 and substances of Classes 4.1 or 5.2 for which marking with three blue cones or three blue lights is prescribed in column (12) of Table A of Chapter 3.2 are on board, no goods of any kind may be loaded or unloaded except at the places designated or permitted for this purpose by the competent authority.

7.1.4.8   *Time and duration of loading and unloading operations*

7.1.4.8.1   Loading and unloading operations of substances or articles of Class 1 and substances of Classes 4.1 or 5.2 for which marking with three blue cones or three blue lights is prescribed in column (12) of Table A of Chapter 3.2 shall not start without permission in writing from the competent authority. This provision also applies to loading or unloading of other goods when substances or articles of Class 1 or substances of Classes 4.1 or 5.2 for which marking with three blue cones or three blue lights is prescribed in column (12) of Table A of Chapter 3.2 are on board.

7.1.4.8.2   Loading and unloading operations of substances or articles of Class 1 and substances of Classes 4.1 or 5.2, for which marking with three blue cones or three blue lights is prescribed in column (12) of Table A of Chapter 3.2, shall be suspended in the event of a storm.

7.1.4.9   *Cargo transhipment operations*

Partial or complete cargo transhipment into another vessel without permission from the competent authority is prohibited outside a cargo transhipment place approved for this purpose.

Note: *For transhipment to means of transport of another mode see 7.1.4.7.1.*

7.1.4.10   *Precautions with respect to foodstuffs, other articles of consumption and animal feeds*

7.1.4.10.1   When special provision 802 is indicated for a dangerous good in column (6) of Table A of Chapter 3.2, precautions shall be taken as follows with respect to foodstuffs, other articles of consumption and animal feeds:

Packages as well as uncleaned empty packagings, including large packagings and intermediate bulk containers (IBCs), bearing labels conforming to models Nos. 6.1 or 6.2, and those bearing labels of Class 9, containing substances of Class 9, UN Nos. 2212, 2315, 2590, 3151, 3152 or 3245, shall not be stacked on or loaded in immediate proximity to packages known to contain foodstuffs, other articles of consumption or animal feeds in the same hold and at places of loading and unloading or trans-shipment.

When these packages, bearing the said labels, are loaded in immediate proximity of packages known to contain foodstuffs, other articles of consumption or animal feeds, they shall be kept apart from the latter:

(a)   by complete partitions which should be as high as the packages bearing the said labels, or

(b)   by packages not bearing labels conforming to models Nos. 6.1, 6.2 or 9 or packages bearing labels of Class 9 but not containing substances of that class, UN Nos. 2212, 2315, 2590, 3151, 3152 or 3245, or

(c)   by a space of at least 0.8 m,

unless the packages bearing said labels are provided with an additional packaging or are completely covered (e.g. by a sheeting, a fibreboard cover or other measures).

7.1.4.11   *Stowage plan*

7.1.4.11.1   The master shall enter on a stowage plan the dangerous goods stowed in the individual holds or on deck. The goods shall be described as in the transport document in accordance with 5.4.1.1.1 (a), (b), (c) and (d).

7.1.4.11.2   Where the dangerous goods are transported in containers, the number of the container shall suffice. In this case, the stowage plan shall contain as an annex a list of all containers with their numbers and the description of the goods contained therein in accordance with 5.4.1.1.1 (a), (b), (c) and (d).

7.1.4.12   *Ventilation*

7.1.4.12.1   During loading or unloading of road vehicles into or from the holds of ro-ro-vessels, there shall be not less than five changes of air per hour based upon the total volume of the empty hold.

7.1.4.12.2   On board vessels carrying dangerous goods only in containers placed in open holds, ventilators do not require to be incorporated but must be on board. Where damage of the container or release of content inside the container is suspected, the holds shall be ventilated so as to reduce the concentration of gases given off by the cargo to less than 10% of the lower explosive limit or in the case of toxic gases to below any significant concentration.

7.1.4.12.3   If tank-containers, portable tanks, MEGCs, tank vehicles or tank wagons are carried in closed holds, such holds shall be permanently ventilated for ensuring five air changes per hour.

7.1.4.13   *Measures to be taken before loading*

The holds and cargo areas shall be cleaned prior to loading. The holds shall be ventilated.

7.1.4.14   *Handling and stowage of the cargo*

7.1.4.14.1   The various components of the cargo shall be stowed such as to prevent them from shifting in relation to one another or to the vessel and such that no damage can be caused by other cargo.

7.1.4.14.1.1   Packages containing dangerous substances and unpackaged dangerous articles shall be secured by suitable means capable of restraining the goods (such as fastening straps, sliding slatboards, adjustable brackets) in a manner that will prevent any movement during carriage which would change the orientation of the packages or cause them to be damaged. When dangerous goods are carried with other goods (e.g. heavy machinery or crates), all goods shall be securely fixed or packed so as to prevent the release of dangerous goods. Movement of packages may also be prevented by filling any voids by the use of dunnage or by blocking and bracing. Where restraints such as banding or straps are used, these shall not be over-tightened to cause damage or deformation of the package. Flexible bulk containers shall be stowed in such way that there are no void spaces between them in the hold. If the flexible bulk containers do not completely fill the hold, adequate measures shall be taken to avoid shifting of cargo.

7.1.4.14.1.2   Packages shall not be stacked unless designed for that purpose. Where different design types of packages that have been designed for stacking are to be loaded together, consideration shall be given to their compatibility for stacking with each other. Where necessary, stacked packages shall be prevented from damaging the package below by the use of load-bearing devices. Flexible bulk containers may be stacked on each other in holds provided that the stacking height does not exceed three high. When flexible bulk containers are fitted with venting devices, the stowage of the flexible bulk containers shall not impede their function.

7.1.4.14.1.3   During loading and unloading, packages containing dangerous goods shall be protected from being damaged.

NOTE: Particular attention shall be paid to the handling of packages during their preparation for carriage, the type of vessel on which they are to be carried and to the method of loading or unloading, so that accidental damage is not caused through dragging or mishandling the packages.

7.1.4.14.1.4   When orientation arrows are required, packages and overpacks shall be oriented in accordance with such markings.

NOTE: Liquid dangerous goods shall be loaded below dry dangerous goods whenever practicable.

7.1.4.14.2   Dangerous goods shall be stowed at a distance of not less than 1 m from the accommodation, the engine rooms, the wheelhouse and any sources of heat.

When the accommodation or wheelhouse is situated above a hold, dangerous goods shall in no case be stowed beneath such accommodation or wheelhouse.

7.1.4.14.3   Packages shall be protected against heat, sunlight and the effects of the weather. This provision does not apply to vehicles, wagons, tank-containers, portable tanks, MEGCs and containers.

Where packages are not enclosed in vehicles, wagons or containers but loaded on deck, they shall be covered with tarpaulins that are not readily flammable.

The ventilation shall not be obstructed.

7.1.4.14.4   The dangerous goods shall be stowed in the holds. However, dangerous goods packed or loaded in:

–   containers having complete sprayproof walls;

–   MEGCs;

- vehicles or wagons having complete sprayproof walls;

- tank-containers or portable tanks;

- tank vehicles or tank wagons;

may be carried on deck in the protected area.

7.1.4.14.5   Packages containing dangerous goods of Classes 3, 4.1, 4.2, 5.1 or 8 may be stowed on deck in the protected area provided that drums are used or that they are contained in containers with complete walls or vehicles or wagons with complete walls. Substances of Class 2 may be stowed on deck in the protected area, provided they are contained in cylinders.

7.1.4.14.6   For seagoing vessels, the stowage requirements set out in 7.1.4.14.1 to 7.1.4.14.5 above and 7.1.4.14.7 below shall be deemed to have been met, if the relevant stowage provisions of the IMDG Code and, in the case of carriage of dangerous goods in bulk, those set out in subsection 9.3 of the IMSBC Code have been complied with.

7.1.4.14.7   *Handling and stowage of radioactive material*

NOTE 1: "Critical group" means a group of members of the public which is reasonably homogeneous with respect to its exposure for a given radiation source and given exposure pathway and is typical of individuals receiving the highest effective dose by the given exposure pathway from the given source.

NOTE 2: "Members of the public" means in a general sense, any individuals in the population except when subject to occupational or medical exposure.

NOTE 3: "Workers" are any persons who work, whether full time, part-time or temporarily, for an employer and who have recognized rights and duties in relation to occupational radiation protection.

7.1.4.14.7.1   *Segregation*

7.1.4.14.7.1.1   Packages, overpacks, containers, tanks and vehicles and wagons containing radioactive material and unpackaged radioactive material shall be segregated during carriage:

(a)   from workers in regularly occupied working areas;

(i)   in accordance with Table A below; or

(ii)   by distances calculated using a dose criterion of 5 mSv in a year and conservative model parameters;

NOTE: Workers subject to individual monitoring for the purposes of radiation protection shall not be considered for the purposes of segregation.

(b)   from members of the critical group of the public, in areas where the public has regular access;

(i)   in accordance with Table A below; or

(ii)   by distances calculated using a dose criterion of 1 mSv in a year and conservative model parameters;

(c)   from undeveloped photographic film and mailbags;

    (i)   in accordance with Table B below; or

    (ii)   by distances calculated using a radiation exposure criterion for undeveloped photographic film due to the transport of radioactive material for 0.1 mSv per consignment of such film; and

NOTE: Mailbags shall be assumed to contain undeveloped film and plates and therefore be separated from radioactive material in the same way.

(d)   from other dangerous goods in accordance with 7.1.4.3.

Table A: Minimum distances between packages of category II-YELLOW or of category III-YELLOW and persons

Sum of transport indexes not more than	Exposure time per year (hours)			
	Areas where members of the public have regular access		Regularly occupied working areas	
	50	250	50	250
	Segregation distance in metres, no shielding material intervening, from:			
2	1	3	0.5	1
4	1.5	4	0.5	1.5
8	2.5	6	1.0	2.5
12	3	7.5	1.0	3
20	4	9.5	1.5	4
30	5	12	2	5
40	5.5	13.5	2.5	5.6
50	6.5	15.5	3	6.5

Table B: Minimum distances between packages of category II-YELLOW or of category III-YELLOW and packages bearing the word "FOTO", or mailbags

Total number of packages not more than		Sum of transport indexes not more than	Journey or storage duration, in hours							
Category			1	2	4	10	24	48	120	240
III-yellow	II-yellow		Minimum distances in metres							
		0.2	0.5	0.5	0.5	0.5	1	1	2	3
		0.5	0.5	0.5	0.5	1	1	2	3	5
	1	1	0.5	0.5	1	1	2	3	5	7
	2	2	0.5	1	1	1.5	3	4	7	9
	4	4	1	1	1.5	3	4	6	9	13
	8	8	1	1.5	2	4	6	8	13	18
1	10	10	1	2	3	4	7	9	14	20
2	20	20	1.5	3	4	6	9	13	20	30
3	30	30	2	3	5	7	11	16	25	35
4	40	40	3	4	5	8	13	18	30	40
5	50	50	3	4	6	9	14	20	32	45

7.1.4.14.7.1.2　Category II-YELLOW or III-YELLOW packages or overpacks shall not be carried in compartments occupied by passengers, except those exclusively reserved for couriers specially authorized to accompany such packages or overpacks.

7.1.4.14.7.1.3　No persons other than the master of the vessel or the driver of the vehicle embarked, persons who are on board for duty reasons and the other members of the crew shall be permitted in vessels carrying packages, overpacks or containers bearing category II-YELLOW or III-YELLOW labels.

7.1.4.14.7.2　*Activity limits*

The total activity in a single hold or compartment of a vessel, or in another conveyance, for carriage of LSA material or SCO articles in Type IP-1, Type IP-2, Type IP-3 or unpackaged, shall not exceed the limits shown in Table C below:

Table C: Conveyance activity limits for LSA material and SCO
in industrial packages or unpackaged

Nature of material or articles	Activity limit for conveyances other than by vessel	Activity limit for a hold or compartment of a vessel
LSA-I	No limit	No limit
LSA-II and LSA-III non-combustible solids	No limit	$100A_2$
LSA-II and LSA-III combustible solids, and all liquids and gases	$100A_2$	$10A_2$
SCO	$100A_2$	$10A_2$

7.1.4.14.7.3　*Stowage during carriage and storage in transit*

7.1.4.14.7.3.1　Consignments shall be securely stowed.

7.1.4.14.7.3.2　Provided that its average surface heat flux does not exceed $15W/m^2$ and that the immediately surrounding cargo is not in bags, a package or overpack may be carried or stored among packaged general cargo without any special stowage provisions except as may be specifically required by the competent authority in an applicable approval certificate.

7.1.4.14.7.3.3　Loading of containers and accumulation of packages, overpacks and containers shall be controlled as follows:

(a)　Except under the conditions of exclusive use, and for consignments of LSA-I material, the total number of packages, overpacks and containers aboard a single conveyance shall be so limited that the total sum of the transport indexes aboard the conveyance does not exceed the values shown in Table D below;

(b)　The radiation level under routine conditions of carriage shall not exceed 2 mSv/h at any point on, and 0.1 mSv/h at 2 m from, the external surface of the conveyance, except for consignments carried under exclusive use, for which the radiation limits around the conveyance are set forth in 7.1.4.14.7.3.5 (b) and (c);

(c)　The total sum of the criticality safety indexes in a container and aboard a conveyance shall not exceed the values shown in Table E below.

Table D: Transport Index limits for containers and conveyances not under exclusive use

Type of container or conveyance	Limit on total sum of transport indexes in a container or aboard a conveyance
Small container	50
Large container	50
Vehicle or wagon	50
Vessel	50

Table E: Criticality Safety Index for containers and vehicles containing fissile material

Type of container or conveyance	Limit on total sum of criticality safety indexes	
	Not under exclusive use	Under exclusive use
Small container	50	n.a.
Large container	50	100
Vehicle or wagon	50	100
Vessel	50	100

7.1.4.14.7.3.4   Any package or overpack having either a transport index greater than 10, or any consignment having a criticality safety index greater than 50, shall be carried only under exclusive use.

7.1.4.14.7.3.5   For consignments under exclusive use in vehicles or wagons, the radiation level shall not exceed:

(a)   10 mSv/h at any point on the external surface of any package or overpack, and may only exceed 2 mSv/h provided that:

(i)   the vehicle or wagon is equipped with an enclosure which, during routine conditions of carriage, prevents the access of unauthorized persons to the interior of the enclosure;

(ii)   provisions are made to secure the package or overpack so that its position within the vehicle or wagon enclosure remains fixed during routine conditions of carriage; and

(iii)   there is no loading or unloading during the shipment;

(b)   2 mSv/h at any point on the outer services of the vehicle or wagon, including the upper and lower surfaces, or, in the case of an open vehicle or wagon, at any point on the vertical planes projected from the outer edges of the vehicle or wagon, on the upper surface of the load, and on the lower external surface of the vehicle or wagon; and

(c)   0.1 mSv/h at any point 2 m from the vertical planes represented by the outer lateral surfaces of the vehicle or wagon, or, if the load is carried in an open vehicle or wagon, at any point 2 m from the vertical planes projected from the outer edges of the vehicle or wagon.

7.1.4.14.7.3.6   Packages or overpacks having a surface radiation area greater than 2 mSv/h, unless being carried in or on a vehicle or wagon under exclusive use and unless they are removed from the vehicle or wagon when on board the vessel shall not be transported by vessel except under special arrangement.

7.4.1.14.7.3.7   The transport of consignments by means of a special use vessel which, by virtue of its design, or by reason of its being chartered, is dedicated to the purpose of carrying radioactive material, shall be excepted from the requirements specified in 7.1.4.14.7.3.3 provided that the following conditions are met:

(a)   A radiation protection programme for the shipment shall be approved by the competent authority of the flag state of the vessel and, when requested, by the competent authority at each port of call of the transit countries;

(b)   Stowage arrangements shall be predetermined for the whole voyage including any consignments to be loaded at ports of call en route; and

(c)   The loading, carriage and unloading of the consignments shall be supervised by persons qualified in the transport of radioactive material.

7.1.4.14.7.4   *Segregation of packages containing fissile material during carriage and storage in transit*

7.1.4.14.7.4.1   Any group of packages, overpacks, and containers containing fissile material stored in transit in any one storage area shall be so limited that the total sum of the criticality safety indexes in the group does not exceed 50. Each group shall be stored so as to maintain a spacing of at least 6 m from other such groups.

7.1.4.14.7.4.2   Where the total sum of the criticality safety indexes on board a vehicle, a wagon or in a container exceeds 50, as permitted in Table E above, storage shall be such as to maintain a spacing of at least 6 m from other groups of packages, overpacks or containers containing fissile material or other vehicles or wagons carrying radioactive material. The space between such groups may be used for other dangerous goods of ADN. The carriage of other goods with consignments under exclusive use is permitted provided that the pertinent provisions have been taken by the consignor and that carriage is not prohibited under other requirements.

7.1.4.14.7.4.3   Fissile material meeting one of the provisions (a) to (f) of 2.2.7.2.3.5 shall meet the following requirements:

(a)   Only one of the provisions (a) to (f) of 2.2.7.2.3.5 is allowed per consignment;

(b)   Only one approved fissile material in packages classified in accordance with 2.2.7.2.3.5 (f) is allowed per consignment unless multiple materials are authorized in the certificate of approval;

(c)   Fissile material in packages classified in accordance with 2.2.7.2.3.5 (c) shall be carried in a consignment with no more than 45 g of fissile nuclides;

(d)   Fissile material in packages classified in accordance with 2.2.7.2.3.5 (d) shall be carried in a consignment with no more than 15 g of fissile nuclides;

(e)   Unpackaged or packaged fissile material classified in accordance with 2.2.7.2.3.5 (e) shall be carried under exclusive use on a vehicle with no more than 45 g of fissile nuclides.

7.1.4.14.7.5   *Damaged or leaking packages, contaminated packagings*

7.1.4.14.7.5.1   If it is evident that a package is damaged or leaking, or if it is suspected that the package may have leaked or been damaged, access to the package shall be restricted and a qualified person shall, as soon as possible, assess the extent of contamination and the resultant radiation level of the package. The scope of the assessment shall include the package, the vehicle, the wagon, the vessel, the adjacent loading and unloading areas, and, if necessary, all other material which has been carried in the vessel. When necessary, additional steps for the protection of persons, property and the environment, in accordance with provisions established by the competent authority, shall be taken to overcome and minimize the consequences of such leakage or damage.

7.1.4.14.7.5.2 Packages damaged or leaking radioactive contents in excess of allowable limits for normal conditions of carriage may be removed to an acceptable interim location under supervision, but shall not be forwarded until repaired or reconditioned and decontaminated.

7.1.4.14.7.5.3 Vehicles, wagons, vessels and equipment used regularly for the carriage of radioactive material shall be periodically checked to determine the level of contamination. The frequency of such checks shall be related to the likelihood of contamination and the extent to which radioactive material is carried.

7.1.4.14.7.5.4 Except as provided in paragraph 7.1.4.14.7.5.6, any vessel, or equipment or part thereof which has become contaminated above the limits specified in 7.1.4.14.7.5.5 in the course of carriage of radioactive material, or which shows a radiation level in excess of 5 µSv/h at the surface, shall be decontaminated as soon as possible by a qualified person and shall not be re-used unless the following conditions are fulfilled:

(a)   the non-fixed contamination shall not exceed the limits specified in 4.1.9.1.2 of ADR;

(b)   the radiation level resulting from the fixed contamination shall not exceed 5 µSv/h at the surface.

7.1.4.14.7.5.5 For the purposes of 7.1.4.14.7.5.4, non-fixed contamination shall not exceed:

–   4 Bq/cm^2 for beta and gamma emitters and low toxicity alpha emitters;

–   0.4 Bq/cm^2 for all other alpha emitters.

These are average limits applicable to any area of 300 cm^2 on any part of the surface.

7.1.4.14.7.5.6 Vessels dedicated to the carriage of radioactive material under exclusive use shall be excepted from the requirements of the previous paragraph 7.1.4.14.7.5.4 solely with regard to its internal surfaces and only for as long as it remains under that specific exclusive use.

7.1.4.14.7.6   *Limitation of the effect of temperature*

7.1.4.14.7.6.1 If the temperature of the accessible outer surfaces of a Type B (U) or Type B (M) package could exceed 50 °C in the shade, carriage is permitted only under exclusive use. As far as practicable, the surface temperature shall be limited to 85 °C. Account may be taken of barriers or screens intended to give protection to transport workers without the barriers or screens being subject to any test.

7.1.4.14.7.6.2 If the average heat flux from the external surfaces of a Type B (U) or B (M) package could exceed 15 W/m^2, the special stowage requirements specified in the competent authority package design approval certificate shall be met.

7.1.4.14.7.7   *Other provisions*

If neither the consignor nor the consignee can be identified or if the consignment cannot be delivered to the consignee and the carrier has no instructions from the consignor the consignment shall be placed in a safe location and the competent authority shall be informed as soon as possible and a request made for instructions on further action.

7.1.4.15   *Measures to be taken after unloading*

7.1.4.15.1   After unloading the holds shall be inspected and cleaned if necessary. In the case of carriage in bulk, this requirement does not apply if the new cargo comprises the same goods as the previous cargo.

7.1.4.15.2   For material of Class 7 see also 7.1.4.14.7.5.

7.1.4.15.3   A cargo transport unit or hold space which has been used to carry infectious substances shall be inspected for release of the substance before re-use. If the infectious substances were released during carriage, the cargo transport unit or hold space shall be decontaminated before it is re-used. Decontamination may be achieved by any means which effectively inactivates the released infectious substance.

7.1.4.16   *Measures to be taken during loading, carriage, unloading and handling of the cargo*

The filling or emptying of receptacles, tank vehicles, tank wagons, intermediate bulk containers (IBCs), large packagings, MEGCs, portable tanks or tank-containers on board the vessel is prohibited without special permission from the competent authority.

7.1.4.17 to 7.1.4.40   *(Reserved)*

7.1.4.41   *Fire and naked light*

The use of fire or naked light is prohibited while substances or articles of Divisions 1.1, 1.2, 1.3, 1.5 or 1.6 of Class 1 are on board and the holds are open or the goods to be loaded are located at a distance of less than 50 m from the vessel.

7.1.4.42 to 7.1.4.50   *(Reserved)*

7.1.4.51   *Electrical equipment*

The use of radiotelephone or radar transmitters is not permitted while substances or articles of Divisions 1.1, 1.2, 1.3, 1.5 or 1.6 of Class 1 are being loaded or unloaded.

This shall not apply to VHF-transmitters of the vessel, in cranes or in the vicinity of the vessel, provided the power of the VHF-transmitter does not exceed 25 W and no part of its aerial is located at a distance less than 2.00 m from the substances or articles mentioned above.

7.1.4.52   *(Reserved)*

7.1.4.53   *Lighting*

If loading or unloading is performed at night or in conditions of poor visibility, effective lighting shall be provided.

If provided from the deck, it shall be effected by properly secured electric lamps which shall be positioned in such a way that they cannot be damaged.

Where these lamps are positioned on deck in the protected area, they shall be of "limited explosion risk" type.

7.1.4.54 to 7.1.4.74   *(Reserved)*

7.1.4.75   *Risk of sparking*

All electrically continuous connections between the vessel and the shore as well as appliances used in the protected area shall be so designed that they do not present a source of ignition.

7.1.4.76   *Synthetic ropes*

During loading or unloading operations, the vessel may be moored by means of synthetic ropes only when steel cables are used to prevent the vessel from going adrift.

Steel cables sheathed in synthetic material or natural fibres are considered as equivalent when the minimum tensile strength required in accordance with the Regulations referred to in 1.1.4.6 is obtained from the steel strands.

However, during loading or unloading of containers, vessels may be moored by means of synthetic ropes.

7.1.4.77   **Possible means of evacuation in case of an emergency**

		Dry cargo bulk (vessel and barge)		Container (vessel and barge) and packaged goods
		Class		Class
		4.1, 4.2, 4.3	5.1, 6.1, 7, 8, 9	All classes
1	Two escape routes inside or outside the protected area in opposite directions	•	•	•
2	One escape route outside the protected area and one safe haven outside the vessel including the escape route towards it at the opposite end	•	•	•
3	One escape route outside the protected area and one safe haven on the vessel at the opposite end	•	•	•
4	One escape route outside the protected area and one life boat at the opposite end	•	•	•
5	One escape route outside the protected cargo area and one escape boat at the opposite end	•	•	•
6	One escape route inside the protected area and one escape route outside the cargo area at the opposite end	•	•	•
7	One escape route inside the protected area and one safe haven outside the vessel in the opposite direction	•	•	•
8	One escape route inside the protected area and one safe haven on the vessel in the opposite direction	•	•	•
9	One escape route inside the protected cargo area and one life boat at the opposite end	•	•	•
10	One escape route inside the protected area and one escape boat at the opposite end	•	•	•
11	One escape route inside or outside the protected cargo area and two safe havens on the vessel at opposite ends	•	•	•
12	One escape route inside or outside the protected area and two safe areas on the vessel at opposite ends	•	•	•
13	One escape route outside the protected area	•	•	•
14	One escape route inside the protected area	•	•	•
15	One or more safe havens outside the vessel, including the escape route towards it	•	•	•
16	One or more safe havens on the vessel		•	•
17	One or more escape boats	•	•	•
18	One escape boat and one evacuation boat	•	•	•
19	One or more evacuation boats		•	•

• = *Possible option.*

        Based on local circumstances, competent authorities may prescribe additional requirements for the availability of means of evacuation.

7.1.4.78 to 7.1.4.99   *(Reserved)*

7.1.5   **Additional requirements concerning the operation of vessels**

7.1.5.0   *Marking*

7.1.5.0.1   Vessels carrying dangerous goods listed in Table A of Chapter 3.2 shall, in accordance with Chapter 3 of the European Code for Inland Waterways (CEVNI), display the markings prescribed in column (12) in this table.

7.1.5.0.2   Vessels carrying the dangerous goods listed in Table A of Chapter 3.2 in packages placed exclusively in containers shall display the number of blue cones or blue lights indicated in column (12) of Table A of Chapter 3.2 where:

–   three blue cones or three blue lights are required, or

–   two blue cones or two blue lights are required, a substance of Class 2 is involved or packing group I is indicated in column (4) of Table A of Chapter 3.2 and the total gross mass of these dangerous goods exceeds 30,000 kg, or

–   one blue cone or one blue light is required, a substance of Class 2 is involved or packing group I is indicated in column (4) of Table A of Chapter 3.2 and the total gross mass of these dangerous goods exceeds 130,000 kg.

7.1.5.0.3   Vessels carrying empty, uncleaned tanks, battery vehicles, battery wagons or MEGCs shall display the marking referred to in column (12) of Table A of Chapter 3.2 if these cargo transport units have contained dangerous goods for which this table prescribes marking.

7.1.5.0.4   Where more than one marking could apply to a vessel, only the marking which includes the greatest number of blue cones or blue lights shall apply, i.e. in the following order of precedence:

–   three blue cones or three blue lights; or

–   two blue cones or two blue lights; or

–   one blue cone or one blue light.

7.1.5.0.5   By derogation from paragraph 7.1.5.0.1, and in accordance with the footnotes to article 3.14 of the European Code for Inland Waterways (CEVNI), the competent authority of a Contracting Party may authorize seagoing vessels temporarily operating in an inland navigation area on the territory of this Contracting Party, the use of the day and night signals prescribed in the Recommendations on the Safe Transport of Dangerous Cargoes and Related Activities in Port Areas adopted by the Maritime Safety Committee of the International Maritime Organization (by night an all-round fixed red light and by day flag "B" of the International Code of Signals), instead of the signals prescribed in 7.1.5.0.1. Contracting Parties which have taken the initiative with respect to the derogation granted shall notify the Executive Secretary of the UNECE, who shall bring this derogation to the attention of the Administrative Committee.

7.1.5.1   *Mode of navigation*

7.1.5.1.1   The competent authorities may impose restrictions on the inclusion of vessels carrying dangerous goods in pushed conveys of large dimension.

7.1.5.1.2   When vessels carry substances or articles of Class 1, or substances of Classes 4.1 or 5.2 for which marking with three blue cones or three blue lights is prescribed in column (12) of Table A of Chapter 3.2, or material of Class 7 of UN Nos. 2912, 2913, 2915, 2916, 2917, 2919, 2977, 2978 or 3321 to 3333, the competent authority may impose restrictions on the dimensions of convoys or side-by-side formations. Nevertheless, the use of a motorized vessel giving temporary towing assistance is permitted.

7.1.5.2 *Vessels under way*

Vessels carrying substances or articles of Class 1, or substances of Classes 4.1 or 5.2 for which marking with three blue cones or three blue lights is prescribed in column (12) of Table A of Chapter 3.2, when under way shall keep not less than 50 m away from any other vessel, if possible.

7.1.5.3 *Mooring*

Vessels shall be moored securely, but in such a way that they can be released quickly in an emergency.

7.1.5.4 *Berthing*

7.1.5.4.1 The distances to be kept by vessels carrying dangerous goods at berth from other vessels shall not be less than the distance prescribed by the European Code for Inland Waterways (CEVNI).

7.1.5.4.2 An expert in accordance with 8.2.1.2 shall be permanently on board berthed vessels for which marking is prescribed in column (12) of Table A of Chapter 3.2.

The competent authority may, however, exempt from this obligation those vessels which are berthed in a harbour basin or in an accepted berthing position.

7.1.5.4.3 Outside the berthing areas specifically designated by the competent authority, the distances to be kept by berthed vessels shall not be less than:

– 100 m   from residential areas, civil engineering structures or storage tanks, if the vessel is required to be marked with one blue cone or one blue light in accordance with the requirements of column (12) of Table A of Chapter 3.2;

– 100 m   from civil engineering structures and storage tanks and 300 m from residential areas if the vessel is required to be marked with two blue cones or two blue lights in accordance with the requirements of column (12) of Table A of Chapter 3.2;

– 500 m   from residential areas, civil engineering structures and storage tanks holding gas or flammable liquids if the vessel is required to be marked with three blue cones or three blue lights in accordance with the requirements of column (12) of Table A of Chapter 3.2.

While waiting in front of locks or bridges, vessels are allowed to keep distances different from and lower than those given above. In no case shall the distance be less than 100 m.

7.1.5.4.4 The competent authority may prescribe distances lower than those given in 7.1.5.4.3 above, especially taking local conditions into account.

7.1.5.5   *Stopping of vessels*

If navigation of a vessel carrying substances and articles of Class 1 or substances of Class 4.1 or 5.2 for which marking with three blue cones or three blue lights is prescribed in column (12) of Table A of Chapter 3.2 threatens to become dangerous owing either to:

–   external factors (bad weather, unfavourable conditions of the waterway, etc.), or

–   the condition of the vessel itself (accident or incident),

the vessel shall be stopped at a suitable berthing area as far away as possible from residential areas, harbours, civil engineering structures or storage tanks for gas or flammable liquids, regardless of the provisions set out in 7.1.5.4.

The competent authority shall be notified without delay.

7.1.5.6 and 7.1.5.7   *(Reserved)*

7.1.5.8   *Reporting duty*

7.1.5.8.1   In the States where the reporting duty is in force, the master of the vessel shall provide information in accordance with paragraph 1.1.4.6.1.

7.1.5.8.2 to 7.1.5.8.4   *(Deleted)*

7.1.5.9 to 7.1.5.99   *(Reserved)*

7.1.6   **Additional requirements**

7.1.6.1 to 7.1.6.10   *(Reserved)*

7.1.6.11   *Carriage in bulk*

The following additional requirements shall be met when they are indicated in column (11) of Table A of Chapter 3.2:

CO01:   The surfaces of holds shall be coated or lined such that they are not readily flammable and not liable to impregnation by the cargo.

CO02:   Any part of the holds and of the hatchway covers which may come into contact with this substance shall consist of metal or of wood having a specific density of not less than 750 kg/m^3 (seasoned wood).

CO03:   The inner surfaces of holds shall be lined or coated so as to prevent corrosion.

ST01:   The substances shall have been stabilized in accordance with the requirements applicable to ammonium nitrate fertilizers set out in the IMSBC Code. Stabilizing shall be certified by the consignor in the transport document.

In those States where this is required, these substances may be carried in bulk only with the approval of the competent authority.

ST02:   These substances may be carried in bulk if the results of the trough test according to subsection 38.2 of the *Manual of Tests and Criteria* show that the self-sustaining decomposition rate is not greater than 25 cm/h.

RA01:   The materials may be carried in bulk provided that:

      (a)   for materials other than natural ores, carriage is under exclusive use and there is no escape of contents out of the vessel and no loss of shielding under normal conditions of transport; or

      (b)   for natural ores, carriage is under exclusive use.

RA02:   The materials may be carried in bulk provided that:

      (a)   they are carried in a vessel so that, under normal conditions of transport, there is no escape of contents or loss of shielding;

      (b)   they are carried under exclusive use if the contamination on the accessible and inaccessible surfaces is greater than 4 Bq/cm^2 (10^{-4} $\mu Ci/cm^2$) for beta and gamma emitters and low toxicity alpha emitters or 0.4 Bq/cm^2 (10^{-5} $\mu Ci/cm^2$) for all other alpha emitters;

      (c)   measures are taken to ensure that radioactive material is not released into the vessel, if it is suspected that non-fixed contamination exists on inaccessible surfaces of more than 4 Bq/cm^2 (10^{-4} $\mu Ci/cm^2$) for beta and gamma emitters and low toxicity alpha emitters or 0.4 Bq/cm^2 (10^{-5} $\mu Ci/cm^2$) for all other alpha emitters.

Surface contaminated objects group (SCO-II) shall not be carried in bulk.

RA03:   *Merged with RA02.*

7.1.6.12   *Ventilation*

The following additional requirements shall be met when they are indicated in column (10) of Table A of Chapter 3.2:

VE01:   Holds containing these substances shall be ventilated with the ventilators operating at full power, where after measurement it has been established that the concentration of gases given off by the cargo exceeds 10% of the lower explosive limit. The measurement shall be carried out immediately after loading. The measurement shall be repeated after one hour for monitoring purposes. The results of the measurement shall be recorded in writing.

VE02:   Holds containing these substances shall be ventilated with the ventilators operating at full power, where after measurement it has been established that the holds are not free from gases given off by the cargo. The measurement shall be carried out immediately after loading. The measurement shall be repeated after one hour for monitoring purposes. The results of the measurement shall be recorded in writing. Alternatively, on vessels only containing these substances in containers in open holds, the holds containing such containers may be ventilated with the ventilation operating at full power only when it is suspected that the holds are not free of gas. Prior to unloading, the unloader shall be informed about this suspicion.

VE03:   Spaces such as holds, accommodation and engine rooms, adjacent to holds containing these substances shall be ventilated.

After unloading, holds having contained these substances shall undergo forced ventilation.

After ventilation, the concentration of gases in these holds shall be measured.

The results of the measurement shall be recorded in writing.

VE04   When aerosols are carried for the purposes of reprocessing or disposal under special provision 327 of chapter 3.3, provisions of VE01 and VE02 are applied.

7.1.6.13   *Measures to be taken before loading*

The following additional requirements shall be met when they are indicated in column (11) of Table A of Chapter 3.2:

LO01:   Before these substances or articles are loaded, it shall be ensured that there are no metal objects in the hold which are not an integral part of the vessel.

LO02:   These substances may be loaded in bulk only if their temperature is not above 55 °C.

LO03:   Before loading these substances in bulk or unpackaged, holds should be made as dry as possible.

LO04:   Any loose organic material shall be removed from holds before loading these substances in bulk.

LO05:   Prior to carriage of pressure receptacles it shall be ensured that the pressure has not risen due to potential hydrogen generation.

7.1.6.14   *Handling and stowage of cargo*

The following additional requirements shall be met when they are indicated in column (11) of Table A of Chapter 3.2:

HA01:   These substances or articles shall be stowed at a distance of not less than 3.00 m from the accommodation, engine rooms, the wheelhouse and from any sources of heat.

HA02:   These substances or articles shall be stowed at a distance of not less than 2.00 m from the vertical planes defined by the sides of the vessel.

HA03:   Any friction, impact, jolting, overturning or dropping shall be prevented during handling of these substances or articles.

All packages loaded in the same hold shall be stowed and wedged as to prevent any jolting or friction during carriage.

Stacking of non-dangerous goods on top of packages containing these substances or articles is prohibited.

Where these substances or articles are loaded together with other goods in the same hold, these substances or articles shall be loaded after, and unloaded before, all the other goods.

There is no need for these substances or articles to be loaded after, and unloaded before, all others if these substances or articles are contained in containers.

While these substances or articles are being loaded or unloaded, no loading or unloading operations shall take place in the other holds and no filling or emptying of fuel tanks shall be allowed. The competent authority may, however, permit exceptions to this provision.

HA04:   *Merged with HA03.*

HA05:   *Merged with HA03.*

HA06:   *Merged with HA03.*

HA07:   It is prohibited to load or unload these substances in bulk or unpackaged if there is a danger that they may get wet because of the prevailing weather conditions.

HA08:   If the packages with these substances are not contained in a container, they shall be placed on gratings and covered with waterproof tarpaulins arranged in such a way that the water drains off to the outside and the air circulation is not hindered.

HA09:   If these substances are carried in bulk they shall not be loaded in the same hold together with flammable substances.

HA10:   These substances shall be stowed on deck in the protected area. For seagoing vessels, the stowage requirements are deemed to be met if the provisions of the IMDG Code are complied with.

7.1.6.15   *(Reserved)*

7.1.6.16   ***Measures to be taken during loading, carriage, unloading and handling of cargo***

The following additional requirements shall be met when they are indicated in column (11) of Table A of Chapter 3.2:

IN01:   After loading and unloading of these substances in bulk or unpackaged and before leaving the cargo transfer site, the concentration of gases in the accommodation, engine rooms and adjacent holds shall be measured by the consignor or consignee using a flammable gas detector.

        Before any person enters a hold and prior to unloading, the concentration of gases shall be measured by the consignee of the cargo.

        The hold shall not be entered or unloading started until the concentration of gases in the airspace above the cargo is below 50% of the lower explosive limit.

        If significant concentrations of gases are found in these spaces, the necessary safety measures shall be taken immediately by the consignor or the consignee.

IN02:   If a hold contains these substances in bulk or unpackaged, the gas concentration shall be measured in all other spaces of the vessel which are used by the crew at least once every eight hours with a toximeter. The results of the measurements shall be recorded in writing.

IN03:   If a hold contains these substances in bulk or unpackaged, the master shall make sure every day by checking the hold bilge wells or pump ducts that no water has entered the hold bilges.

        Water which has entered the hold bilges shall be removed immediately.

7.1.6.17 to 7.1.9.99   *(Reserved)*

CHAPTER 7.2

TANK VESSELS

7.2.0   **General requirements**

7.2.0.1   The provisions of 7.2.0 to 7.2.5 are applicable to tank vessels.

7.2.0.2 to 7.2.0.99   *(Reserved)*

7.2.1   **Mode of carriage of goods**

7.2.1.1 to 7.2.1.20   *(Reserved)*

7.2.1.21   *Carriage in cargo tanks*

7.2.1.21.1   Substances, their assignment to the various types of tank vessels and the special conditions for their carriage in these tank vessels, are listed in Table C of Chapter 3.2.

7.2.1.21.2   Substances, which according to column (6) of Table C of Chapter 3.2, have to be carried in a tank vessel of type N, open, may also be carried in a tank vessel of type N, open, with flame-arresters; type N, closed; types C or G provided that all conditions of carriage prescribed for tank vessels of type N, open, as well as all other conditions of carriage required for these substances in Table C of Chapter 3.2 are met.

7.2.1.21.3   Substances which, according to column (6) of Table C of Chapter 3.2 have to be carried in a tank vessel of type N, open, with flame-arresters, may also be carried in tank vessels of type N, closed, and types C or G provided that all conditions of carriage prescribed for tank vessels of type N, open, with flame arresters, as well as all other conditions of carriage required for these substances in Table C of Chapter 3.2 are met.

7.2.1.21.4   Substances which, according to column (6) of Table C of Chapter 3.2 have to be carried in a tank vessel of type N, closed, may also be carried in tank vessels of type C or G provided that all conditions of carriage prescribed for tank vessels of type N, closed, as well as all other conditions of carriage required for these substances in Table C of Chapter 3.2 are met.

7.2.1.21.5   Substances which, according to column (6) of Table C of Chapter 3.2 have to be carried in tank vessels of type C may also be carried in tank vessels of type G provided that all conditions of carriage prescribed for tank vessels of type C as well as all other conditions of carriage required for these substances in Table C of Chapter 3.2 are met.

7.2.1.21.6   Oily and greasy wastes resulting from the operation of the vessel may only be carried in fire-resistant receptacles, fitted with a lid, or in cargo tanks.

7.2.1.21.7   A substance which according to column (8) of Table C of Chapter 3.2 must be carried in cargo tank type 2 (integral cargo tank), may also be carried in a cargo tank type 1 (independent cargo tank) or cargo tank type 3 (cargo tank with walls distinct from the outer hull) of the vessel type prescribed in Table C or a vessel type prescribed in 7.2.1.21.2 to 7.2.1.21.5, provided that all other conditions of carriage required for this substance by Table C of Chapter 3.2 are met.

7.2.1.21.8   A substance which according to column (8) of Table C of Chapter 3.2 must be carried in cargo tank type 3 (cargo tank with walls distinct from the outer hull), may also be carried in a cargo tank type 1 (independent cargo tank) of the vessel type prescribed in Table C or a vessel type prescribed in 7.2.1.21.2 to 7.2.1.21.5 or in a type C vessel with cargo tank type 2 (integral cargo tank), provided that at least the conditions of carriage concerning the prescribed N type are met and all other conditions of carriage required for this substance by Table C of Chapter 3.2 or 7.2.1.21.2 to 7.2.1.21.5 are met.

7.2.1.22 to 7.2.1.99   *(Reserved)*

7.2.2   **Requirements applicable to vessels**

7.2.2.0   *Permitted vessels*

NOTE 1: The relief pressure of the safety valves or of the high-velocity vent valves shall be indicated in the certificate of approval (see 8.6.1.3).

NOTE 2: The design pressure and the test pressure of cargo tanks shall be indicated in the certificate of the recognised classification society prescribed in 9.3.1.8.1 or 9.3.2.8.1 or 9.3.3.8.1.

NOTE 3: Where a vessel carries cargo tanks with different valve-relief pressures, the relief pressure of each tank shall be indicated in the certificate of approval and the design and test pressures of each tank shall be indicated in the certificate of the recognised classification society.

7.2.2.0.1   Dangerous substances may be carried in tank vessels of Types G, C or N in accordance with the requirements of sections 9.3.1, 9.3.2 or 9.3.3 respectively. The type of tank vessel to be used is specified in column (6) of Table C in chapter 3.2 and in 7.2.1.21.

NOTE: The substances accepted for carriage in the individual vessel are listed in the vessel substance list to be drawn up by the recognised classification society (see 1.16.1.2.5).

7.2.2.1 to 7.2.2.4   *(Reserved)*

7.2.2.5   *Instructions for the use of devices and installations*

Where specific safety rules have to be complied with when using any device or installation, instructions for the use of the particular device or installation shall be readily available for consultation at appropriate places on board in the language normally spoken on board, and also, if that language is not English, French or German, in English, French or German unless agreements concluded between the countries concerned in the transport operation provide otherwise.

7.2.2.6   *Gas detection system*

The sensors of the gas detection system shall be set at not more than 20% of the lower explosive limit of the substances allowed for carriage in the vessel.

The system shall have been approved by the competent authority or a recognized classification society.

7.2.2.7 to 7.2.2.18   *(Reserved)*

7.2.2.19   *Pushed convoys and side-by-side formations*

7.2.2.19.1   Where at least one vessel of a convoy or side-by-side formation is required to be in possession of a certificate of approval for the carriage of dangerous goods, all vessels of such convoy or side-by-side formation shall be provided with an appropriate certificate of approval.

Vessels not carrying dangerous goods shall comply with the provisions of 7.1.2.19.

7.2.2.19.2   For the purposes of the application of this Chapter, the entire pushed convoy or side-by-side formation shall be deemed to be a single vessel.

7.2.2.19.3   When a pushed convoy or a side-by-side formation comprises a tank vessel carrying dangerous substances, vessels used for propulsion shall meet the requirements of the following paragraphs:

1.16.1.1, 1.16.1.2, 1.16.1.3, 7.2.2.5, 8.1.4, 8.1.5, 8.1.6.1, 8.1.6.3, 8.1.7, 9.3.3.0.1, 9.3.3.0.3 (d), 9.3.3.0.5, 9.3.3.10.1, 9.3.3.10.2, 9.3.3.12.4, 9.3.3.12.6, 9.3.3.16, 9.3.3.17.1 to 9.3.3.17.4, 9.3.3.31.1 to 9.3.3.31.5, 9.3.3.32.2, 9.3.3.34.1, 9.3.3.34.2, 9.3.3.40.1 (however, one single fire or ballast pump shall be sufficient), 9.3.3.40.2, 9.3.3.41, 9.3.3.50.1 (c), 9.3.3.50.2, 9.3.3.51, 9.3.3.52.3 to 9.3.3.52.6, 9.3.3.56.5, 9.3.3.71 and 9.3.3.74.

Vessels moving only type N open tank vessels do not have to meet the requirements of paragraphs 9.3.3.10.1, 9.3.3.10.2 and 9.3.3.12.6. In this case the following entry shall be made in the certificate of approval or provisional certificate of approval under number 5, permitted derogations: "Derogation from 9.3.3.10.1, 9.3.3.10.2 and 9.3.3.12.6; the vessel may only move tank vessels of type N open".

7.2.2.20   *(Reserved)*

7.2.2.21   *Safety and control equipment*

It shall be possible to interrupt loading or unloading of substances of Class 2 and substances assigned to UN Nos. 1280 and 2983 of Class 3 by means of switches installed at two locations on the vessel (fore and aft) and at two locations ashore (directly at the access to the vessel and at an appropriate distance on shore). Interruption of loading and unloading shall be effected by the means of a quick action stop valve which shall be directly fitted to the flexible connecting hose between the vessel and the shore facility.

The system of disconnection shall be designed in accordance with the closed circuit principle.

7.2.2.22   *Cargo tank openings*

When substances for which a type C vessel is required in column (6) of Table C of Chapter 3.2 are carried, the high-velocity vent valves shall be set so that blowing-off does not normally occur while the vessel is under way.

7.2.2.23 to 7.2.2.99   *(Reserved)*

7.2.3   General service requirements

7.2.3.1   *Access to cargo tanks, residual cargo tanks, cargo pump-rooms below deck, cofferdams, double-hull spaces, double bottoms and hold spaces; inspections*

7.2.3.1.1   The cofferdams shall be empty. They shall be inspected once a day in order to ascertain that they are dry (except for condensation water).

7.2.3.1.2   Access to the cargo tanks, residual cargo tanks, cofferdams, double-hull spaces, double bottoms and hold spaces is not permitted except for carrying out inspections or cleaning operations.

7.2.3.1.3   Access to the double-hull spaces and the double bottoms is not permitted while the vessel is under way.

7.2.3.1.4   When the gas concentration or oxygen content has to be measured before entry into cargo tanks, residual cargo tanks, cargo pump-rooms below deck, cofferdams, double-hull spaces, double bottoms or hold spaces, the results of these measurements shall be recorded in writing.

The measurement may only be effected by persons equipped with breathing apparatus suited to the substance carried.

Entry into these spaces is not permitted for the purpose of measuring.

7.2.3.1.5   Before any person enters cargo tanks, the cargo pump-rooms below deck, cofferdams, double-hull spaces, double bottoms or hold spaces:

(a)   When dangerous substances of Classes 2, 3, 4.1, 6.1, 8 or 9 for which a flammable gas detector is required in column (18) of Table C of Chapter 3.2 are carried on board the vessel, it shall be established, by means of this device that the gas concentration in these cargo tanks, cargo pump-rooms below deck, cofferdams, double-hull spaces, double bottoms or hold spaces is not more than 50% of the lower explosive limit of the cargo. For the cargo pump-rooms below deck this may be determined by means of the permanent gas detection system;

(b)   When dangerous substances of Classes 2, 3, 4.1, 6.1, 8 or 9 for which a toximeter is required in column (18) of Table C of Chapter 3.2 are carried on board the vessel, it shall be established, by means of this device that the cargo tanks, cargo pump-rooms below deck, cofferdams, double-hull spaces, double bottoms or hold spaces do not contain any significant concentration of toxic gases.

7.2.3.1.6   Entry into empty cargo tanks, the cargo pump-rooms below deck, cofferdams, double-hull spaces, double bottoms and hold spaces is not permitted, except where:

–   there is no lack of oxygen and no measurable amount of dangerous substances in dangerous concentrations; or

–   the person entering the spaces wears a self-contained breathing apparatus and other necessary protective and rescue equipment, and is secured by a line. Entry into these spaces is only permitted if this operation is supervised by a second person for whom the same equipment is readily at hand. Another two persons capable of giving assistance in an emergency shall be on the vessel within calling distance. If a rescue winch has been installed, only one other person is sufficient.

7.2.3.2   *Cargo pump-rooms below deck*

7.2.3.2.1   When carrying dangerous substances of classes 3, 4.1, 6.1, 8 or 9, the cargo pump-rooms below deck shall be inspected daily so as to ascertain that there are no leaks. The bilges and the drip pans shall be kept free from products.

7.2.3.2.2   When the gas detection system is activated, the loading and unloading operations shall be stopped immediately. All shut-off devices shall be closed and the cargo pump-rooms shall be evacuated immediately. All entrances shall be closed. The loading or unloading operations shall not be continued except when the damage has been repaired or the fault eliminated.

7.2.3.3 to 7.2.3.5   *(Reserved)*

7.2.3.6   *Gas detection system*

The gas detection system shall be maintained and calibrated in accordance with the instructions of the manufacturer.

7.2.3.7   *Gas-freeing of empty cargo tanks*

7.2.3.7.0   Gas-freeing of empty or unloaded cargo tanks is permitted under the conditions below but only if it is not prohibited on the basis of international or domestic legal requirements.

7.2.3.7.1   Empty or unloaded cargo tanks having previously contained dangerous substances of Class 2 or Class 3, with a classification code including the letter "T" in column (3b) of Table C of Chapter 3.2, Class 6.1 or packing group I of Class 8, may only be gas-freed by either competent persons according to sub-section 8.2.1.2 or companies approved by the competent authority for that purpose. Gas-freeing may be carried out only at the locations approved by the competent authority.

7.2.3.7.2   Gas-freeing of empty or unloaded cargo tanks having contained dangerous goods other than those referred to under 7.2.3.7.1 above, may be carried out while the vessel is underway or at locations approved by the competent authority by means of suitable venting equipment with the tank lids closed and by leading the gas/air mixtures through flame-arresters capable of withstanding steady burning. In normal conditions of operation, the gas concentration in the vented mixture at the outlet shall be less than 50% of the lower explosive limit. The suitable venting equipment may be used for gas-freeing by extraction only when a flame-arrester is fitted immediately before the ventilation fan on the extraction side. The gas concentration shall be measured once each hour during the two first hours after the beginning of the gas-freeing operation by forced ventilation or by extraction, by an expert referred to in 7.2.3.15. The results of these measurements shall be recorded in writing.

Gas-freeing is, however, prohibited within the area of locks including their lay-bys.

7.2.3.7.3   Where gas-freeing of cargo tanks having previously contained the dangerous goods referred to in 7.2.3.7.1 above is not practicable at the locations designated or approved for this purpose by the competent authority, gas-freeing may be carried out while the vessel is under way, provided that:

–   the requirements of 7.2.3.7.2 are complied with; the concentration of dangerous substances in the vented mixture at the outlet shall, however, be not more than 10% of the lower explosive limit;

–   there is no risk involved for the crew;

–   any entrances or openings of spaces connected to the outside are closed; this provision does not apply to the air supply openings of the engine room and overpressure ventilation systems;

–   any member of the crew working on deck is wearing suitable protective equipment;

–   it is not carried out within the area of locks including their lay-bys, under bridges or within densely populated areas.

7.2.3.7.4   Gas-freeing operations shall be interrupted during a thunderstorm or when, due to unfavourable wind conditions, dangerous concentrations of gases are to be expected outside the cargo area in front of accommodation, the wheelhouse and service spaces. The critical state is reached as soon as concentrations of more than 20% of the lower explosive limit have been detected in those areas by measurements by means of portable equipment.

7.2.3.7.5   The marking prescribed in column (19) of Table C of Chapter 3.2 may be withdrawn by the master when, after gas-freeing of the cargo tanks, it has been ascertained, using the equipment described in column (18) of Table C of Chapter 3.2, that the cargo tanks no longer contain flammable gases in concentrations of more than 20% of the lower explosive limit or do not contain any significant concentration of toxic gases.

7.2.3.7.6   Before taking measures which could cause hazards as described in section 8.3.5, cargo tanks and pipes in the cargo area shall be cleaned and gas-freed. The result of the gas-freeing shall be documented in a gas-free certificate. The condition of being gas-free may only be declared and certified by a person approved by the competent authority.

7.2.3.8 to 7.2.3.11   *(Reserved)*

7.2.3.12   *Ventilation*

7.2.3.12.1   While the machinery in the service spaces is operating, the extension ducts connected to the air inlets, if any, shall be in the upright position; otherwise the inlets shall be closed. This provision does not apply to air inlets of service spaces outside the cargo area, provided the inlets without extension duct are located not less than 0.50 m above the deck.

7.2.3.12.2   The ventilation of pump rooms shall be in operation:

–   at least 30 minutes before entry and during occupation;

–   during loading, unloading and gas-freeing; and

–   after the gas detection system has been activated.

7.2.3.13 and 7.2.3.14   *(Reserved)*

7.2.3.15   *Expert on board the vessel*

When dangerous substances are carried, the responsible master shall at the same time be an expert according to 8.2.1.2. In addition this expert shall be:

–   An expert as referred to in 8.2.1.5 when dangerous goods are carried for which a type G tank vessel is prescribed in column (6) of Table C of Chapter 3.2; and

–   An expert as referred to in 8.2.1.7 when dangerous goods are carried for which a type C tank vessel is prescribed in column (6) of Table C of Chapter 3.2.

NOTE: Which master of the vessel's crew is the responsible master shall be determined and documented on board by the carrier. If there is no such determination, the requirement applies to every master.

By derogation from this, for the loading and unloading of dangerous goods in a tank barge, it is sufficient that the person who is responsible for loading and unloading and for ballasting of the tank barge has the expertise required according to 8.2.1.2.

During the carriage of goods for which a type C tank vessel is prescribed in column (6) of Table C of Chapter 3.2 and cargo tank type 1 in column (8), an expert referred to in 8.2.1.5 for carriage in type G vessels is sufficient.

7.2.3.16 to 7.2.3.19   *(Reserved)*

7.2.3.20   *Water ballast*

7.2.3.20.1   Cofferdams and hold spaces containing insulated cargo tanks shall not be filled with water. Double-hull spaces, double bottoms and hold spaces which do not contain insulated cargo tanks may be filled with ballast water provided:

–   this has been taken into account in the intact and damage stability calculations; and

–   the filling is not prohibited in column (20) of Table C of Chapter 3.2.

If the water in the ballast tanks and compartments leads to the vessel no longer respecting these stability criteria:

—   fixed level indicators shall be installed; or

—   the filling level of the ballast tanks and compartments shall be checked daily before departure and during operations.

In case of the existence of level indicators, ballast tanks may also be partially filled. Otherwise they shall be completely full or empty.

7.2.3.20.2   (*Deleted*)

7.2.3.21   (*Reserved*)

7.2.3.22   *Entrances to hold spaces, cargo pump-rooms below deck and cofferdams, openings of cargo tanks and residual cargo tanks; closing devices*

The cargo tanks, residual cargo tanks and entrances to cargo pump-rooms below deck, cofferdams and hold spaces shall remain closed. This requirement shall not apply to cargo pump-rooms on board oil separator and supply vessels or to the other exceptions set out in this Part.

7.2.3.23 and 7.2.3.24   (*Reserved*)

7.2.3.25   *Connections between pipes*

7.2.3.25.1   Connecting two or more of the following groups of pipes is prohibited:

(a)   piping for loading and unloading;

(b)   pipes for ballasting and draining cargo tanks, cofferdams, hold spaces, double-hull spaces and double bottoms;

(c)   pipes located outside the cargo area.

7.2.3.25.2   The provision of 7.2.3.25.1 above does not apply to removable pipe connections between cofferdam pipes and

—   piping for loading and unloading;

—   pipes located outside the cargo area while the cofferdams have to be filled with water in an emergency.

In these cases the connections shall be designed so as to prevent water from being drawn from the cargo tanks. The cofferdams shall be emptied only by means of ejectors or an independent system within the cargo area.

7.2.3.25.3   The provisions of 7.2.3.25.1 (b) and (c) above do not apply to:

—   pipes intended for ballasting and draining double-hull spaces and double bottoms which do not have a common boundary with the cargo tanks;

—   pipes intended for ballasting hold spaces where the pipes of the fire-fighting system within the cargo area are used for this purpose. Double-hull and double bottom spaces and hold spaces shall be stripped only by means of ejectors or an independent system within the cargo area.

7.2.3.26 and 7.2.3.27   *(Reserved)*

7.2.3.28   *Refrigeration system*

For the carriage of refrigerated substances, an instruction shall be on board mentioning the permissible maximum loading temperature in relation to the capacity of the refrigeration system and the insulation design of the cargo tanks.

7.2.3.29   Lifeboats

7.2.3.29.1   The lifeboat required in accordance with the Regulations referred to in 1.1.4.6 shall be stowed outside the cargo area. The lifeboat may, however, be stowed in the cargo area provided an easily accessible collective life-saving appliance conforming to the Regulations referred to in 1.1.4.6 is available within the accommodation area.

7.2.3.29.2   7.2.3.29.1 above does not apply to oil separator or supply vessels.

7.2.3.30   *(Reserved)*

7.2.3.31   Engines

7.2.3.31.1   The use of engines running on fuels having a flash-point below 55° C (e.g. petrol engines) is prohibited. This requirement does not apply to the outboard motors of lifeboats.

7.2.3.31.2   The carriage of power-driven conveyances such as passenger cars and motor boats in the cargo area is prohibited.

7.2.3.32   Oil fuel tanks

Double bottoms with a height of at least 0.60 m may be used as oil fuel tanks, provided they have been constructed in accordance with Part 9.

7.2.3.33 to 7.2.3.40   *(Reserved)*

7.2.3.41   Fire and naked light

7.2.3.41.1   The use of fire or naked light is prohibited.

This provision does not apply to the accommodation and the wheelhouse.

7.2.3.41.2   Heating, cooking and refrigerating appliances shall not be fuelled with liquid fuels, liquid gas or solid fuels.

Cooking and refrigerating appliances may only be used in the accommodation and in the wheelhouse.

7.2.3.41.3   Heating appliances or boilers fuelled with liquid fuels having a flash-point above 55° C which are installed in the engine room or in another suitable space may, however, be used.

7.2.3.42   Cargo heating system

7.2.3.42.1   Heating of the cargo is not permitted except where there is risk of solidification of the cargo or where the cargo, because of its viscosity, cannot be unloaded in the usual manner.

In general, a liquid shall not be heated up to a temperature above its flash-point.

Special provisions are included in column 20 of Table C of Chapter 3.2.

7.2.3.42.2   Cargo tanks containing substances which are heated during transport shall be equipped with devices for measuring the temperature of the cargo.

7.2.3.42.3   During unloading, the cargo heating system may be used provided that the space where it has been installed meets in all respects the provisions of 9.3.2.52.3 or 9.3.3.52.3.

7.2.3.42.4   The provisions of 7.2.3.42.3 above do not apply when the cargo heating system is supplied with steam from shore and only the circulation pump is in operation, as well as when the flash-point of the cargo being unloaded is not less than 60° C.

7.2.3.43   *(Reserved)*

7.2.3.44   *Cleaning operations*

The use of liquids having a flash-point below 55° C for cleaning purposes is permitted only in the cargo area.

7.2.3.45 to 7.2.3.50   *(Reserved)*

7.2.3.51   *Electrical installations*

7.2.3.51.1   The electrical installations shall be properly maintained in a faultless condition.

7.2.3.51.2   The use of movable electric cables is prohibited in the cargo area.

This provision does not apply to:

– intrinsically safe electric circuits;

– electric cables for connecting signal lights or gangway lighting, provided the socket is permanently fitted to the vessel close to the signal mast or gangway;

– electric cables for connecting submerged pumps on board oil separator vessels.

7.2.3.51.3   The sockets for connecting the signal lights and gangway lighting or for submerged pumps on board oil separator vessels shall not be live except when the signal lights or the gangway lighting or the submerged pumps on board oil separator vessels are switched on.

Connecting or disconnecting shall not be possible except when the sockets are not live.

7.2.3.52 to 7.2.3.99   *(Reserved)*

7.2.4   **Additional requirements concerning loading, carriage, unloading and other handling of cargo**

7.2.4.1   *Limitation of the quantities carried*

7.2.4.1.1   The carriage of packages in the cargo area is prohibited. This prohibition does not apply to:

– residual cargo, washing water, cargo residues and slops contained in not more than six approved receptacles for residual products and receptacles for slops having a maximum individual capacity of not more than 2 m³. These receptacles for residual products shall meet the requirements of international regulations applicable to the substance concerned. The receptacles for residual products and the receptacles for slops shall be properly secured in the cargo area and comply with the provisions of 9.3.2.26.4 or 9.3.3.26.4 concerning them;

– to cargo samples, up to a maximum of 30, of substances accepted for carriage in the tank vessel, where the maximum contents are 500 ml per receptacle. Receptacles shall meet the packing requirements referred to in Part 4 of ADR and shall be placed on board, at a specific point in the cargo area, such that under normal conditions of carriage they cannot break or be punctured and their contents cannot spill in the hold space. Fragile receptacles shall be suitably padded.

7.2.4.1.2 On board oil separator vessel receptacles with a maximum capacity of 2.00 m³ oily and greasy wastes resulting from the operation of vessels may be placed in the cargo area provided that these receptacles are properly secured.

7.2.4.1.3 On board supply vessels or other vessels delivering products for the operation of vessels, packages of dangerous goods and non-dangerous goods may be carried in the cargo area up to a gross quantity of 5,000 kg provided that this possibility is mentioned in the certificate of approval. The packages shall be properly secured and shall be protected against heat, sun and bad weather.

7.2.4.1.4 On board supply vessels or other vessels delivering products for the operation of vessels, the number of cargo samples referred to in 7.2.4.1.1 may be increased from 30 to a maximum of 500.

7.2.4.2 *Reception of oily and greasy wastes resulting from the operation of vessels and delivery of products for the operation of vessels*

7.2.4.2.1 The reception of unpackaged liquid oily and greasy wastes resulting from the operation of vessels may only be effected by suction.

7.2.4.2.2 The landing and reception of oily and greasy wastes may not take place during the loading and unloading of substances for which protection against explosion is required in column (17) of Table C of Chapter 3.2 nor during the gas-freeing of tank vessels. This requirement does not apply to oil separator vessels provided that the provisions for protection against explosion applicable to the dangerous substance are complied with.

7.2.4.2.3 Berthing and handing over of products for the operation of vessels shall not take place during the loading or unloading of substances for which protection against explosions is required in column (17) of Table C of Chapter 3.2 nor during the gas-freeing of tank vessels. This requirement does not apply to supply vessels provided that the provisions for protection against explosion applicable to the dangerous substance are complied with.

7.2.4.2.4 The competent authority may issue derogations to the requirements of 7.2.4.2.1 and 7.2.4.2.2 above. During unloading it may also issue derogations to 7.2.4.2.3 above.

7.2.4.3 to 7.2.4.6 *(Reserved)*

7.2.4.7 *Places of loading and unloading*

7.2.4.7.1 Tank vessels shall be loaded, unloaded or gas-freed only at the places designated or approved for this purpose by the competent authority.

7.2.4.7.2 The reception of unpackaged oily and greasy liquid wastes resulting from the operation of vessels and the handing over of products for the operation of vessels shall not be taken to be loading or unloading within the meaning of 7.2.4.7.1 above.

7.2.4.8 *(Reserved)*

7.2.4.9   *Cargo transfer operations*

Partial or complete cargo transfer into another vessel without permission from the competent authority is prohibited outside a cargo transfer place approved for this purpose.

NOTE: For transhipment to means of transport of another mode see 7.2.4.7.1.

7.2.4.10   *Checklist*

7.2.4.10.1   Loading or unloading shall start only once a checklist conforming with section 8.6.3 of ADN has been completed for the cargo in question and questions 1 to 19 of the list have been checked off with an "X". Irrelevant questions should be deleted. The list shall be completed, after the pipes intended for the handling are connected and prior to the handling, in duplicate and signed by the master or a person mandated by him and the person responsible for the handling at the shore facilities. If a positive response to all the questions is not possible, loading or unloading is only permitted with the prior consent of the competent authority.

The competent authority may accept that, until 31 December 2016 at the latest, by derogation from 8.6.3 a control list containing question 4 in the version in force until 31 December 2014 be used.

7.2.4.10.2   The list shall conform to the model in 8.6.3.

7.2.4.10.3   The checklist shall be printed at least in languages understood by the master and the person responsible for the handling at the shore facilities.

7.2.4.10.4   The provisions of 7.2.4.10.1 to 7.2.4.10.3 above shall not apply to the reception of oily and greasy wastes by oil separator vessels nor to the handing over of products for the operation of vessels by supply vessels.

7.2.4.11   *Loading plan*

7.2.4.11.1   (*Deleted*)

7.2.4.11.2   The master shall enter on a cargo stowage plan the goods carried in the individual cargo tanks. The goods shall be described as in the transport document (information according to 5.4.1.1.2 (a) to (d)).

7.2.4.12   *Registration during the voyage*

The following particulars shall immediately be entered in the register referred to in 8.1.11:

Loading: Place of loading and loading berth, date and time, UN number or identification number of the substance, proper shipping name of the substance, the class and packing group if any;

Unloading: Place of unloading and unloading berth, date and time;

Gas-freeing of UN No. 1203 petrol: Gas-freeing place and facility or sector, date and time.

These particulars shall be provided for each cargo tank.

7.2.4.13   **Measures to be taken before loading**

7.2.4.13.1   When residues of the previous cargo may cause dangerous reactions with the next cargo, any such residues shall be properly removed.

Substances which react dangerously with other dangerous goods shall be separated by a cofferdam, an empty space, a pump-room, an empty cargo tank or a cargo tank loaded with a substance which does not react with the cargo.

Where an empty, uncleaned cargo tank, or a cargo tank containing cargo residues of a substance liable to react dangerously with other dangerous goods, this separation is not required if the master has taken appropriate measures to avoid a dangerous reaction.

If the vessel is equipped with piping for loading and unloading below the deck passing through the cargo tanks, the mixed loading or carriage of substances likely to react dangerously with each other is prohibited.

7.2.4.13.2   Before the start of loading operations, any prescribed safety and control devices and any items of equipment shall, if possible, be checked and controlled for proper functioning.

7.2.4.13.3   Before the start of loading operations the overflow control device switch shall be connected to the shore installation.

7.2.4.14   *Cargo handling and stowage*

Dangerous goods shall be loaded in the cargo area in cargo tanks, in cargo residue tanks or in packages permitted under 7.2.4.1.1.

7.2.4.15   *Measures to be taken after unloading (stripping system)*

7.2.4.15.1   If the provisions listed in 1.1.4.6.1 foresee the application of a stripping system, the cargo tanks and the cargo piping shall be emptied by means of the stripping system in accordance with the conditions laid down in the testing procedure after each unloading operation. This provision need not be complied with if the new cargo is the same as the previous cargo or a different cargo, the carriage of which does not require a prior cleaning of the cargo tanks.

Residual cargo shall be discharged ashore by means of the equipment provided for that effect (article 7.04 Nr. 1 and appendix II model 1 of CDNI) or shall be stored in the vessel's own tank for residual products or in receptacles for residual products according to 7.2.4.1.1.

7.2.4.15.2   During the filling of the receptacle for residual products, released gases shall be safely evacuated.

7.2.4.15.3   The gas-freeing of cargo tanks and piping for loading and unloading shall be carried out in compliance with the conditions of 7.2.3.7.

7.2.4.16   *Measures to be taken during loading, carriage, unloading and handling*

7.2.4.16.1   The loading rate and the maximum operational pressure of the cargo pumps shall be determined in agreement with the personnel of the shore installation.

7.2.4.16.2   All safety or control devices required in the cargo tanks shall remain switched on. During carriage this provision is only applicable for the installations mentioned in 9.3.1.21.1 (e) and (f), 9.3.2.21.1 (e) and (f) or 9.3.3.21.1 (e) and (f).

In the event of a failure of a safety or control device, loading or unloading shall be suspended immediately.

When a cargo pump-room is located below deck, the prescribed safety and control devices in the cargo pump-room shall remain permanently switched on.

Any failure of the gas detection system shall be immediately signalled in the wheelhouse and on deck by a visual and audible warning.

7.2.4.16.3 The shut-off devices of the loading and unloading piping as well as of the pipes of the stripping systems shall remain closed except during loading, unloading, stripping, cleaning or gas-freeing operations.

7.2.4.16.4 If the vessel is fitted with a transverse bulkhead according to 9.3.1.25.3, 9.3.2.25.3 or 9.3.3.25.3, the doors in this bulkhead shall be closed during loading and unloading.

7.2.4.16.5 Receptacles intended for recovering possible liquid spillage shall be placed under connections to shore installations used for loading and unloading. Before coupling and after uncoupling the connections and in between if necessary, the receptacles shall be emptied. These requirements shall not apply to the carriage of substance of Class 2.

7.2.4.16.6 In case of recovery of the gas-air mixture from shore into the vessel, the pressure at the connection point shall not be more than the opening pressure of the high velocity vent valve.

7.2.4.16.7 When a tank vessel conforms to 9.3.2.25.5 (d) or 9.3.3.22.5 (d), the individual cargo tanks shall be closed off during transport and opened during loading, unloading and gas-freeing.

7.2.4.16.8 Persons entering the premises located in the cargo area below deck during loading or unloading shall wear the PP equipment referred to in 8.1.5 if this equipment is prescribed in column (18) of Table C of Chapter 3.2.

Persons connecting or disconnecting the loading and unloading piping or the venting piping, or taking samples, carrying out measurements, replacing the flame arrester plate stack or relieving pressure in cargo tanks shall wear the PP equipment referred to in 8.1.5 if this equipment is prescribed in column (18) of Table C of Chapter 3.2. They shall also wear protective equipment A if a toximeter (TOX) is prescribed in column (18) of Table C of Chapter 3.2.

7.2.4.16.9 (a) During loading or unloading in a closed tank vessel of substances for which an open type N vessel with a flame arrester is sufficient according to columns (6) and (7) of Table C of Chapter 3.2, the cargo tanks may be opened using the safe pressure-relief device referred to in 9.3.2.22.4 (a) or 9.3.3.22.4 (a).

(b) During loading or unloading in a closed tank vessel of substances for which an open type N vessel is sufficient according to columns (6) and (7) of Table C of Chapter 3.2, the cargo tanks may be opened using the safe pressure-relief device referred to in 9.3.2.22.4 (a) or 9.3.3.22.4 (a) or using another suitable opening in the venting piping if any accumulation of water and its penetration into the cargo tanks is prevented and the opening is appropriately closed again after loading or unloading.

7.2.4.16.10 7.2.4.16.9 shall not apply when the cargo tanks contain gases or vapour from substances for the carriage of which a closed-type tank vessel is required in column (7) of Table C of Chapter 3.2.

7.2.4.16.11 The connection closure referred to in 9.3.1.21.1 (g), 9.3.2.21.1 (g) or 9.3.3.21.1 (g) can be opened only after a gastight connection has been made to the closed or partly closed sampling device.

7.2.4.16.12 For substances requiring protection against explosions according to column (17) of Table C of Chapter 3.2, the connection of the venting piping to the shore installation shall be such that the vessel is protected against detonations and the passage of flames from the shore. The protection of the vessel against detonations and the passage of flames from the shore is not required when the cargo tanks are inerted in accordance with 7.2.4.18.

7.2.4.16.13   For the carriage of substances of UN No. 2448, or of goods of Class 5.1 or 8, the bulwark ports, openings in the foot rail, etc., shall not be closed off. Nor shall they be closed off, during the voyage, in the event of carriage of other dangerous goods.

7.2.4.16.14   If supervision is required in column (20) of Table C of Chapter 3.2 for substances of Classes 2 or 6.1, loading and unloading shall be carried out under the supervision of a person who is not a member of the crew and has been mandated for the task by the consignor or the consignee.

7.2.4.16.15   The initial cargo throughput established in the loading instructions shall be such as to ensure that no electrostatic charge exists at the start of loading.

7.2.4.16.16   Measures to be taken before loading refrigerated liquefied gases

Unless the temperature of the cargo is controlled in accordance with 9.3.1.24.1 (a) or 9.3.1.24.1 (c) guaranteeing the use of the maximal boil-off in any service conditions, the holding time has to be determined by the master or another person on his behalf before loading and validated by the master or another person on his behalf during loading and shall be documented on board.

7.2.4.16.17   Determination of the holding time

A table, approved by the recognized classification society that certified the vessel, giving the relation between holding time and filling conditions, incorporating the parameters below shall be kept on board.

The holding time of the cargo shall be determined on the basis of the following parameters:

–   The heat transmission coefficient as defined in 9.3.1.27.9;

–   The set pressure of the safety valves;

–   The initial filling conditions (temperature of cargo during loading and degree of filling);

–   The ambient temperatures as given in 9.3.1.24.2;

–   When using the boil-off vapours, the minimum guaranteed use of the boil-off vapours (that is the amount of boil-off vapours used under any service conditions), may be taken into account.

Adequate safety margin

To leave an adequate margin to ensure safety, the holding time is at least three times the expected duration of the journey of the vessel, including the following:

–   To ensure safety for short journeys of (as expected) no more than 5 days, the minimum holding time for any vessel with refrigerated liquefied gases is 15 days.

–   For long journeys of (as expected) more than 10 days, the minimum holding time shall be 30 days, adding two days for each day the journeys takes more than 10 days.

As soon as it becomes clear that the cargo will not be unloaded within the holding time, the master shall inform the nearest emergency services according to 1.4.1.2.

7.2.4.17   *Closing of windows and doors*

7.2.4.17.1   During loading, unloading and gas-freeing operations, all entrances or openings of spaces which are accessible from the deck and all openings of spaces facing the outside shall remain closed.

This provision does not apply to:

–   air intakes of running engines;

–   ventilation inlets of engine rooms while the engines are running;

–   air intakes of the overpressure ventilation system referred to in 9.3.1.52.3, 9.3.2.52.3 or 9.3.3.52.3;

–   air intakes of air conditioning in installations if these openings are fitted with a gas detection system referred to in 9.3.1.52.3, 9.3.2.52.3 or 9.3.3.52.3.

These entrances and openings may only be opened when necessary and for a short time, after the master has given his permission.

7.2.4.17.2   After the loading, unloading and gas-freeing operations, the spaces which are accessible from the deck shall be ventilated.

7.2.4.17.3   The provisions of 7.2.4.17.1 and 7.2.4.17.2 above shall not apply to the reception of oily and greasy wastes resulting from the operation of vessels nor to the handing over of products for the operation of vessels.

7.2.4.18   *Blanketing of the cargo and inerting*

7.2.4.18.1   In cargo tanks and the corresponding piping, inerting in the gaseous phase or blanketing of the cargo may be necessary. Inerting and blanketing of the cargo are defined as follows:

–   Inerting: cargo tanks and the corresponding piping and other spaces for which this process is prescribed in column (20) of Table C of Chapter 3.2 are filled with gases or vapours which prevent combustion, do not react with the cargo and maintain this state;

–   Blanketing of the cargo: spaces in the cargo tanks above the cargo and the corresponding piping are filled with a liquid, gas or vapour so that the cargo is separated from the air and this state is maintained.

7.2.4.18.2   For certain substances the requirements for inerting and blanketing of the cargo in cargo tanks, in the corresponding piping and in adjacent empty spaces are given in column (20) of Table C of Chapter 3.2.

7.2.4.18.3   *(Reserved)*

7.2.4.18.4   Inerting or blanketing of flammable cargoes shall be carried out in such a way as to reduce the electrostatic charge as far as possible when the inerting agent is added.

7.2.4.19   *(Deleted)*

7.2.4.20   *(Reserved)*

7.2.4.21   *Filling of cargo tanks*

7.2.4.21.1   The degree of filling given in column (11) of Table C of Chapter 3.2 or calculated in accordance with 7.2.4.21.3 for the individual cargo tank shall not be exceeded.

7.2.4.21.2   The provisions of 7.2.4.21.1 above do not apply to cargo tanks the contents of which are maintained at the filling temperature during carriage by means of heating equipment. In this case calculation of the degree of filling at the beginning of carriage and control of the temperature shall be such that, during carriage, the maximum allowable degree of filling is not exceeded.

7.2.4.21.3   For carriage of substances having a relative density higher than that stated in the certificate of approval, the maximum permissible degree of filling of the cargo tanks shall be calculated in accordance with the following formula:

maximum permissible degree of filling (%) = a * 100/b

a = relative density stated in the certificate of approval,

b = relative density of the substance.

The degree of filling given in column (11) of Table C of Chapter 3.2 shall, however, not be exceeded.

NOTE: Furthermore, the requirements concerning stability, longitudinal strength and the deepest permissible draught of the vessel shall be observed when filling the cargo tanks.

7.2.4.21.4   If the degree of filling of 97.5% is exceeded a technical installation shall be authorized to pump off the overflow. During such an operation an automatic visual alarm shall be activated on deck.

7.2.4.22   *Opening of openings of cargo tanks*

7.2.4.22.1   Opening of cargo tanks apertures shall be permitted only after the tanks have been relieved of pressure.

7.2.4.22.2   Opening of sampling outlets and ullage openings and opening of the housing of the flame arrester shall not be permitted except for the purpose of inspecting or cleaning empty cargo tanks.

When in column (17) of Table C of Chapter 3.2 anti-explosion protection is required, the opening of cargo tank covers or of the housing of the flame arrester for the purpose of mounting or removing the flame arrester plate stack in unloaded cargo tanks shall be permitted only if the cargo tanks in question have been gas-freed and the concentration of flammable gases in the tanks is less than 10% of the lower explosive limit.

7.2.4.22.3   Sampling shall be permitted only if a device prescribed in column (13) of Table C of Chapter 3.2 or a device ensuring a higher level of safety is used.

Opening of sampling outlets and ullage openings of cargo tanks loaded with substances for which marking with one or two blue cones or one or two blue lights is prescribed in column (19) of Table C of Chapter 3.2 shall be permitted only when loading has been interrupted for not less than 10 minutes.

7.2.4.22.4   The sampling receptacles including all accessories such as ropes, etc., shall consist of electrostatically conductive material and shall, during sampling, be electrically connected to the vessel's hull.

7.2.4.22.5   The duration of opening shall be limited to the time necessary for control, cleaning, replacing the flame arrester, gauging or sampling.

7.2.4.22.6   Pressure relief of cargo tanks is permitted only when carried out by means of the device for safe pressure relief prescribed in 9.3.2.22.4 (a) or 9.3.3.22.4 (a).

7.2.4.22.7 The provisions of 7.2.4.22.1 to 7.2.4.22.6 above shall not apply to oil separator or supply vessels.

7.2.4.23 *(Reserved)*

7.2.4.24 *Simultaneous loading and unloading*

During loading or unloading of cargo tanks, no other cargo shall be loaded or unloaded. The competent authority may grant exceptions during unloading.

7.2.4.25 *Cargo piping*

7.2.4.25.1 Loading and unloading as well as stripping of cargo tanks shall be carried out by means of the fixed cargo piping of the vessel.

The metal fittings of the connections to the shore piping shall be electrically earthed so as to prevent the accumulation of electrostatic charges.

7.2.4.25.2 The loading and unloading piping shall not be extended by pipes or hose assemblies fore or aft beyond the cofferdams.

This requirement shall not apply to hose assemblies used for the reception of oily and greasy wastes resulting from the operation of vessels and the delivery of products for the operation of vessels.

7.2.4.25.3 The shut-off devices of the loading and unloading cargo piping shall not be open except as necessary during loading, unloading or gas-freeing operations.

7.2.4.25.4 The liquid remaining in the piping shall be completely drained into the cargo tanks, if possible, or safely removed. This requirement shall not apply to supply vessels.

7.2.4.25.5 The gas/air mixtures shall be returned ashore through a vapour return piping during loading operations when a closed type vessel is required in column (7) of Table C of Chapter 3.2.

7.2.4.25.6 When substances of Class 2 are carried the requirements of 7.2.4.25.4 shall be deemed to have been satisfied if the piping for loading and unloading have been purged with the cargo gas or with nitrogen.

7.2.4.26 and 7.2.4.27 *(Reserved)*

7.2.4.28 *Water-spray system*

7.2.4.28.1 If a gas or vapour water-spray system is required in column (9) of Table C of Chapter 3.2, it shall be kept ready for operation during loading, unloading and carriage. If a water-spray system is required to cool the tank-deck, it shall be kept ready for operation during the carriage.

7.2.4.28.2 When water-spraying is required in column (9) of Table C of Chapter 3.2 and the pressure of the gaseous phase in the cargo tanks may reach 80% of the relief pressure of the high velocity vent valves, the master shall take all measures compatible with safety to prevent the pressure from reaching that value. He shall in particular activate the water-spray system.

7.2.4.28.3 If a water-spray system is required in column (9) of Table C of Chapter 3.2 and remark 23 is indicated in column (20) of Table C of Chapter 3.2, the instrument measuring the internal pressure shall activate an alarm when the internal pressure reaches 40 kPa (0.4 bar). The water-spray system shall immediately be activated and remain in operation until the internal pressure drops to 30 kPa (0.3 bar).

7.2.4.29   *Transport of refrigerated liquefied gases*

During loading or unloading the drip tray as mentioned in 9.3.1.21.11 shall be placed under the shore connection of the piping for loading and unloading in use, and a water film as mentioned in 9.3.1.21.11 shall be activated.

7.2.4.30 to 7.2.4.39   *(Reserved)*

7.2.4.40   *Fire-extinguishing arrangements*

During loading and unloading, the fire extinguishing systems, the fire main with hydrants complete with couplings and jet/spray nozzles or with couplings and hose assemblies with couplings and jet/spray nozzles shall be kept ready for operation in the cargo area on deck.

The freezing of fire-mains and hydrants shall be prevented.

7.2.4.41   *Fire or naked light*

During loading, unloading or gas-freeing operations fires and naked lights are prohibited on board the vessel.

However, the provisions of 7.2.3.42.3 and 7.2.3.42.4 are applicable.

7.2.4.42   *Cargo heating system*

The maximum allowable temperature for carriage indicated in column (20) of Table C of Chapter 3.2 shall not be exceeded.

7.2.4.43 to 7.2.4.50   *(Reserved)*

7.2.4.51   *Electrical installations*

7.2.4.51.1   During loading, unloading or gas-freeing operations, only electrical equipment conforming to the rules for construction in Part 9 or which are installed in spaces complying with the conditions of 9.3.1.52.3, 9.3.2.52.3 or 9.3.3.52.3, may be used. All other electrical equipment marked in red shall be switched off.

7.2.4.51.2   Electrical equipment which has been switched off by the device referred to in 9.3.1.52.3, 9.3.2.52.3 or 9.3.3.52.3 shall only be switched on after the gas-free condition has been established in these spaces.

7.2.4.51.3   Equipment for active cathodic corrosion protection shall be disconnected before berthing and may not be re-connected until after the departure of the vessel, at earliest.

7.2.4.52   *(Reserved)*

7.2.4.53   *Lighting*

If loading or unloading is performed at night or in conditions of poor visibility, effective lighting shall be provided. If provided from the deck, it shall be effected by properly secured electric lamps which shall be positioned in such a way that they cannot be damaged. Where these lamps are positioned in the cargo area, they shall be of the "certified safe" type.

7.2.4.54 to 7.2.4.59   *(Reserved)*

7.2.4.60 *Special equipment*

The shower and the eye and face bath prescribed in the rules for construction shall be kept ready in all weather conditions for use during loading and unloading operations and cargo transfer operations by pumping.

7.2.4.61 to 7.2.4.73 *(Reserved)*

7.2.4.74 *Prohibition of smoking, fire and naked light*

The prohibition of smoking does not apply in accommodation or wheelhouses conforming to the provisions of 9.3.1.52.3, 9.3.2.52.3 or 9.3.3.52.3.

7.2.4.75 *Risk of sparking*

All electrical connections between the vessel and the shore shall be so designed that they do not present a source of ignition.

7.2.4.76 *Synthetic ropes*

During loading and unloading operations, the vessel may be moored by means of synthetic ropes only when steel cables are used to prevent the vessel from going adrift.

Steel cables sheathed in synthetic material or natural fibres are considered as equivalent when the minimum tensile strength required in accordance with the Regulations referred to in 1.1.4.6 is obtained from the steel strands.

Oil separator vessels may, however, be moored by means of appropriate synthetic ropes during the reception of oily and greasy wastes resulting from the operation of vessels, as may supply vessels and other vessels during the delivery of products for the operation of vessels.

7.2.4.77 Possible means of evacuation in case of an emergency

		Tank vessel/tank barge				
		Class				
		2, 3 packing group I, II and rest of III	3 packing group III (UN No. 1202: second and third entries in table C), 4.1	5.1, 6.1	8	9
1	Two escape routes inside or outside the cargo area in opposite directions	•	•	•	•	•
2	One escape route outside the cargo area and one safe haven outside the vessel including the escape route towards it from the opposite end	•	•	•	•	•
3	One escape route outside the cargo area and one safe haven on the vessel at the opposite end	•	•	•**	•	•
4	One escape route outside the cargo area and one life boat at the opposite end		•		•	•
5	One escape route outside the cargo area and one escape boat at the opposite end	•	•	•	•	•
6	One escape route inside the cargo area and one escape route outside the cargo area at the opposite end	•	•	•	•	•
7	One escape route inside the cargo area and one safe haven outside the vessel in the opposite direction	•	•	•	•	•
8	One escape route inside the cargo area and one safe haven on the vessel in the opposite direction	•	•	•**	•	•
9	One escape route inside the cargo area and one life boat at the opposite end		•		•	•
10	One escape route inside the cargo area and one escape boat at the opposite end	•	•	•	•	•
11	One escape route inside or outside the cargo area and two safe havens on the vessel at opposite ends	•	•	•**	•	•
12	One escape route inside or outside the cargo area and two safe areas on the vessel at opposite ends	•	•	•**	•	•
13	One escape route outside the cargo area		•		*•	•
14	One escape route inside the cargo area		•		*•	•
15	One or more safe havens outside the vessel, including the escape route towards it	•	•	•	*•	•

• = Possible option.

* = Not accepted in case of classification codes TFC, CF or CFT.

**= Not accepted if there is a risk that oxidizing substances in combination with flammable liquids may cause an explosion.

Based on local circumstances, competent authorities may prescribe additional requirements for the availability of means of evacuation.

7.2.4.78 to 7.2.4.99   *(Reserved)*

7.2.5   **Additional requirements concerning the operation of vessels**

7.2.5.0   *Marking*

7.2.5.0.1   Vessels carrying dangerous goods listed in Table C of Chapter 3.2 shall display the number of blue cones or blue lights indicated in column (19) and in accordance with CEVNI. When because of the cargo carried no marking with blue cones or blue lights is prescribed but the concentration of flammable gases in the cargo tanks is higher than 20% of the lower explosion limit, the number of blue cones or blue lights to be carried is determined by the last cargo for which this marking was required.

7.2.5.0.2   When more than one marking should apply to a vessel, the first of the options below shall apply:

–   two blue cones or two blue lights; or

–   one blue cone or one blue light.

7.2.5.0.3   By derogation from 7.2.5.0.1 above, and in accordance with the footnotes to article 3.14 of the CEVNI, the competent authority of a Contracting Party may authorize seagoing vessels temporarily operating in an inland navigation area on the territory of this Contracting Party, the use of the day and night signals prescribed in the Recommendations on the Safe Transport of Dangerous Cargoes and Related Activities in Port Areas adopted by the Maritime Safety Committee of the International Maritime Organization (by night an all-round fixed red light and by day flag "B" of the International Code of Signals), instead of the signals prescribed in 7.2.5.0.1. The competent authority which has taken the initiative with respect to the derogation granted shall notify the Executive Secretary of the UNECE, who shall bring this derogation to the attention of the Administrative Committee.

7.2.5.1   *Mode of navigation*

The competent authorities may impose restrictions on the inclusion of tank vessels in pushed convoys of large dimension.

7.2.5.2   *(Reserved)*

7.2.5.3   *Mooring*

Vessels shall be moored securely, but in such a way that electrical power cables and hose assemblies are not subject to tensile strain and the vessels can be released quickly in an emergency.

7.2.5.4   *Berthing*

7.2.5.4.1   The distances from other vessels to be kept by berthed vessels carrying dangerous goods shall be not less than those prescribed by the Regulations referred to in 1.1.4.6.

7.2.5.4.2   An expert, as required by 7.2.3.15 shall be permanently on board berthed vessels carrying dangerous substances. The competent authority may, however, exempt from this obligation those vessels which are berthed in the harbour basin or in a permitted berthing position.

7.2.5.4.3   Outside the berthing areas specifically designated by the competent authority, the distances to be kept by berthed vessels shall not be less than:

–   100 m   from residential areas, civil engineering structures or storage tanks, if the vessel is required to be marked with one blue cone or blue light in accordance with column (19) of Table C of Chapter 3.2;

–   100 m   from civil engineering structures and storage tanks; and 300 m from residential areas if the vessel is required to be marked with two blue cones or two blue lights in accordance with column (19) of Table C of Chapter 3.2.

While waiting in front of locks or bridges, vessels are allowed to keep distances less than those given above. In no case shall the distance be less than 100 m.

7.2.5.4.4   The competent authority may prescribe distances less than those given in 7.2.5.4.3 above.

7.2.5.5 to 7.2.5.7   *(Reserved)*

7.2.5.8   *Reporting duty*

7.2.5.8.1   In the States where the reporting duty is in force, the master of the vessel shall provide information in accordance with paragraph 1.1.4.6.1.

7.2.5.8.2 to 7.2.5.8.4   *(Deleted)*

7.2.5.9 to 7.2.9.99   *(Reserved)*

PART 8

Provisions for vessel crews, equipment, operation and documentation

CHAPTER 8.1

GENERAL REQUIREMENTS APPLICABLE TO VESSELS AND EQUIPMENT

8.1.1   (*Reserved*)

8.1.2   Documents

8.1.2.1   In addition to the documents required by other regulations, the following documents shall be kept on board:

(a)   The vessel's certificate of approval referred to in 1.16.1.1 or the vessel's provisional certificate of approval referred to in 1.16.1.3 and the annex referred to in 1.16.1.4;

(b)   Transport documents referred to in 5.4.1 for all dangerous goods on board and, where necessary the large container, vehicle or wagon packing certificate (see 5.4.2);

(c)   The instructions in writing prescribed in 5.4.3;

(d)   A copy of the ADN with the latest version of its annexed Regulations which may be a copy which can be consulted by electronic means at any time;

(e)   The inspection certificate of the insulation resistance of the electrical installations prescribed in 8.1.7;

(f)   The inspection certificate of the fire–extinguishing hoses prescribed in 8.1.6.1;

(g)   A book in which all required measurement results are recorded;

(h)   A copy of the relevant text of the special authorizations referred to in 1.5 if the transport operation is performed under this/these special authorization(s);

(i)   Means of identification, which include a photograph, for each crew member, in accordance with 1.10.1.4; and

(j)   (*Deleted*).

8.1.2.2   In addition to the documents prescribed in 8.1.2.1, the following documents shall be carried on board dry cargo vessels:

(a)   The stowage plan prescribed in 7.1.4.11;

(b)   The ADN specialized knowledge certificate prescribed in 8.2.1.2;

(c)   For vessels complying with the additional requirements for double–hull vessels:

–   a damage–control plan;

–   the documents concerning intact stability as well as all conditions of intact stability taken into account for the damaged stability calculation in a form the master understands;

–   the certificate of the recognized classification society (see 9.1.0.88 or 9.2.0.88);

(d)   The inspection certificates concerning the fixed fire extinguishing systems prescribed in 9.1.0.40.2.9.

8.1.2.3   In addition to the documents prescribed in 8.1.2.1, the following documents shall be carried on board tank vessels:

(a)   The cargo stowage plan prescribed in 7.2.4.11.2;

(b)   The ADN specialized knowledge certificate prescribed in 7.2.3.15;

(c)   For vessels which have to conform to the conditions of damage–control (see 9.3.1.15, 9.3.2.15 or 9.3.3.15)

–   a damage–control plan;

–   the documents concerning intact stability as well as all conditions of intact stability taken into account for the damaged stability calculation in a form the master understands; the stability booklet and the proof of the loading instrument having been approved by the recognized classification society;

(d)   The documents concerning the electrical installations prescribed in 9.3.1.50, 9.3.2.50 or 9.3.3.50;

(e)   The certificate of class issued by the recognized classification society prescribed in 9.3.1.8.1, 9.3.2.8.1 or 9.3.3.8.1;

(f)   The flammable gas detector certificate prescribed in 9.3.1.8.3, 9.3.2.8.3 or 9.3.3.8.3;

(g)   The vessel substance list prescribed in 1.16.1.2.5;

(h)   The inspection certificate for the hose assemblies for loading and unloading prescribed in 8.1.6.2;

(i)   The instructions relating to the loading and unloading flows prescribed in 9.3.2.25.9 or 9.3.3.25.9;

(j)   *(Deleted);*

(k)   In the event of the carriage of goods having a melting point $\geq 0°$ C, heating instructions;

(l)   The inspection certificate for the pressure relief and vacuum relief valves prescribed in 8.1.6.5, except for open type N tank vessels, or open type N vessels with flame–arresters;

(m)   The registration document referred to in 8.1.11;

(n)   For the carriage of refrigerated substances, the instruction required in 7.2.3.28;

(o)   The certificate concerning the refrigeration system, prescribed in 9.3.1.27.10, 9.3.2.27.10 or 9.3.3.27.10;

(p)   The inspection certificates concerning the fixed fire extinguishing systems prescribed in 9.3.1.40.2.9, 9.3.2.40.2.9 or 9.3.3.40.2.9; and

(q)   When transporting refrigerated liquefied gases and the temperature is not controlled in accordance with 9.3.1.24.1 (a) and 9.3.1.24.1 (c), the determination of the holding time (7.2.4.16.16, 7.2.4.16.17). The heat transmission coefficient shall be documented and kept on board.

8.1.2.4   The instructions in writing referred to in 5.4.3 shall be handed to the master before loading. They shall be kept readily at hand in the wheelhouse.

On board dry cargo vessels, the transport documents shall be handed to the master before loading and on board tank vessels they shall be handed to him after loading and before the journey commences.

8.1.2.5   (*Reserved*)

8.1.2.6   The presence on board of the certificate of approval is not required in the case of pusher barges which are not carrying dangerous goods, provided that the following additional particulars are indicated, in identical lettering, on the plate furnished by CEVNI:

Number of the certificate of approval: …

issued by: …

valid until: …

The barge–owner shall thereafter keep the certificate of approval and the annex covered by 1.16.1.4 in his possession.

The similarity of the particulars on the plate and those contained in the certificate of approval shall be certified by a competent authority which shall affix its stamp to the plate.

8.1.2.7   The presence on board of the certificate of approval is not required in the case of dry cargo barges or tank barges carrying dangerous goods provided that the plate furnished by CEVNI is supplemented by a second metal or plastic plate reproducing by photo–optical means a copy of the entire certificate of approval. A photo–optical copy of the annex referred to in 1.16.1.4 is not required.

The barge–owner shall thereafter keep the certificate of approval and the annex referred to in 1.16.1.4 in his possession.

The similarity of the particulars on the plate and the certificate of approval shall be certified by a competent authority which shall affix its stamp to the plate.

8.1.2.8   All documents shall be on board in a language the master is able to read and understand. If that language is not English, French or German, all documents, with the exception of the copy of ADN with its annexed Regulations and those for which the Regulations include special provisions concerning languages, shall be on board also in English, French or German, unless agreements concluded between the countries concerned in the transport operation provide otherwise.

8.1.2.9   8.1.2.1 (b), 8.1.2.1 (g), 8.1.2.4 and 8.1.2.5 do not apply to oil separator vessels or supply vessels. 8.1.2.1 (c) does not apply to oil separator vessels.

8.1.3   (*Reserved*)

8.1.4   Fire–extinguishing arrangements

In addition to the fire–extinguishing appliances prescribed in the Regulations referred to in 1.1.4.6, each vessel shall be equipped with at least two additional hand fire–extinguishers having the same capacity. The fire–extinguishing agent contained in these additional hand fire–extinguishers shall be suitable for fighting fires involving the dangerous goods carried.

8.1.5   **Special equipment**

8.1.5.1   Insofar as the provisions of Chapter 3.2, Tables A or C require, the following equipment shall be available on board:

PP: for each member of the crew, a pair of protective goggles, a pair of protective gloves, a protective suit and a suitable pair of protective shoes (or protective boots, if necessary). On board tank vessels, protective boots are required in all cases;

EP: a suitable escape device for each person on board;

EX: a flammable gas detector with the instructions for its use;

TOX: a toximeter with the instructions for its use;

A: a breathing apparatus ambient air–dependent.

8.1.5.2   (*Reserved*)

8.1.5.3   For pushed convoys or side–by–side formations under way, it shall be sufficient, however, if the pusher tug or the vessel propelling the formation is equipped with the special equipment referred to in 8.1.5.1 above, when this is required in Chapter 3.2, Tables A or C.

8.1.6   **Checking and inspection of equipment**

8.1.6.1   Hand fire–extinguishers and fire–extinguishing hoses shall be inspected at least once every two years by persons authorized for this purpose by the competent authority. Proof of inspection shall be affixed to the hand fire–extinguishers. A certificate concerning this inspection shall be carried on board. A certificate concerning the inspection of fire extinguishing hoses shall be carried on board.

8.1.6.2   Hose assemblies used for loading, unloading or delivering products for the operation of the vessel and residual cargo shall comply with European standard EN 12115:2011–04 (Rubber and thermoplastics hoses and hose assemblies) or EN 13765:2010–08 (Thermoplastic multilayer (non–vulcanized) hoses and hose assemblies) or EN ISO 10380:2003–10 (Corrugated metal hoses and hose assemblies). They shall be checked and inspected in accordance with table A.1 of standard EN 12115:2011–04 or table K.1 of standard EN 13765:2010–08 or paragraph 7 of standard EN ISO 10380:2003–10 at least once a year, according to the manufacturer's instructions, by persons authorized for this purpose by the competent authority. A certificate concerning this inspection shall be carried on board.

8.1.6.3   The special equipment referred to in 8.1.5.1 and the gas detection system shall be checked and inspected in accordance with the instructions of the manufacturer by the manufacturer concerned or by persons authorized for this purpose by the competent authority. A certificate concerning this inspection shall be carried on board.

8.1.6.4   The measuring instruments prescribed in 8.1.5.1 shall be checked each time before use by the user in accordance with the instructions for use.

8.1.6.5   The pressure relief and vacuum relief valves prescribed in 9.3.1.22, 9.3.2.22, 9.3.2.26.4, 9.3.3.22 and 9.3.3.26.4 shall be inspected on each renewal of the certificate of approval by the manufacturer or by a firm approved by the manufacturer. A certificate concerning this inspection shall be carried on board.

8.1.6.6   (*Deleted*)

8.1.7   **Electrical installations**

The insulation resistance of the electrical installations, the earthing and the certified safe type electrical equipment and the conformity of the documents required in 9.3.1.50.1, 9.3.2.50.1 or 9.3.3.50.1 with the circumstances on board shall be inspected whenever the certificate of approval is renewed and, in addition, within the third year from the date of issue of the certificate of approval by a person authorized for this purpose by the competent authority. An appropriate inspection certificate shall be kept on board.

8.1.8   *(Deleted)*

8.1.9   *(Deleted)*

8.1.10   *(Deleted)*

8.1.11   **Register of operations during carriage relating to the carriage of UN 1203**

Tank vessels accepted for the carriage of UN No. 1203 petrol shall have on board a register of operations during carriage. This register may consist of other documents containing the information required. This register or these other documents shall be kept on board for not less than three months and cover at least the last three cargoes.

CHAPTER 8.2

REQUIREMENTS CONCERNING TRAINING

8.2.1   **General requirements concerning training of experts**

8.2.1.1   An expert shall not be less than 18 years of age.

8.2.1.2   An expert is a person who has a special knowledge of the ADN. Proof of this knowledge shall be furnished by means of a certificate from a competent authority or from an agency recognized by the competent authority.

This certificate shall be issued to persons who, after training, have passed a qualifying ADN examination.

8.2.1.3   The experts referred to in 8.2.1.2 shall take part in a basic training course. Training shall take place in the context of classes approved by the competent authority. The primordial objective of the training is to make the experts aware of the hazards of the carriage of dangerous goods and provide them with the necessary basic knowledge to reduce the dangers of an incident to a minimum, to enable them to take the necessary measures to ensure their own safety, general safety and the protection of the environment and to limit the consequences of the incident. This training, which shall include individual practical exercises, takes the form of a basic course; it shall cover at least the objectives referred to in 8.2.2.3.1.1 and in 8.2.2.3.1.2 or 8.2.2.3.1.3.

8.2.1.4   After five years, the certificate shall be renewed by the competent authority or by a body recognized by it if the expert furnishes proof , of successful completion of a refresher course taken in the last year prior to the expiry of the certificate, covering at least the objectives referred to in 8.2.2.3.1.1 and in 8.2.2.3.1.2 or 8.2.2.3.1.3 and comprising current new developments in particular. A refresher course shall be considered to have been successfully completed if a final written test conducted by the course organizer under 8.2.2.2 has been passed. The test can be retaken as often as desired during the validity of the certificate. The new period of validity shall begin on the expiry date of the certificate; if the test is passed more than one year before the date of expiry of the certificate, it shall begin on the date of the certificate of participation in the course.

8.2.1.5   Experts for the carriage of gases shall take part in a specialization course covering at least the objectives referred to in 8.2.2.3.3.1. Training shall take place in the context of classes approved by the competent authority. An expert certificate shall be issued to persons who, after training, have successfully passed an examination concerning the carriage of gases and have produced evidence of not less than one year's work on board a type G vessel during a period of two years prior to or following the examination.

8.2.1.6   After five years, the certificate shall be renewed by the competent authority or by a body recognized by it if the expert on the carriage of gases furnishes proof:

–   that during the year preceding the expiry of the certificate, he has participated in a refresher specialization course covering at least the objectives referred to in 8.2.2.3.3.1 and comprising current new developments in particular, or

–   that during the previous two years he has performed a period of work of not less than one year on board a type G tank vessel.

When the refresher specialization training course is taken in the year preceding the date of expiry of the certificate, the new period of validity shall begin on the expiry date of the preceding certificate, but in other cases it shall begin on the date of certification of participation in the course.

| 8.2.1.7 | Experts for the carriage of chemicals shall take part in a specialization course covering at least the objectives referred to in 8.2.2.3.3.2. Training shall take place in the context of classes approved by the competent authority. An expert certificate shall be issued to persons who, after training, have successfully passed an examination concerning the carriage of chemicals and have produced evidence of not less than one year's work on board a type C vessel during a period of two years prior to or following the examination. |

8.2.1.8 After five years, the certificate shall be renewed by the competent authority or by a body recognized by it if the expert on the carriage of chemicals furnishes proof:,

– that during the year preceding the expiry of the certificate, he has participated in a refresher specialization course covering at least the objectives referred to in 8.2.2.3.3.2 and comprising current new developments in particular, or

– that during the previous two years he had performed a period of work of not less than one year on board a type C tank vessel.

When the refresher specialization training course is taken in the year preceding the date of expiry of the certificate, the new period of validity shall begin on the expiry date of the preceding certificate, but in other cases it shall begin on the date of certification of participation in the course.

8.2.1.9 The document attesting training and experience in accordance with the requirements of Chapter V of the STCW Code on Training and Qualifications of Masters, Officers and Ratings of Tankers carrying LPG/LNG shall be equivalent to the certificate referred to in 8.2.1.5, provided it has been recognized by a competent authority. No more than five years shall have passed since the date of issue or renewal of such a document.

8.2.1.10 The document attesting training and experience in accordance with Chapter V of the STCW Code on Training and Qualifications of Masters, Officers and Ratings of Tankers carrying chemicals in bulk shall be equivalent to the certificate referred to in 8.2.1.7, provided it has been recognized by a competent authority. No more than five years shall have passed since the date of issue or renewal of such a document.

8.2.1.11 The certificate shall conform to the model in 8.6.2.

8.2.2 **Special requirements for the training of experts**

8.2.2.1 Theoretical knowledge and practical abilities shall be acquired as a result of training in theory and practical exercises. The theoretical knowledge shall be tested by an examination. During the refresher course exercises and tests shall ensure that the participant takes an active role in the training.

8.2.2.2 The training organizer shall ensure that training instructors have a good knowledge of the subject and shall take into account the latest developments concerning the Regulations and the requirements for training in the transport of dangerous goods. Teaching shall relate closely to practice. In accordance with the approval, the teaching syllabus shall be drawn up on the basis of the objectives referred to in 8.2.2.3.1.1 to 8.2.2.3.1.3 and in 8.2.2.3.3.1 or 8.2.2.3.3.2. Basic training and their refresher courses shall comprise individual practical exercises (see 8.2.2.3.1.1).

8.2.2.3 *Organization of training*

Initial basic training and the refresher courses shall be organized in the context of basic courses (see 8.2.2.3.1) and if necessary specialization courses (see 8.2.2.3.3). The courses referred to in 8.2.2.3.1 may comprise three variants: transport of dry cargo, transport in tank vessels and a combination of transport of dry cargo and transport in tank vessels.

8.2.2.3.1 *Basic course*

Basic course on the transport of dry cargo

Prior training:	none
Knowledge:	ADN in general, except Chapter 3.2, Table C, Chapters 7.2 and 9.3
Authorized for:	dry cargo vessel
Training:	general 8.2.2.3.1.1 and dry cargo vessels 8.2.2.3.1.2

Basic course on transport by tank vessels

Prior training:	none
Knowledge:	ADN in general, except Chapter 3.2, Tables A and B, Chapters 7.1, 9.1, 9.2 and sections 9.3.1 and 9.3.2
Authorized for:	tank vessels for the transport of substances for which a type N tank vessel is prescribed
Training:	general 8.2.2.3.1.1 and tank vessels 8.2.2.3.1.3

Basic course – combination of transport of dry cargo and transport in tank vessels

Prior training:	none
Knowledge:	ADN in general, except sections 9.3.1 and 9.3.2
Authorized for:	dry cargo vessels and tank vessels for the transport of substances for which a type N tank vessel is prescribed
Training:	general 8.2.2.3.1.1, dry cargo vessels 8.2.2.3.1.2 and tank vessels 8.2.2.3.1.3

8.2.2.3.1.1 The general part of the basic training course shall comprise at least the following objectives:

General:

– Objectives and structure of ADN.

Construction and equipment:

– Construction and equipment of vessels subject to ADN.

Measurement techniques:

– Measurements of toxicity, oxygen content, explosivity.

Knowledge of products:

– Classification and hazard characteristics of the dangerous goods.

Loading, unloading and transport:

– Loading, unloading, general service requirements and requirements relating to transport.

Documents:

– Documents which must be on board during transport.

Hazards and measures of prevention:

– General safety measures.

Practical exercises:

– Practical exercises, in particular with respect to entry into spaces, use of fire–extinguishers, fire–fighting equipment and personal protective equipment as well as flammable gas detectors, oxygen meters and toximeters.

Stability:

– parameters of relevance to stability;

– heeling moments;

– exemplary calculations;

– damage stability, intermediate states and final state of flooding;

– influence of free surfaces;

– evaluation of stability on the basis of existing stability criteria (text of Regulations);

– evaluation of intact stability with the help of the lever arm curve

– application of loading instruments;

– use of loading instruments;

– application of the stability booklet according to 9.3.13.3.

8.2.2.3.1.2 The "dry cargo vessels" part of the basic training course shall comprise at least the following objectives:

Construction and equipment:

– Construction and equipment of dry cargo vessels.

Treatment of holds and adjacent spaces:

– degassing, cleaning, maintenance,

– ventilation of holds and spaces outside the protected area.

Loading, unloading and transport:

– loading, unloading, general service and transport requirements,

– labelling of packages.

Documents:

– documents which must be on board during transport.

Hazards and measures of prevention:

– general safety measures,

– personal protective and safety equipment.

8.2.2.3.1.3   The "tank vessel" part of the basic training course shall comprise at least the following objectives:

Construction and equipment:

– construction and equipment of tank vessels,

– ventilation,

– loading and unloading systems.

Treatment of cargo tanks and adjacent spaces:

– degassing, cleaning, maintenance,

– heating and cooling of cargo,

– handling of receptacles for residual products.

Measurement and sampling techniques:

– measurements of toxicity, oxygen content and explosivity,

– sampling.

Loading, unloading and transport:

– loading, unloading, general service and transport requirements.

Documents:

– documents which must be on board during transport.

Hazards and measures of prevention:

– prevention and general safety measures,

– spark formation,

– personal protective and safety equipment,

– fires and fire–fighting.

8.2.2.3.2   *Refresher training courses*

Refresher training course on transport of dry cargo

Prior training:	valid ADN "dry cargo vessels" or combined "dry cargo vessels/tank vessels" certificate
Knowledge:	ADN in general, except Chapter 3.2, Table C, Chapters 7.2 and 9.3
Authorized for:	dry cargo vessel
Training:	general 8.2.2.3.1.1 and dry cargo vessels 8.2.2.3.1.2

Refresher training course on transport in tank vessels

Prior training:	valid ADN "tank vessels" or combined "dry cargo vessels/tank vessels" certificate
Knowledge:	ADN in general, except Chapter 3.2, Tables A and B, Chapters 7.1, 9.1 and 9.2 and sections 9.3.1 and 9.3.2
Authorized for:	tank vessels for the transport of substances for which a type N tank vessel is prescribed
Training:	general 8.2.2.3.1.1 and tank vessels 8.2.2.3.1.3

Refresher training course – combination of transport of dry cargo and transport in tank vessels

Prior training:	valid ADN combined "dry cargo vessels and tank vessels" certificate
Knowledge:	ADN in general, including sections 9.3.1 and 9.3.2
Authorized for:	dry cargo vessels and tank vessels for the transport of substances for which a type N tank vessel is prescribed
Training:	general 8.2.2.3.1.1, dry cargo vessels 8.2.2.3.1.2 and tank vessels 8.2.2.3.1.3

8.2.2.3.3   *Specialization courses*

Specialization course on gases

Prior training:	valid ADN "tank vessels" or combined "dry cargo vessels/tank vessels" certificate
Knowledge:	ADN, in particular knowledge relating to loading, transport, unloading and handling of gases
Authorization for:	tank vessels for the transport of substances for which a type G tank vessel is required and transport in type G of substances for which a type C is required with cargo tank design 1 required in column (7) of Table C of Chapter 3.2.
Training:	gases 8.2.2.3.3.1

Specialization course on chemicals

Prior training: valid ADN "tank vessels" or combined "dry cargo vessels/tank vessels" certificate

Knowledge: ADN, in particular knowledge relating to loading, transport, unloading and handling of chemicals

Authorized for: tank vessels for the transport of substances for which a type C tank vessel is required

Training: chemicals 8.2.2.3.3.2

8.2.2.3.3.1 The specialization course on gases shall comprise at least the following objectives:

Knowledge of physics and chemistry:

– laws of gases, e.g. Boyle, Gay–Lussac and fundamental law

– partial pressures and mixtures, e.g. definitions and simple calculations, pressure increase and gas release from cargo tanks

– Avogadro's number and calculation of masses of ideal gas and application of the mass formula

– mass density, relative density and volume of liquids, e.g. mass density, relative density, volume in terms of temperature increase and maximum degree of filling

– critical pressure and temperature

– polymerization, e.g. theoretical and practical questions, conditions of carriage

– vaporization, condensation, e.g. definition, liquid volume and vapour volume ratio

– mixtures, e.g. vapour pressure, composition and hazard characteristics

– chemical bonds and formulae.

Practice:

– flushing of cargo tanks, e.g. flushing in the event of a change of cargo, addition of air to the cargo, methods of flushing (degassing) before entering cargo tanks

– sampling

– danger of explosion

– health risks

– gas concentration measures, e.g. which apparatus to use and how to use it

– monitoring of closed spaces and entry to these spaces

– certificates for degassing and permitted work

– degree of filling and over–filling

– safety installations

–   pumps and compressors

–   handling refrigerated liquefied gases.

Emergency measures:

–   physical injury, e.g. substances on the skin, breathing in gas, assistance

–   irregularities relating to the cargo, e.g. leak in a connection, over-filling, polymerization and hazards in the vicinity of the vessel.

8.2.2.3.3.2   The specialization course on chemicals shall comprise at least the following objectives:

Knowledge of physics and chemistry:

–   chemical products, e.g. molecules, atoms, physical state, acids, bases, oxidation

–   mass density, relative density, pressure and volume of liquids, e.g. mass density, relative density, volume and pressure in terms of temperature increase, maximum degree of filling

–   critical temperature

–   polymerization, e.g. theoretical and practical questions, conditions of carriage

–   mixtures, e.g. vapour pressure, composition and hazard characteristics

–   chemical bonds and formulae.

Practice:

–   cleaning of cargo tanks, e.g. gas freeing, washing, residual cargo and receptacles for residual products

–   loading and unloading, e.g. venting piping systems, rapid closing devices, effects of temperature

–   sampling

–   danger of explosion

–   health risks

–   gas concentration measures, e.g. which apparatus to use and how to use it

–   monitoring of closed spaces and entry to these spaces

–   certificates for degassing and permitted work

–   degree of filling and over–filling

–   safety installations

–   pumps and compressors.

Emergency measures:

–   physical injury, e.g. contact with the cargo, breathing in gas, assistance

–   irregularities relating to the cargo, e.g. leak in a connection, over–filling, polymerization and hazards in the vicinity of the vessel.

8.2.2.3.4   *Refresher courses*

Refresher course on gases

Prior training:	valid ADN "gases" and "tank vessels" certificate or combined "dry cargo/tank vessels" certificate;
Knowledge:	ADN, in particular, loading, transport, unloading and handling of gases;
Authorization for:	tank vessels for the transport of substances for which a type G tank vessel is required and transport in type G of substances for which a type C is required with cargo tank design 1 required in column (7) of Table C of Chapter 3.2.
Training:	gases 8.2.2.3.3.1.

Refresher course on chemicals

Prior training:	valid ADN "chemicals" and "tank vessels" certificate or combined "dry cargo/tank vessels" certificate;
Knowledge:	ADN, in particular, loading, transport, unloading and handling of gases;
Authorization for:	tank vessels for the transport of substances for which a type C tank vessel is required;
Training:	chemicals 8.2.2.3.3.2.

8.2.2.4   ***Planning of refresher and specialization courses***

The following minimum periods of training shall be observed:

Basic "dry cargo vessels course"	32 lessons of 45 minutes each
Basic "tank vessels" course	32 lessons of 45 minutes each
Basic combined course	40 lessons of 45 minutes each
Specialization course on gases	16 lessons of 45 minutes each
Specialization course on chemicals	16 lessons of 45 minutes each

Each day of training may comprise not more than eight lessons.

If the theoretical training is by correspondence, equivalences to the above–mentioned lessons shall be determined. Training by correspondence shall be completed within a period of nine months.

Approximately 30% of basic training shall be devoted to practical exercises. Practical exercises shall, where possible, be undertaken during the period of theoretical training; in any event, they shall be completed not later than three months following the completion of theoretical training.

8.2.2.5   *Planning of refresher course*

The refresher course shall take place before the expiry of the deadline referred to in 8.2.1.4, 8.2.1.6 or 8.2.1.8.

The following minimum periods of training shall be observed:

Basic refresher course:

–	dry cargo vessels	16 lessons of 45 minutes each
–	tank vessels	16 lessons of 45 minutes each
–	combined dry cargo vessels and tank vessels	16 lessons of 45 minutes each
Specialization refresher course on gases		8 lessons of 45 minutes each
Specialization refresher course on chemicals		8 lessons of 45 minutes each

Each day of training may comprise not more than eight lessons.

Approximately 30% of basic training shall be devoted to practical exercises. Practical exercises shall, where possible, be undertaken during the period of theoretical training; in any event, they shall be completed not later than three months following the completion of theoretical training. The proportion of stability training in the refresher course shall amount to at least 2 lessons.

8.2.2.6   *Approval of training courses*

8.2.2.6.1   Training courses shall be approved by the competent authority.

8.2.2.6.2   Approval shall be granted only on written application.

8.2.2.6.3   Applications for approval shall be accompanied by:

(a)   the detailed course curriculum showing the course topics and the length of time to be devoted to them, as well as the teaching methods envisaged;

(b)   the roster of training instructors, listing their qualifications and the subjects to be taught by each one;

(c)   information on classrooms and teaching materials, as well as on the facilities available for practical exercises;

(d)   enrolment requirements, e.g. the number of participants;

(e)   detailed plan for final tests, including, if necessary, the infrastructure and organisation of electronic examinations in accordance with 8.2.2.7.1.7, if these are to be carried out.

8.2.2.6.4   The competent authority shall be responsible for monitoring training courses and examinations.

8.2.2.6.5   The approval comprises the following conditions, <u>inter alia</u>:

(a)   training courses shall conform to the information accompanying the application for approval;

(b)   the competent authority may send inspectors to attend training courses and examinations;

(c)   the timetables for the various training courses shall be notified in advance to the competent authority.

Approval shall be granted in writing for a limited period. It may be withdrawn in the event of failure to comply with the conditions of approval.

8.2.2.6.6   The approval document shall indicate whether the course in question is a basic training course, a specialization course or a refresher course.

8.2.2.6.7   If, after approval is granted, the training body wishes to change conditions affecting the approval, it shall seek the prior agreement of the competent authority. This provision shall apply in particular to amendments to syllabuses.

8.2.2.6.8   Training courses shall take account of the current developments in the various subjects taught. The course organizer shall be responsible for ensuring that recent developments are brought to the attention of, and properly understood by, training instructors.

8.2.2.7   *Examinations and final tests*

8.2.2.7.0   The examination shall be organized by the competent authority or by an examining body designated by the competent authority. The examining body shall not be a training provider.

The examining body shall be designated in writing. This approval may be of limited duration and should be based on the following criteria:

–   Competence of the examining body;

–   Specifications of the form of the examinations the examining body is proposing, including, if necessary, the infrastructure and organisation of electronic examinations in accordance with 8.2.2.7.1.7, if these are to be carried out;

–   Measures intended to ensure that examinations are impartial;

–   Independence of the body from all natural or legal persons employing ADN experts.

8.2.2.7.1   *Basic training courses*

8.2.2.7.1.1   After initial training an ADN basic training examination shall be taken. This examination shall be held either immediately after the training or within six months following the completion of such training.

8.2.2.7.1.2   In the examination the candidate shall furnish evidence that, in accordance with the basic training course, he has the knowledge, understanding and capabilities required of an expert on board a vessel.

8.2.2.7.1.3   The Administrative Committee shall establish a catalogue of questions comprising the objectives set out in 8.2.2.3.1.1 to 8.2.2.3.1.3 and a directive on the use of the catalogue of questions.[1] The examination questions shall be selected from this list. The candidate shall not have advance knowledge of the questions selected.

8.2.2.7.1.4   The model attached to the directive on the use of the catalogue of questions is to be used to compile the examination questions.

8.2.2.7.1.5   The examination shall be written. Candidates shall be asked 30 questions. The examination shall last 60 minutes. It is deemed to have been passed if at least 25 of the 30 questions have been answered correctly.

8.2.2.7.1.6   The competent authority or an examining body designated by the competent authority shall invigilate every examination. Any manipulation and deception shall be ruled out as far as possible. Authentication of candidates shall be ensured.

The use in the written test of documentation other than the texts of regulations on dangerous goods, CEVNI and related police regulations, is not permitted. Non–programmable pocket calculators are authorized for use during specialization courses and shall be supplied by the competent authority or by the examining body designated by the competent authority.

Examination documents (questions and answers) shall be recorded and kept as a print–out or electronically as a file.

8.2.2.7.1.7   Written examinations may be performed, in whole or in part, as electronic examinations, where the answers are recorded and evaluated using electronic data processing (EDP) processes, provided the following conditions are met:

(a)   The hardware and software shall be checked and accepted by the competent authority or by the examining body designated by the competent authority.

(b)   Electronic media may be used only if provided by the competent authority or by the examining body designated by the competent authority.

(c)   Proper technical functioning shall be ensured. Arrangements as to whether and how the examination can be continued shall be made in the case of a failure of the devices and applications. No aids shall be available on the input devices (e.g. electronic search function); the electronic data processing equipment provided shall not allow the candidates to communicate with any other device during the examination.

(d)   There shall be no means of a candidate introducing further data to the electronic media provided; the candidate may only answer the questions posed.

(e)   The final inputs of each candidate shall be logged. The determination of the results shall be transparent.

8.2.2.7.2   *Specialization course on gases and chemicals*

8.2.2.7.2.1   Candidates who are successful in the ADN basic training examination may apply for enrolment in a "gases" and/or "chemicals" specialization course, to be followed by an examination. The examination shall be based on the Administrative Committee's list of questions.

---

[1]   *Note by the secretariat: the catalogue of questions and the directive for its application are available on the website of the secretariat of the United Nations Economic Commission for Europe (http://www.unece.org/trans/danger/publi/adn/catalog_of_questions.html.*

8.2.2.7.2.2   During the examination the candidate shall furnish proof that, in accordance with the "gases" and/or "chemicals" specialization course, he has the knowledge, understanding and capabilities required of the expert on board vessels carrying gases or chemicals, respectively.

8.2.2.7.2.3   The Administrative Committee shall prepare a catalogue of questions for the examination, comprising the objectives set out in 8.2.2.3.3.1 or 8.2.2.3.3.2 and a directive on the use of the catalogue of questions[1]. The examination questions shall be selected from the list. The candidate shall not have advance knowledge of the questions selected.

8.2.2.7.2.4   The model attached to the directive on the use of the catalogue of questions is to be used to compile the examination questions.

8.2.2.7.2.5   The examination shall be written.

The candidate is to be asked 30 multiple–choice questions and one substantive question. The examination shall last a total of 150 minutes, of which 60 minutes for the multiple–choice questions and 90 minutes for the substantive questions.

The examination shall be marked out of a total of 60, of which 30 marks will go to the multiple–choice questions (one mark per question) and 30 to the substantive question (the distribution of marks is left to the appreciation of the competent authority). A total of 44 marks must be achieved to pass. However, not less than 20 marks must be obtained in each part. If the candidate obtains 44 but does not achieve 20 in one part, the part in question may be resat once.

The provisions of 8.2.2.7.1.6 and 8.2.2.7.1.7 shall apply by analogy.

8.2.2.7.3   *Refresher training course*

8.2.2.7.3.1   At the end of the refresher course in accordance with paragraph 8.2.1.4, the course organizer shall conduct a test.

8.2.2.7.3.2   The test shall be in writing. Candidates shall be asked 20 multiple–choice questions. At the end of every refresher course, a fresh question paper shall be prepared. The test shall last 40 minutes. It shall be deemed to have been passed if at least 16 of the 20 questions have been answered correctly.

8.2.2.7.3.3   The provisions of 8.2.2.7.1.2, 8.2.2.7.1.3, 8.2.2.1.7.6 and 8.2.2.1.7.7 shall apply to the administration of the tests (outside the provisions of the directive on the use of the catalogue of questions for examining authorities and bodies).

8.2.2.7.3.4   The course organizer shall deliver to successful candidates a written certificate for presentation to the competent authority under paragraph 8.2.2.8.

---

[1]   *Note by the secretariat: the catalogue of questions and the directive for its application are available on the website of the secretariat of the United Nations Economic Commission for Europe (http://www.unece.org/trans/danger/publi/adn/catalog_of_questions.html).*

8.2.2.7.3.5 The course organizer shall keep test papers of candidates for five years from the date of the test.

8.2.2.8 *ADN specialized knowledge certificate*

8.2.2.8.1 The issue and renewal of the ADN specialized knowledge certificate conforming to 8.6.2 shall be the responsibility of the competent authority or a body authorized by the competent authority.

Certificates shall be issued to:

– candidates who have attended a basic or specialized training course and have passed the examination;

– candidates who have taken part in a refresher course.

Candidates who have obtained the "gases" and/or "chemicals" specialized training certificate shall be issued with a new certificate containing all the certificates relating to the basic and specialized training courses. The validity of the new certificate shall be five years as from the date of the basic training examination.

If the refresher course was not fully completed before the expiry of the period of validity of the certificate, a new certificate shall not be issued until the candidate has completed a further initial basic training course and passed an examination referred to in 8.2.2.7 above.

If a new certificate is issued following attendance at a specialized or refresher course, and the previous certificate was issued by another competent authority or by a body authorized by another competent authority, the previous certificate shall be retained and returned to the authority or body that issued it.

8.2.2.8.2 Contracting Parties shall provide the UNECE secretariat with an example of the national model for any certificate intended for issue in accordance with this section, along with examples of models for certificates which are still valid. A Contracting Party may additionally provide explanatory notes. The UNECE secretariat shall make the information received available to all Contracting Parties.

CHAPTER 8.3

MISCELLANEOUS REQUIREMENTS TO BE COMPLIED WITH

BY THE CREW OF THE VESSEL

8.3.1   **Persons authorized on board**

8.3.1.1   Unless otherwise provided for in Part 7, only the following persons are authorized to be on board:

(a)   members of the crew;

(b)   persons who, although not being members of the crew, normally live on board; and

(c)   persons who are on board for duty reasons.

8.3.1.2   The persons referred to in 8.3.1.1 (b) are not authorized to remain in the protected area of dry cargo vessels or in the cargo area of tank vessels except for short periods.

8.3.1.3   When the vessel is required to carry two blue cones or two blue lights in accordance with column (19) of Table C of Chapter 3.2, persons under 14 years of age are not permitted on board.

8.3.2   **Portable lamps**

On board dry cargo vessels, the only portable lamps permitted in the protected area are lamps having their own source of power.

On board tank vessels, the only portable lamps permitted in the cargo area and on the deck outside the cargo area are lamps having their own source of power.

They shall be of the certified safe type.

8.3.3   **Admittance on board**

No unauthorized person shall be permitted on board. This prohibition shall be displayed on notice boards at appropriate places.

8.3.4   **Prohibition on smoking, fire and naked light**

Smoking on board the vessel is prohibited. The prohibition of smoking also applies to electronic cigarettes and other similar devices. This prohibition shall be displayed on notice boards at appropriate places.

This prohibition does not apply to the accommodation or the wheelhouse provided their windows, doors, skylights and hatches are closed.

8.3.5   **Danger caused by work on board**

No repair or maintenance work requiring the use of an open flame or electric current or liable to cause sparks may be carried out

–   on board dry cargo vessels in the protected area or on the deck less than 3 m forward or aft of that area;

–   on board tank vessels.

This requirement does not apply:

when dry cargo vessels are furnished with an authorization from the competent authority or a certificate attesting to the totally gas–free condition of the protected area;

when tank vessels are furnished with an authorization from the competent authority or a certificate attesting to the totally gas–free condition of the vessel;

– to berthing operations.

Such work on board tank vessels may be undertaken without permission in the service spaces outside the cargo area, provided the doors and openings are closed and the vessel is not being loaded, unloaded or gas–freed.

The use of chromium vanadium steel screwdrivers and wrenches or screwdrivers and wrenches of equivalent material from the point of view of spark formation is permitted.

CHAPTER 8.4

(Reserved)

CHAPTER 8.5

(Reserved)

(Reserved)

CHAPTER 8.6

DOCUMENTS

8.6.1   **Certificate of approval**

8.6.1.1   *Model for a certificate of approval for dry cargo vessels*

1

Competent authority: ....

Space reserved for the emblem and name of the State

ADN certificate of approval No.:

1.   Name of vessel ....

2.   Official number ....

3.   Type of vessel ....

4.   Additional requirements: vessel referred to in 7.1.2.19.1[1]

                                 vessel referred to in 7.2.2.19.3[1]

                                 The vessel complies with the additional rules of construction referred to in 9.1.0.80 to 9.1.0.95/ 9.2.0.80 to 9.2.0.95 for double hull vessels[1]

5.   Permitted derogations[1]:   ....
         ....
         ....
         ....
         ....

6.   The validity of this certificate of approval expires on .... (date)

7.   The previous certificate of approval No. ....was issued on ....
     by .... (competent authority)

8.   The vessel is approved for the carriage of dangerous goods based on:
     – inspection on[1] (date).....................................
     – The inspection report of a recognized classification society [1]
         (name of the classification society) .... (date).........................
     – The inspection report of a recognized inspection body [1]
         (name of the inspection body) .... (date).........................

9.   Subject to permitted equivalence:[1] ....
         ....
         ....
         ....

10.   Subject to special authorizations:[1] ....
         ....
         ....
         ....

11.   Issued at: .... on ....
                 (place)   (date)

12.   (Stamp)

                                       ....
                                       (competent authority)

                                       ....
                                       (signature)

---

[1] Delete as appropriate

Extension of the validity of the certificate of approval

13.   The validity of this certificate is extended under Chapter 1.16 of ADN

        until ....
                (date)

14.   .... on ....
        (place)   (date)

15.   (Stamp)

                                                        ....
                                                        (competent authority)

                                                        ....
                                                        (signature)

1

Competent authority: ....

Space reserved for the emblem and name of the State

ADN provisional certificate of approval No: ....

1.   Name of vessel ....
2.   Official number ....
3.   Type of vessel ....
4.   Additional requirements:

vessel referred to in 7.1.2.19.1[1]

vessel referred to in 7.2.2.19.3[1]

The vessel complies with the additional rules of construction

referred to in 9.1.0.80 to 9.1.0.95/9.2.0.80. to 9.2.0.95 for

double hull vessels[1]

5.   Permitted derogations[1]: ....

....

....

6.   The provisional certificate of approval is valid...

6.1   until ....[1]

6.2   for a single journey from .... to ....[1]

7.   Issued at .... on ....

(place)   (date)

8.   (Stamp)   ....

(competent authority)

....

(signature)

....

_____

[1] Delete as appropriate.

NOTE: *This model provisional certificate of approval may be replaced by a single certificate model combining a provisional certificate of inspection and the provisional certificate of approval, provided that this single certificate model contains the same particulars as the model above and is approved by the competent authority.*

1

Competent authority: ....

Space reserved for the emblem and name of the State

ADN certificate of approval No.:

1.	Name of vessel	....
2.	Official number	....
3.	Type of vessel	....
4.	Type of tank vessel	....

5.   Cargo tank design

       1.   Pressure cargo tanks[1,2]
       2.   Closed cargo tanks[1,2]
       3.   Open cargo tanks with flame arresters[1,2]
       4.   Open cargo tanks[1,2]

6.   Types of cargo tanks

       1.   Independent cargo tanks[1,2]
       2.   Integral cargo tanks[1,2]
       3.   Cargo tank wall distinct from the hull[1,2]

7.   Opening pressure of high–velocity vent valves/safety valves .... kPa[1,2]

8.   Additional equipment:

- Sampling device
  connection for a sampling device...........yes/no[1,2]
  sampling opening ....…. yes/no[1,2]
- Water–spray system .... yes/no[1,2]
  Internal pressure alarm 40 kPa .... yes/no[1,2]
- Cargo heating system:
  possibility of cargo heating from shore ..... yes/no[1,2]
  cargo heating installation on board .... yes/no[1,2]
- Cargo refrigeration system .... yes/no[1,2]
- Inerting facilities ....…………….. yes/no[1,2]
- Cargo pump–room below deck ....….. yes/no[1]
- Ventilation system ensuring an overpressure   yes/no[1]
- Venting piping according to ....… .………
  piping and installation heated ....…… yes/no[1,2]
- Conforms to the rules of construction resulting from the remark(s) …….. of column (20) of Table C of Chapter 3.2[1,2]

9.   Electrical equipment:

- Temperature class: ....…
- Explosion group: ....…

10.   Loading/unloading rate: .... m^3/h [1] or see loading instructions on loading and unloading[1]

11.   Permitted relative density: ....

12.   Additional observations [1]...
....
....

---

[1] Delete as appropriate.
[2] If the tanks are not all of the same type, see page 3.

13.   The validity of this certificate of approval expires on .... (date)

14.   The previous certificate of approval No. .... was issued on ....
      by .... (competent authority)

15.   The vessel is approved for the carriage of the dangerous goods entered in the vessel substance
      list according to 1.16.1.2.5 based on:
      – Inspection on[1] (date)..
      – The inspection report of a recognized classification society [1]
         (name of the classification society) .... (date)...........................
      – The inspection report of a recognized inspection body [1]
         (name of the inspection body) .... (date)...........................

16.   Subjected to permitted equivalence:[1]
      ....
      ....

17.   Subject to special authorizations:[1]
      ....
      ....

18.   Issued at: ....   on ....
                  (place)   (date)

19.   (Stamp)   ....
                                                        (competent authority)

                                                       ....
                                                        (signature)

_____
[1] Delete as appropriate

Extension of the validity of the certificate of approval

20.   The validity of this certificate is extended under Chapter 1.16 of ADN

      Until ....
      (date)

21.   ....   on ....
      (place)   (date)

22.   (Stamp)   ....
                                                        (competent authority)

                                                       ....
                                                        (signature)

If the cargo tanks of the vessel are not all of the same type or the same design or the equipment is not the same, their type, their design and their equipment shall be indicated below:

Cargo tank number	1	2	3	4	5	6	7	8	9	10	11	12
pressure cargo tank												
closed cargo tank												
open cargo tank with flame arrester												
open cargo tank												
independent cargo tank												
integral cargo tank												
cargo tank wall distinct from the hull												
opening pressure of the high–velocity vent valve/safety valve in kPa												
connection for a sampling device												
sampling opening												
water–spray system												
internal pressure alarm 40 kPa ……..												
possibility of cargo heating from shore												
cargo heating installation on board												
cargo refrigeration installation												
inerting facilities												
venting piping according to 9.3.2.22.5 or 9.3.3.22.5												
venting piping and heated installation												
Conforms to the rules of construction resulting from the remark(s) …...of column (20) of Table C of Chapter 3.2												

1

Competent authority: ....

Space reserved for the emblem and name of the State

ADN provisional certificate of approval No: ....

1.   Name of vessel...

2.   Official number..

3.   Type of vessel..

4.   Type of tank vessel ....

5.   Cargo tank design
        1.   Pressure cargo tanks [1 2]
        2.   Closed cargo tanks [1 2]
        3.   Open cargo tanks with flame arresters [1 2]
        4.   Open cargo tanks [1 2]

6.   Types of cargo tanks
        1.   Independent cargo tanks [1 2]
        2.   Integral cargo tanks [1 2]
        3.   Cargo tank wall distinct from the hull [1 2]

7.   Opening pressure of high–velocity vent valves/safety valves .... kPa [1 2]

8.   Additional equipment:
   - Sampling device
     connection for a sampling device ....   yes/no[1 2]
     sampling opening ....   yes/no[1 2]
   - Water–spray system ....   yes/no[1 2]
         Internal pressure alarm 40 kPa ....   yes/no[1 2]
   - Cargo heating system:
     possibility of cargo heating from shore ....   yes/no[1 2]
     cargo heating installation on board ....   yes/no[1 2]
   - Cargo refrigeration system ....   yes/no[1 2]
   - Inerting facilities ....   yes/no[1 2]
   - Cargo pump–room below deck ....   yes/no[1]
   - Ventilation system ensuring an overpressure.........   yes/no[1]
   - Venting piping according to ....
     piping and installation heated ....   yes/no[1 2]
   - Conforms to the rules of construction resulting from the remark(s) .... of column (20) of Table C of Chapter 3.2[1 2]

9.   Electrical equipment:
   - Temperature class: ....
   - Explosion group: ....

10.   Loading/unloading rate .... m^3/h[1] or see loading instructions[1] or see instructions on loading and unloading[1].

11.   Permitted relative density: ....

12. Additional observations: [1] ....
....

13. The provisional certificate of approval is valid..
     13.1   until [1]..

     13.2   for a single journey from [1]................. to ....

14. Issued at .... on ....
     (place)   (date)

15. (Stamp)   ....
                                   (competent authority)

                                 ....
                                   (signature)

---

[1] Delete as appropriate.

NOTE: *This model provisional certificate of approval may be replaced by a single certificate model combining a provisional certificate of inspection and the provisional certificate of approval, provided that this single certificate model contains the same particulars as the model above and is approved by the competent authority.*

If the cargo tanks of the vessel are not all of the same type or the same design or the equipment is not the same, their type, their design and their equipment shall be indicated below:

Cargo tank number	1	2	3	4	5	6	7	8	9	10	11	12
pressure cargo tank												
closed cargo tank												
open cargo tank with flame arrester												
open cargo tank												
independent cargo tank												
integral cargo tank												
cargo tank wall distinct from the hull												
opening pressure of the high–velocity vent valve/safety valve in kPa												
connection for a sampling device												
sampling opening												
water–spray system												
internal pressure alarm 40 kPa ……..												
possibility of cargo heating from shore												
cargo heating installation on board												
cargo refrigeration installation												
inerting facilities												
venting piping according to 9.3.2.22.5 or 9.3.3.22.5												
venting piping and heated installation												
Conforms to the rules of construction resulting from the remark(s) …….. of column (20) of Table C of Chapter 3.2												

8.6.1.5 Annex to the certificate of approval and provisional certificate of approval according to 1.16.1.3.1 (a)

Annex to the certificate of approval

1.   Official number ....
2.   Type of vessel ....
3.   Transitional provisions applicable as from

ADN certificate of approval No.:	Competent authority	Issued on	Valid until	Stamp and signature]

ADN certificate of approval No.:	Competent authority	Issued on	Valid until	Stamp and signature

8.6.2 Certificate of special knowledge of ADN according to 8.2.1.2, 8.2.1.5 or 8.2.1.7

(Format: A6, Colour: orange)

No. of certificate: ………………………………

(Space reserved for the emblem of State, competent authority)

Name……………………………………………

First name(s): ……………………………………..

Born on: ……………………………………………

Nationality: ………………………………………

ADN certificate

of special knowledge of ADN

Signature of holder: ………………………………..

The holder of this certificate has special knowledge of ADN.

The holder of this certificate has participated in an 8–lesson stability training.

The certificate is valid for special knowledge of ADN according to
8.2.1.3 (dry cargo vessels)*
8.2.1.3 (tanks vessels)*
8.2.1.5*
8.2.1.7*

until: ………………………………………………

Issued by: …………………………………………

Date: ………………………………………………

(Stamp)

Signature: …………………………………………

_____
* Delete as appropriate.

(*Recto*)   (*Verso*)

1

ADN Checklist

concerning the observance of safety provisions and the implementation of the necessary measures for loading/unloading

- **Particulars of vessel**

  ………………………………………..   No. ……………………………………………...
  (name of vessel)   (official number)

  ………………………………………..
  (vessel type)

- **Particulars of loading or unloading operations**

  ………………………………………...   …………………………………………………...
  (shore loading or unloading installation)   (place)

  ………………………………………...   …………………………………………………...
  (date)   (time)

- **Particulars of the cargo as indicated in the transport document**

Quantity m³	Proper shipping name***	UN Number or Identification number	Dangers* …………….	Packing Group
……………...	…………………………………	………………..	…………….	…………….
……………...	…………………………………	………………..	…………….	…………….
……………...	…………………………………	………………..	…………….	…………….

- **Particulars of last cargo****

Proper shipping name ***	UN Number or Identification number	Dangers* …………….	Packing Group
…………………………………………	………………..	…………….	…………….
…………………………………………	………………..	…………….	…………….
…………………………………………	………………..	…………….	…………….

Dangers indicated in column (5) of Table C, as relevant (as mentioned in the transport document in accordance with 5.4.1.1.2 (c)).
**To be filled in only if vessel is to be loaded.
*** The proper shipping name given in column (2) of Table C of Chapter 3.2, supplemented, when applicable, by the technical name in parenthesis.

Loading/unloading rate (not to be filled in if vessel is to be loaded with gas or have gas unloaded)

Proper shipping name**	Cargo tank number	agreed rate of loading/unloading					
		start		half way		end	
		rate m^3/h	quantity m^3	rate m^3/h	quantity m^3	rate m^3/h	quantity m^3
....	....	....	....	....	....	....	....
....	....	....	....	....	....	....	....
....	....	....	....	....	....	....	....

Will the cargo piping be drained after loading or unloading by stripping or by blowing residual quantities to the shore installation/to the vessel?*

by blowing*
by stripping*

If drained by blowing, how?

....
(e.g. air, inert gas, sleeve)

.... kPa
(permissible maximum pressure in the cargo tank)

....................................litres
(estimated residual quantity)

Questions to the master or the person mandated by him and the person in charge at the loading/unloading place

Loading/unloading may only be started after all questions on the checklist have been checked off by "X", i.e. answered with YES and the list has been signed by both persons.

Non–applicable questions have to be deleted.

If not all questions can be answered with YES, loading/unloading is only allowed with consent of the competent authority.

---

* *Delete as appropriate.*
** *The proper shipping name given in column (2) of Table C of Chapter 3.2, supplemented, when applicable, by the technical name in parenthesis.*

		vessel	**3** loading/ unloading place
1.	Is the vessel permitted to carry this cargo?	O*	O*
2.	(*Reserved*)		
3.	Is the vessel well moored in view of local circumstances?	O	–
4.	Have suitable means in accordance with 7.2.4.77 been provided for leaving the vessel, including in cases of emergency?	O	O
5.	Are the escape routes and the loading/unloading place adequately lighted?	O	O
6.	Vessel/shore connection		
	6.1 Is the piping for loading or unloading between vessel and shore in satisfactory condition?	–	O
	Is it correctly connected?	–	O
	6.2 Are all the connecting flanges fitted with suitable gaskets?	–	O
	6.3 Are all the connecting bolts fitted and tightened?	O	O
	6.4 Are the shoreside loading arms free to move in all directions and do the hose assemblies have enough room for easy movement?	–	O
7.	Are all flanges of the connections of the piping for loading and unloading and of the venting piping not in use, correctly blanked off?	O	O
8.	Are suitable means of collecting leakages placed under the pipe connections which are in use and are they empty??	O	O
9.	Are the movable connecting pieces between the ballast and bilge piping on the one hand and the piping for loading and unloading on the other hand disconnected?	O	–
10.	Is continuous and suitable supervision of loading/unloading ensured for the whole period of the operation?	O	O
11.	Is communication between vessel and shore ensured?	O	O
12.1	For the loading of the vessel, is the venting piping, where required, or if it exists, connected with the vapour return piping?	O	O
12.2	Is it ensured that the shore installation is such that the pressure at the connecting point cannot exceed the opening pressure of the high–velocity vent valves (pressure at connecting point __ kPa)?	–	O*
12.3	When anti–explosion protection is required in Chapter 3.2, Table C, column (17) does the shore installation ensure that its vapour return piping is such that the vessel is protected against detonations and flame fronts from the shore.	–	O
13.	Is it known what actions are to be taken in the event of an "Emergency–stop" and an "Alarm"?	O	O

* *To be filled in only if vessel is to be loaded.*

		vessel	loading/ unloading place **4**
14.	Check on the most important operational requirements:		
	– Are the required fire extinguishing systems and appliances operational?	O	O
	– Have all valves and other closing devices been checked for correct open – or closed position?	O	O
	– Has smoking been generally prohibited?	O	O
	– Are the flame operated heating applications on board turned off?	O	–
	– Is the voltage cut off from the radar installations?	O	–
	– Is all electrical equipment marked red switched off?	O	–
	– Are all windows and doors closed?	O	–
15.1	Has the starting working pressure of the vessel's cargo discharge pump been adjusted to the permissible working pressure of the shore installation? (agreed pressure __ kPa)	O	–
15.2	Has the starting working pressure of the shore pump been adjusted to the permissible working pressure of the on–board installation? (agreed pressure __ kPa)	–	O
16.	Is the liquid level alarm–installation operational?	O	–
17.	Is the following system plugged in, in working order and tested?		
	Overflow prevention device ☐ when loading ☐ when unloading	O	O
	Device for switching off the on–board pump from the shore facility (only when unloading the vessel)	O	O
18.	To be filled in only in the case of loading or unloading of substances for the carriage of which a vessel of the closed type or a vessel of the open type with flame arrester is required.		
	Are the cargo tank hatches and cargo tank inspection, gauging and sampling openings closed or protected by flame arresters in good condition?	O	–
19.	When transporting refrigerated liquefied gases, has the holding time been determined according to 7.2.4.16.16, and is known and documented on board?	O**	O**

Checked, filled in and signed

for the vessel:	for the installation of loading and unloading:
....	....
(name in capital letters)	(name in capital letters)
....	....
(signature)	(signature)

**   *To be filled in only if the vessel is to be loaded.*

Explanation

Question 3

"Well moored" means that the vessel is fastened to the pier or the cargo transfer station in such a way that, without intervention of a third person, movements of the vessel in any direction that could hamper the operation of the cargo transfer gear will be prevented. Established or predictable variations of the water–level at that location and special factors have to be taken into account.

Question 4

It must be possible to board or escape from the vessel at any time. If there is none or only one protected escape route available at the shoreside for a quick escape from the vessel in case of emergency, a suitable means of escape has to be provided on the vessel side if required in accordance with 7.1.4.77 and 7.2.4.77.

Question 6

A valid inspection certificate for the hose assemblies must be available on board. The material of the piping for loading and unloading must be able to withstand the expected loads and be suitable for cargo transfer of the respective substances. The piping for loading and unloading between vessel and shore must be placed so that it cannot be damaged by ordinary movements of the vessel during the loading and unloading process or by variations of the water. In addition, all flanged joints must be fitted with appropriate gaskets and sufficient bolt connections in order to exclude the possibility of leakage.

Question 10

Loading/unloading must be supervised on board and ashore so that dangers which may occur in the vicinity of piping for loading and unloading between vessel and shore can be recognized immediately. When supervision is effected by additional technical means it must be agreed between the shore installation and the vessel how it is to be ensured.

Question 11

For a safe loading/unloading operation good communications between vessel and shore are required. For this purpose telephone and radio equipment may be used only if of an explosion protected type and located within reach of the supervisor.

Question 13

Before the start of the loading/unloading operation the representative of the shore installation and the master or the person mandated by him must agree on the applicable procedure. The specific properties of the substances to be loaded/unloaded have to be taken into account.

Question 17

To prevent backflow from the shore, it is also necessary to activate the overflow prevention device on the vessel under certain circumstances when unloading. It is obligatory during loading and optional during unloading. Delete this item if it is not necessary during unloading.

8.6.4 *(Deleted)*

PART 9

Rules for construction

CHAPTER 9.1

RULES FOR CONSTRUCTION OF DRY CARGO VESSELS

9.1.0   **Rules for construction applicable to dry cargo vessels**

Provisions of 9.1.0.0 to 9.1.0.79 apply to dry cargo vessels.

9.1.0.0   *Materials of construction*

The vessel's hull shall be constructed of shipbuilding steel or other metal, provided that this metal has at least equivalent mechanical properties and resistance to the effects of temperature and fire.

9.1.0.1   *Vessel record*

NOTE: For the purpose of this paragraph, the term "owner" has the same meaning as in 1.16.0.

The vessel record shall be retained by the owner who shall be able to provide this documentation at the request of the competent authority and the recognized classification society.

The vessel record shall be maintained and updated throughout the life of the vessel and shall be retained for 6 months after the vessel is taken out of service.

Should a change of owner occur during the life of the vessel the vessel record shall be transferred to the new owner.

Copies of the vessel record or all necessary documents shall be made available on request to the competent authority for the issuance of the certificate of approval and for the recognized classification society or inspection body for first inspection, periodic inspection, special inspection or exceptional checks.

9.1.0.2 to 9.1.0.10   *(Reserved)*

9.1.0.11   **Holds**

9.1.0.11.1   (a)   Each hold shall be bounded fore and aft by watertight metal bulkheads.

             (b)   The holds shall have no common bulkhead with the oil fuel tanks.

9.1.0.11.2   The bottom of the holds shall be such as to permit them to be cleaned and dried.

9.1.0.11.3   The hatchway covers shall be spraytight and weathertight or be covered by waterproof tarpaulins.

Tarpaulins used to cover the holds shall not readily ignite.

9.1.0.11.4   No heating appliances shall be installed in the holds.

9.1.0.12   *Ventilation*

9.1.0.12.1   It must be possible to ventilate each hold by means of two mutually independent extraction ventilators having a capacity of not less than five changes of air per hour based on the volume of the empty hold. The ventilator fan shall be designed so that no sparks may be emitted on contact of the impeller blades with the housing and no static electricity may be generated. The extraction ducts shall be positioned at the extreme ends of the hold and extend down to not more than 50 mm above the bottom. The extraction of gases and vapours through the duct shall also be ensured for carriage in bulk.

If the extraction ducts are movable they shall be suitable for the ventilator assembly and capable of being firmly fixed. Protection shall be ensured against bad weather and spray. The air intake shall be ensured during ventilation.

9.1.0.12.2   The ventilation system of a hold shall be arranged so that dangerous gases cannot penetrate into the accommodation, wheelhouse or engine rooms.

9.1.0.12.3   Ventilation shall be provided for the accommodation and for service spaces.

9.1.0.13 to 9.1.0.16   (*Reserved*)

9.1.0.17   *Accommodation and service spaces*

9.1.0.17.1   The accommodation shall be separated from the holds by metal bulkheads having no openings.

9.1.0.17.2   Gastight closing appliances shall be provided for openings in the accommodation and wheelhouse facing the holds.

9.1.0.17.3   No entrances or openings of the engine rooms and service spaces shall face the protected area.

9.1.0.18 and 9.1.0.19   (*Reserved*)

9.1.0.20   *Water ballast*

The double-hull spaces and double bottoms may be arranged for being filled with water ballast.

9.1.0.21 to 9.1.0.30   (*Reserved*)

9.1.0.31   *Engines*

9.1.0.31.1   Only internal combustion engines running on fuel having a flashpoint above 55 °C are allowed.

9.1.0.31.2   The air vents in the engine rooms and the air intakes of the engines which do not take air in directly from the engine room shall be located not less than 2.00 m from the protected area.

9.1.0.31.3   Sparking shall not be possible in the protected area.

9.1.0.32   *Oil fuel tanks*

9.1.0.32.1   Double bottoms within the hold area may be arranged as oil fuel tanks provided their depth is not less than 0.6 m. Oil fuel pipes and openings to such tanks are not permitted in the holds.

9.1.0.32.2   The air pipes of all oil fuel tanks shall be led to 0.50 m above the open deck. Their open ends and the open ends of the overflow pipes leaking to the deck shall be fitted with a protective device consisting of a gauze grid or by a perforated plate.

9.1.0.33   (*Reserved*)

9.1.0.34 *Exhaust pipes*

9.1.0.34.1   Exhausts shall be evacuated from the vessel into the open air either upwards through an exhaust pipe or through the shell plating. The exhaust outlet shall be located not less than 2.00 m from the hatchway openings. The exhaust pipes of engines shall be arranged so that the exhausts are led away from the vessel. The exhaust pipes shall not be located within the protected area.

9.1.0.34.2   Exhaust pipes shall be provided with a device preventing the escape of sparks, e.g. spark arresters.

9.1.0.35 *Stripping installation*

The stripping pumps intended for the holds shall be located in the protected area. This requirement shall not apply when stripping is effected by eductors.

9.1.0.36 to 9.1.0.39   (*Reserved*)

9.1.0.40 *Fire-extinguishing arrangements*

9.1.0.40.1   A fire-extinguishing system shall be installed on the vessel. This system shall comply with the following requirements:

–   It shall be supplied by two independent fire or ballast pumps one of which shall be ready for use at any time. These pumps and their means of propulsion and electrical equipment shall not be installed in the same space;

–   It shall be provided with a water main fitted with at least three hydrants in the protected area above deck. Three suitable and sufficiently long hoses with jet/spray nozzles having a diameter of not less than 12 mm shall be provided. Alternatively one or more of the hose assemblies may be substituted by directable jet/spray nozzles having a diameter of not less than 12 mm. It shall be possible to reach any point of the deck in the protected area simultaneously with at least two jets of water which do not emanate from the same hydrant. A spring-loaded non-return valve shall be fitted to ensure that no gases can escape through the fire-extinguishing system into the accommodation or service spaces outside the protected area;

–   The capacity of the system shall be at least sufficient for a jet of water to reach a distance of not less than the vessel's breadth from any location on board with two spray nozzles being used at the same time.;

–   The water supply system shall be capable of being put into operation from the wheelhouse and from the deck;

–   Measures shall be taken to prevent the freezing of fire-mains and hydrants.

A single fire or ballast pump shall suffice on board pushed barges without their own means of propulsion.

9.1.0.40.2   In addition, the engine rooms shall be provided with a permanently fixed fire-extinguishing system meeting the following requirements:

9.1.0.40.2.1 *Extinguishing agents*

For the protection of spaces in engine rooms, boiler rooms and pump rooms, only permanently fixed fire-extinguishing systems using the following extinguishing agents are permitted:

(a) CO_2 (carbon dioxide);

(b)   HFC 227 ea (heptafluoropropane);

(c)   IG-541 (52% nitrogen, 40% argon, 8% carbon dioxide);

(d)   FK-5-1-12 (dodecafluoro 2–methylpentane–3–one).

Other extinguishing agents are permitted only on the basis of recommendations by the Administrative Committee.

9.1.0.40.2.2 *Ventilation, air extraction*

(a)   The combustion air required by the combustion engines which ensure propulsion should not come from spaces protected by permanently fixed fire-extinguishing systems. This requirement is not mandatory if the vessel has two independent main engine rooms with a gastight separation or if, in addition to the main engine room, there is a separate engine room installed with a bow thruster that can independently ensure propulsion in the event of a fire in the main engine room.

(b)   All forced ventilation systems in the space to be protected shall be shut down automatically as soon as the fire-extinguishing system is activated.

(c)   All openings in the space to be protected which permit air to enter or gas to escape shall be fitted with devices enabling them to be closed rapidly. It shall be clear whether they are open or closed.

(d)   Air escaping from the pressure–relief valves of the pressurised air tanks installed in the engine rooms shall be evacuated to the open air.

(e)   Overpressure or negative pressure caused by the diffusion of the extinguishing agent shall not destroy the constituent elements of the space to be protected. It shall be possible to ensure the safe equalisation of pressure.

(f)   Protected spaces shall be provided with a means of extracting the extinguishing agent. If extraction devices are installed, it shall not be possible to start them up during extinguishing.

9.1.0.40.2.3 *Fire alarm system*

The space to be protected shall be monitored by an appropriate fire alarm system. The alarm signal shall be audible in the wheelhouse, the accommodation and the space to be protected.

9.1.0.40.2.4 *Piping system*

(a)   The extinguishing agent shall be routed to and distributed in the space to be protected by means of a permanent piping system. Piping installed in the space to be protected and their fittings shall be made of steel. This shall not apply to the connecting nozzles of tanks and compensators provided that the materials used have equivalent fire–retardant properties. Piping shall be protected against corrosion both internally and externally.

(b)   The discharge nozzles shall be so arranged as to ensure the regular diffusion of the extinguishing agent. In particular, the extinguishing agent must also be effective beneath the floor.

9.1.0.40.2.5   *Triggering device*

(a)   Automatically activated fire-extinguishing systems are not permitted.

(b)   It shall be possible to activate the fire-extinguishing system from a suitable point located outside the space to be protected.

(c)   Triggering devices shall be so installed that they can be activated in the event of a fire and so that the risk of their breakdown in the event of a fire or an explosion in the space to be protected is reduced as far as possible.

Systems which are not mechanically activated shall be supplied from two energy sources independent of each other. These energy sources shall be located outside the space to be protected. The control lines located in the space to be protected shall be so designed as to remain capable of operating in the event of a fire for a minimum of 30 minutes. The electrical installations are deemed to meet this requirement if they conform to the IEC 60331–21:1999 standard.

When the triggering devices are so placed as not to be visible, the component concealing them shall carry the "Fire-fighting system" symbol, each side being not less than 10 cm in length, with the following text in red letters on a white ground:

Fire-extinguishing system

(d)   If the fire-extinguishing system is intended to protect several spaces, it shall comprise a separate and clearly–marked triggering device for each space;

(e)   The instructions shall be posted alongside all triggering devices and shall be clearly visible and indelible. The instructions shall be in a language the master can read and understand and if this language is not English, French or German, they shall be in English, French or German. They shall include information concerning:

(i)   the activation of the fire-extinguishing system;

(ii)   the need to ensure that all persons have left the space to be protected;

(iii)   The correct behaviour of the crew in the event of activation and when accessing the space to be protected following activation or diffusion, in particular in respect of the possible presence of dangerous substances;

(iv)   the correct behaviour of the crew in the event of the failure of the fire-extinguishing system to function properly.

(f)   The instructions shall mention that prior to the activation of the fire-extinguishing system, combustion engines installed in the space and aspirating air from the space to be protected, shall be shut down.

9.1.0.40.2.6   *Alarm device*

(a)   Permanently fixed fire-extinguishing systems shall be fitted with an audible and visual alarm device;

(b)   The alarm device shall be set off automatically as soon as the fire-extinguishing system is first activated. The alarm device shall function for an appropriate period of time before the extinguishing agent is released; it shall not be possible to turn it off;

(c)   Alarm signals shall be clearly visible in the spaces to be protected and their access points and be clearly audible under operating conditions corresponding to the highest possible sound level. It shall be possible to distinguish them clearly from all other sound and visual signals in the space to be protected;

(d)   Sound alarms shall also be clearly audible in adjoining spaces, with the communicating doors shut, and under operating conditions corresponding to the highest possible sound level;

(e)   If the alarm device is not intrinsically protected against short circuits, broken wires and drops in voltage, it shall be possible to monitor its operation;

(f)   A sign with the following text in red letters on a white ground shall be clearly posted at the entrance to any space the extinguishing agent may reach:

Warning, fire-extinguishing system!
Leave this space immediately when the … (description) alarm is activated!

9.1.0.40.2.7   *Pressurised tanks, fittings and piping*

(a)   Pressurised tanks, fittings and piping shall conform to the requirements of the competent authority or, if there are no such requirements, to those of a recognized classification society.

(b)   Pressurised tanks shall be installed in accordance with the manufacturer's instructions.

(c)   Pressurised tanks, fittings and piping shall not be installed in the accommodation.

(d)   The temperature of cabinets and storage spaces for pressurised tanks shall not exceed 50 °C.

(e)   Cabinets or storage spaces on deck shall be securely stowed and shall have vents so placed that in the event of a pressurised tank not being gastight, the escaping gas cannot penetrate into the vessel. Direct connections with other spaces are not permitted.

9.1.0.40.2.8   *Quantity of extinguishing agent*

If the quantity of extinguishing agent is intended for more than one space, the quantity of extinguishing agent available does not need to be greater than the quantity required for the largest of the spaces thus protected.

9.1.0.40.2.9   *Installation, maintenance, monitoring and documents*

(a)   The mounting or modification of the system shall only be performed by a company specialised in fire-extinguishing systems. The instructions (product data sheet, safety data sheet) provided by the manufacturer of the extinguishing agent or the system shall be followed.

(b)   The system shall be inspected by an expert:

    (i)   before being brought into service;

    (ii)   each time it is put back into service after activation;

    (iii)   after every modification or repair;

    (iv)   regularly, not less than every two years.

(c)   During the inspection, the expert is required to check that the system conforms to the requirements of 9.1.0.40.2.

(d)   The inspection shall include, as a minimum:

    (i)   an external inspection of the entire system;

    (ii)   an inspection to ensure that the piping is leakproof;

    (iii)   an inspection to ensure that the control and activation systems are in good working order;

    (iv)   an inspection of the pressure and contents of tanks;

    (v)   an inspection to ensure that the means of closing the space to be protected are leakproof;

    (vi)   an inspection of the fire alarm system;

    (vii)   an inspection of the alarm device.

(e)   The person performing the inspection shall establish, sign and date a certificate of inspection.

(f)   The number of permanently fixed fire-extinguishing systems shall be mentioned in the vessel certificate.

9.1.0.40.2.10   *Fire-extinguishing system operating with CO$_2$*

In addition to the requirements contained in 9.1.0.40.2.1 to 9.1.0.40.2.9, fire-extinguishing systems using CO$_2$ as an extinguishing agent shall conform to the following provisions:

(a)   Tanks of CO$_2$ shall be placed in a gastight space or cabinet separated from other spaces. The doors of such storage spaces and cabinets shall open outwards; they shall be capable of being locked and shall carry on the outside the symbol "Warning: general danger," not less than 5 cm high and "CO$_2$" in the same colours and the same size;

(b)   Storage cabinets or spaces for CO$_2$ tanks located below deck shall only be accessible from the outside. These spaces shall have an artificial ventilation system with extractor hoods and shall be completely independent of the other ventilation systems on board;

(c)   The level of filling of CO$_2$ tanks shall not exceed 0.75 kg/l. The volume of depressurised CO$_2$ shall be taken to be 0.56 m^3/kg;

(d) The concentration of CO_2 in the space to be protected shall be not less than 40% of the gross volume of the space. This quantity shall be released within 120 seconds. It shall be possible to monitor whether diffusion is proceeding correctly;

(e) The opening of the tank valves and the control of the diffusing valve shall correspond to two different operations;

(f) The appropriate period of time mentioned in 9.1.0.40.2.6 (b) shall be not less than 20 seconds. A reliable installation shall ensure the timing of the diffusion of CO_2.

9.1.0.40.2.11   *Fire-extinguishing system operating with HFC-227 ea (heptafluoropropane)*

In addition to the requirements of 9.1.0.40.2.1 to 9.1.0.40.2.9, fire-extinguishing systems using HFC-227 ea as an extinguishing agent shall conform to the following provisions:

(a) Where there are several spaces with different gross volumes, each space shall be equipped with its own fire-extinguishing system;

(b) Every tank containing HFC-227 ea placed in the space to be protected shall be fitted with a device to prevent overpressure. This device shall ensure that the contents of the tank are safely diffused in the space to be protected if the tank is subjected to fire, when the fire-extinguishing system has not been brought into service;

(c) Every tank shall be fitted with a device permitting control of the gas pressure;

(d) The level of filling of tanks shall not exceed 1.15 kg/l. The specific volume of depressurised HFC-227 ea shall be taken to be 0.1374 m^3/kg;

(e) The concentration of HFC-227 ea in the space to be protected shall be not less than 8% of the gross volume of the space. This quantity shall be released within 10 seconds;

(f) Tanks of HFC-227 ea shall be fitted with a pressure monitoring device which triggers an audible and visual alarm in the wheelhouse in the event of an unscheduled loss of propellant gas. Where there is no wheelhouse, the alarm shall be triggered outside the space to be protected;

(g) After discharge, the concentration in the space to be protected shall not exceed 10.5% (volume);

(h) The fire-extinguishing system shall not comprise aluminium parts.

9.1.0.40.2.12   *Fire-extinguishing system operating with IG-541*

In addition to the requirements of 9.1.0.40.2.1 to 9.1.0.40.2.9, fire-extinguishing systems using IG-541 as an extinguishing agent shall conform to the following provisions:

(a) Where there are several spaces with different gross volumes, every space shall be equipped with its own fire-extinguishing system;

(b) Every tank containing IG-541 placed in the space to be protected shall be fitted with a device to prevent overpressure. This device shall ensure that the contents of the tank are safely diffused in the space to be protected if the tank is subjected to fire, when the fire-extinguishing system has not been brought into service;

(c) Each tank shall be fitted with a device for checking the contents;

(d) The filling pressure of the tanks shall not exceed 200 bar at a temperature of +15 °C;

(e) The concentration of IG-541 in the space to be protected shall be not less than 44% and not more than 50% of the gross volume of the space. This quantity shall be released within 120 seconds.

9.1.0.40.2.13 *Fire-extinguishing system operating with FK-5-1-12*

In addition to the requirements of 9.1.0.40.2.1 to 9.1.0.40.2.9, fire-extinguishing systems using FK-5-1-12 as an extinguishing agent shall comply with the following provisions:

(a) Where there are several spaces with different gross volumes, every space shall be equipped with its own fire-extinguishing system;

(b) Every tank containing FK-5-1-12 placed in the space to be protected shall be fitted with a device to prevent overpressure. This device shall ensure that the contents of the tank are safely diffused in the space to be protected if the tank is subjected to fire, when the fire-extinguishing system has not been brought into service;

(c) Every tank shall be fitted with a device permitting control of the gas pressure;

(d) The level of filling of tanks shall not exceed 1.00 kg/l. The specific volume of depressurized FK-5-1-12 shall be taken to be 0.0719 m^3/kg;

(e) The volume of FK-5-1-12 in the space to be protected shall be not less than 5.5% of the gross volume of the space. This quantity shall be released within 10 seconds;

(f) Tanks of FK-5-1-12 shall be fitted with a pressure monitoring device which triggers an audible and visual alarm in the wheelhouse in the event of an unscheduled loss of extinguishing agent. Where there is no wheelhouse, the alarm shall be triggered outside the space to be protected;

(g) After discharge, the concentration in the space to be protected shall not exceed 10.0%.

9.1.0.40.2.14 *Fixed fire-extinguishing system for physical protection*

In order to ensure physical protection in the engine rooms, boiler rooms and pump rooms, permanently fixed fire-extinguishing systems are accepted solely on the basis of recommendations by the Administrative Committee.

9.1.0.40.3 The two hand fire–extinguishers referred to in 8.1.4 shall be located in the protected area or in proximity to it.

9.1.0.40.4 The fire-extinguishing agent in the permanently fixed fire-extinguishing system shall be suitable and sufficient for fighting fires.

9.1.0.41 **Fire and naked light**

9.1.0.41.1 The outlets of funnels shall be located not less than 2 m from the hatchway openings. Arrangements shall be provided to prevent the escape of sparks and the entry of water.

9.1.0.41.2 Heating, cooking and refrigerating appliances shall not be fuelled with liquid fuels, liquid gas or solid fuels. The installation in the engine room or other separate space of heating appliances fuelled with liquid fuel having a flashpoint above 55 °C is, however, permitted.

Cooking and refrigerating appliances are permitted only in wheelhouses with metal floor and in the accommodation.

9.1.0.41.3   Electric lighting appliances are only permitted outside the accommodation and the wheelhouse.

9.1.0.42 to 9.1.0.51   (*Reserved*)

9.1.0.52   *Type and location of electrical equipment*

9.1.0.52.1   It shall be possible to isolate the electrical equipment in the protected area by means of centrally located switches except where:

–   in the holds it is of a certified safe type corresponding at least to temperature class T4 and explosion group II B; and

–   in the protected area on the deck it is of the limited explosion risk type.

The corresponding electrical circuits shall have control lamps to indicate whether or not the circuits are live.

The switches shall be protected against unintended unauthorized operation. The sockets used in this area shall be so designed as to prevent connections being made except when they are not live. Submerged pumps installed or used in the holds shall be of the certified safe type at least for temperature class T4 and explosion group II B.

9.1.0.52.2   Electric motors for hold ventilators which are arranged in the air flow shall be of the certified safe type.

9.1.0.52.3   Sockets for the connection of signal lights and gangway lighting shall be solidly fitted to the vessel close to the signal mast or the gangway. Sockets intended to supply the submerged pumps, hold ventilators and containers shall be permanently fitted to the vessel in the vicinity of the hatches.

9.1.0.52.4   Accumulators shall be located outside the protected area.

9.1.0.53 to 9.1.0.55   (*Reserved*)

9.1.0.56   *Electric cables*

9.1.0.56.1   Cables and sockets in the protected area shall be protected against mechanical damage.

9.1.0.56.2   Movable cables are prohibited in the protected area, except for intrinsically safe electric circuits or for the supply of signal lights and gangway lighting, for containers, for submerged pumps, hold ventilators and for electrically operated cover gantries.

9.1.0.56.3   For movable cables permitted in accordance with 9.1.0.56.2 above, only rubber–sheathed cables of type H07 RN-F in accordance with standard IEC–60 245–4:1994 or cables of at least equivalent design having conductors with a cross-section of not less than 1.5 mm^2, shall be used. These cables shall be as short as possible and installed so that damage is not likely to occur.

9.1.0.57 to 9.1.0.69   (*Reserved*)

9.1.0.70   *Metal wires, masts*

All metal wires passing over the holds and all masts shall be earthed, unless they are electrically bonded to the metal hull of the vessel through their installation.

9.1.0.71   *Admittance on board*

The notice boards displaying the prohibition of admittance in accordance with 8.3.3 shall be clearly legible from either side of the vessel.

9.1.0.72 and 9.1.0.73   (*Reserved*)

9.1.0.74   *Prohibition of smoking, fire and naked light*

9.1.0.74.1   The notice boards displaying the prohibition of smoking in accordance with 8.3.4 shall be clearly legible from either side of the vessel.

9.1.0.74.2   Notice boards indicating the circumstances under which the prohibition applies shall be fitted near the entrances to the spaces where smoking or the use of fire or naked light is not always prohibited.

9.1.0.74.3   Ashtrays shall be provided close to each exit of the accommodation and the wheelhouse.

9.1.0.75 to 9.1.0.79   (*Reserved*)

9.1.0.80   *Additional rules applicable to double-hull vessels*

The rules of 9.1.0.88 to 9.1.0.99 are applicable to double-hull vessels intended to carry dangerous goods of Classes 2, 3, 4.1, 4.2, 4.3, 5.1, 5.2, 6.1, 7, 8 or 9, except those for which label No. 1 is prescribed in column (5) of Table A of Chapter 3.2, in quantities exceeding those of 7.1.4.1.1.

9.1.0.81 to 9.1.0.87   (*Reserved*)

9.1.0.88   *Classification*

9.1.0.88.1   Double-hull vessels intended to carry dangerous goods of Classes 2, 3, 4.1, 4.2, 4.3, 5.1, 5.2, 6.1, 7, 8 or 9 except those for which label No. 1 is prescribed in column (5) of Table A of Chapter 3.2, in quantities exceeding those referred to in 7.1.4.1.1 shall be built or transformed under survey of a recognised classification society in accordance with the rules established by that classification society to its highest class. This shall be confirmed by the classification society by the issue of an appropriate certificate.

9.1.0.88.2   Continuation of class is not required.

9.1.0.88.3   Future conversions and major repairs to the hull shall be carried out under survey of this classification society.

9.1.0.89 and 9.1.0.90   (*Reserved*)

9.1.0.91   *Holds*

9.1.0.91.1   The vessel shall be built as a double-hull vessel with double-hull spaces and double bottom within the protected area.

9.1.0.91.2   The distance between the sides of the vessel and the longitudinal bulkheads of the hold shall be not less than 0.80 m. Regardless of the requirements relating to the width of walkways on deck, a reduction of this distance to 0.60 m is permitted, provided that, compared with the scantlings specified in the rules for construction published by a recognised classification society, the following reinforcements have been made:

(a)   Where the vessel's sides are constructed according to the longitudinal framing system, the frame spacing shall not exceed 0.60 m.

The longitudinals shall be supported by web frames with lightening holes similar to the floors in the double bottom and spaced not more than 1.80 m apart. These intervals may be increased if the construction is correspondingly reinforced;

(b)   Where the vessel's sides are constructed according to the transverse framing system, either:

–   two longitudinal side shell stringers shall be fitted. The distance between the two stringers and between the uppermost stringer and the gangboard shall not exceed 0.80 m. The depth of the stringers shall be at least equal to that of the transverse frames and the cross-section of the face plate shall be not less than 15 cm^2.

The longitudinal stringers shall be supported by web frames with lightening holes similar to plate floors in the double bottom and spaced not more than 3.60 m apart. The transverse shell frames and the hold bulkhead vertical stiffeners shall be connected at the bilge by a bracket plate with a height of not less than 0.90 m and thickness equal to the thickness of the floors; or

–   web frames with lightening holes similar to the double bottom plate floors shall be arranged on each transverse frame;

(c)   The gangboards shall be supported by transverse bulkheads or cross-ties spaced not more than 32 m apart.

As an alternative to compliance with the requirements of (c) above, a proof by calculation, issued by a recognised classification society confirming that additional reinforcements have been fitted in the double-hull spaces and that the vessel's transverse strength may be regarded as satisfactory.

9.1.0.91.3   The depth of the double bottom shall be at least 0.50 m. The depth below the suction wells may, however, be locally reduced, but the space between the bottom of the suction well and the bottom of the vessel floor shall be at least 0.40 m. If spaces are between 0.40 m and 0.49 m, the surface area of the suction well shall not exceed 0.5 m^2.

The capacity of the suction wells must not exceed 0.120 m^3.

9.1.0.92   *Emergency exit*

Spaces the entrances or exits of which are partly or fully immersed in damaged condition shall be provided with an emergency exit not less than 0.10 m above the waterline. This does not apply to forepeak and afterpeak.

9.1.0.93   *Stability (general)*

9.1.0.93.1   Proof of sufficient stability shall be furnished including stability in the damaged condition.

9.1.0.93.2   The basic values for the stability calculation – the vessel's lightweight and the location of the centre of gravity – shall be determined either by means of an inclining experiment or by detailed mass and moment calculation. In the latter case the lightweight shall be checked by means of a lightweight test with a resulting difference of not more than ± 5% between the mass determined by the calculation and the displacement determined by the draught readings.

9.1.0.93.3   Proof of sufficient intact stability shall be furnished for all stages of loading and unloading and for the final loading condition.

Floatability after damage shall be proved for the most unfavourable loading condition. For this purpose calculated proof of sufficient stability shall be established for critical intermediate stages of flooding and for the final stage of flooding. Negative values of stability in intermediate stages of flooding may be accepted only if the continued range of curve of righting lever in damaged condition indicates adequate positive values of stability.

9.1.0.94   *Stability (intact)*

9.1.0.94.1   The requirements for intact stability resulting from the damaged stability calculation shall be fully complied with.

9.1.0.94.2   For the carriage of containers, proof of sufficient stability shall also be furnished in accordance with the provisions of the Regulations referred to in 1.1.4.6.

9.1.0.94.3   The most stringent of the requirements of 9.1.0.94.1 and 9.1.0.94.2 above shall prevail for the vessel.

9.1.0.95   *Stability (damaged condition)*

9.1.0.95.1   The following assumptions shall be taken into consideration for the damaged condition:

(a)   The extent of side damage is as follows:

longitudinal extent: at least 0.10 L, but not less than 5.00 m;
transverse extent: 0.59 m inboard from the vessel's side at right angles to the centreline at the level corresponding to the maximum draught;
vertical extent: from the baseline upwards without limit;

(b)   The extent of bottom damage is as follows:

longitudinal extent: at least 0.10 L, but not less than 5.00 m;
transverse extent: 3.00 m;
vertical extent: from the base 0.49 m upwards, the sump excepted;

(c)   Any bulkheads within the damaged area shall be assumed damaged, which means that the location of bulkheads shall be chosen so as to ensure that the vessel remains afloat after the flooding of two or more adjacent compartments in the longitudinal direction.

The following provisions are applicable:

–   For bottom damage also two adjacent athwartships compartments shall be assumed as flooded;

–   The lower edge of any openings that cannot be closed watertight (e.g. doors, windows, access hatchways) shall, at the final stage of flooding, be not less than 0.10 m above the damage waterline;

–   In general, permeability shall be assumed to be 95%. Where an average permeability of less than 95% is calculated for any compartment, this calculated value may be used.

However, the following minimum values shall be used:

–   engine rooms: 85%

–   accommodation: 95%

–   double bottoms, oil fuel tanks, ballast tanks, etc.,
depending on whether, according to their
function, they have to be assumed as full or
empty for the vessel floating at the maximum
permissible draught: 0% or 95%

For the main engine room only the one–compartment standard needs to be taken into account, i.e. the end bulkheads of the engine room shall be assumed as not damaged.

9.1.0.95.2   At the stage of equilibrium (final stage of flooding) the angle of heel shall not exceed 12°. Non-watertight openings shall not be immersed before reaching the stage of equilibrium. If such openings are immersed before that stage, the corresponding spaces shall be considered as flooded for the purpose of stability calculation.

The positive range of the righting lever curve beyond the position of equilibrium shall have a righting lever of ≥ 0.05 m in association with an area under the curve of ≥ 0.0065 m.rad. The minimum values of stability shall be satisfied up to immersion of the first non-weathertight opening and in any event up to an angle of heel $\leq 27°$. If non-weathertight openings are immersed before that stage, the corresponding spaces shall be considered as flooded for the purposes of stability calculation.

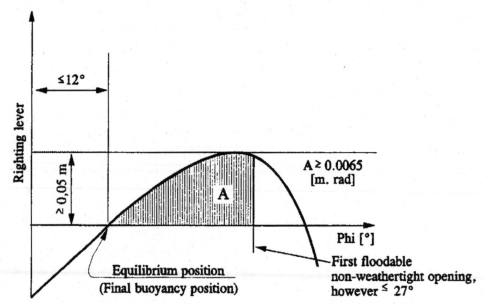

9.1.0.95.3   Inland navigation vessels carrying containers which have not been secured shall satisfy the following damage stability criteria:

At the stage of equilibrium (final stage of flooding) the angle of heel shall not exceed 5°. Non-watertight openings shall not be immersed before reaching the stage of equilibrium. If such openings are immersed before that stage, the corresponding spaces shall be considered as flooded for the purpose of stability calculation;

The positive range of the righting lever curve beyond the position of equilibrium shall have an area under the curve of ≥ 0.0065 m.rad. The minimum values of stability shall be satisfied up to immersion of the first non-weathertight opening and in any event up to an angle of heel $\leq 10°$. If non-weathertight openings are immersed before that stage, the corresponding spaces shall be considered as flooded for the purposes of stability calculation.

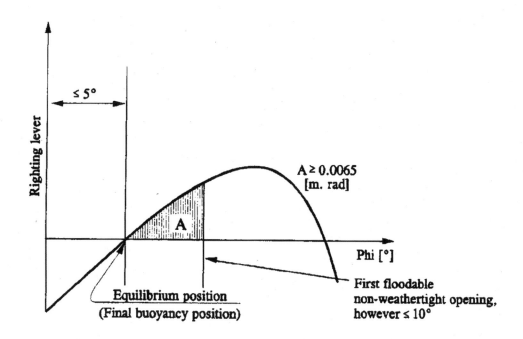

9.1.0.95.4   If openings through which undamaged compartments may become additionally flooded are capable of being closed watertight, the closing devices shall be appropriately marked.

9.1.0.95.5   Where cross- or down-flooding openings are provided for reduction of unsymmetrical flooding, the time for equalisation shall not exceed 15 minutes if during the intermediate stages of flooding sufficient stability has been proved.

9.1.0.96 to 9.1.0.99   (*Reserved*)

CHAPTER 9.2

RULES FOR CONSTRUCTION APPLICABLE TO SEAGOING

VESSELS WHICH COMPLY WITH THE REQUIREMENTS OF THE SOLAS 74 CONVENTION, CHAPTER II-2, REGULATION 19 OR SOLAS 74, CHAPTER II-2, REGULATION 54

9.2.0   The requirements of 9.2.0.0 to 9.2.0.79 are applicable to seagoing vessels which comply with the following requirements:

– SOLAS 74, Chapter II-2, Regulation 19 in its amended version; or

– SOLAS 74, Chapter II-2, Regulation 54 in its amended version in accordance with the resolutions mentioned in Chapter II-2, Regulation 1, paragraph 2.1, provided that the vessel was constructed before 1 July 2002.

Seagoing vessels which do not comply with the above–mentioned requirements of the SOLAS 74 Convention shall meet the requirements of 9.1.0.0 to 9.1.0.79.

9.2.0.0   *Materials of construction*

The vessel's hull shall be constructed of shipbuilding steel or other metal, provided that this metal has at least equivalent mechanical properties and resistance to the effects of temperature and fire.

9.2.0.1 to 9.2.0.19   (*Reserved*)

9.2.0.20   *Water ballast*

The double-hull spaces and double bottoms may be arranged for being filled with water ballast.

9.2.0.21 to 9.2.0.30   (*Reserved*)

9.2.0.31   *Engines*

9.2.0.31.1   Only internal combustion engines running on a fuel having a flashpoint above 60 °C, are allowed.

9.2.0.31.2   Ventilation inlets of the engine rooms and the air intakes of the engines which do not take air in directly from the engine room shall be located not less than 2 m from the protected area.

9.2.0.31.3   Sparking shall not be possible in the protected area.

9.2.0.32 and 9.2.0.33   (*Reserved*)

9.2.0.34   *Exhaust pipes*

9.2.0.34.1   Exhausts shall be evacuated from the vessel into the open–air either upwards through an exhaust pipe or through the shell plating. The exhaust outlet shall be located not less than 2.00 m from the hatchway openings. The exhaust pipes of engines shall be arranged so that the exhausts are led away from the vessel. The exhaust pipes shall not be located within the protected area.

9.2.0.34.2   Exhaust pipes shall be provided with a device preventing the escape of sparks, e.g. spark arresters.

9.2.0.35 to 9.2.0.40   (*Reserved*)

9.2.0.41   *Fire and naked light*

9.2.0.41.1   The outlets of funnels shall be located not less than 2.00 m from the hatchway openings. Arrangements shall be provided to prevent the escape of sparks and the entry of water.

9.2.0.41.2   Heating, cooking and refrigerating appliances shall not be fuelled with liquid fuels, liquid gas or solid fuels. The installation in the engine room or other separate space of heating appliances fuelled with liquid fuel having a flashpoint above 55 °C shall, however, be permitted.

Cooking and refrigerating appliances are permitted only in wheelhouses with metal floor and in the accommodation.

9.2.0.41.3   Electric lighting appliances are only permitted outside the accommodation and the wheelhouse.

9.2.0.42 to 9.2.0.70   (*Reserved*)

9.2.0.71   *Admittance on board*

The notice boards displaying the prohibition of admittance in accordance with 8.3.3 shall be clearly legible from either side of the vessel.

9.2.0.72 and 9.2.0.73   (*Reserved*)

9.2.0.74   *Prohibition of smoking, fire and naked light*

9.2.0.74.1   The notice boards displaying the prohibition of smoking in accordance with 8.3.4 shall be clearly legible from either side of the vessel.

9.2.0.74.2   Notice boards indicating the circumstances under which the prohibition applies shall be fitted near the entrances to the spaces where smoking or the use of fire or naked light is not always prohibited.

9.2.0.74.3   Ashtrays shall be provided close to each exit of the wheelhouse.

9.2.0.75 to 9.2.0.79   (*Reserved*)

9.2.0.80   *Additional rules applicable to double-hull vessels*

The rules of 9.2.0.88 to 9.2.0.99 are applicable to double-hull vessels intended to carry dangerous goods of Classes 2, 3, 4.1, 4.2, 4.3, 5.1, 5.2, 6.1, 7, 8 or 9, except those for which label No. 1 is prescribed in column (5) of Table A of Chapter 3.2, in quantities exceeding those of 7.1.4.1.1.

9.2.0.81 to 9.2.0.87   (*Reserved*)

9.2.0.88   *Classification*

9.2.0.88.1   Double-hull vessels intended to carry dangerous goods of Classes 2, 3, 4.1, 4.2, 4.3, 5.1, 5.2, 6.1, 7, 8 or 9 except those for which label No. 1 is prescribed in column (5) of Table A of Chapter 3.2, in quantities exceeding those referred to in 7.1.4.1, shall be built under survey of a recognised classification society in accordance with the rules established by that classification society to its highest class. This shall be confirmed by the classification society by the issue of an appropriate certificate.

9.2.0.88.2 The vessel's highest class shall be continued.

9.2.0.89 and 9.2.0.90 (*Reserved*)

9.2.0.91 *Holds*

9.2.0.91.1 The vessel shall be built as a double-hull vessel with double–wall spaces and double bottom within the protected area.

9.2.0.91.2 The distance between the sides of the vessel and the longitudinal bulkheads of the hold shall be not less than 0.80 m. A locally reduced distance at the vessel's ends shall be permitted, provided the smallest distance between vessel's side and the longitudinal bulkhead (measured perpendicular to the side) is not less than 0.60 m. The sufficient structural strength of the vessel (longitudinal, transverse and local strength) shall be confirmed by the certificate of class.

9.2.0.91.3 The depth of the double bottom shall be not less than 0.50 m.

The depth below the suction wells may however be locally reduced to 0.40 m, provided the suction well has a capacity of not more than 0.03 m^3.

9.2.0.92 (*Reserved*)

9.2.0.93 *Stability (general)*

9.2.0.93.1 Proof of sufficient stability shall be furnished including stability in the damaged condition.

9.2.0.93.2 The basic values for the stability calculation – the vessel's lightweight and the location of the centre of gravity – shall be determined either by means of an inclining experiment or by detailed mass and moment calculation. In the latter case the lightweight shall be checked by means of a lightweight test with a resulting difference of not more than ± 5% between the mass determined by the calculation and the displacement determined by the draught readings.

9.2.0.93.3 Proof of sufficient intact stability shall be furnished for all stages of loading and unloading and for the final loading condition.

Floatability after damage shall be proved for the most unfavourable loading condition. For this purpose calculated proof of sufficient stability shall be established for critical intermediate stages of flooding and for the final stage of flooding. Negative values of stability in intermediate stages of flooding may be accepted only if the continued range of curve of righting lever in damaged condition indicates adequate positive values of stability.

9.2.0.94 *Stability (intact)*

9.2.0.94.1 The requirements for intact stability resulting from the damaged stability calculation shall be fully complied with.

9.2.0.94.2 For the carriage of containers, additional proof of sufficient stability shall be furnished in accordance with the requirements of the Regulations referred to in 1.1.4.6.

9.2.0.94.3 The most stringent of the requirements of 9.2.0.94.1 and 9.2.0.94.2 shall prevail for the vessel.

9.2.0.94.4   For seagoing vessels the provisions of 9.2.0.94.2 above may be regarded as having been complied with if the stability conforms to Resolution A.749 (18) of the International Maritime Organization and the stability documents have been checked by the competent authority. This applies only when all containers are secured as usual on seagoing vessels and a relevant stability document has been approved by the competent authority.

9.2.0.95   *Stability (damaged condition)*

9.2.0.95.1   The following assumptions shall be taken into consideration for the damaged condition:

(a)   The extent of side damage is as follows:

longitudinal extent:	at least 0.10 L, but not less than 5.00 m;
transverse extent:	0.59 m inboard from the vessel's side at right angles to the centreline at the level corresponding to the maximum draught;
vertical extent:	from the baseline upwards without limit;

(b)   The extent of bottom damage is as follows:

longitudinal extent:	at least 0.10 L, but not less than 5.00 m;
transverse extent:	3.00 m;
vertical extent:	from the base 0.49 m upwards, the sump excepted;

(c)   Any bulkheads within the damaged area shall be assumed damaged, which means that the location of bulkheads shall be chosen so that the vessel will remain afloat after flooding of two or more adjacent compartments in the longitudinal direction.

The following provisions are applicable:

–   For bottom damage, adjacent athwartship compartments shall also be assumed as flooded;

–   The lower edge of any openings that cannot be closed watertight (e.g. doors, windows, access hatchways) shall, at the final stage of flooding, be not less than 0.10 m above the damage waterline;

–   In general, permeability shall be assumed to be 95%. Where an average permeability of less than 95% is calculated for any compartment, this calculated value may be used.

However, the following minimum values shall be used:

–   engine rooms   85%

–   accommodation   95%

–   double bottoms, oil fuel tanks, ballast tanks, etc.,

depending on whether according to their function,

they have to be assumed as full or empty for the

vessel floating at the maximum permissible draught   0% or 95%

For the main engine room only the one–compartment standard needs to be taken into account. (Consequently, the end bulkheads of the engine room shall be assumed as not damaged.)

9.2.0.95.2   At the stage of equilibrium (final stage of flooding) the angle of heel shall not exceed 12°. Non-watertight openings shall not be immersed before reaching the stage of equilibrium. If such openings are immersed before that stage, the corresponding spaces shall be considered as flooded for the purpose of stability calculation.

The positive range of the righting lever curve beyond the position of equilibrium shall have a righting lever of ≥ 0.05 m in association with an area under the curve of ≥ 0.0065 m.rad. The minimum values of stability shall be satisfied up to immersion of the first non-weathertight opening and in any event up to an angle of heel ≤ 27°. If non-weathertight openings are immersed before that stage, the corresponding spaces shall be considered as flooded for the purposes of stability calculation.

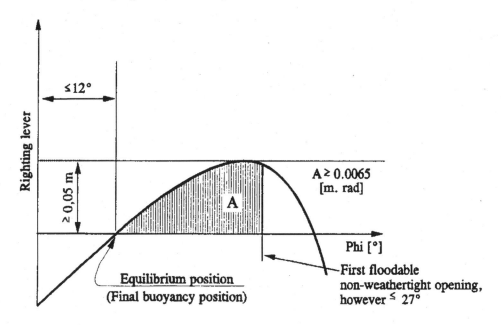

9.2.0.95.3   If openings through which undamaged compartments may become additionally flooded are capable of being closed watertight, the closing devices shall be appropriately marked.

9.2.0.95.4   Where cross- or down-flooding openings are provided for reduction of unsymmetrical flooding, the time for equalisation shall not exceed 15 minutes if during the intermediate stages of flooding sufficient stability has been proved.

9.2.0.96 to 9.2.0.99   (*Reserved*)

CHAPTER 9.3

RULES FOR CONSTRUCTION OF TANK VESSELS

9.3.1 **Rules for construction of type G tank vessels**

The rules for construction of 9.3.1.0 to 9.3.1.99 apply to type G tank vessels.

9.3.1.0 *Materials of construction*

9.3.1.0.1 (a) The vessel's hull and the cargo tanks shall be constructed of shipbuilding steel or other at least equivalent metal.

The independent cargo tanks may also be constructed of other materials, provided these have at least equivalent mechanical properties and resistance against the effects of temperature and fire.

(b) Every part of the vessel including any installation and equipment which may come into contact with the cargo shall consist of materials which can neither be dangerously affected by the cargo nor cause decomposition of the cargo or react with it so as to form harmful or hazardous products. In case it has not been possible to examine this during classification and inspection of the vessel a relevant reservation shall be entered in the vessel substance list according to 1.16.1.2.5.

9.3.1.0.2 Except where explicitly permitted in 9.3.1.0.3 below or in the certificate of approval, the use of wood, aluminium alloys or plastic materials within the cargo area is prohibited.

9.3.1.0.3 (a) The use of wood, aluminium alloys or plastic materials within the cargo area is only permitted for:

– gangways and external ladders;

– movable items of equipment;

– chocking of cargo tanks which are independent of the vessel's hull and chocking of installations and equipment;

– masts and similar round timber;

– engine parts;

– parts of the electrical installation;

– lids of boxes which are placed on the deck.

(b) The use of wood or plastic materials within the cargo area is only permitted for:

– supports and stops of any kind.

(c)   The use of plastic materials or rubber within the cargo area is only permitted for:

–   all kinds of gaskets (e.g. for dome or hatch covers);

–   electric cables;

–   hose assemblies for loading and unloading;

–   insulation of cargo tanks and of piping for loading and unloading;

–   photo–optical copies of the certificate of approval according to 8.1.2.6 or 8.1.2.7.

(d)   All permanently fitted materials in the accommodation or wheelhouse, with the exception of furniture, shall not readily ignite. They shall not evolve fumes or toxic gases in dangerous quantities, if involved in a fire.

9.3.1.0.4   The paint used in the cargo area shall not be liable to produce sparks in case of impact.

9.3.1.0.5   The use of plastic material for vessel's boats is permitted only if the material does not readily ignite.

9.3.1.1   *Vessel record*

NOTE: For the purpose of this paragraph, the term "owner" has the same meaning as in 1.16.0.

The vessel record shall be retained by the owner who shall be able to provide this documentation at the request of the competent authority and the recognized classification society.

The vessel record shall be maintained and updated throughout the life of the vessel and shall be retained for 6 months after the vessel is taken out of service.

Should a change of owner occur during the life of the vessel the vessel record shall be transferred to the new owner.

Copies of the vessel record or all necessary documents shall be made available on request to the competent authority for the issuance of the certificate of approval and for the recognized classification society or inspection body for first inspection, periodic inspection, special inspection or exceptional checks.

9.3.1.2 to 9.3.1.7   (*Reserved*)

9.3.1.8   *Classification*

9.3.1.8.1   The tank vessel shall be built under the survey of a recognised classification society and be classed in its highest class.

The vessel's highest class shall be continued. This shall be confirmed by an appropriate certificate issued by the recognized classification society (certificate of class).

The certificate of class shall confirm that the vessel is in conformity with its own additionally applicable rules and regulations that are relevant for the intended use of the vessel.

The design pressure and the test pressure of cargo tanks shall be entered in the certificate.

If a vessel has cargo tanks with different valve opening pressures, the design and test pressures of each tank shall be entered in the certificate.

The recognized classification society shall draw up a vessel substance list mentioning all the dangerous goods accepted for carriage by the tank vessel (see also 1.16.1.2.5).

9.3.1.8.2   The cargo pump-rooms shall be inspected by a recognised classification society whenever the certificate of approval has to be renewed as well as during the third year of validity of the certificate of approval. The inspection shall comprise at least:

–   an inspection of the whole system for its condition, for corrosion, leakage or conversion works which have not been approved;

–   a checking of the condition of the gas detection system in the cargo pump-rooms.

Inspection certificates signed by the recognised classification society with respect to the inspection of the cargo pump-rooms shall be kept on board. The inspection certificates shall at least include particulars of the above inspection and the results obtained as well as the date of the inspection.

9.3.1.8.3   The condition of the gas detection system referred to in 9.3.1.52.3 shall be checked by a recognised classification society whenever the certificate of approval has to be renewed and during the third year of validity of the certificate of approval. A certificate signed by the recognised classification society shall be kept on board.

9.3.1.9   (*Reserved*)

9.3.1.10   *Protection against the penetration of gases*

9.3.1.10.1   The vessel shall be designed so as to prevent gases from penetrating into the accommodation and the service spaces.

9.3.1.10.2   Outside the cargo area, the lower edges of door-openings in the sidewalls of superstructures and the coamings of access hatches to under-deck spaces shall have a height of not less than 0.50 m above the deck

This requirement need not be complied with if the wall of the superstructures facing the cargo area extends from one side of the ship to the other and has doors the sills of which have a height of not less than 0.50 m. The height of this wall shall not be less than 2.00 m. In this case, the lower edges of door-openings in the sidewalls of superstructures and the coamings of access hatches behind this wall shall have a height of not less than 0.10 m. The sills of engine room doors and the coamings of its access hatches shall, however, always have a height of not less than 0.50 m.

9.3.1.10.3   In the cargo area, the lower edges of door-openings in the sidewalls of superstructures shall have a height of not less than 0.50 m above the deck and the sills of hatches and ventilation openings of premises located under the deck shall have a height of not less than 0.50 m above the deck. This requirement does not apply to access openings to double-hull and double bottom spaces.

9.3.1.10.4   The bulwarks, foot-rails, etc shall be provided with sufficiently large openings which are located directly above the deck.

9.3.1.11 *Hold spaces and cargo tanks*

9.3.1.11.1   (a)   The maximum permissible capacity of a cargo tank shall be determined in accordance with the following table:

$L \times B \times H$ (m³)	Maximum permissible capacity of a cargo tank (m³)
up to 600	$L \times B \times H \times 0.3$
600 to 3 750	$180 + (L \times B \times H - 600) \times 0.0635$
> 3 750	380

Alternative constructions in accordance with 9.3.4 are permitted.

In the table above $L \times B \times H$ is the product of the main dimensions of the tank vessel in metres (according to the measurement certificate), where:

L = overall length of the hull in m;

B = extreme breadth of the hull in m;

H = shortest vertical distance between the top of the keel and the lowest point of the deck at the side of the vessel (moulded depth) within the cargo area in m;

where:

For trunk vessels, H shall be replaced by H', where H' shall be obtained from the following formula:

$$H' = H + \left(ht \times \frac{bt}{B} \times \frac{lt}{L} \right)$$

where:

ht = trunk height (distance between trunk deck and main deck measured on trunk side at L/2) in m;

bt = trunk breadth in m;

lt = trunk length in m;

    (b)   Pressure tanks whose ratio of length to diameter exceeds 7 are prohibited.

    (c)   The pressure tanks shall be designed for a cargo temperature of + 40 °C.

9.3.1.11.2   (a)   In the cargo area, the hull shall be designed as follows:[1]

       –   as a double-hull and double bottom vessel. The internal distance between the sideplatings of the vessel and the longitudinal bulkheads shall not be less than 0.80 m, the height of the double bottom shall be not less than 0.60 m, the cargo tanks shall be supported by saddles extending between the tanks to not less than 20° below the horizontal centreline of the cargo tanks.

---

[1]   *For a different design of the hull in the cargo area, proof shall be furnished by way of calculation that in the event of a lateral collision with another vessel having a straight bow, an energy of 22 MJ can be absorbed without any rupture of the cargo tanks and the piping leading to the cargo tanks. Alternative constructions in accordance with 9.3.4 are permitted.*

Refrigerated cargo tanks and cargo tanks used for the transport of refrigerated liquefied gases shall be installed only in hold spaces bounded by double-hull spaces and double–bottom. Cargo tank fastenings shall meet the requirements of a recognised classification society; or

– as a single–hull vessel with the sideplatings of the vessel between gangboard and top of floor plates provided with side stringers at regular intervals of not more than 0.60 m which are supported by web frames spaced at intervals of not more than 2.00 m. The side stringers and the web frames shall have a height of not less than 10% of the depth, however, not less than 0.30 m. The side stringers and web frames shall be fitted with a face plate made of flat steel and having a cross-section of not less that 7.5 cm^2 and 15 cm^2, respectively.

The distance between the sideplating of the vessel and the cargo tanks shall be not less than 0.80 m and between the bottom and the cargo tanks not less than 0.60 m. The depth below the suction wells may be reduced to 0.50 m.

The lateral distance between the suction well of the cargo tanks and the bottom structure shall be not less than 0.10 m.

The cargo tank supports and fastenings should extend to not less than 10° below the horizontal centreline of the cargo tanks.

(b) The cargo tanks shall be fixed so that they cannot float.

(c) The capacity of a suction well shall be limited to not more than 0.10 m^3. For pressure cargo tanks, however, the capacity of a suction well may be of 0.20 m^3.

(d) Side–struts linking or supporting the load-bearing components of the sides of the vessel with the load-bearing components of the longitudinal walls of cargo tanks and side–struts linking the load-bearing components of the vessel's bottom with the tank-bottom are prohibited.

(e) Cargo tanks intended to contain products at a temperature below -10°C shall be suitably insulated to ensure that the temperature of the vessel's structure does not fall below the minimum allowable material design temperature. The insulation material shall be resistant to flame spread.

9.3.1.11.3   (a) The hold spaces shall be separated from the accommodation, engine rooms and service spaces outside the cargo area below deck by bulkheads provided with a Class A-60 fire protection insulation according to SOLAS 74, Chapter II-2, Regulation 3. A space of not less than 0.20 m shall be provided between the cargo tanks and the end bulkheads of the hold spaces. Where the cargo tanks have plane end bulkheads this space shall be not less than 0.50 m.

(b) The hold spaces and cargo tanks shall be capable of being inspected.

(c) All spaces in the cargo area shall be capable of being ventilated. Means for checking their gas-free condition shall be provided.

9.3.1.11.4   The bulkheads bounding the hold spaces shall be watertight. The cargo tanks and the bulkheads bounding the cargo area shall have no openings or penetrations below deck.

The bulkhead between the engine room and the service spaces within the cargo area or between the engine room and a hold space may be fitted with penetrations provided that they conform to the requirements of 9.3.1.17.5.

9.3.1.11.5   Double-hull spaces and double bottoms in the cargo area shall be arranged for being filled with ballast water only. Double bottoms may, however, be used as oil fuel tanks, provided they comply with the requirements of 9.3.1.32.

9.3.1.11.6   (a)   A space in the cargo area below deck may be arranged as a service space, provided that the bulkhead bounding the service space extends vertically to the bottom and the bulkhead not facing the cargo area extends from one side of the vessel to the other in one frame plane. This service space shall only be accessible from the deck.

              (b)   The service space shall be watertight with the exception of its access hatches and ventilation inlets.

              (c)   No piping for loading or unloading shall be fitted within the service space referred to under (a) above.

                     Piping for loading and unloading may be fitted in the cargo pump-rooms below deck only when they conform to the provisions of 9.3.1.17.6.

9.3.1.11.7   Where service spaces are located in the cargo area under deck, they shall be arranged so as to be easily accessible and to permit persons wearing protective clothing and breathing apparatus to safely operate the service equipment contained therein. They shall be designed so as to allow injured or unconscious personnel to be removed from such spaces without difficulty, if necessary by means of fixed equipment.

9.3.1.11.8   Hold spaces and other accessible spaces within the cargo area shall be arranged so as to ensure that they may be completely inspected and cleaned in an appropriate manner. The dimensions of openings, except for those of double-hull spaces and double bottoms which do not have a wall adjoining the cargo tanks, shall be sufficient to allow a person wearing breathing apparatus to enter or leave the space without difficulty. These openings shall have a minimum cross-sectional area of 0.36 m^2 and a minimum side length of 0.50 m. They shall be designed so as to allow injured or unconscious persons to be removed from the bottom of such spaces without difficulties, if necessary by means of fixed equipment. In these spaces the distance between the reinforcements shall not be less than 0.50 m. In double bottoms this distance may be reduced to 0.45 m.

              Cargo tanks may have circular openings with a diameter of not less than 0.68 m.

9.3.1.11.9   In case the vessel has insulated cargo tanks, the hold spaces shall only contain dry air to protect the insulation of the cargo tanks against moisture.

9.3.1.12   *Ventilation*

9.3.1.12.1   Each hold space shall have two openings the dimensions and location of which shall be such as to permit effective ventilation of any part of the hold space. If there are no such openings, it shall be possible to fill the hold spaces with inert gas or dry air.

9.3.1.12.2   Double-hull spaces and double bottoms within the cargo area which are not arranged for being filled with ballast water and cofferdams between engine rooms and pump-rooms, if they exist, shall be provided with ventilation systems.

9.3.1.12.3   Any service spaces located in the cargo area below deck shall be provided with a system of forced ventilation with sufficient power for ensuring at least 20 changes of air per hour based on the volume of the space.

              The ventilation exhaust ducts shall extend down to 50 mm above the bottom of the service space. The air shall be supplied through a duct at the top of the service space. The air inlets shall be located not less than 2.00 m above the deck, at a distance of not less than 2.00 m from tank openings and 6.00 m from the outlets of safety valves.

The extension pipes, which may be necessary, may be of the hinged type.

9.3.1.12.4   Ventilation of accommodation and service spaces shall be possible.

9.3.1.12.5   Ventilators used in the cargo area shall be designed so that no sparks may be emitted on contact of the impeller blades with the housing and no static electricity may be generated.

9.3.1.12.6   Notice boards shall be fitted at the ventilation inlets indicating the conditions when they shall be closed. All ventilation inlets of accommodation and service spaces leading outside shall be fitted with fire flaps. Such ventilation inlets shall be located not less than 2.00 m from the cargo area.

Ventilation inlets of service spaces in the cargo area may be located within such area.

9.3.1.13   *Stability (general)*

9.3.1.13.1   Proof of sufficient stability shall be furnished including for stability in damaged condition.

9.3.1.13.2   The basic values for the stability calculation – the vessel's lightweight and location of the centre of gravity – shall be determined either by means of an inclining experiment or by detailed mass and moment calculation. In the latter case the lightweight of the vessel shall be checked by means of a lightweight test with a tolerance limit of ± 5% between the mass determined by calculation and the displacement determined by the draught readings.

9.3.1.13.3   Proof of sufficient intact stability shall be furnished for all stages of loading and unloading and for the final loading condition for all the relative densities of the substances transported contained in the vessel substance list according to 1.16.1.2.5.

For every loading case, taking account of the actual fillings and floating position of cargo tanks, ballast tanks and compartments, drinking water and sewage tanks and tanks containing products for the operation of the vessel, the vessel shall comply with the intact and damage stability requirements.

Intermediate stages during operations shall also be taken into consideration.

The proof of sufficient stability shall be shown for every operating, loading and ballast condition in the stability booklet, to be approved by the recognized classification society, which classes the vessel. If it is unpractical to pre-calculate the operating, loading and ballast conditions, a loading instrument approved by the recognised classification society which classes the vessel shall be installed and used which contains the contents of the stability booklet.

NOTE: A stability booklet shall be worded in a form comprehensible for the responsible master and containing the following details:

General description of the vessel:

–   *General arrangement and capacity plans indicating the assigned use of compartments and spaces (cargo tanks, stores, accommodation, etc.);*

–   *A sketch indicating the position of the draught marks referring to the vessel's perpendiculars;*

–   *A scheme for ballast/bilge pumping and overflow prevention systems;*

–   *Hydrostatic curves or tables corresponding to the design trim, and, if significant trim angles are foreseen during the normal operation of the vessel, curves or tables corresponding to such range of trim are to be introduced;*

- *Cross curves or tables of stability calculated on a free trimming basis, for the ranges of displacement and trim anticipated in normal operating conditions, with an indication of the volumes which have been considered buoyant;*

- *Tank sounding tables or curves showing capacities, centres of gravity, and free surface data for all cargo tanks, ballast tanks and compartments, drinking water and sewage water tanks and tanks containing products for the operation of the vessel;*

- *Lightship data (weight and centre of gravity) resulting from an inclining test or deadweight measurement in combination with a detailed mass balance or other acceptable measures. Where the above–mentioned information is derived from a sister vessel, the reference to this sister vessel shall be clearly indicated, and a copy of the approved inclining test report relevant to this sister vessel shall be included;*

- *A copy of the approved test report shall be included in the stability booklet;*

- *Operating loading conditions with all relevant details, such as:*

  - *Lightship data, tank fillings, stores, crew and other relevant items on board (mass and centre of gravity for each item, free surface moments for liquid loads);*

  - *Draughts amidships and at perpendiculars;*

  - *Metacentric height corrected for free surfaces effect;*

  - *Righting lever values and curve;*

  - *Longitudinal bending moments and shear forces at read–out points;*

  - *Information about openings (location, type of tightness, means of closure); and*

  - *Information for the master.*

- *Calculation of the influence of ballast water on stability with information on whether fixed level gauges for ballast tanks and compartments have to be installed or the ballast tanks, or compartments shall only be completely full or completely empty when underway.*

9.3.1.13.4   Floatability after damage shall be proved for the most unfavourable loading condition. For this purpose, calculated proof of sufficient stability shall be established for critical intermediate stages of flooding and for the final stage of flooding.

9.3.1.14   ***Stability (intact)***

9.3.1.14.1   The requirements for intact stability resulting from the damaged stability calculation shall be fully complied with.

9.3.1.14.2   For vessels with cargo tanks of more than 0.70 B in width, proof shall be furnished that the following stability requirements have been complied with:

(a)   In the positive area of the righting lever curve up to immersion of the first non-watertight opening there shall be a righting lever (GZ) of not less than 0.10 m;

(b)   The surface of the positive area of the righting lever curve up to immersion of the first non-watertight opening and in any event up to an angle of heel < 27° shall not be less than 0.024 m.rad;

(c)   The metacentric height (GM) shall be not less than 0.10 m.

These conditions shall be met bearing in mind the influence of all free surfaces in tanks for all stages of loading and unloading.

9.3.1.14.3   The most stringent requirement of 9.3.1.14.1 and 9.3.1.14.2 is applicable to the vessel.

9.3.1.15   *Stability (damaged condition)*

9.3.1.15.1   The following assumptions shall be taken into consideration for the damaged condition:

(a)   The extent of side damage is as follows:

longitudinal extent:	at least 0.10 L, but not less than 5.00 m;
transverse extent:	0.79 m inboard from the vessel's side at right angles to the centreline at the level corresponding to the maximum draught, or when applicable, the distance allowed by section 9.3.4, reduced by 0.01 m;
vertical extent:	from the base line upwards without limit;

(b)   The extent of bottom damage is as follows:

longitudinal extent:	at least 0.10 L, but not less than 5.00 m;
transverse extent:	3.00 m;
vertical extent:	from the base 0.59 m upwards, the well excepted;

(c)   Any bulkheads within the damaged area shall be assumed damaged, which means that the location of bulkheads shall be chosen so as to ensure that the vessel remains afloat after the flooding of two or more adjacent compartments in the longitudinal direction.

The following provisions are applicable:

–   For bottom damage, adjacent athwartship compartments shall also be assumed as flooded;

–   The lower edge of any non-watertight openings (e.g. doors, windows, access hatchways) shall, at the final stage of flooding, be not less than 0.10 m above the damage waterline;

–   In general, permeability shall be assumed to be 95%. Where an average permeability of less than 95% is calculated for any compartment, this calculated value obtained may be used.

However, the following minimum values shall be used:

- engine rooms: 85%;

- accommodation: 95%;

- double bottoms, oil fuel tanks, ballast tanks,

  etc., depending on whether, according

  to their function, they have to be assumed

  as full or empty for the vessel floating

  at the maximum permissible draught: 0% or 95%.

For the main engine room only the one–compartment standard need be taken into account, i.e. the end bulkheads of the engine room shall be assumed as not damaged.

9.3.1.15.2   For the intermediate stage of flooding the following criteria have to be fulfilled:

GZ >= 0.03m

Range of positive GZ: 5°.

At the stage of equilibrium (final stage of flooding), the angle of heel shall not exceed 12°. Non-watertight openings shall not be flooded before reaching the stage of equilibrium. If such openings are immersed before that stage, the corresponding spaces shall be considered as flooded for the purpose of stability calculation.

The positive range of the righting lever curve beyond the stage of equilibrium shall have a righting level of $\geq$ 0.05 m in association with an area under the curve of $\geq$ 0.0065 m.rad. The minimum values of stability shall be satisfied up to immersion of the first non-weathertight opening and in any event up to an angle of heel $\leq$ 27°. If non-watertight openings are immersed before that stage, the corresponding spaces shall be considered as flooded for the purpose of stability calculation.

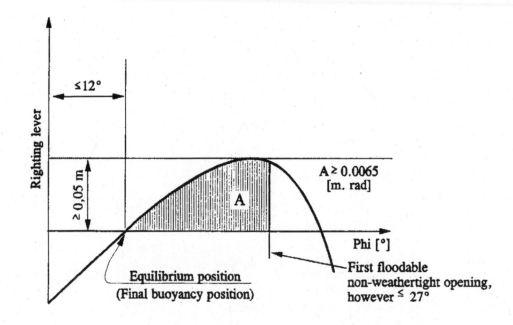

-434-

9.3.1.15.3   If openings through which undamaged compartments may additionally become flooded are capable of being closed watertight, the closing appliances shall be marked accordingly.

9.3.1.15.4   When cross- or down-flooding openings are provided for reduction of unsymmetrical flooding, the time for equalisation shall not exceed 15 minutes, if during the intermediate stages of flooding sufficient stability has been proved.

9.3.1.16   *Engine rooms*

9.3.1.16.1   Internal combustion engines for the vessel's propulsion as well as internal combustion engines for auxiliary machinery shall be located outside the cargo area. Entrances and other openings of engine rooms shall be at a distance of not less than 2.00 m from the cargo area.

9.3.1.16.2   The engine room shall be accessible from the deck; the entrances shall not face the cargo area. When the doors are not located in a recess whose depth is at least equal to the door width, the hinges shall face the cargo area.

9.3.1.17   *Accommodation and service spaces*

9.3.1.17.1   Accommodation spaces and the wheelhouse shall be located outside the cargo area forward of the fore vertical plane or abaft the aft vertical plane bounding the part of the cargo area below deck. Windows of the wheelhouse which are located not less than 1.00 m above the bottom of the wheelhouse may tilt forward.

9.3.1.17.2   Entrances to spaces and openings of superstructures shall not face the cargo area. Doors opening outward and not located in a recess the depth of which is at least equal to the width of the doors shall have their hinges facing the cargo area.

9.3.1.17.3   Entrances from the deck and openings of spaces facing the weather shall be capable of being closed. The following instruction shall be displayed at the entrance of such spaces:

**Do not open during loading, unloading and degassing
without the permission of the master.
Close immediately.**

9.3.1.17.4   Entrances and windows of superstructures and accommodation spaces which can be opened as well as other openings of these spaces shall be located not less than 2.00 m from the cargo area. No wheelhouse doors and windows shall be located within 2.00 m from the cargo area, except where there is no direct connection between the wheelhouse and the accommodation.

9.3.1.17.5   (a)   Driving shafts of the bilge or ballast pumps may penetrate through the bulkhead between the service space and the engine room, provided the arrangement of the service space is in compliance with 9.3.1.11.6.

(b)   The penetration of the shaft through the bulkhead shall be gastight and shall have been approved by a recognised classification society.

(c)   The necessary operating instructions shall be displayed.

(d)   Penetrations through the bulkhead between the engine room and the service space in the cargo area, and the bulkhead between the engine room and the hold spaces may be provided for electrical cables, hydraulic lines and piping for measuring, control and alarm systems, provided that the penetrations have been approved by a recognised classification society. The penetrations shall be gastight. Penetrations through a bulkhead with an "A-60" fire protection insulation according to SOLAS 74, Chapter II-2, Regulation 3, shall have an equivalent fire protection.

(e)   Pipes may pass through the bulkhead between the engine room and the service space in the cargo area provided that these are pipes between the mechanical equipment in the engine room and the service space which do not have any openings within the service space and which are provided with shut-off devices at the bulkhead in the engine room.

(f)   Notwithstanding 9.3.1.11.4, pipes from the engine room may pass through the service space in the cargo area or a cofferdam or a hold space or a double-hull space to the outside provided that within the service space or cofferdam or hold space or double-hull space they are of the thick-walled type and have no flanges or openings.

(g)   Where a driving shaft of auxiliary machinery penetrates through a wall located above the deck the penetration shall be gastight.

9.3.1.17.6   A service space located within the cargo area below deck shall not be used as a cargo pump-room for the vessel's own gas discharging system, e.g. compressors or the compressor/heat exchanger/pump combination, except where:

–   the pump-room is separated from the engine room or from service spaces outside the cargo area by a cofferdam or a bulkhead with an "A-60" fire protection insulation according to SOLAS 74, Chapter II-2, Regulation 3, or by a service space or a hold space;

–   the "A-60" bulkhead required above does not include penetrations referred to in 9.3.1.17.5 (a);

–   ventilation exhaust outlets are located not less than 6.00 m from entrances and openings of the accommodation and service spaces;

–   the access hatches and ventilation inlets can be closed from the outside;

–   all piping for loading and unloading (at the suction side and delivery side) are led through the deck above the pump-room. The necessary operation of the control devices in the pump-room, starting of pumps or compressors and necessary control of the liquid flow rate shall be effected from the deck;

–   the system is fully integrated in the gas and liquid piping system;

–   the cargo pump-room is provided with a permanent gas detection system which automatically indicates the presence of explosive gases or lack of oxygen by means of direct-measuring sensors and which actuates a visual and audible alarm when the gas concentration has reached 20% of the lower explosive limit. The sensors of this system shall be placed at suitable positions at the bottom and directly below the deck.

        Measurement shall be continuous.

        The audible and visual alarms are installed in the wheelhouse and in the cargo pump-room and, when the alarm is actuated, the loading and unloading system is shut down. Failure of the gas detection system shall be immediately signalled in the wheelhouse and on deck by means of audible and visual alarms;

–   the ventilation system prescribed in 9.3.1.12.3 has a capacity of not less than 30 changes of air per hour based on the total volume of the service space.

9.3.1.17.7   The following instruction shall be displayed at the entrance of the cargo pump-room:

Before entering the cargo pump-room check whether
it is free from gases and contains sufficient oxygen.
Do not open doors and entrance openings without
the permission of the master.
Leave immediately in the event of alarm.

9.3.1.18   *Inerting facility*

In cases in which inerting or blanketing of the cargo is prescribed, the vessel shall be equipped with an inerting system.

This system shall be capable of maintaining a permanent minimum pressure of 7 kPa (0.07 bar) in the spaces to be inerted. In addition, the inerting system shall not increase the pressure in the cargo tank to a pressure greater than that at which the pressure valve is regulated. The set pressure of the vacuum-relief valve shall be 3.5 kPa (0.035 bar).

A sufficient quantity of inert gas for loading or unloading shall be carried or produced on board if it is not possible to obtain it on shore. In addition, a sufficient quantity of inert gas to offset normal losses occurring during carriage shall be on board.

The premises to be inerted shall be equipped with connections for introducing the inert gas and monitoring systems so as to ensure the correct atmosphere on a permanent basis.

When the pressure or the concentration of inert gas in the gaseous phase falls below a given value, this monitoring system shall activate an audible and visible alarm in the wheelhouse. When the wheelhouse is unoccupied, the alarm shall also be perceptible in a location occupied by a crew member.

9.3.1.19 and 9.3.1.20   (*Reserved*)

9.3.1.21   *Safety and control installations*

9.3.1.21.1   Cargo tanks shall be provided with the following equipment:

(a)   *(Reserved)*;

(b)   a level gauge;

(c)   a level alarm device which is activated at the latest when a degree of filling of 86% is reached;

(d)   a high level sensor for actuating the facility against overflowing at the latest when a degree of filling of 97.5% is reached;

(e)   an instrument for measuring the pressure of the gas phase in the cargo tank;

(f)   an instrument for measuring the temperature of the cargo;

(g)   a connection for a closed-type sampling device.

9.3.1.21.2   When the degree of filling in per cent is determined, an error of not more than 0.5% is permitted. It shall be calculated on the basis of the total cargo tank capacity including the expansion trunk.

9.3.1.21.3   The level gauge shall allow readings from the control position of the shut-off devices of the particular cargo tank. The permissible maximum filling levels of 91%, 95% and 97%, as given in the list of substances, shall be marked on each level gauge.

Permanent reading of the overpressure and vacuum shall be possible from a location from which loading or unloading operations may be interrupted. The permissible maximum overpressure and vacuum shall be marked on each level gauge.

Readings shall be possible in all weather conditions.

9.3.1.21.4   The level alarm device shall give a visual and audible warning on board when actuated. The level alarm device shall be independent of the level gauge.

9.3.1.21.5   (a)   The high level sensor referred to in 9.3.1.21.1 (d) shall give a visual and audible alarm on board and at the same time actuate an electrical contact which in the form of a binary signal interrupts the electric current loop provided and fed by the shore facility, thus initiating measures at the shore facility against overflowing during loading operations.

The signal shall be transmitted to the shore facility via a watertight two–pin plug of a connector device in accordance with standard EN 60309–2:1999 + A1:2007 + A2:2012 for direct current of 40 to 50 volts, identification colour white, position of the nose 10 h.

The plug shall be permanently fitted to the vessel close to the shore connections of the loading and unloading piping.

The high level sensor shall also be capable of switching off the vessel's own discharging pump.

The high level sensor shall be independent of the level alarm device, but it may be connected to the level gauge.

(b)   During discharging by means of the on-board pump, it shall be possible for the shore facility to switch it off. For this purpose, an independent intrinsically safe power line, fed by the vessel, shall be switched off by the shore facility by means of an electrical contact.

It shall be possible for the binary signal of the shore facility to be transmitted via a watertight two–pole socket or a connector device in accordance with standard EN 60309-2:1999 + A1:2007 + A2:2012, for direct current of 40 to 50 volts, identification colour white, position of the nose 10 h.

This socket shall be permanently fitted to the vessel close to the shore connections of the unloading piping.

9.3.1.21.6   The visual and audible signals given by the level alarm device shall be clearly distinguishable from those of the high level sensor.

The visual alarm shall be visible at each control position on deck of the cargo tank stop valves. It shall be possible to easily check the functioning of the sensors and electric circuits or these shall be of the "failsafe" design.

9.3.1.21.7   When the pressure or the temperature exceeds a set value, the instruments for measuring the pressure and the temperature of the cargo shall activate a visual and an audible alarm in the wheelhouse. When the wheelhouse is unoccupied the alarm shall also be perceptible in a location occupied by a crew member.

When the pressure exceeds a set value during loading or unloading, the instrument for measuring the pressure shall simultaneously initiate an electrical contact which, by means of the plug referred to in 9.3.1.21.5 above, enables measures to be taken to interrupt the loading and unloading operation. When the vessel's own discharge pump is used, it shall be switched off automatically. The sensor for the alarms referred to above may be connected to the alarm installation.

9.3.1.21.8   Where the control elements of the shut-off devices of the cargo tanks are located in a control room, it shall be possible to stop the loading pumps and read the level gauges in the control room, and the visual and audible warning given by the level alarm device, the high level sensor referred to in 9.3.1.21.1 (d) and the instruments for measuring the pressure and temperature of the cargo shall be noticeable in the control room and on deck.

Satisfactory monitoring of the cargo area shall be ensured from the control room.

9.3.1.21.9   The vessel shall be so equipped that loading or unloading operations can be interrupted by means of switches, i.e. the quick-action stop valve located on the flexible vessel–to–shore connecting line must be capable of being closed. The switches shall be placed at two points on the vessel (fore and aft).

The interruption systems shall be designed according to the quiescent current principle.

9.3.1.21.10   When refrigerated substances are carried the opening pressure of the safety system shall be determined by the design of the cargo tanks. In the event of the transport of substances that must be carried in a refrigerated state the opening pressure of the safety system shall be not less than 25 kPa (0.25 bar) greater than the maximum pressure calculated according to 9.3.1.27.

9.3.1.21.11   On vessels certified to carry refrigerated liquefied gases the following protective measures shall be provided in the cargo area:

–   Drips trays shall be installed under the shore connections of the piping for loading and unloading through which the loading and unloading operation is carried out. They must be made of materials which are able to resist the temperature of the cargo and be insulated from the deck. The drip trays shall have a sufficient volume and an overboard drain;

–   A water spray system to cover:

1.   exposed cargo tank domes and exposed parts of cargo tanks;

2.   exposed on–deck storage vessels for flammable or toxic products;

3.   parts of the cargo deck area where a leakage may occur.

The capacity of the water spray system shall be such that when all spray nozzles are in operation, the outflow is of 300 liters per square meter of cargo deck area per hour. The system shall be capable of being put into operation from the wheelhouse and from the deck;

–   A water film around the shore connection of the piping for loading and unloading in use to protect the deck and the shipside in the way of the shore connection of the piping for loading and unloading in use during connecting and disconnecting the loading arm or hose. The water film shall have sufficient capacity. The system shall be capable of being put into operation from the wheelhouse and from the deck.

9.3.1.21.12   Vessels carrying refrigerated liquefied gases shall have on board, for the purpose of preventing damage to the cargo tanks during loading and the piping for loading and unloading during loading and unloading, a written instruction for pre–cooling. This instruction shall be applied before the vessel is put into operation and after long–term maintenance.

9.3.1.22   *Cargo tank openings*

9.3.1.22.1   (a)   Cargo tank openings shall be located on deck in the cargo area.

(b)   Cargo tank openings with a cross-section greater than 0.10 m^2 shall be located not less than 0.50 m above the deck.

9.3.1.22.2   Cargo tank openings shall be fitted with gastight closures which comply with the provisions of 9.3.1.23.1.

9.3.1.22.3   The exhaust outlets of the pressure relief valves shall be located not less than 2.00 m above the deck at a distance of not less than 6.00 m from the accommodation and from the service spaces located outside the cargo area. This height may be reduced when within a radius of 1.00 m round the pressure relief valve outlet there is no equipment, no work is being carried out and signs indicate the area.

9.3.1.22.4   The closing devices normally used in loading and unloading operations shall not be capable of producing sparks when operated.

9.3.1.22.5   Each tank in which refrigerated substances are carried shall be equipped with a safety system to prevent unauthorized vacuum or overpressure.

9.3.1.23   *Pressure test*

9.3.1.23.1   Cargo tanks and piping for loading and unloading shall comply with the provisions concerning pressure vessels which have been established by the competent authority or a recognised classification society for the substances carried.

9.3.1.23.2   Any cofferdams shall be subjected to initial tests before being put into service and thereafter at the prescribed intervals.

The test pressure shall be not less than 10 kPa (0.10 bar) gauge pressure.

9.3.1.23.3   The maximum intervals for the periodic tests referred to in 9.3.1.23.2 above shall be 11 years.

9.3.1.24   *Regulation of cargo pressure and temperature*

9.3.1.24.1   Unless the entire cargo system is designed to resist the full effective vapour pressure of the cargo at the upper limits of the ambient design temperatures, the pressure of the tanks shall be kept below the permissible maximum set pressure of the safety valves, by one or more of the following means:

(a)   a system for the regulation of cargo tank pressure using mechanical refrigeration;

(b)   a system ensuring safety in the event of the heating or increase in pressure of the cargo. The insulation or the design pressure of the cargo tank, or the combination of these two elements, shall be such as to leave an adequate margin for the operating period and the temperatures expected; in each case the system shall be deemed acceptable by a recognized classification society and shall ensure safety for a minimum time of three times the operation period;

(c)   For UN No. 1972 only, and when the use of LNG as fuel is authorized according to 1.5.3.2, a system for the regulation of cargo tank pressure whereby the boil-off vapours are utilized as fuel;

(d)   other systems deemed acceptable by a recognized classification society.

9.3.1.24.2   The systems prescribed in 9.3.1.24.1 shall be constructed, installed and tested to the satisfaction of the recognized classification society. The materials used in their construction shall be compatible with the cargoes to be carried. For normal service, the upper ambient design temperature limits shall be:

air: +30° C;
water: +20° C.

9.3.1.24.3   The cargo storage system shall be capable of resisting the full vapour pressure of the cargo at the upper limits of the ambient design temperatures, whatever the system adopted to deal with the boil-off gas. This requirement is indicated by remark 37 in column (20) of Table C of Chapter 3.2.

9.3.1.25   *Pumps and piping*

9.3.1.25.1   Pumps, compressors and accessory loading and unloading piping shall be placed in the cargo area. Cargo pumps and compressors shall be capable of being shut down from the cargo area and, in addition, from a position outside the cargo area. Cargo pumps and compressors situated on deck shall be located not less than 6.00 m from entrances to, or openings of, the accommodation and service spaces outside the cargo area.

9.3.1.25.2   (a)   Piping for loading and unloading shall be independent of any other piping of the vessel. No cargo piping shall be located below deck, except those inside the cargo tanks and in the service spaces intended for the installation of the vessel's own gas discharging system.

(b)   *(Reserved)*

(c)   Piping for loading and unloading shall be clearly distinguishable from other piping, e.g. by means of colour marking.

(d)   The piping for loading and unloading on deck, the venting piping with the exception of the shore connections but including the safety valves, and the valves shall be located within the longitudinal line formed by the outer boundaries of the domes and not less than one quarter of the vessel's breadth from the outer shell. This requirement does not apply to the relief pipes situated behind the safety valves. If there is, however, only one dome athwartships, these pipes and their valves shall be located at a distance not less than 2.70 m from the shell.

Where cargo tanks are placed side by side, all the connections to the domes shall be located on the inner side of the domes. The external connections may be located on the fore and aft centre line of the dome. The shut-off devices shall be located directly at the dome or as close as possible to it. The shut-off devices of the loading and unloading piping shall be duplicated, one of the devices being constituted by a remote-controlled quick-action stop device. When the inside diameter of a shut-off device is less than 50 mm this device may be regarded as a safety device against bursts in the piping.

(e)   The shore connections shall be located not less than 6.00 m from the entrances to or openings of, the accommodation and service spaces outside the cargo area.

(f)   Each shore connection of the venting piping and shore connections of the piping for loading and unloading, through which the loading or unloading operation is carried out, shall be fitted with a shut-off device and a quick-action stop valve. However, each shore connection shall be fitted with a blind flange when it is not in operation.

(g)   Piping for loading and unloading, and venting piping, shall not have flexible connections fitted with sliding seals.

For transport of refrigerated liquefied gases

(h)   The piping for loading and unloading and cargo tanks shall be protected from excessive stresses due to thermal movement and from movements of the tank and hull structure.

(i)   Where necessary, piping for loading and unloading shall be thermally insulated from the adjacent hull structure to prevent the temperature of the hull falling below the design temperature of the hull material.

(j)   All piping for loading and unloading, which may be closed off at each end when containing liquid (residue), shall be provided with safety valves. These safety valves shall discharge into the cargo tanks and shall be protected against inadvertent closing.

9.3.1.25.3   The distance referred to in 9.3.1.25.1 and 9.3.1.25.2 (e) may be reduced to 3.00 m if a transverse bulkhead complying with 9.3.1.10.2 is situated at the end of the cargo area. The openings shall be provided with doors.

The following notice shall be displayed on the doors:

**Do not open during loading and unloading without
the permission of the master.
Close immediately.**

9.3.1.25.4   Every component of the piping for loading and unloading shall be electrically connected to the hull.

9.3.1.25.5   The stop valves or other shut-off devices of the piping for loading and unloading shall indicate whether they are open or shut.

9.3.1.25.6   The piping for loading and unloading shall have, at the test pressure, the required elasticity, leakproofness and resistance to pressure.

9.3.1.25.7   The piping for unloading shall be fitted with pressure gauges at the inlet and outlet of the pump.

Reading of the pressure gauges shall be possible from the control position of the vessel's own gas discharging system. The maximum permissible overpressure or vacuum shall be indicated by a measuring device.

Readings shall be possible in all weather conditions.

9.3.1.25.8   Use of the cargo piping for ballasting purposes shall not be possible.

9.3.1.25.9   *(Reserved)*

9.3.1.25.10   Compressed air generated outside the cargo area or wheelhouse can be used in the cargo area subject to the installation of a spring-loaded non-return valve to ensure that no gases can escape from the cargo area through the compressed air system into accommodation or service spaces outside the cargo area.

9.3.1.26   *(Reserved)*

9.3.1.27   *Refrigeration system*

9.3.1.27.1   The refrigeration system referred to in 9.3.1.24.1 (a) shall be composed of one or more units capable of keeping the pressure and temperature of the cargo at the upper limits of the ambient design temperatures at the prescribed level. Unless another means of regulating cargo pressure and temperature deemed satisfactory by a recognized classification society is provided, provision shall be made for one or more stand-by units with an output at least equal to that of the largest prescribed unit. A stand-by unit shall include a compressor, its engine, its control system and all necessary accessories to enable it to operate independently of the units normally used. Provision shall be made for a stand-by heat-exchanger unless the system's normal heat-exchanger has a surplus capacity equal to at least 25% of the largest prescribed capacity. It is not necessary to make provision for separate piping.

Cargo tanks, piping and accessories shall be insulated so that, in the event of a failure of all cargo refrigeration systems, the entire cargo remains for at least 52 hours in a condition not causing the safety valves to open.

9.3.1.27.2   The security devices and the connecting lines from the refrigeration system shall be connected to the cargo tanks above the liquid phase of the cargo when the tanks are filled to their maximum permissible degree of filling. They shall remain within the gaseous phase, even if the vessel has a list up to 12 degrees.

9.3.1.27.3   When several refrigerated cargoes with a potentially dangerous chemical reaction are carried simultaneously, particular care shall be given to the refrigeration systems so as to prevent any mixing of the cargoes. For the carriage of such cargoes, separate refrigeration systems, each including the full stand-by unit referred to in 9.3.1.27.1, shall be provided for each cargo. When, however, refrigeration is ensured by an indirect or combined system and no leak in the heat exchangers can under any foreseeable circumstances lead to the mixing of cargoes, no provision need be made for separate refrigeration units for the different cargoes.

9.3.1.27.4   When several refrigerated cargoes are not soluble in each other under conditions of carriage such that their vapour pressures are added together in the event of mixing, particular care shall be given to the refrigeration systems to prevent any mixing of the cargoes.

9.3.1.27.5   When the refrigeration systems require water for cooling, a sufficient quantity shall be supplied by a pump or pumps used exclusively for the purpose. This pump or pumps shall have at least two suction pipes, leading from two water intakes, one to port, the other to starboard. Provision shall be made for a stand-by pump with a satisfactory flow; this may be a pump used for other purposes provided that its use for supplying water for cooling does not impair any other essential service.

9.3.1.27.6   The refrigeration system may take one of the following forms:

(a)   Direct system: the cargo vapours are compressed, condensed and returned to the cargo tanks. This system shall not be used for certain cargoes specified in Table C of Chapter 3.2. This requirement is indicated by remark 35 in column (20) of Table C of Chapter 3.2;

(b)   Indirect system: the cargo or the cargo vapours are cooled or condensed by means of a coolant without being compressed;

(c)   Combined system: the cargo vapours are compressed and condensed in a cargo/coolant heat-exchanger and returned to the cargo tanks. This system shall not be used for certain cargoes specified in Table C of Chapter 3.2. This requirement is indicated by remark 36 in column (20) of Table C of Chapter 3.2.

9.3.1.27.7 All primary and secondary coolant fluids shall be compatible with each other and with the cargo with which they may come into contact. Heat exchange may take place either at a distance from the cargo tank, or by using cooling coils attached to the inside or the outside of the cargo tank.

9.3.1.27.8 When the refrigeration system is installed in a separate service space, this service space shall meet the requirements of 9.3.1.17.6.

9.3.1.27.9 For all cargo systems, the heat transmission coefficient as used for the determination of the holding time (7.2.4.16.16 and 7.2.4.16.17) shall be determined by calculation. Upon completion of the vessel, the correctness of the calculation shall be checked by means of a heat balance test. The calculation and test shall be performed under supervision by the recognized classification society which classified the vessel.

The heat transmission coefficient shall be documented and kept on board. The heat transmission coefficient shall be verified at every renewal of the certificate of approval.

9.3.1.27.10 A certificate from a recognized classification society stating that 9.3.1.24.1 to 9.3.1.24.3, 9.3.1.27.1 and 9.3.1.27.4 above have been complied with shall be submitted together with the application for issue or renewal of the certificate of approval.

9.3.1.28 *Water-spray system*

When water-spraying is required in column (9) of Table C of Chapter 3.2 a water-spray system shall be installed in the cargo area on deck for the purpose of reducing gases given off by the cargo by spraying water.

The system shall be fitted with a connection device for supply from the shore. The spray nozzles shall be so installed that released gases are precipitated safely. The system shall be capable of being put into operation from the wheelhouse and from the deck. The capacity of the water-spray system shall be such that when all the spray nozzles are in operation, the outflow is of 50 litres per square metre of cargo deck area and per hour.

9.3.1.29 and 9.3.1.30   (*Reserved*)

9.3.1.31 *Engines*

9.3.1.31.1 Only internal combustion engines running on fuel with a flashpoint of more than 55 °C are allowed.

9.3.1.31.2 Ventilation inlets of the engine room and, when the engines do not take in air directly from the engine room, the air intakes of the engines shall be located not less than 2.00 m from the cargo area.

9.3.1.31.3 Sparking shall not be possible within the cargo area.

9.3.1.31.4 The surface temperature of the outer parts of engines used during loading or unloading operations, as well as that of their air inlets and exhaust ducts shall not exceed the allowable temperature according to the temperature class of the substances carried. This provision does not apply to engines installed in service spaces provided the provisions of 9.3.1.52.3 are fully complied with.

9.3.1.31.5 The ventilation in the closed engine room shall be designed so that, at an ambient temperature of 20 °C, the average temperature in the engine room does not exceed 40 °C.

9.3.1.32   *Oil fuel tanks*

9.3.1.32.1   When the vessel is fitted with hold spaces and double bottoms, double bottoms within the cargo area may be arranged as oil fuel tanks, provided their depth is not less than 0.6 m.

Oil fuel pipes and openings of such tanks are not permitted in the hold space.

9.3.1.32.2   Open ends of air pipes of all oil fuel tanks shall extend to not less than 0.5 m above the open deck. The open ends and the open ends of overflow pipes leading to the deck shall be fitted with a protective device consisting of a gauze diaphragm or a perforated plate.

9.3.1.33   (*Reserved*)

9.3.1.34   *Exhaust pipes*

9.3.1.34.1   Exhausts shall be evacuated from the vessel into the open air either upwards through an exhaust pipe or through the shell plating. The exhaust outlet shall be located not less than 2 m from the cargo area. The exhaust pipes of engines shall be arranged so that the exhausts are led away from the vessel. The exhaust pipes shall not be located within the cargo area.

9.3.1.34.2   Exhaust pipes of engines shall be provided with a device preventing the escape of sparks, e.g. spark arresters.

9.3.1.35   *Bilge pumping and ballasting arrangements*

9.3.1.35.1   Bilge and ballast pumps for spaces within the cargo area shall be installed within such area.

This provision does not apply to:

–   double-hull spaces and double bottoms which do not have a common boundary wall with the cargo tanks;

–   cofferdams and hold spaces where ballasting is carried out using the piping of the fire-fighting system in the cargo area and bilge–pumping is performed using eductors.

9.3.1.35.2   Where the double bottom is used as a liquid oil fuel tank, it shall not be connected to the bilge piping system.

9.3.1.35.3   Where the ballast pump is installed in the cargo area, the standpipe and its outboard connection for suction of ballast water shall be located within the cargo area.

9.3.1.35.4   It shall be possible for an under-deck pump-room to be stripped in an emergency using a system located in the cargo area and independent of any other system. This stripping system shall be located outside the pump-room.

9.3.1.36 to 9.3.1.39   (*Reserved*)

9.3.1.40   *Fire-extinguishing arrangements*

9.3.1.40.1   A fire-extinguishing system shall be installed on the vessel.

This system shall comply with the following requirements:

–   It shall be supplied by two independent fire or ballast pumps, one of which shall be ready for use at any time. These pumps and their means of propulsion and electrical equipment shall not be installed in the same space;

–   It shall be provided with a water main fitted with at least three hydrants in the cargo area above deck. Three suitable and sufficiently long hoses with jet/spray nozzles having a diameter of not less than 12 mm shall be provided. Alternatively one or more of the hose assemblies may be substituted by directable jet/spray nozzles having a diameter of not less than 12 mm. It shall be possible to reach any point of the deck in the cargo area simultaneously with at least two jets of water which do not emanate from the same hydrant.

A spring-loaded non-return valve shall be fitted to ensure that no gases can escape through the fire-extinguishing system into the accommodation or service spaces outside the cargo area or wheelhouse;

–   The capacity of the system shall be at least sufficient for a jet of water to have a minimum reach of not less than the vessel's breadth from any location on board with two spray nozzles being used at the same time;

–   The water supply system shall be capable of being put into operation from the wheelhouse and from the deck;

–   Measures shall be taken to prevent the freezing of fire-mains and hydrants.

9.3.1.40.2   In addition the engine rooms, the cargo pump-room and all spaces containing special equipment (switchboards, compressors, etc.) for the refrigerant equipment if any, shall be provided with a permanently fixed fire-extinguishing system meeting the following requirements:

9.3.1.40.2.1   *Extinguishing agents*

For the protection of spaces in engine rooms, boiler rooms and pump rooms, only permanently fixed fire-extinguishing systems using the following extinguishing agents are permitted:

(a) CO_2 (carbon dioxide);

(b)   HFC 227 ea (heptafluoropropane);

(c)   IG-541 (52% nitrogen, 40% argon, 8% carbon dioxide).

(d)   FK-5-1-12 (dodecafluoro 2–methylpentane–3–one).

Other extinguishing agents are permitted only on the basis of recommendations by the Administrative Committee.

9.3.1.40.2.2   *Ventilation, air extraction*

(a)   The combustion air required by the combustion engines which ensure propulsion should not come from spaces protected by permanently fixed fire-extinguishing systems. This requirement is not mandatory if the vessel has two independent main engine rooms with a gastight separation or if, in addition to the main engine room, there is a separate engine room installed with a bow thruster that can independently ensure propulsion in the event of a fire in the main engine room.

(b)   All forced ventilation systems in the space to be protected shall be shut down automatically as soon as the fire-extinguishing system is activated.

(c)   All openings in the space to be protected which permit air to enter or gas to escape shall be fitted with devices enabling them to be closed rapidly. It shall be clear whether they are open or closed.

(d)   Air escaping from the pressure–relief valves of the pressurised air tanks installed in the engine rooms shall be evacuated to the open air.

(e)   Overpressure or negative pressure caused by the diffusion of the extinguishing agent shall not destroy the constituent elements of the space to be protected. It shall be possible to ensure the safe equalisation of pressure.

(f)   Protected spaces shall be provided with a means of extracting the extinguishing agent. If extraction devices are installed, it shall not be possible to start them up during extinguishing.

9.3.1.40.2.3   *Fire alarm system*

The space to be protected shall be monitored by an appropriate fire alarm system. The alarm signal shall be audible in the wheelhouse, the accommodation and the space to be protected.

9.3.1.40.2.4   *Piping system*

(a)   The extinguishing agent shall be routed to and distributed in the space to be protected by means of a permanent piping system. Piping installed in the space to be protected and their fittings shall be made of steel. This shall not apply to the connecting nozzles of tanks and compensators provided that the materials used have equivalent fire–retardant properties. Piping shall be protected against corrosion both internally and externally.

(b)   The discharge nozzles shall be so arranged as to ensure the regular diffusion of the extinguishing agent. In particular, the extinguishing agent must also be effective beneath the floor.

9.3.1.40.2.5   *Triggering device*

(a)   Automatically activated fire-extinguishing systems are not permitted.

(b)   It shall be possible to activate the fire-extinguishing system from a suitable point located outside the space to be protected.

(c)   Triggering devices shall be so installed that they can be activated in the event of a fire and so that the risk of their breakdown in the event of a fire or an explosion in the space to be protected is reduced as far as possible.

Systems which are not mechanically activated shall be supplied from two energy sources independent of each other. These energy sources shall be located outside the space to be protected. The control lines located in the space to be protected shall be so designed as to remain capable of operating in the event of a fire for a minimum of 30 minutes. The electrical installations are deemed to meet this requirement if they conform to the IEC 60331–21:1999 standard.

When the triggering devices are so placed as not to be visible, the component concealing them shall carry the "Fire-fighting system" symbol, each side being not less than 10 cm in length, with the following text in red letters on a white ground:

Fire-extinguishing system

(d)   If the fire-extinguishing system is intended to protect several spaces, it shall comprise a separate and clearly–marked triggering device for each space.

(e) The instructions shall be posted alongside all triggering devices and shall be clearly visible and indelible. The instructions shall be in a language the master can read and understand and if this language is not English, French or German, they shall be in English, French or German. They shall include information concerning:

(i) the activation of the fire-extinguishing system;

(ii) the need to ensure that all persons have left the space to be protected;

(iii) The correct behaviour of the crew in the event of activation and when accessing the space to be protected following activation or diffusion, in particular in respect of the possible presence of dangerous substances;

(iv) the correct behaviour of the crew in the event of the failure of the fire-extinguishing system to function properly.

(f) The instructions shall mention that prior to the activation of the fire-extinguishing system, combustion engines installed in the space and aspirating air from the space to be protected, shall be shut down.

9.3.1.40.2.6 *Alarm device*

(a) Permanently fixed fire-extinguishing systems shall be fitted with an audible and visual alarm device.

(b) The alarm device shall be set off automatically as soon as the fire-extinguishing system is first activated. The alarm device shall function for an appropriate period of time before the extinguishing agent is released; it shall not be possible to turn it off.

(c) Alarm signals shall be clearly visible in the spaces to be protected and their access points and be clearly audible under operating conditions corresponding to the highest possible sound level. It shall be possible to distinguish them clearly from all other sound and visual signals in the space to be protected.

(d) Sound alarms shall also be clearly audible in adjoining spaces, with the communicating doors shut, and under operating conditions corresponding to the highest possible sound level.

(e) If the alarm device is not intrinsically protected against short circuits, broken wires and drops in voltage, it shall be possible to monitor its operation.

(f) A sign with the following text in red letters on a white ground shall be clearly posted at the entrance to any space the extinguishing agent may reach:

Warning, fire-extinguishing system!
Leave this space immediately when the ... (description) alarm is activated!

9.3.1.40.2.7 *Pressurised tanks, fittings and piping*

(a) Pressurised tanks, fittings and piping shall conform to the requirements of the competent authority or, if there are no such requirements, to those of a recognized classification society.

(b) Pressurised tanks shall be installed in accordance with the manufacturer's instructions.

(c) Pressurised tanks, fittings and piping shall not be installed in the accommodation.

(d) The temperature of cabinets and storage spaces for pressurised tanks shall not exceed 50 °C.

(e) Cabinets or storage spaces on deck shall be securely stowed and shall have vents so placed that in the event of a pressurised tank not being gastight, the escaping gas cannot penetrate into the vessel. Direct connections with other spaces are not permitted.

9.3.1.40.2.8 *Quantity of extinguishing agent*

If the quantity of extinguishing agent is intended for more than one space, the quantity of extinguishing agent available does not need to be greater than the quantity required for the largest of the spaces thus protected.

9.3.1.40.2.9 *Installation, maintenance, monitoring and documents*

(a) The mounting or modification of the system shall only be performed by a company specialised in fire-extinguishing systems. The instructions (product data sheet, safety data sheet) provided by the manufacturer of the extinguishing agent or the system shall be followed.

(b) The system shall be inspected by an expert:

(i) before being brought into service;

(ii) each time it is put back into service after activation;

(iii) after every modification or repair;

(iv) regularly, not less than every two years.

(c) During the inspection, the expert is required to check that the system conforms to the requirements of 9.3.1.40.2.

(d) The inspection shall include, as a minimum:

(i) an external inspection of the entire system;

(ii) an inspection to ensure that the piping is leakproof;

(iii) an inspection to ensure that the control and activation systems are in good working order;

(iv) an inspection of the pressure and contents of tanks;

(v) an inspection to ensure that the means of closing the space to be protected are leakproof;

(vi) an inspection of the fire alarm system;

(vii) an inspection of the alarm device.

(e) The person performing the inspection shall establish, sign and date a certificate of inspection.

(f) The number of permanently fixed fire-extinguishing systems shall be mentioned in the vessel certificate.

9.3.1.40.2.10   *Fire-extinguishing system operating with CO₂*

In addition to the requirements contained in 9.3.1.40.2.1 to 9.3.1.40.2.9, fire-extinguishing systems using CO_2 as an extinguishing agent shall conform to the following provisions:

(a)   Tanks of CO_2 shall be placed in a gastight space or cabinet separated from other spaces. The doors of such storage spaces and cabinets shall open outwards; they shall be capable of being locked and shall carry on the outside the symbol "Warning: general danger", not less than 5 cm high and "CO_2" in the same colours and the same size;

(b)   Storage cabinets or spaces for CO_2 tanks located below deck shall only be accessible from the outside. These spaces shall have an artificial ventilation system with extractor hoods and shall be completely independent of the other ventilation systems on board;

(c)   The level of filling of CO_2 tanks shall not exceed 0.75 kg/l. The volume of depressurised CO_2 shall be taken to be 0.56 m³/kg;

(d)   The concentration of CO_2 in the space to be protected shall be not less than 40% of the gross volume of the space. This quantity shall be released within 120 seconds. It shall be possible to monitor whether diffusion is proceeding correctly;

(e)   The opening of the tank valves and the control of the diffusing valve shall correspond to two different operations;

(f)   The appropriate period of time mentioned in 9.3.1.40.2.6 (b) shall be not less than 20 seconds. A reliable installation shall ensure the timing of the diffusion of CO_2.

9.3.1.40.2.11   *Fire-extinguishing system operating with HFC-227 ea (heptafluoropropane)*

In addition to the requirements of 9.3.1.40.2.1 to 9.3.1.40.2.9, fire-extinguishing systems using HFC-227 ea as an extinguishing agent shall conform to the following provisions:

(a)   Where there are several spaces with different gross volumes, each space shall be equipped with its own fire-extinguishing system;

(b)   Every tank containing HFC-227 ea placed in the space to be protected shall be fitted with a device to prevent overpressure. This device shall ensure that the contents of the tank are safely diffused in the space to be protected if the tank is subjected to fire, when the fire-extinguishing system has not been brought into service;

(c)   Every tank shall be fitted with a device permitting control of the gas pressure;

(d)   The level of filling of tanks shall not exceed 1.15 kg/l. The specific volume of depressurised HFC-227 ea shall be taken to be 0.1374 m³/kg;

(e)   The concentration of HFC-227 ea in the space to be protected shall be not less than 8% of the gross volume of the space. This quantity shall be released within 10 seconds;

(f)   Tanks of HFC-227 ea shall be fitted with a pressure monitoring device which triggers an audible and visual alarm in the wheelhouse in the event of an unscheduled loss of propellant gas. Where there is no wheelhouse, the alarm shall be triggered outside the space to be protected;

(g)   After discharge, the concentration in the space to be protected shall not exceed 10.5% (volume);

(h)   The fire-extinguishing system shall not comprise aluminium parts.

9.3.1.40.2.12   *Fire-extinguishing system operating with IG-541*

In addition to the requirements of 9.3.1.40.2.1 to 9.3.1.40.2.9, fire-extinguishing systems using IG-541 as an extinguishing agent shall conform to the following provisions:

(a)   Where there are several spaces with different gross volumes, every space shall be equipped with its own fire-extinguishing system;

(b)   Every tank containing IG-541 placed in the space to be protected shall be fitted with a device to prevent overpressure. This device shall ensure that the contents of the tank are safely diffused in the space to be protected if the tank is subjected to fire, when the fire-extinguishing system has not been brought into service;

(c)   Each tank shall be fitted with a device for checking the contents;

(d)   The filling pressure of the tanks shall not exceed 200 bar at a temperature of +15 °C;

(e)   The concentration of IG-541 in the space to be protected shall be not less than 44% and not more than 50% of the gross volume of the space. This quantity shall be released within 120 seconds.

9.3.1.40.2.13   *Fire-extinguishing system operating with FK-5-1-12*

In addition to the requirements of 9.3.1.40.2.1 to 9.3.1.40.2.9, fire-extinguishing systems using FK-5-1-12 as an extinguishing agent shall comply with the following provisions:

(a)   Where there are several spaces with different gross volumes, every space shall be equipped with its own fire-extinguishing system;

(b)   Every tank containing FK-5-1-12 placed in the space to be protected shall be fitted with a device to prevent overpressure. This device shall ensure that the contents of the tank are safely diffused in the space to be protected if the tank is subjected to fire, when the fire-extinguishing system has not been brought into service;

(c)   Every tank shall be fitted with a device permitting control of the gas pressure;

(d)   The level of filling of tanks shall not exceed 1.00 kg/l. The specific volume of depressurized FK-5-1-12 shall be taken to be 0.0719 m^3/kg;

(e)   The volume of FK-5-1-12 in the space to be protected shall be not less than 5.5% of the gross volume of the space. This quantity shall be released within 10 seconds;

(f)   Tanks of FK-5-1-12 shall be fitted with a pressure monitoring device which triggers an audible and visual alarm in the wheelhouse in the event of an unscheduled loss of extinguishing agent. Where there is no wheelhouse, the alarm shall be triggered outside the space to be protected;

(g)   After discharge, the concentration in the space to be protected shall not exceed 10.0%.

9.3.1.40.2.14   *Fixed fire-extinguishing system for physical protection*

In order to ensure physical protection in the engine rooms, boiler rooms and pump rooms, permanently fixed fire-extinguishing systems are accepted solely on the basis of recommendations by the Administrative Committee.

9.3.1.40.3   The two hand fire–extinguishers referred to in 8.1.4 shall be located in the cargo area.

9.3.1.40.4   The fire-extinguishing agent and the quantity contained in the permanently fixed fire-extinguishing system shall be suitable and sufficient for fighting fires.

9.3.1.41   ***Fire and naked light***

9.3.1.41.1   The outlets of funnels shall be located not less than 2.00 m from the cargo area. Arrangements shall be provided to prevent the escape of sparks and the entry of water.

9.3.1.41.2   Heating, cooking and refrigerating appliances shall not be fuelled with liquid fuels, liquid gas or solid fuels.

The installation in the engine room or in another separate space of heating appliances fuelled with liquid fuel having a flash–point above 55 °C is, however, permitted.

Cooking and refrigerating appliances are permitted only in the accommodation.

9.3.1.41.3   Only electrical lighting appliances are permitted.

9.3.1.42 to 9.3.1.49   *(Reserved)*

9.3.1.50   ***Documents concerning electrical installations***

9.3.1.50.1   In addition to the documents required by the Regulations referred to in 1.1.4.6, the following documents shall be on board:

(a)   a drawing indicating the boundaries of the cargo area and the location of the electrical equipment installed in this area;

(b)   a list of the electrical equipment referred to in (a) above including the following particulars:

machine or appliance, location, type of protection, type of protection against explosion, testing body and approval number;

(c)   a list of or general plan indicating the electrical equipment outside the cargo area which may be operated during loading, unloading or gas-freeing. All other electrical equipment shall be marked in red. See 9.3.1.52.3 and 9.3.1.52.4.

9.3.1.50.2   The documents listed above shall bear the stamp of the competent authority issuing the certificate of approval.

9.3.1.51   *Electrical installations*

9.3.1.51.1   Only distribution systems without return connection to the hull are permitted.

This provision does not apply to:

–   active cathodic corrosion protection;

–   local installations outside the cargo area (e.g. connections of starters of diesel engines);

–   the device for checking the insulation level referred to in 9.3.1.51.2 below.

9.3.1.51.2   Every insulated distribution network shall be fitted with an automatic device with a visual and audible alarm for checking the insulation level.

9.3.1.51.3   For the selection of electrical equipment to be used in zones presenting an explosion risk, the explosion groups and temperature classes assigned to the substances carried in the list of substances shall be taken into consideration (See columns (15) and (16) of Table C of Chapter 3.2).

9.3.1.52   *Type and location of electrical equipment*

9.3.1.52.1   (a)   Only the following equipment may be installed in cargo tanks and piping for loading and unloading (comparable to zone 0):

–   measuring, regulation and alarm devices of the EEx (ia) type of protection.

(b)   Only the following equipment may be installed in the cofferdams, double-hull spaces, double bottoms and hold spaces (comparable to zone 1):

–   measuring, regulation and alarm devices of the certified safe type;

–   lighting appliances of the "flame-proof enclosure" or "apparatus protected by pressurization" type of protection;

–   hermetically sealed echo sounding devices the cables of which are led through thick-walled steel tubes with gastight connections up to the main deck;

–   cables for the active cathodic protection of the shell plating in protective steel tubes such as those provided for echo sounding devices;

The following equipment may be installed only in double-hull spaces and double bottoms if used for ballasting:

–   Permanently fixed submerged pumps with temperature monitoring, of the certified safe type.

(c)   Only the following equipment may be installed in the service spaces in the cargo area below deck (comparable to zone 1):

–   measuring, regulation and alarm devices of the certified safe type;

–   lighting appliances of the "flame-proof enclosure" or "apparatus protected by pressurization" type of protection;

–   motors driving essential equipment such as ballast pumps with temperature monitoring; they shall be of the certified safe type.

(d)   The control and protective equipment of the electrical equipment referred to in (a), (b) and (c) above shall be located outside the cargo area if they are not intrinsically safe.

(e)   The electrical equipment in the cargo area on deck (comparable to zone 1) shall be of the certified safe type.

9.3.1.52.2   Accumulators shall be located outside the cargo area.

9.3.1.52.3   (a)   Electrical equipment used during loading, unloading and gas-freeing during berthing and which are located outside the cargo area (comparable to zone 2) shall be at least of the "limited explosion risk" type.

(b)   This provision does not apply to:

(i)   lighting installations in the accommodation, except for switches near entrances to accommodation;

(ii)   radiotelephone installations in the accommodation or the wheelhouse;

(iii)   mobile and fixed telephone installations in the accommodation or the wheelhouse;

(iv)   electrical installations in the accommodation, the wheelhouse or the service spaces outside the cargo areas if:

1.   These spaces are fitted with a ventilation system ensuring an overpressure of 0.1 kPa (0.001 bar) and none of the windows is capable of being opened; the air intakes of the ventilation system located as far away as possible, however, not less than 6.00 m from the cargo area and not less than 2.00 m above the deck;

2.   The spaces are fitted with a gas detection system with sensors:

–   at the suction inlets of the ventilation system;

–   directly at the top edge of the sill of the entrance doors of the accommodation and service spaces when the cargo in the gas phase is heavier than air; otherwise sensors shall be fitted close to the ceiling;

3.   The gas concentration measurement is continuous;

4.   When the gas concentration reaches 20% of the lower explosive limit, the ventilators shall be switched off. In such a case and when the overpressure is not maintained or in the event of failure of the gas detection system, the electrical installations which do not comply with (a) above, shall be switched off. These operations shall be performed immediately and automatically and activate the emergency lighting in the accommodation, the wheelhouse and the service spaces, which shall comply at least with the "limited explosion risk" type. The switching-off shall be indicated in the accommodation and wheelhouse by visual and audible signals;

5.   The ventilation system, the gas detection system and the alarm of the switch-off device fully comply with the requirements of (a) above;

6.   The automatic switch-off device is set so that no automatic switching-off may occur while the vessel is under way.

(v)   Inland AIS (automatic identification systems) stations in the accommodation and in the wheelhouse if no part of an aerial for electronic apparatus is situated above the cargo area and if no part of a VHF antenna for AIS stations is situated within 2 m from the cargo area.

9.3.1.52.4   The electrical equipment which does not meet the requirements set out in 9.3.1.52.3 above together with its switches shall be marked in red. The disconnection of such equipment shall be operated from a centralised location on board.

9.3.1.52.5   An electric generator which is permanently driven by an engine and which does not meet the requirements of 9.3.1.52.3 above, shall be fitted with a switch capable of shutting down the excitation of the generator. A notice board with the operating instructions shall be displayed near the switch.

9.3.1.52.6   Sockets for the connection of signal lights and gangway lighting shall be permanently fitted to the vessel close to the signal mast or the gangway. Connecting and disconnecting shall not be possible except when the sockets are not live.

9.3.1.52.7   The failure of the power supply for the safety and control equipment shall be immediately indicated by visual and audible signals at the locations where the alarms are usually actuated.

9.3.1.53   *Earthing*

9.3.1.53.1   The metal parts of electrical appliances in the cargo area which are not live as well as protective metal tubes or metal sheaths of cables in normal service shall be earthed, unless they are so arranged that they are automatically earthed by bonding to the metal structure of the vessel.

9.3.1.53.2   The provisions of 9.3.1.53.1 above apply also to equipment having service voltages of less than 50 V.

9.3.1.53.3   Independent cargo tanks shall be earthed.

9.3.1.53.4   Receptacles for residual products shall be capable of being earthed.

9.3.1.54 and 9.3.1.55   (*Reserved*)

9.3.1.56   *Electrical cables*

9.3.1.56.1   All cables in the cargo area shall have a metallic sheath.

9.3.1.56.2   Cables and sockets in the cargo area shall be protected against mechanical damage.

9.3.1.56.3   Movable cables are prohibited in the cargo area, except for intrinsically safe electric circuits or for the supply of signal lights and gangway lighting.

9.3.1.56.4   Cables of intrinsically safe circuits shall only be used for such circuits and shall be separated from other cables not intended for being used in such circuits (e.g. they shall not be installed together in the same string of cables and they shall not be fixed by the same cable clamps).

9.3.1.56.5   For movable cables intended for signal lights and gangway lighting, only sheathed cables of type H 07 RN-F in accordance with standard IEC 60 245–4:1994 or cables of at least equivalent design having conductors with a cross-section of not less than 1.5 mm^2 shall be used.

These cables shall be as short as possible and installed so that damage is not likely to occur.

9.3.1.56.6 The cables required for the electrical equipment referred to in 9.3.1.52.1 (b) and (c) are accepted in cofferdams, double-hull spaces, double bottoms, hold spaces and service spaces below deck.

9.3.1.57 to 9.3.1.59 (*Reserved*)

9.3.1.60 *Special equipment*

A shower and an eye and face bath shall be provided on the vessel at a location which is directly accessible from the cargo area.

9.3.1.61 to 9.3.1.70 (*Reserved*)

9.3.1.71 *Admittance on board*

The notice boards displaying the prohibition of admittance in accordance with 8.3.3 shall be clearly legible from either side of the vessel.

9.3.1.72 and 9.3.1.73 (*Reserved*)

9.3.1.74 *Prohibition of smoking, fire or naked light*

9.3.1.74.1 The notice boards displaying the prohibition of smoking in accordance with 8.3.4 shall be clearly legible from either side of the vessel.

9.3.1.74.2 Notice boards indicating the circumstances under which the prohibition is applicable shall be fitted near the entrances to the spaces where smoking or the use of fire or naked light is not always prohibited.

9.3.1.74.3 Ashtrays shall be provided close to each exit of the accommodation and the wheelhouse.

9.3.1.75 to 9.3.1.91 (*Reserved*)

9.3.1.92 *Emergency exit*

Spaces the entrances or exits of which are likely to become partly or completely immersed in the damaged condition shall have an emergency exit which is situated not less than 0.10 m above the damage waterline. This does not apply to forepeak and afterpeak.

9.3.1.93 to 9.3.1.99 (*Reserved*)

9.3.2 **Rules for construction of type C tank vessels**

The rules for construction of 9.3.2.0 to 9.3.2.99 apply to type C tank vessels.

9.3.2.0 *Materials of construction*

9.3.2.0.1 (a) The vessel's hull and the cargo tanks shall be constructed of shipbuilding steel or other at least equivalent metal.

The independent cargo tanks may also be constructed of other materials, provided these have at least equivalent mechanical properties and resistance against the effects of temperature and fire.

(b) Every part of the vessel including any installation and equipment which may come into contact with the cargo shall consist of materials which can neither be dangerously affected by the cargo nor cause decomposition of the cargo or react with it so as to form harmful or hazardous products. In case it has not been possible to examine this during classification and inspection of the vessel a relevant reservation shall be entered in the vessel substance list according to 1.16.1.2.5.

(c) Venting piping shall be protected against corrosion.

9.3.2.0.2 Except where explicitly permitted in 9.3.2.0.3 below or in the certificate of approval, the use of wood, aluminium alloys or plastic materials within the cargo area is prohibited.

9.3.2.0.3 (a) The use of wood, aluminium alloys or plastic materials within the cargo area is only permitted for:

– gangways and external ladders;

– movable items of equipment (aluminium gauging rods are, however permitted, provided that they are fitted with brass feet or protected in another way to avoid sparking);

– chocking of cargo tanks which are independent of the vessel's hull and chocking of installations and equipment;

– masts and similar round timber;

– engine parts;

– parts of the electrical installation;

– loading and unloading appliances;

– lids of boxes which are placed on the deck.

(b) The use of wood or plastic materials within the cargo area is only permitted for:

– supports and stops of any kind.

(c) The use of plastic materials or rubber within the cargo area is only permitted for:

– coating of cargo tanks and of piping for loading and unloading;

– all kinds of gaskets (e.g. for dome or hatch covers);

– electric cables;

– hose assemblies for loading and unloading;

– insulation of cargo tanks and of piping for loading and unloading;

– photo–optical copies of the certificate of approval according to 8.1.2.6 or 8.1.2.7.

(d) All permanently fitted materials in the accommodation or wheelhouse, with the exception of furniture, shall not readily ignite. They shall not evolve fumes or toxic gases in dangerous quantities, if involved in a fire.

9.3.2.0.4 The paint used in the cargo area shall not be liable to produce sparks in case of impact.

9.3.2.0.5 The use of plastic material for vessel's boats is permitted only if the material does not readily ignite.

9.3.2.1 **Vessel record**

NOTE: For the purpose of this paragraph, the term "owner" has the same meaning as in 1.16.0.

The vessel record shall be retained by the owner who shall be able to provide this documentation at the request of the competent authority and the recognized classification society.

The vessel record shall be maintained and updated throughout the life of the vessel and shall be retained for 6 months after the vessel is taken out of service.

Should a change of owner occur during the life of the vessel the vessel record shall be transferred to the new owner.

Copies of the vessel record or all necessary documents shall be made available on request to the competent authority for the issuance of the certificate of approval and for the recognized classification society or inspection body for first inspection, periodic inspection, special inspection or exceptional checks.

9.3.2.2 to 9.3.2.7   (*Reserved*)

9.3.2.8 *Classification*

9.3.2.8.1 The tank vessel shall be built under the survey of a recognised classification society and be classed in its highest class.

The vessel's highest class shall be continued. This shall be confirmed by an appropriate certificate issued by the recognized classification society (certificate of class).

The design pressure and the test pressure of cargo tanks shall be entered in the certificate.

If a vessel has cargo tanks with different valve opening pressures, the design and test pressures of each tank shall be entered in the certificate.

The recognized classification society shall draw up a vessel substance list mentioning all the dangerous goods accepted for carriage by the tank vessel (see also 1.16.1.2.5).

9.3.2.8.2 The cargo pump-rooms shall be inspected by a recognised classification society whenever the certificate of approval has to be renewed as well as during the third year of validity of the certificate of approval. The inspection shall comprise at least:

–   an inspection of the whole system for its condition, for corrosion, leakage or conversion works which have not been approved;

–   a checking of the condition of the gas detection system in the cargo pump-rooms.

Inspection certificates signed by the recognised classification society with respect to the inspection of the cargo pump-rooms shall be kept on board. The inspection certificates shall at least include particulars of the above inspection and the results obtained as well as the date of the inspection.

9.3.2.8.3   The condition of the gas detection system referred to in 9.3.2.52.3 shall be checked by a recognised classification society whenever the certificate of approval has to be renewed and during the third year of validity of the certificate of approval. A certificate signed by the recognised classification society shall be kept on board.

9.3.2.9   (*Reserved*)

9.3.2.10   *Protection against the penetration of gases*

9.3.2.10.1   The vessel shall be designed so as to prevent gases from penetrating into the accommodation and the service spaces.

9.3.2.10.2   Outside the cargo area, the lower edges of door-openings in the sidewalls of superstructures and the coamings of access hatches to under-deck spaces shall have a height of not less than 0.50 m above the deck.

This requirement need not be complied with if the wall of the superstructures facing the cargo area extends from one side of the ship to the other and has doors the sills of which have a height of not less than 0.50 m. The height of this wall shall be not less than 2.00 m. In this case, the lower edges of door-openings in the sidewalls of superstructures and of coamings of access hatches behind this wall shall have a height of not less than 0.10 m. The sills of engine–room doors and the coamings of its access hatches shall, however, always have a height of not less than 0.50 m.

9.3.2.10.3   In the cargo area, the lower edges of door-openings in the sidewalls of superstructures shall have a height of not less than 0.50 m above the deck and the sills of hatches and ventilation openings of premises located under the deck shall have a height of not less than 0.50 m above the deck. This requirement does not apply to access openings to double-hull and double bottom spaces.

9.3.2.10.4   The bulwarks, foot-rails, etc. shall be provided with sufficiently large openings which are located directly above the deck.

9.3.2.11   *Hold spaces and cargo tanks*

9.3.2.11.1   (a)   The maximum permissible capacity of a cargo tank shall be determined in accordance with the following table:

$L \times B \times H$ (m^3)	Maximum permissible capacity of a cargo tank (m^3)
up to 600	$L \times B \times H \times 0.3$
600 to 3 750	$180 + (L \times B \times H - 600) \times 0.0635$
> 3 750	380

Alternative constructions in accordance with 9.3.4 are permitted.

In the table above $L \times B \times H$ is the product of the main dimensions of the tank vessel in metres (according to the measurement certificate), where:

L   =   overall length of the hull in m;

B   =   extreme breadth of the hull in m;

H   =   shortest vertical distance in m between the top of the keel and the lowest point of the deck at the side of the vessel (moulded depth) within the cargo area.

(b)   The relative density of the substances to be carried shall be taken into consideration in the design of the cargo tanks. The maximum relative density shall be indicated in the certificate of approval;

(c)   When the vessel is provided with pressure cargo tanks, these tanks shall be designed for a working pressure of 400 kPa (4 bar);

(d)   For vessels with a length of not more than 50.00 m, the length of a cargo tank shall not exceed 10.00 m; and

For vessels with a length of more than 50.00 m, the length of a cargo tank shall not exceed 0.20 L;

This provision does not apply to vessels with independent built–in cylindrical tanks having a length to diameter ratio ≤ 7.

9.3.2.11.2   (a)   In the cargo area (except cofferdams) the vessel shall be designed as a flush–deck double-hull vessel, with double-hull spaces and double bottoms, but without a trunk;

Cargo tanks independent of the vessel's hull and refrigerated cargo tanks may only be installed in a hold space which is bounded by double-hull spaces and double bottoms in accordance with 9.3.2.11.7 below. The cargo tanks shall not extend beyond the deck;

Refrigerated cargo tank fastenings shall meet the requirements of a recognised classification society;

(b)   The cargo tanks independent of the vessel's hull shall be fixed so that they cannot float;

(c)   The capacity of a suction well shall be limited to not more than 0.10 m^3;

(d)   Side–struts linking or supporting the load-bearing components of the sides of the vessel with the load-bearing components of the longitudinal walls of cargo tanks and side–struts linking the load-bearing components of the vessel's bottom with the tank-bottom are prohibited;

(e)   A local recess in the cargo deck, contained on all sides, with a depth greater than 0.1 m, designed to house the loading and unloading pump, is permitted if it fulfils the following conditions:

–   The recess shall not be greater than 1 m in depth;

–   The recess shall be located not less than 6 m from entrances and openings to accommodation and service spaces outside the cargo area;

–   The recess shall be located at a minimum distance from the side plating equal to one quarter of the vessel's breadth;

–   All pipes linking the recess to the cargo tanks shall be fitted with shut-off devices fitted directly on the bulkhead;

–   All the controls required for the equipment located in the recess shall be activated from the deck;

–   If the recess is deeper than 0.5 m, it shall be provided with a permanent gas detection system which automatically indicates the presence of explosive gases by means of direct-measuring sensors and actuates a visual and audible alarm when the gas concentration has reached 20% of the lower explosion limit. The sensors of this system shall be placed at suitable positions at the bottom of the recess. Measurement shall be continuous;

–   Visual and audible alarms shall be installed in the wheelhouse and on deck and, when the alarm is actuated, the vessel loading and unloading system shall be shut down. Failure of the gas detection system shall be immediately signalled in the wheelhouse and on deck by means of visual and audible alarms;

–   It shall be possible to drain the recess using a system installed on deck in the cargo area and independent of any other system;

–   The recess shall be provided with a level alarm device which activates the draining system and triggers a visual and audible alarm in the wheelhouse when liquid accumulates at the bottom;

–   When the recess is located above the cofferdam, the engine room bulkhead shall have an 'A-60' fire protection insulation according to SOLAS 74, Chapter II-2, Regulation 3;

–   When the cargo area is fitted with a water-spray system, electrical equipment located in the recess shall be protected against infiltration of water;

–   Pipes connecting the recess to the hull shall not pass through the cargo tanks.

9.3.2.11.3   (a)   The cargo tanks shall be separated by cofferdams of at least 0.60 m in width from the accommodation, engine rooms and service spaces outside the cargo area below deck or, if there are no such accommodation, engine rooms and service spaces, from the vessel's ends. Where the cargo tanks are installed in a hold space, a space of not less than 0.50 m shall be provided between such tanks and the end bulkheads of the hold space. In this case an end bulkhead meeting at least the definition for Class "A-60" according to SOLAS 74, Chapter II-2, Regulation 3, shall be deemed equivalent to a cofferdam. For pressure cargo tanks, the 0.50 m distance may be reduced to 0.20 m;

(b)   Hold spaces, cofferdams and cargo tanks shall be capable of being inspected;

(c)   All spaces in the cargo area shall be capable of being ventilated. Means for checking their gas-free condition shall be provided.

9.3.2.11.4   The bulkheads bounding the cargo tanks, cofferdams and hold spaces shall be watertight. The cargo tanks and the bulkheads bounding the cargo area shall have no openings or penetrations below deck.

The bulkhead between the engine room and the cofferdam or service space in the cargo area or between the engine room and a hold space may be fitted with penetrations provided that they conform to the provisions of 9.3.2.17.5.

The bulkhead between the cargo tank and the cargo pump-room below deck may be fitted with penetrations provided that they conform to the provisions of 9.3.2.17.6. The bulkheads between the cargo tanks may be fitted with penetrations provided that the loading or unloading piping are fitted with shut-off devices in the cargo tank from which they come. These shut-off devices shall be operable from the deck.

9.3.2.11.5   Double-hull spaces and double bottoms in the cargo area shall be arranged for being filled with ballast water only. Double bottoms may, however, be used as oil fuel tanks, provided they comply with the provisions of 9.3.2.32.

9.3.2.11.6   (a)   A cofferdam, the centre part of a cofferdam or another space below deck in the cargo area may be arranged as a service space, provided the bulkheads bounding the service space extend vertically to the bottom. This service space shall only be accessible from the deck;

              (b)   The service space shall be watertight with the exception of its access hatches and ventilation inlets;

              (c)   No piping for loading and unloading shall be fitted within the service space referred to under (a) above;

                     Piping for loading and unloading may be fitted in the cargo pump-rooms below deck only when they conform to the provisions of 9.3.2.17.6.

9.3.2.11.7   For double-hull construction with the cargo tanks integrated in the vessel's structure, the distance between the side wall of the vessel and the longitudinal bulkhead of the cargo tanks shall be not less than 1.00 m. A distance of 0.80 m may however be permitted, provided that, compared with the scantling requirements specified in the rules for construction of a recognised classification society, the following reinforcements have been made:

              (a)   25% increase in the thickness of the deck stringer plate;

              (b)   15% increase in the side plating thickness;

              (c)   Arrangement of a longitudinal framing system at the vessel's side, where depth of the longitudinals shall be not less than 0.15 m and the longitudinals shall have a face plate with the cross-sectional area of at least 7.0 cm^2;

              (d)   The stringer or longitudinal framing systems shall be supported by web frames, and like bottom girders fitted with lightening holes, at a maximum spacing of 1.80 m. These distances may be increased if the longitudinals are strengthened accordingly.

              When a vessel is built according to the transverse framing system, a longitudinal stringer system shall be arranged instead of (c) above. The distance between the longitudinal stringers shall not exceed 0.80 m and their depth shall be not less than 0.15 m, provided they are completely welded to the frames. The cross-sectional area of the facebar or faceplate shall be not less than 7.0 cm^2 as in (c) above. Where cut-outs are arranged in the stringer at the connection with the frames, the web depth of the stringer shall be increased with the depth of cut-outs.

              The mean depth of the double bottoms shall be not less than 0.70 m. It shall, however, never be less than 0.60 m.

              The depth below the suction wells may be reduced to 0.50 m.

              Alternative constructions in accordance with 9.3.4 are permitted.

9.3.2.11.8   When a vessel is built with cargo tanks located in a hold space or refrigerated cargo tanks, the distance between the double walls of the hold space shall be not less than 0.80 m and the depth of the double bottom shall be not less than 0.60 m.

9.3.2.11.9   Where service spaces are located in the cargo area under deck, they shall be arranged so as to be easily accessible and to permit persons wearing protective clothing and breathing apparatus to safely operate the service equipment contained therein. They shall be designed so as to allow injured or unconscious personnel to be removed from such spaces without difficulties, if necessary by means of fixed equipment.

9.3.2.11.10   Cofferdams, double-hull spaces, double bottoms, cargo tanks, hold spaces and other accessible spaces within the cargo area shall be arranged so that they may be completely inspected and cleaned in an appropriate manner. The dimensions of openings except for those of double-hull spaces and double bottoms which do not have a wall adjoining the cargo tanks shall be sufficient to allow a person wearing breathing apparatus to enter or leave the space without difficulties. These openings shall have a minimum cross-sectional area of 0.36 m^2 and a minimum side length of 0.50 m. They shall be designed so as to allow an injured or unconscious person to be removed from the bottom of such a space without difficulties, if necessary by means of fixed equipment. In these spaces the distance between the reinforcements shall not be less than 0.50 m. In double bottoms this distance may be reduced to 0.45 m.

Cargo tanks may have circular openings with a diameter of not less than 0.68 m.

9.3.2.12   *Ventilation*

9.3.2.12.1   Each hold space shall have two openings the dimensions and location of which shall be such as to permit effective ventilation of any part of the hold space. If there are no such openings, it shall be possible to fill the hold spaces with inert gas or dry air.

9.3.2.12.2   Double-hull spaces and double bottoms within the cargo area which are not arranged for being filled with ballast water, hold spaces and cofferdams shall be provided with ventilation systems.

9.3.2.12.3   Any service spaces located in the cargo area below deck shall be provided with a system of forced ventilation with sufficient power for ensuring at least 20 changes of air per hour based on the volume of the space.

The ventilation exhaust ducts shall extend down to 50 mm above the bottom of the service space. The air shall be supplied through a duct at the top of the service space. The air inlets shall be located not less than 2.00 m above the deck, at a distance of not less than 2.00 m from tank openings and 6.00 m from the outlets of safety valves.

The extension pipes, which may be necessary, may be of the hinged type.

9.3.2.12.4   Ventilation of accommodation and service spaces shall be possible.

9.3.2.12.5   Ventilators used in the cargo area shall be designed so that no sparks may be emitted on contact of the impeller blades with the housing and no static electricity may be generated.

9.3.2.12.6   Notice boards shall be fitted at the ventilation inlets indicating the conditions when they shall be closed. Any ventilation inlets of accommodation and service spaces leading outside shall be fitted with fire flaps. Such ventilation inlets shall be located not less than 2.00 m from the cargo area.

Ventilation inlets of service spaces in the cargo area may be located within such area.

9.3.2.12.7   The flame-arresters prescribed in 9.3.2.20.4, 9.3.2.22.4, 9.3.2.22.5 and 9.3.2.26.4 shall be of a type approved for this purpose by the competent authority.

9.3.2.13   *Stability (general)*

9.3.2.13.1   Proof of sufficient stability shall be furnished including for stability in damaged condition.

9.3.2.13.2   The basic values for the stability calculation - the vessel's lightweight and location of the centre of gravity - shall be determined either by means of an inclining experiment or by detailed mass and moment calculation. In the latter case the lightweight of the vessel shall be checked by means of a lightweight test with a tolerance limit of ± 5% between the mass determined by calculation and the displacement determined by the draught readings.

9.3.2.13.3   Proof of sufficient intact stability shall be furnished for all stages of loading and unloading and for the final loading condition for all the relative densities of the substances transported contained in the vessel substance list according to 1.16.1.2.5.

For every loading operation, taking account of the actual fillings and floating position of cargo tanks, ballast tanks and compartments, drinking water and sewage tanks and tanks containing products for the operation of the vessel, the vessel shall comply with the intact and damage stability requirements.

Intermediate stages during operations shall also be taken into consideration.

The proof of sufficient stability shall be shown for every operating, loading and ballast condition in the stability booklet, to be approved by the recognized classification society, which classes the vessel. If it is unpractical to pre-calculate the operating, loading and ballast conditions, a loading instrument approved by the recognised classification society which classes the vessel shall be installed and used which contains the contents of the stability booklet.

NOTE: A stability booklet shall be worded in a form comprehensible for the responsible master and containing the following details:

General description of the vessel:

— *General arrangement and capacity plans indicating the assigned use of compartments and spaces (cargo tanks, stores, accommodation, etc.);*

— *A sketch indicating the position of the draught marks referring to the vessel's perpendiculars;*

— *A scheme for ballast/bilge pumping and overflow prevention systems;*

— *Hydrostatic curves or tables corresponding to the design trim, and, if significant trim angles are foreseen during the normal operation of the vessel, curves or tables corresponding to such range of trim are to be introduced;*

— *Cross curves or tables of stability calculated on a free trimming basis, for the ranges of displacement and trim anticipated in normal operating conditions, with an indication of the volumes which have been considered buoyant;*

— *Tank sounding tables or curves showing capacities, centres of gravity, and free surface data for all cargo tanks, ballast tanks and compartments, drinking water and sewage water tanks and tanks containing products for the operation of the vessel;*

— *Lightship data (weight and centre of gravity) resulting from an inclining test or deadweight measurement in combination with a detailed mass balance or other acceptable measures. Where the above-mentioned information is derived from a sister vessel, the reference to this sister vessel shall be clearly indicated, and a copy of the approved inclining test report relevant to this sister vessel shall be included;*

–   *A copy of the approved test report shall be included in the stability booklet;*

–   *Operating loading conditions with all relevant details, such as:*

  –   *Lightship data, tank fillings, stores, crew and other relevant items on board (mass and centre of gravity for each item, free surface moments for liquid loads);*
  –   *Draughts amidships and at perpendiculars;*
  –   *Metacentric height corrected for free surfaces effect;*
  –   *Righting lever values and curve;*
  –   *Longitudinal bending moments and shear forces at read–out points;*
  –   *Information about openings (location, type of tightness, means of closure); and*
  –   *Information for the master;*

–   *Calculation of the influence of ballast water on stability with information on whether fixed level gauges for ballast tanks and compartments have to be installed or whether the ballast tanks or compartments shall be completely full or completely empty when underway.*

9.3.2.13.4   Floatability after damage shall be proved for the most unfavourable loading condition. For this purpose, calculated proof of sufficient stability shall be established for critical intermediate stages of flooding and for the final stage of flooding.

9.3.2.14   *Stability (intact)*

9.3.2.14.1   The requirements for intact stability resulting from the damage stability calculation shall be fully complied with.

9.3.2.14.2   For vessels with cargo tanks of more than 0.70 B in width, proof shall be furnished that the following stability requirements have been complied with:

(a)   In the positive area of the righting lever curve up to immersion of the first non-watertight opening there shall be a righting lever (GZ) of not less than 0.10 m;

(b)   The surface of the positive area of the righting lever curve up to immersion of the first non-watertight opening and in any event up to an angle of heel $\leq 27°$ shall not be less than 0.024 m.rad;

(c)   The metacentric height (GM) shall be not less than 0.10 m.

These conditions shall be met bearing in mind the influence of all free surfaces in tanks for all stages of loading and unloading.

9.3.2.14.3   The most stringent requirement of 9.3.2.14.1 and 9.3.2.14.2 is applicable to the vessel.

9.3.2.15   *Stability (damaged condition)*

9.3.2.15.1   The following assumptions shall be taken into consideration for the damaged condition:

(a)   The extent of side damage is as follows:

longitudinal extent:	at least 0.10 L, but not less than 5.00 m;
transverse extent:	0.79 m inboard from the vessel's side at right angles to the centreline at the level corresponding to the maximum draught, or when applicable, the distance allowed by section 9.3.4, reduced by 0.01 m;
vertical extent:	from the base line upwards without limit.

(b)   The extent of bottom damage is as follows:

longitudinal extent: at least 0.10 L, but not less than 5.00 m;
transverse extent: 3.00 m;
vertical extent: from the base 0.59 m upwards, the sump excepted.

(c)   Any bulkheads within the damaged area shall be assumed damaged, which means that the location of bulkheads shall be chosen so as to ensure that the vessel remains afloat after the flooding of two or more adjacent compartments in the longitudinal direction.

The following provisions are applicable:

–   For bottom damage, adjacent athwartship compartments shall also be assumed as flooded;

–   The lower edge of any non-watertight openings (e.g. doors, windows, access hatchways) shall, at the final stage of flooding, be not less than 0.10 m above the damage waterline;

–   In general, permeability shall be assumed to be 95%. Where an average permeability of less than 95% is calculated for any compartment, this calculated value obtained may be used.

However, the following minimum values shall be used:

–   engine rooms: 85%;

–   accommodation: 95%;

–   double bottoms, oil fuel tanks, ballast tanks,
        etc., depending on whether, according to
        their function, they have to be assumed
        as full or empty for the vessel floating at
        the maximum permissible draught: 0% or 95%.

For the main engine room only the one–compartment standard need be taken into account, i.e. the end bulkheads of the engine room shall be assumed as not damaged.

9.3.2.15.2   For the intermediate stage of flooding the following criteria have to be fulfilled:

GZ >= 0.03m
Range of positive GZ: 5°.

At the stage of equilibrium (final stage of flooding), the angle of heel shall not exceed 12°. Non-watertight openings shall not be flooded before reaching the stage of equilibrium. If such openings are immersed before that stage, the corresponding spaces shall be considered as flooded for the purpose of the stability calculation.

The positive range of the righting lever curve beyond the stage of equilibrium shall have a righting lever of ≥ 0.05 m in association with an area under the curve of ≥ 0.0065 m.rad. The minimum values of stability shall be satisfied up to immersion of the first non-watertight opening and in any event up to an angle of heel $\leq 27°$. If non-watertight openings are immersed before that stage, the corresponding spaces shall be considered as flooded for the purposes of stability calculation.

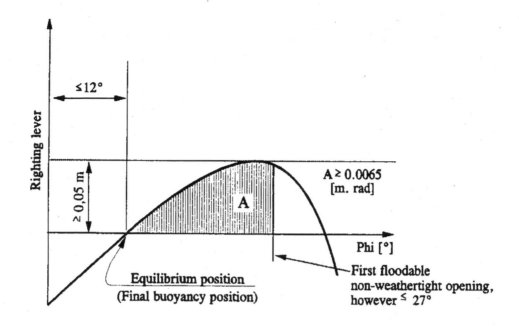

9.3.2.15.3   If openings through which undamaged compartments may additionally become flooded are capable of being closed watertight, the closing appliances shall be marked accordingly.

9.3.2.15.4   Where cross- or down-flooding openings are provided for reduction of unsymmetrical flooding, the time for equalisation shall not exceed 15 minutes, if during the intermediate stages of flooding sufficient stability has been proved.

9.3.2.16   Engine rooms

9.3.2.16.1   Internal combustion engines for the vessel's propulsion as well as internal combustion engines for auxiliary machinery shall be located outside the cargo area. Entrances and other openings of engine rooms shall be at a distance of not less than 2.00 m from the cargo area.

9.3.2.16.2   The engine rooms shall be accessible from the deck; the entrances shall not face the cargo area. Where the doors are not located in a recess whose depth is at least equal to the door width, the hinges shall face the cargo area.

9.3.2.17   Accommodation and service spaces

9.3.2.17.1   Accommodation spaces and the wheelhouse shall be located outside the cargo area forward of the fore vertical plane or abaft the aft vertical plane bounding the part of the cargo area below deck. Windows of the wheelhouse which are located not less than 1.00 m above the bottom of the wheelhouse may tilt forward.

9.3.2.17.2   Entrances to spaces and openings of superstructures shall not face the cargo area. Doors opening outward and not located in a recess the depth of which is at least equal to the width of the doors shall have their hinges face the cargo area.

9.3.2.17.3   Entrances from the deck and openings of spaces facing the weather shall be capable of being closed. The following instruction shall be displayed at the entrance of such spaces:

**Do not open during loading, unloading and degassing
without the permission of the master.
Close immediately.**

9.3.2.17.4   Entrances and windows of superstructures and accommodation spaces which can be opened as well as other openings of these spaces shall be located not less than 2.00 m from the cargo area. No wheelhouse doors and windows shall be located within 2.00 m from the cargo area, except where there is no direct connection between the wheelhouse and the accommodation.

9.3.2.17.5   (a)   Driving shafts of the bilge or ballast pumps in the cargo area may penetrate through the bulkhead between the service space and the engine room, provided the arrangement of the service space is in compliance with 9.3.2.11.6.

(b)   The penetration of the shaft through the bulkhead shall be gastight and shall have been approved by a recognised classification society.

(c)   The necessary operating instructions shall be displayed.

(d)   Penetrations through the bulkhead between the engine room and the service space in the cargo area and the bulkhead between the engine room and the hold spaces may be provided for electrical cables, hydraulic and piping for measuring, control and alarm systems, provided that the penetration have been approved by a recognised classification society. The penetrations shall be gastight. Penetrations through a bulkhead with an "A-60" fire protection insulation according to SOLAS 74, Chapter II-2, Regulation 3, shall have an equivalent fire protection.

(e)   Pipes may penetrate the bulkhead between the engine room and the service space in the cargo area provided that these are pipes between the mechanical equipment in the engine room and the service space which do not have any openings within the service space and which are provided with shut-off devices at the bulkhead in the engine room.

(f)   Notwithstanding 9.3.2.11.4, pipes from the engine room may pass through the service space in the cargo area or a cofferdam or a hold space or a double-hull space to the outside provided that within the service space or cofferdam or hold space or double-hull space they are of the thick-walled type and have no flanges or openings.

(g)   Where a driving shaft of auxiliary machinery penetrates through a wall located above the deck the penetration shall be gastight.

9.3.2.17.6   A service space located within the cargo area below deck shall not be used as a cargo pump-room for the loading and unloading system, except where:

–   the pump room is separated from the engine room or from service spaces outside the cargo area by a cofferdam or a bulkhead with an "A-60" fire protection insulation according to SOLAS 74, Chapter II-2, Regulation 3, or by a service space or a hold space;

–   the "A-60" bulkhead required above does not include penetrations referred to in 9.3.2.17.5 (a);

–   ventilation exhaust outlets are located not less than 6.00 m from entrances and openings of the accommodation and service spaces outside the cargo area;

–   the access hatches and ventilation inlets can be closed from the outside;

–   all piping for loading and unloading as well as those of stripping systems are provided with shut-off devices at the pump suction side in the cargo pump-room immediately at the bulkhead. The necessary operation of the control devices in the pump-room, starting of pumps and necessary control of the liquid flow rate shall be effected from the deck;

–   the bilge of the cargo pump-room is equipped with a gauging device for measuring the filling level which activates a visual and audible alarm in the wheelhouse when liquid is accumulating in the cargo pump-room bilge;

–   the cargo pump-room is provided with a permanent gas–detection system which automatically indicates the presence of explosive gases or lack of oxygen by means of direct-measuring sensors and which actuates a visual and audible alarm when the gas concentration has reached 20% of the lower explosive limit. The sensors of this system shall be placed at suitable positions at the bottom and directly below the deck.

Measurement shall be continuous.

The audible and visual alarms are installed in the wheelhouse and in the cargo pump-room and, when the alarm is actuated, the loading and unloading system is shut down. Failure of the gas detection system shall be immediately signalled in the wheelhouse and on deck by means of audible and visual alarms;

–   the ventilation system prescribed in 9.3.2.12.3 has a capacity of not less than 30 changes of air per hour based on the total volume of the service space.

9.3.2.17.7   The following instruction shall be displayed at the entrance of the cargo pump-room:

**Before entering the cargo pump-room check whether
it is free from gases and contains sufficient oxygen.
Do not open doors and entrance openings without
the permission of the master.
Leave immediately in the event of alarm.**

9.3.2.18   *Inerting facility*

In cases in which inerting or blanketing of the cargo is prescribed, the vessel shall be equipped with an inerting system.

This system shall be capable of maintaining a permanent minimum pressure of 7 kPa (0.07 bar) in the spaces to be inerted. In addition, the inerting system shall not increase the pressure in the cargo tank to a pressure greater than that at which the pressure valve is regulated. The set pressure of the vacuum-relief valve shall be 3.5 kPa (0.035 bar).

A sufficient quantity of inert gas for loading or unloading shall be carried or produced on board if it is not possible to obtain it on shore. In addition, a sufficient quantity of inert gas to offset normal losses occurring during carriage shall be on board.

The premises to be inerted shall be equipped with connections for introducing the inert gas and monitoring systems so as to ensure the correct atmosphere on a permanent basis.

When the pressure or the concentration of inert gas in the gaseous phase falls below a given value, this monitoring system shall activate an audible and visible alarm in the wheelhouse. When the wheelhouse is unoccupied, the alarm shall also be perceptible in a location occupied by a crew member.

9.3.2.19   (*Reserved*)

9.3.2.20 *Arrangement of cofferdams*

9.3.2.20.1   Cofferdams or cofferdam compartments remaining once a service space has been arranged in accordance with 9.3.2.11.6 shall be accessible through an access hatch.

9.3.2.20.2   Cofferdams shall be capable of being filled with water and emptied by means of a pump. Filling shall be effected within 30 minutes. These requirements are not applicable when the bulkhead between the engine room and the cofferdam comprises fire-protection insulation "A-60" in accordance with SOLAS 74, Chapter II-2, Regulation 3, or has been fitted out as a service space. The cofferdams shall not be fitted with inlet valves.

9.3.2.20.3   No fixed pipe shall permit connection between a cofferdam and other piping of the vessel outside the cargo area.

9.3.2.20.4   When the list of substances on the vessel according to 1.16.1.2.5 contains substances for which protection against explosion is required in column (17) of Table C of Chapter 3.2, the ventilation openings of cofferdams shall be fitted with a flame-arrester withstanding a deflagration.

9.3.2.21   *Safety and control installations*

9.3.2.21.1   Cargo tanks shall be provided with the following equipment:

(a)   a mark inside the tank indicating the liquid level of 95%;

(b)   a level gauge;

(c)   a level alarm device which is activated at the latest when a degree of filling of 90% is reached;

(d)   a high level sensor for actuating the facility against overflowing at the latest when a degree of filling of 97.5% is reached;

(e)   an instrument for measuring the pressure of the vapour phase inside the cargo tank;

(f)   an instrument for measuring the temperature of the cargo, if in column (9) of Table C of Chapter 3.2 a heating installation is required, or if a maximum temperature is indicated in column (20) of that list;

(g)   a connection for a closed-type or partly closed-type sampling device, and/or at least one sampling opening as required in column (13) of Table C of Chapter 3.2.

9.3.2.21.2   When the degree of filling in per cent is determined, an error of not more than 0.5% is permitted. It shall be calculated on the basis of the total cargo tank capacity including the expansion trunk.

9.3.2.21.3   The level gauge shall allow readings from the control position of the shut-off devices of the particular cargo tank. The permissible maximum filling levels of 95% and 97%, as given in the list of substances, shall be marked on each level gauge.

Permanent reading of the overpressure and vacuum shall be possible from a location from which loading or unloading operations may be interrupted. The permissible maximum overpressure and vacuum shall be marked on each level gauge.

Readings shall be possible in all weather conditions.

9.3.2.21.4   The level alarm device shall give a visual and audible warning on board when actuated. The level alarm device shall be independent of the level gauge.

9.3.2.21.5   (a)   The high level sensor referred to in 9.3.2.21.1 (d) above shall give a visual and audible alarm on board and at the same time actuate an electrical contact which in the form of a binary signal interrupts the electric current loop provided and fed by the shore facility, thus initiating measures at the shore facility against overflowing during loading operations.

                 The signal shall be transmitted to the shore facility via a watertight two–pin plug of a connector device in accordance with standard EN 60309-2:1999 + A1:2007 + A2:2012 for direct current of 40 to 50 volts, identification colour white, position of the nose 10 h.

                 The plug shall be permanently fitted to the vessel close to the shore connections of the loading and unloading piping.

                 The high level sensor shall also be capable of switching off the vessel's own discharging pump. The high level sensor shall be independent of the level alarm device, but it may be connected to the level gauge.

       (b)   During discharging by means of the on-board pump, it shall be possible for the shore facility to switch it off. For this purpose, an independent intrinsically safe power line, fed by the vessel, shall be switched off by the shore facility by means of an electrical contact.

                 It shall be possible for the binary signal of the shore facility to be transmitted via a watertight two–pole socket or a connector device in accordance with standard EN 60309-2:1999 + A1:2007 + A2:2012, for direct current of 40 to 50 volts, identification colour white, position of the nose 10 h.

                 This socket shall be permanently fitted to the vessel close to the shore connections of the unloading piping.

       (c)   Vessels which may be delivering products required for operation of vessels shall be equipped with a transhipment facility compatible with European standard EN 12827:1999 and a rapid closing device enabling refuelling to be interrupted. It shall be possible to actuate this rapid closing device by means of an electrical signal from the overflow prevention system. The electrical circuits actuating the rapid closing device shall be secured according to the quiescent current principle or other appropriate error detection measures. The state of operation of electrical circuits which cannot be controlled using the quiescent current principle shall be capable of being easily checked.

                 It shall be possible to actuate the rapid closing device independently of the electrical signal.

                 The rapid closing device shall actuate a visual and audible alarm on board.

9.3.2.21.6   The visual and audible signals given by the level alarm device shall be clearly distinguishable from those of the high level sensor.

           The visual alarm shall be visible at each control position on deck of the cargo tank stop valves. It shall be possible to easily check the functioning of the sensors and electric circuits or these shall be "intrinsically safe apparatus".

9.3.2.21.7   When the pressure or temperature exceeds a set value, instruments for measuring the vacuum or overpressure of the gaseous phase in the cargo tank or the temperature of the cargo, shall activate a visual and audible alarm in the wheelhouse. When the wheelhouse is unoccupied the alarm shall also be perceptible in a location occupied by a crew member.

When the pressure exceeds the set value during loading and unloading, the instrument for measuring the pressure shall, by means of the plug referred to in 9.3.2.21.5 above, initiate immediately an electrical contact which shall put into effect measures to interrupt the loading or unloading operation. If the vessel's own discharge pump is used, it shall be switched off automatically.

The instrument for measuring the overpressure or vacuum shall activate the alarm at latest when an overpressure is reached equal to 1.15 times the opening pressure of the pressure relief device, or a vacuum pressure equal to the construction vacuum pressure but not exceeding 5 kPa (0.05 bar). The maximum allowable temperature is indicated in column (20) of Table C of Chapter 3.2. The sensors for the alarms mentioned in this paragraph may be connected to the alarm device of the sensor.

When it is prescribed in column (20) of Table C of Chapter 3.2, the instrument for measuring the overpressure of the gaseous phase shall activate a visible and audible alarm in the wheelhouse when the overpressure exceeds 40 kPa (0.4 bar) during the voyage. When the wheelhouse is unoccupied, the alarm shall also be perceptible in a location occupied by a crew member.

9.3.2.21.8   Where the control elements of the shut-off devices of the cargo tanks are located in a control room, it shall be possible to stop the loading pumps and read the level gauges in the control room, and the visual and audible warning given by the level alarm device, the high level sensor referred to in 9.3.2.21.1 (d) and the instruments for measuring the pressure and temperature of the cargo shall be noticeable in the control room and on deck.

Satisfactory monitoring of the cargo area shall be ensured from the control room.

9.3.2.21.9   The vessel shall be so equipped that loading or unloading operations can be interrupted by means of switches, i.e. the quick-action stop valve located on the flexible vessel–to–shore connecting line must be capable of being closed. The switch shall be placed at two points on the vessel (fore and aft).

This provision applies only when prescribed in column (20) of Table C of Chapter 3.2.

The interruption system shall be designed according to the quiescent current principle.

9.3.2.21.10   When refrigerated substances are carried the opening pressure of the safety system shall be determined by the design of the cargo tanks. In the event of the transport of substances that must be carried in a refrigerated state the opening pressure of the safety system shall be not less than 25 kPa (0.25 bar) greater than the maximum pressure calculated according to 9.3.2.27.

9.3.2.22   _Cargo tank openings_

9.3.2.22.1   (a)   Cargo tank openings shall be located on deck in the cargo area.

              (b)   Cargo tank openings with a cross-section of more than 0.10 m² and openings of safety devices for preventing overpressures shall be located not less than 0.50 m above deck.

9.3.2.22.2   Cargo tank openings shall be fitted with gastight closures capable of withstanding the test pressure in accordance with 9.3.2.23.2

9.3.2.22.3   Closures which are normally used during loading or unloading operations shall not cause sparking when operated.

9.3.2.22.4   (a)   Each cargo tank or group of cargo tanks connected to a common venting piping shall be fitted with:

- safety devices for preventing unacceptable overpressures or vacuums. When anti-explosion protection is required in column (17) of Table C of Chapter 3.2, the vacuum valve shall be fitted with a flame arrester capable of withstanding a deflagration and the pressure–relief valve with a high-velocity vent valve capable of withstanding steady burning.

  The gases shall be discharged upwards. The opening pressure of the high-velocity vent valve and the opening pressure of the vacuum valve shall be indelibly indicated on the valves;

- a connection for the safe return ashore of gases expelled during loading;

- a device for the safe depressurization of the tanks. When the list of substances on the vessel according to 1.16.1.2.5 contains substances for which protection against explosion is required in column (17) of Table C of Chapter 3.2, this device shall include at least a flame arrester capable of withstanding steady burning and a stop valve which clearly indicates whether it is open or shut.

(b) The outlets of high-velocity vent valves shall be located not less than 2.00 m above the deck and at a distance of not less than 6.00 m from the accommodation and from the service spaces outside the cargo area. This height may be reduced when within a radius of 1.00 m round the outlet of the high-velocity vent valve, there is no equipment, no work is being carried out and signs indicate the area. The setting of the high-velocity vent valves shall be such that during the transport operation they do not blow off until the maximum permissible working pressure of the cargo tanks is reached.

9.3.2.22.5   (a) Insofar as anti-explosion protection is prescribed in column (17) of Table C of Chapter 3.2, venting piping connecting two or more cargo tanks shall be fitted, at the connection to each cargo tank, with a flame arrester with a fixed or spring-loaded plate stack, capable of withstanding a detonation. This equipment may consist of:

(i) a flame arrester fitted with a fixed plate stack, where each cargo tank is fitted with a vacuum valve capable of withstanding a deflagration and a high-velocity vent valve capable of withstanding steady burning;

(ii) a flame arrester fitted with a spring-loaded plate stack, where each cargo tank is fitted with a vacuum valve capable of withstanding a deflagration;

(iii) a flame arrester with a fixed or spring-loaded plate stack;

(iv) a flame arrester with a fixed plate stack, where the pressure–measuring device is fitted with an alarm system in accordance with 9.3.2.21.7;

(v) (*Deleted*).

When a fire-fighting installation is permanently mounted on deck in the cargo area and can be brought into service from the deck and from the wheelhouse, flame arresters need not be required for individual cargo tanks.

Only substances which do not mix and which do not react dangerously with each other may be carried simultaneously in cargo tanks connected to a common venting piping;

or

(b) Insofar as anti-explosion protection is prescribed in column (17) of Table C of Chapter 3.2, venting piping connecting two or more cargo tanks shall be fitted, at the connection to each cargo tank, with a pressure/vacuum relief valve incorporating a flame arrester capable of withstanding a detonation/deflagration.

Only substances which do not mix and which do not react dangerously with each other may be carried simultaneously in cargo tanks connected to a common venting piping;

or

(c) Insofar as anti-explosion protection is prescribed in column (17) of Table C of Chapter 3.2, an independent venting piping for each cargo tank, fitted with a vacuum valve incorporating a flame arrester capable of withstanding a deflagration and a high velocity vent valve incorporating a flame arrester capable of withstanding steady burning. Several different substances may be carried simultaneously;

or

(d) Insofar as anti-explosion protection is prescribed in column (17) of Table C of Chapter 3.2, venting piping connecting two or more cargo tanks shall be fitted, at the connection to each cargo tank, with a shut-off device capable of withstanding a detonation, where each cargo tank is fitted with a vacuum valve capable of withstanding a deflagration and a high-velocity vent valve capable of withstanding steady burning.

Only substances which do not mix and which do not react dangerously with each other may be carried simultaneously in cargo tanks connected to a common venting piping.

9.3.2.23 *Pressure tests*

9.3.2.23.1 The cargo tanks, residual cargo tanks, cofferdams, piping for loading and unloading shall be subjected to initial tests before being put into service and thereafter at prescribed intervals.

Where a heating system is provided inside the cargo tanks, the heating coils shall be subjected to initial tests before being put into service and thereafter at prescribed intervals.

9.3.2.23.2 The test pressure for the cargo tanks and residual cargo tanks shall be not less than 1.3 times the construction pressure. The test pressure for the cofferdams and open cargo tanks shall be not less than 10 kPa (0.10 bar) gauge pressure.

9.3.2.23.3 The test pressure for piping for loading and unloading shall be not less than 1,000 kPa (10 bar) gauge pressure.

9.3.2.23.4 The maximum intervals for the periodic tests shall be 11 years.

9.3.2.23.5 The procedure for pressure tests shall comply with the provisions established by the competent authority or a recognised classification society.

9.3.2.24 *Regulation of cargo pressure and temperature*

9.3.2.24.1 Unless the entire cargo system is designed to resist the full effective vapour pressure of the cargo at the upper limits of the ambient design temperatures, the pressure of the tanks shall be kept below the permissible maximum set pressure of the safety valves, by one or more of the following means:

(a) a system for the regulation of cargo tank pressure using mechanical refrigeration;

(b)   a system ensuring safety in the event of the heating or increase in pressure of the cargo. The insulation or the design pressure of the cargo tank, or the combination of these two elements, shall be such as to leave an adequate margin for the operating period and the temperatures expected; in each case the system shall be deemed acceptable by a recognised classification society and shall ensure safety for a minimum time of three times the operation period;

(c)   other systems deemed acceptable by a recognised classification society.

9.3.2.24.2   The systems prescribed in 9.3.2.24.1 shall be constructed, installed and tested to the satisfaction of the recognised classification society. The materials used in their construction shall be compatible with the cargoes to be carried. For normal service, the upper ambient design temperature limits shall be:

air: +30° C;
water: +20° C.

9.3.2.24.3   The cargo storage system shall be capable of resisting the full vapour pressure of the cargo at the upper limits of the ambient design temperatures, whatever the system adopted to deal with the boil-off gas. This requirement is indicated by remark 37 in column (20) of Table C of Chapter 3.2.

9.3.2.25   *Pumps and piping*

9.3.2.25.1   Pumps, compressors and accessory loading and unloading piping shall be placed in the cargo area. Cargo pumps shall be capable of being shut down from the cargo area and, in addition, from a position outside the cargo area. Cargo pumps situated on deck shall be located not less than 6.00 m from entrances to, or openings of, the accommodation and service spaces outside the cargo area.

9.3.2.25.2   (a)   Piping for loading and unloading shall be independent of any other piping of the vessel. No cargo piping shall be located below deck, except those inside the cargo tanks and inside the cargo pump-room.

(b)   The piping for loading and unloading shall be arranged so that, after loading or unloading operations, the liquid remaining in these pipes may be safely removed and may flow either into the vessel's tanks or the tanks ashore.

(c)   Piping for loading and unloading shall be clearly distinguishable from other piping, e.g. by means of colour marking.

(d)   The piping for loading and unloading located on deck, with the exception of the shore connections, shall be located not less than a quarter of the vessel's breadth from the outer shell.

(e)   The shore connections shall be located not less than 6.00 m from the entrances to, or openings of, the accommodation and service spaces outside the cargo area.

(f)   Each shore connection of the venting piping and shore connections of the piping for loading and unloading, through which the loading or unloading operation is carried out, shall be fitted with a shut-off device. However, each shore connection shall be fitted with a blind flange when it is not in operation.

(g)   (*Deleted*)

(h) The flanges and stuffing boxes shall be provided with a spray protection device.

(i) Piping for loading and unloading, and venting piping, shall not have flexible connections fitted with sliding seals.

9.3.2.25.3 The distance referred to in 9.3.2.25.1 and 9.3.2.25.2 (e) may be reduced to 3.00 m if a transverse bulkhead complying with 9.3.2.10.2 is situated at the end of the cargo area. The openings shall be provided with doors.

The following notice shall be displayed on the doors:

**Do not open during loading and unloading without
the permission of the master.
Close immediately.**

9.3.2.25.4 (a) Every component of the piping for loading and unloading shall be electrically connected to the hull.

(b) The piping for loading shall extend down to the bottom of the cargo tanks.

9.3.2.25.5 The stop valves or other shut-off devices of the piping for loading and unloading shall indicate whether they are open or shut.

9.3.2.25.6 The piping for loading and unloading shall have, at the test pressure, the required elasticity, leakproofness and resistance to pressure.

9.3.2.25.7 The piping for loading and unloading shall be fitted with pressure gauges at the outlet of the pumps. The permissible maximum overpressure or vacuum value shall be indicated on each measuring device. Readings shall be possible in all weather conditions.

9.3.2.25.8 (a) When piping for loading and unloading are used for supplying the cargo tanks with washing or ballast water, the suctions of these pipes shall be located within the cargo area but outside the cargo tanks.

Pumps for tank washing systems with associated connections may be located outside the cargo area, provided the discharge side of the system is arranged in such a way that the suction is not possible through that part.

A spring-loaded non-return valve shall be provided to prevent any gases from being expelled from the cargo area through the tank washing system.

(b) A non-return valve shall be fitted at the junction between the water suction pipe and the cargo loading pipe.

9.3.2.25.9 The permissible loading and unloading flows shall be calculated.

Calculations concern the permissible maximum loading and unloading flow for each cargo tank or each group of cargo tanks, taking into account the design of the ventilation system. These calculations shall take into consideration the fact that in the event of an unforeseen cut-off of the vapour return piping of the shore facility, the safety devices of the cargo tanks will prevent pressure in the cargo tanks from exceeding the following values:

over-pressure: 115% of the opening pressure of the high-velocity vent valve;

vacuum pressure: not more than the construction vacuum pressure but not exceeding 5 kPa (0.05 bar).

The main factors to be considered are the following:

1. Dimensions of the ventilation system of the cargo tanks;

2. Gas formation during loading: multiply the largest loading flow by a factor of not less than 1.25;

3. Density of the vapour mixture of the cargo based on 50% volume vapour and 50% volume air;

4. Loss of pressure through ventilation pipes, valves and fittings. Account will be taken of a 30% clogging of the mesh of the flame-arrester;

5. Chocking pressure of the safety valves.

The permissible maximum loading and unloading flows for each cargo tank or for each group of cargo tanks shall be given in an on-board instruction.

9.3.2.25.10   Compressed air generated outside the cargo area or wheelhouse can be used in the cargo area subject to the installation of a spring-loaded non-return valve to ensure that no gases can escape from the cargo area through the compressed air system into accommodation or service spaces outside the cargo area.

9.3.2.25.11   If the vessel is carrying several dangerous substances liable to react dangerously with each other, a separate pump with its own piping for loading and unloading shall be installed for each substance. The piping shall not pass through a cargo tank containing dangerous substances with which the substance in question is liable to react.

9.3.2.26   *Tanks and receptacles for residual products and receptacles for slops*

9.3.2.26.1   If vessels are provided with a tank for residual products, it shall comply with the provisions of 9.3.2.26.3 and 9.3.2.26.4. Receptacles for residual products and receptacles for slops shall be located only in the cargo area. During the filling of the receptacles for residual products, means for collecting any leakage shall be placed under the filling connections.

9.3.2.26.2   Receptacles for slops shall be fire resistant and shall be capable of being closed with lids (drums with removable heads, code 1A2, ADR). The receptacles for slops shall be marked and be easy to handle.

9.3.2.26.3   The maximum capacity of a tank for residual products is 30 m^3.

9.3.2.26.4   The tank for residual products shall be equipped with:

–   pressure-relief and vacuum relief valves.

The high velocity vent valve shall be so regulated as not to open during carriage. This condition is met when the opening pressure of the valve meets the conditions set out in column (10) of Table C of Chapter 3.2;

When anti-explosion protection is required in column (17) of Table C of Chapter 3.2, the vacuum-relief valve shall be capable of withstanding deflagrations and the high-velocity vent valve shall withstand steady burning;

–   a level indicator;

–   connections with shut-off devices, for pipes and hose assemblies.

Receptacles for residual products shall be equipped with:

– a connection enabling gases released during filling to be evacuated safely;

– a possibility of indicating the degree of filling;

– connections with shut-off devices, for pipes and hose assemblies.

Receptacles for residual products shall be connected to the venting piping of cargo tanks only for the time necessary to fill them in accordance with 7.2.4.15.2.

Receptacles for residual products and receptacles for slops placed on the deck shall be located at a minimum distance from the hull equal to one quarter of the vessel's breadth.

9.3.2.27 *Refrigeration system*

9.3.2.27.1 The refrigeration system referred to in 9.3.2.24.1 (a) shall be composed of one or more units capable of keeping the pressure and temperature of the cargo at the upper limits of the ambient design temperatures at the prescribed level. Unless another means of regulating cargo pressure and temperature deemed satisfactory by a recognised classification society is provided, provision shall be made for one or more stand-by units with an output at least equal to that of the largest prescribed unit. A stand-by unit shall include a compressor, its engine, its control system and all necessary accessories to enable it to operate independently of the units normally used. Provision shall be made for a stand-by heat-exchanger unless the system's normal heat-exchanger has a surplus capacity equal to at least 25% of the largest prescribed capacity. It is not necessary to make provision for separate piping.

Cargo tanks, piping and accessories shall be insulated so that, in the event of a failure of all cargo refrigeration systems, the entire cargo remains for at least 52 hours in a condition not causing the safety valves to open.

9.3.2.27.2 The security devices and the connecting lines from the refrigeration system shall be connected to the cargo tanks above the liquid phase of the cargo when the tanks are filled to their maximum permissible degree of filling. They shall remain within the gaseous phase, even if the vessel has a list up to 12 degrees.

9.3.2.27.3 When several refrigerated cargoes with a potentially dangerous chemical reaction are carried simultaneously, particular care shall be given to the refrigeration systems so as to prevent any mixing of the cargoes. For the carriage of such cargoes, separate refrigeration systems, each including the full stand-by unit referred to in 9.3.2.27.1, shall be provided for each cargo. When, however, refrigeration is ensured by an indirect or combined system and no leak in the heat exchangers can under any foreseeable circumstances lead to the mixing of cargoes, no provision need be made for separate refrigeration units for the different cargoes.

9.3.2.27.4 When several refrigerated cargoes are not soluble in each other under conditions of carriage such that their vapour pressures are added together in the event of mixing, particular care shall be given to the refrigeration systems to prevent any mixing of the cargoes.

9.3.2.27.5 When the refrigeration systems require water for cooling, a sufficient quantity shall be supplied by a pump or pumps used exclusively for the purpose. This pump or pumps shall have at least two suction pipes, leading from two water intakes, one to port, the other to starboard. Provision shall be made for a stand-by pump with a satisfactory flow; this may be a pump used for other purposes provided that its use for supplying water for cooling does not impair any other essential service.

9.3.2.27.6   The refrigeration system may take one of the following forms:

(a)   Direct system: the cargo vapours are compressed, condensed and returned to the cargo tanks. This system shall not be used for certain cargoes specified in Table C of Chapter 3.2. This requirement is indicated by remark 35 in column (20) of Table C of Chapter 3.2;

(b)   Indirect system: the cargo or the cargo vapours are cooled or condensed by means of a coolant without being compressed;

(c)   Combined system: the cargo vapours are compressed and condensed in a cargo/coolant heat-exchanger and returned to the cargo tanks. This system shall not be used for certain cargoes specified in Table C of Chapter 3.2. This requirement is indicated by remark 36 in column (20) of Table C of Chapter 3.2.

9.3.2.27.7   All primary and secondary coolant fluids shall be compatible with each other and with the cargo with which they may come into contact. Heat exchange may take place either at a distance from the cargo tank, or by using cooling coils attached to the inside or the outside of the cargo tank.

9.3.2.27.8   When the refrigeration system is installed in a separate service space, this service space shall meet the requirements of 9.3.2.17.6.

9.3.2.27.9   For all cargo systems, the heat transmission coefficient as used for the determination of the holding time (7.2.4.16.16 and 7.2.4.16.17) shall be determined by calculation. Upon completion of the vessel, the correctness of the calculation shall be checked by means of a heat balance test. The calculation and test shall be performed under supervision by the recognized classification society which classified the vessel.

The heat transmission coefficient shall be documented and kept on board. The heat transmission coefficient shall be verified at every renewal of the certificate of approval.

9.3.2.27.10   A certificate from a recognised classification society stating that 9.3.2.24.1 to 9.3.2.24.3, 9.3.2.27.1 and 9.3.2.27.4 above have been complied with shall be submitted together with the application for issue or renewal of the certificate of approval.

9.3.2.28   *Water-spray system*

When water-spraying is required in column (9) of Table C of Chapter 3.2, a water-spray system shall be installed in the cargo area on deck to enable gas emissions from loading to be precipitated and to cool the tops of cargo tanks by spraying water over the whole surface so as to avoid safely the activation of the high-velocity vent valve at 50 kPa (0.5 bar).

The gas precipitation system shall be fitted with a connection device for supply from a shore installation.

The spray nozzles shall be so installed that the entire cargo deck area is covered and the gases released are precipitated safely.

The system shall be capable of being put into operation from the wheelhouse and from the deck. Its capacity shall be such that when all the spray nozzles are in operation, the outflow is not less than 50 litres per square metre of deck area and per hour.

9.3.2.29 and 9.3.2.30   (*Reserved*)

9.3.2.31   *Engines*

9.3.2.31.1   Only internal combustion engines running on fuel with a flashpoint of more than 55° C are allowed.

9.3.2.31.2.   Ventilation inlets of the engine room, and when the engines do not take in air directly from the engine room, air intakes of the engines shall be located not less than 2.00 m from the cargo area.

9.3.2.31.3   Sparking shall not be possible within the cargo area.

9.3.2.31.4   The surface temperature of the outer parts of engines used during loading or unloading operations, as well as that of their air inlets and exhaust ducts shall not exceed the allowable temperature according to the temperature class of the substances carried. This provision does not apply to engines installed in service spaces provided the provisions of 9.3.2.52.3 are fully complied with.

9.3.2.31.5   The ventilation in the closed engine room shall be designed so that, at an ambient temperature of 20 °C, the average temperature in the engine room does not exceed 40° C.

9.3.2.32   *Oil fuel tanks*

9.3.2.32.1   Where the vessel is provided with hold spaces, the double bottoms within these spaces may be arranged as oil fuel tanks, provided their depth is not less than 0.6 m.

Oil fuel pipes and openings of such tanks are not permitted in the hold space.

9.3.2.32.2   The open ends of the air pipes of all oil fuel tanks shall extend to not less than 0.5 m above the open deck. Their open ends and the open ends of overflow pipes leading to the deck shall be fitted with a protective device consisting of a gauze diaphragm or a perforated plate.

9.3.2.33   (*Reserved*)

9.3.2.34   *Exhaust pipes*

9.3.2.34.1   Exhausts shall be evacuated from the vessel into the open air either upwards through an exhaust pipe or through the shell plating. The exhaust outlet shall be located not less than 2.00 m from the cargo area. The exhaust pipes of engines shall be arranged so that the exhausts are led away from the vessel. The exhaust pipes shall not be located within the cargo area.

9.3.2.34.2   Exhaust pipes shall be provided with a device preventing the escape of sparks, e.g. spark arresters.

9.3.2.35   *Bilge pumping and ballasting arrangements*

9.3.2.35.1   Bilge and ballast pumps for spaces within the cargo area shall be installed within such area.

This provision does not apply to:

–   double-hull spaces and double bottoms which do not have a common boundary wall with the cargo tanks;

–   cofferdams, double-hull spaces, hold spaces and double bottoms where ballasting is carried out using the piping of the fire-fighting system in the cargo area and bilge-pumping is performed using eductors.

9.3.2.35.2   Where the double bottom is used as a liquid oil fuel tank, it shall not be connected to the bilge piping system.

9.3.2.35.3   Where the ballast pump is installed in the cargo area, the standpipe and its outboard connection for suction of ballast water shall be located within the cargo area but outside the cargo tanks.

9.3.2.35.4   A cargo pump-room below deck shall be capable of being drained in an emergency by an installation located in the cargo area and independent from any other installation. This installation shall be provided outside the cargo pump-room.

9.3.2.36 to 9.3.2.39   (*Reserved*)

9.3.2.40   *Fire-extinguishing arrangements*

9.3.2.40.1   A fire-extinguishing system shall be installed on the vessel. This system shall comply with the following requirements:

–   It shall be supplied by two independent fire or ballast pumps, one of which shall be ready for use at any time. These pumps and their means of propulsion and electrical equipment shall not be installed in the same space;

–   It shall be provided with a water main fitted with at least three hydrants in the cargo area or wheelhouse above deck. Three suitable and sufficiently long hoses with jet/spray nozzles having a diameter of not less than 12 mm shall be provided. Alternatively one or more of the hose assemblies may be substituted by directable jet/spray nozzles having a diameter of not less than 12 mm. It shall be possible to reach any point of the deck in the cargo area simultaneously with at least two jets of water which do not emanate from the same hydrant.

A spring-loaded non-return valve shall be fitted to ensure that no gases can escape through the fire-extinguishing system into the accommodation or service spaces outside the cargo area;

–   The capacity of the system shall be at least sufficient for a jet of water to have a minimum reach of not less than the vessel's breadth from any location on board with two spray nozzles being used at the same time;

–   The water supply system shall be capable of being put into operation from the wheelhouse and from the deck;

–   Measures shall be taken to prevent the freezing of fire-mains and hydrants.

9.3.2.40.2   In addition, the engine rooms, the pump-room and all spaces containing essential equipment (switchboards, compressors, etc.) for the refrigeration equipment, if any, shall be provided with a permanently fixed fire-extinguishing system meeting the following requirements:

9.3.2.40.2.1   *Extinguishing agents*

For the protection of spaces in engine rooms, boiler rooms and pump rooms, only permanently fixed fire-extinguishing systems using the following extinguishing agents are permitted:

(a) CO_2 (carbon dioxide);

(b)   HFC 227 ea (heptafluoropropane);

(c)   IG-541 (52% nitrogen, 40% argon, 8% carbon dioxide).

(d)   FK-5-1-12 (dodecafluoro 2-methylpentane-3-one).

Other extinguishing agents are permitted only on the basis of recommendations by the Administrative Committee.

9.3.2.40.2.2 *Ventilation, air extraction*

(a)   The combustion air required by the combustion engines which ensure propulsion should not come from spaces protected by permanently fixed fire-extinguishing systems. This requirement is not mandatory if the vessel has two independent main engine rooms with a gastight separation or if, in addition to the main engine room, there is a separate engine room installed with a bow thruster that can independently ensure propulsion in the event of a fire in the main engine room.

(b)   All forced ventilation systems in the space to be protected shall be shut down automatically as soon as the fire-extinguishing system is activated.

(c)   All openings in the space to be protected which permit air to enter or gas to escape shall be fitted with devices enabling them to be closed rapidly. It shall be clear whether they are open or closed.

(d)   Air escaping from the pressure–relief valves of the pressurised air tanks installed in the engine rooms shall be evacuated to the open air.

(e)   Overpressure or negative pressure caused by the diffusion of the extinguishing agent shall not destroy the constituent elements of the space to be protected. It shall be possible to ensure the safe equalisation of pressure.

(f)   Protected spaces shall be provided with a means of extracting the extinguishing agent. If extraction devices are installed, it shall not be possible to start them up during extinguishing.

9.3.2.40.2.3 *Fire alarm system*

The space to be protected shall be monitored by an appropriate fire alarm system. The alarm signal shall be audible in the wheelhouse, the accommodation and the space to be protected.

9.3.2.40.2.4 *Piping system*

(a)   The extinguishing agent shall be routed to and distributed in the space to be protected by means of a permanent piping system. Piping installed in the space to be protected and their fittings shall be made of steel. This shall not apply to the connecting nozzles of tanks and compensators provided that the materials used have equivalent fire-retardant properties. Piping shall be protected against corrosion both internally and externally.

(b)   The discharge nozzles shall be so arranged as to ensure the regular diffusion of the extinguishing agent. In particular, the extinguishing agent must also be effective beneath the floor.

9.3.2.40.2.5 *Triggering device*

(a)   Automatically activated fire-extinguishing systems are not permitted.

(b)   It shall be possible to activate the fire-extinguishing system from a suitable point located outside the space to be protected.

(c)   Triggering devices shall be so installed that they can be activated in the event of a fire and so that the risk of their breakdown in the event of a fire or an explosion in the space to be protected is reduced as far as possible.

Systems which are not mechanically activated shall be supplied from two energy sources independent of each other. These energy sources shall be located outside the space to be protected. The control lines located in the space to be protected shall be so designed as to remain capable of operating in the event of a fire for a minimum of 30 minutes. The electrical installations are deemed to meet this requirement if they conform to the IEC 60331–21:1999 standard.

When the triggering devices are so placed as not to be visible, the component concealing them shall carry the "Fire-fighting system" symbol, each side being not less than 10 cm in length, with the following text in red letters on a white ground:

Fire-extinguishing system

(d)   If the fire-extinguishing system is intended to protect several spaces, it shall comprise a separate and clearly–marked triggering device for each space.

(e)   The instructions shall be posted alongside all triggering devices and shall be clearly visible and indelible. The instructions shall be in a language the master can read and understand and if this language is not English, French or German, they shall be in English, French or German. They shall include information concerning:

(i)   the activation of the fire-extinguishing system;

(ii)   the need to ensure that all persons have left the space to be protected;

(iii)   The correct behaviour of the crew in the event of activation and when accessing the space to be protected following activation or diffusion, in particular in respect of the possible presence of dangerous substances;

(iv)   the correct behaviour of the crew in the event of the failure of the fire-extinguishing system to function properly.

(f)   The instructions shall mention that prior to the activation of the fire-extinguishing system, combustion engines installed in the space and aspirating air from the space to be protected, shall be shut down.

9.3.2.40.2.6 *Alarm device*

(a)   Permanently fixed fire-extinguishing systems shall be fitted with an audible and visual alarm device.

(b)   The alarm device shall be set off automatically as soon as the fire-extinguishing system is first activated. The alarm device shall function for an appropriate period of time before the extinguishing agent is released; it shall not be possible to turn it off.

(c)   Alarm signals shall be clearly visible in the spaces to be protected and their access points and be clearly audible under operating conditions corresponding to the highest possible sound level. It shall be possible to distinguish them clearly from all other sound and visual signals in the space to be protected.

(d)   Sound alarms shall also be clearly audible in adjoining spaces, with the communicating doors shut, and under operating conditions corresponding to the highest possible sound level.

(e)   If the alarm device is not intrinsically protected against short circuits, broken wires and drops in voltage, it shall be possible to monitor its operation.

(f)   A sign with the following text in red letters on a white ground shall be clearly posted at the entrance to any space the extinguishing agent may reach:

Warning, fire-extinguishing system!
Leave this space immediately when the … (description) alarm is activated!

9.3.2.40.2.7   *Pressurised tanks, fittings and piping*

(a)   Pressurised tanks, fittings and piping shall conform to the requirements of the competent authority or, if there are no such requirements, to those of a recognized classification society.

(b)   Pressurised tanks shall be installed in accordance with the manufacturer's instructions.

(c)   Pressurised tanks, fittings and piping shall not be installed in the accommodation.

(d)   The temperature of cabinets and storage spaces for pressurised tanks shall not exceed 50 °C.

(e)   Cabinets or storage spaces on deck shall be securely stowed and shall have vents so placed that in the event of a pressurised tank not being gastight, the escaping gas cannot penetrate into the vessel. Direct connections with other spaces are not permitted.

9.3.2.40.2.8   *Quantity of extinguishing agent*

If the quantity of extinguishing agent is intended for more than one space, the quantity of extinguishing agent available does not need to be greater than the quantity required for the largest of the spaces thus protected.

9.3.2.40.2.9   *Installation, maintenance, monitoring and documents*

(a)   The mounting or modification of the system shall only be performed by a company specialised in fire-extinguishing systems. The instructions (product data sheet, safety data sheet) provided by the manufacturer of the extinguishing agent or the system shall be followed.

(b)   The system shall be inspected by an expert:

(i)   before being brought into service;

(ii)   each time it is put back into service after activation;

(iii)   after every modification or repair;

(iv)   regularly, not less than every two years.

(c)   During the inspection, the expert is required to check that the system conforms to the requirements of 9.3.2.40.2.

(d)   The inspection shall include, as a minimum:

(i)   an external inspection of the entire system;

(ii)   an inspection to ensure that the piping is leakproof;

(iii)   an inspection to ensure that the control and activation systems are in good working order;

(iv)   an inspection of the pressure and contents of tanks;

(v)   an inspection to ensure that the means of closing the space to be protected are leakproof;

(vi)   an inspection of the fire alarm system;

(vii)   an inspection of the alarm device.

(e)   The person performing the inspection shall establish, sign and date a certificate of inspection.

(f)   The number of permanently fixed fire-extinguishing systems shall be mentioned in the vessel certificate.

9.3.2.40.2.10   *Fire-extinguishing system operating with CO_2*

In addition to the requirements contained in 9.3.2.40.2.1 to 9.3.2.40.2.9, fire-extinguishing systems using CO_2 as an extinguishing agent shall conform to the following provisions:

(a)   Tanks of CO_2 shall be placed in a gastight space or cabinet separated from other spaces. The doors of such storage spaces and cabinets shall open outwards; they shall be capable of being locked and shall carry on the outside the symbol "Warning: danger", not less than 5 cm high and "CO_2" in the same colours and the same size;

(b)   Storage cabinets or spaces for CO_2 tanks located below deck shall only be accessible from the outside. These spaces shall have an artificial ventilation system with extractor hoods and shall be completely independent of the other ventilation systems on board;

(c)   The level of filling of CO_2 tanks shall not exceed 0.75 kg/l. The volume of depressurised CO_2 shall be taken to be 0.56 m^3/kg;

(d)   The concentration of CO_2 in the space to be protected shall be not less than 40% of the gross volume of the space. This quantity shall be released within 120 seconds. It shall be possible to monitor whether diffusion is proceeding correctly;

(e)   The opening of the tank valves and the control of the diffusing valve shall correspond to two different operations;

(f)   The appropriate period of time mentioned in 9.3.2.40.2.6 (b) shall be not less than 20 seconds. A reliable installation shall ensure the timing of the diffusion of CO_2.

9.3.2.40.2.11   *Fire-extinguishing system operating with HFC-227 ea (heptafluoropropane)*

In addition to the requirements of 9.3.2.40.2.1 to 9.3.2.40.2.9, fire-extinguishing systems using HFC-227 ea as an extinguishing agent shall conform to the following provisions:

(a)   Where there are several spaces with different gross volumes, each space shall be equipped with its own fire-extinguishing system;

(b)   Every tank containing HFC-227 ea placed in the space to be protected shall be fitted with a device to prevent overpressure. This device shall ensure that the contents of the tank are safely diffused in the space to be protected if the tank is subjected to fire, when the fire-extinguishing system has not been brought into service;

(c)   Every tank shall be fitted with a device permitting control of the gas pressure;

(d)   The level of filling of tanks shall not exceed 1.15 kg/l. The specific volume of depressurised HFC-227 ea shall be taken to be 0.1374 m^3/kg;

(e)   The concentration of HFC-227 ea in the space to be protected shall be not less than 8% of the gross volume of the space. This quantity shall be released within 10 seconds;

(f)   Tanks of HFC-227 ea shall be fitted with a pressure monitoring device which triggers an audible and visual alarm in the wheelhouse in the event of an unscheduled loss of propellant gas. Where there is no wheelhouse, the alarm shall be triggered outside the space to be protected;

(g)   After discharge, the concentration in the space to be protected shall not exceed 10.5% (volume);

(h)   The fire-extinguishing system shall not comprise aluminium parts.

9.3.2.40.2.12   *Fire-extinguishing system operating with IG-541*

In addition to the requirements of 9.3.2.40.2.1 to 9.3.2.40.2.9, fire-extinguishing systems using IG-541 as an extinguishing agent shall conform to the following provisions:

(a)   Where there are several spaces with different gross volumes, every space shall be equipped with its own fire-extinguishing system;

(b)   Every tank containing IG-541 placed in the space to be protected shall be fitted with a device to prevent overpressure. This device shall ensure that the contents of the tank are safely diffused in the space to be protected if the tank is subjected to fire, when the fire-extinguishing system has not been brought into service;

(c)   Each tank shall be fitted with a device for checking the contents;

(d)   The filling pressure of the tanks shall not exceed 200 bar at a temperature of +15 °C;

(e)   The concentration of IG-541 in the space to be protected shall be not less than 44% and not more than 50% of the gross volume of the space. This quantity shall be released within 120 seconds.

9.3.2.40.2.13   *Fire-extinguishing system operating with FK-5-1-12*

In addition to the requirements of 9.3.2.40.2.1 to 9.3.2.40.2.9, fire-extinguishing systems using FK-5-1-12 as an extinguishing agent shall comply with the following provisions:

(a)   Where there are several spaces with different gross volumes, every space shall be equipped with its own fire-extinguishing system;

(b)   Every tank containing FK-5-1-12 placed in the space to be protected shall be fitted with a device to prevent overpressure. This device shall ensure that the contents of the tank are safely diffused in the space to be protected if the tank is subjected to fire, when the fire-extinguishing system has not been brought into service;

(c)   Every tank shall be fitted with a device permitting control of the gas pressure;

(d)   The level of filling of tanks shall not exceed 1.00 kg/l. The specific volume of depressurized FK-5-1-12 shall be taken to be 0.0719 m^3/kg;

(e)   The volume of FK-5-1-12 in the space to be protected shall be not less than 5.5% of the gross volume of the space. This quantity shall be released within 10 seconds;

(f)   Tanks of FK-5-1-12 shall be fitted with a pressure monitoring device which triggers an audible and visual alarm in the wheelhouse in the event of an unscheduled loss of extinguishing agent. Where there is no wheelhouse, the alarm shall be triggered outside the space to be protected;

(g)   After discharge, the concentration in the space to be protected shall not exceed 10.0%.

9.3.2.40.2.14   *Fixed fire-extinguishing system for physical protection*

In order to ensure physical protection in the engine rooms, boiler rooms and pump rooms, permanently fixed fire-extinguishing systems are accepted solely on the basis of recommendations by the Administrative Committee.

9.3.2.40.3   The two hand fire–extinguishers referred to in 8.1.4 shall be located in the cargo area.

9.3.2.40.4   The fire-extinguishing agent and the quantity contained in the permanently fixed fire-extinguishing system shall be suitable and sufficient for fighting fires.

9.3.2.41   ***Fire and naked light***

9.3.2.41.1   The outlets of funnels shall be located not less than 2.00 m from the cargo area. Arrangements shall be provided to prevent the escape of sparks and the entry of water.

9.3.2.41.2   Heating, cooking and refrigerating appliances shall not be fuelled with liquid fuels, liquid gas or solid fuels.

The installation in the engine room or in another separate space of heating appliances fuelled with liquid fuel having a flash–point above 55 °C is, however, permitted.

Cooking and refrigerating appliances are permitted only in the accommodation.

9.3.2.41.3   Only electrical lighting appliances are permitted.

9.3.2.42 *Cargo heating system*

9.3.2.42.1   Boilers which are used for heating the cargo shall be fuelled with a liquid fuel having a flashpoint of more than 55 °C. They shall be placed either in the engine room or in another separate space below deck and outside the cargo area, which is accessible from the deck or from the engine room.

9.3.2.42.2   The cargo heating system shall be designed so that the cargo cannot penetrate into the boiler in the case of a leak in the heating coils. A cargo heating system with artificial draught shall be ignited electrically.

9.3.2.42.3   The ventilation system of the engine room shall be designed taking into account the air required for the boiler.

9.3.2.42.4   Where the cargo heating system is used during loading, unloading or gas-freeing, the service space which contains this system shall fully comply with the requirements of 9.3.2.52.3. This requirement does not apply to the inlets of the ventilation system. These inlets shall be located at a minimum distance of 2 m from the cargo area and 6 m from the openings of cargo tanks or residual cargo tanks, loading pumps situated on deck, openings of high velocity vent valves, pressure relief devices and shore connections of loading and unloading piping and must be located not less than 2 m above the deck.

The requirements of 9.3.2.52.3 are not applicable to the unloading of substances having a flash point of 60 °C or more when the temperature of the product is at least 15 K lower at the flash point.

9.3.2.43 to 9.3.2.49   *(Reserved)*

9.3.2.50   *Documents concerning electrical installations*

9.3.2.50.1   In addition to the documents required in accordance with the Regulations referred to in 1.1.4.6, the following documents shall be on board:

(a)   a drawing indicating the boundaries of the cargo area and the location of the electrical equipment installed in this area;

(b)   a list of the electrical equipment referred to in (a) above including the following particulars:

machine or appliance, location, type of protection, type of protection against explosion, testing body and approval number;

(c)   a list of or general plan indicating the electrical equipment outside the cargo area which may be operated during loading, unloading or gas-freeing. All other electrical equipment shall be marked in red. See 9.3.2.52.3 and 9.3.2.52.4.

9.3.2.50.2   The documents listed above shall bear the stamp of the competent authority issuing the certificate of approval.

9.3.2.51   *Electrical installations*

9.3.2.51.1   Only distribution systems without return connection to the hull are permitted:

This provision does not apply to:

–   active cathodic corrosion protection;

–   local installations outside the cargo area (e.g. connections of starters of diesel engines);

–   the device for checking the insulation level referred to in 9.3.2.51.2 below.

9.3.2.51.2   Every insulated distribution network shall be fitted with an automatic device with a visual and audible alarm for checking the insulation level.

9.3.2.51.3   For the selection of electrical equipment to be used in zones presenting an explosion risk, the explosion groups and temperature classes assigned to the substances carried in accordance with columns (15) and (16) of Table C of Chapter 3.2 shall be taken into consideration.

9.3.2.52   *Type and location of electrical equipment*

9.3.2.52.1   (a)   Only the following equipment may be installed in cargo tanks, residual cargo tanks and piping for loading and unloading (comparable to zone 0):

–   measuring, regulation and alarm devices of the EEx (ia) type of protection.

(b)   Only the following equipment may be installed in the cofferdams, double-hull spaces, double bottoms and hold spaces (comparable to zone 1):

–   measuring, regulation and alarm devices of the certified safe type;

–   lighting appliances of the "flame-proof enclosure" or "pressurised enclosure" type of protection;

–   hermetically sealed echo sounding devices the cables of which are led through thick-walled steel tubes with gastight connections up to the main deck;

–   cables for the active cathodic protection of the shell plating in protective steel tubes such as those provided for echo sounding devices.

The following equipment may be installed only in double-hull spaces and double bottoms if used for ballasting:

–   Permanently fixed submerged pumps with temperature monitoring, of the certified safe type.

(c)   Only the following equipment may be installed in the service spaces in the cargo area below deck (comparable to zone 1):

–   measuring, regulation and alarm devices of the certified safe type;

–   lighting appliances of the "flame-proof enclosure" or "apparatus protected by pressurization" type of protection;

–   motors driving essential equipment such as ballast pumps with temperature monitoring; they shall be of the certified safe type.

(d)   The control and protective equipment of the electrical equipment referred to in paragraphs (a), (b) and (c) above shall be located outside the cargo area if they are not intrinsically safe.

(e)   The electrical equipment in the cargo area on deck (comparable to zone 1) shall be of the certified safe type.

9.3.2.52.2   Accumulators shall be located outside the cargo area.

9.3.2.52.3   (a)   Electrical equipment used during loading, unloading and gas-freeing during berthing and which are located outside the cargo area shall (comparable to zone 2) be at least of the "limited explosion risk" type.

(b)   This provision does not apply to:

(i)   lighting installations in the accommodation, except for switches near entrances to accommodation;

(ii)   radiotelephone installations in the accommodation or the wheelhouse;

(iii)   mobile and fixed telephone installations in the accommodation or the wheelhouse;

(iv)   electrical installations in the accommodation, the wheelhouse or the service spaces outside the cargo areas if:

1.   These spaces are fitted with a ventilation system ensuring an overpressure of 0.1 kPa (0.001 bar) and none of the windows is capable of being opened; the air intakes of the ventilation system shall be located as far away as possible, however, not less than 6.00 m from the cargo area and not less than 2.00 m above the deck;

2.   The spaces are fitted with a gas detection system with sensors:

–   at the suction inlets of the ventilation system;

–   directly at the top edge of the sill of the entrance doors of the accommodation and service spaces;

3.   The gas concentration measurement is continuous;

4.   When the gas concentration reaches 20% of the lower explosive limit, the ventilators are switched off. In such a case and when the overpressure is not maintained or in the event of failure of the gas detection system, the electrical installations which do not comply with (a) above, shall be switched off. These operations shall be performed immediately and automatically and activate the emergency lighting in the accommodation, the wheelhouse and the service spaces, which shall comply at least with the "limited explosion risk" type. The switching-off shall be indicated in the accommodation and wheelhouse by visual and audible signals;

5.   The ventilation system, the gas detection system and the alarm of the switch-off device fully comply with the requirements of (a) above;

6.   The automatic switching-off device is set so that no automatic switch off may occur while the vessel is under way.

(v)   Inland AIS (automatic identification systems) stations in the accommodation and in the wheelhouse if no part of an aerial for electronic apparatus is situated above the cargo area and if no part of a VHF antenna for AIS stations is situated within 2 m from the cargo area.

9.3.2.52.4   The electrical equipment which does not meet the requirements set out in 9.3.2.52.3 above together with its switches shall be marked in red. The disconnection of such equipment shall be operated from a centralised location on board.

9.3.2.52.5 An electric generator which is permanently driven by an engine and which does not meet the requirements of 9.3.2.52.3 above, shall be fitted with a switch capable of shutting down the excitation of the generator. A notice board with the operating instructions shall be displayed near the switch.

9.3.2.52.6 Sockets for the connection of signal lights and gangway lighting shall be permanently fitted to the vessel close to the signal mast or the gangway. Connecting and disconnecting shall not be possible except when the sockets are not live.

9.3.2.52.7 The failure of the power supply for the safety and control equipment shall be immediately indicated by visual and audible signals at the locations where the alarms are usually actuated.

9.3.2.53 *Earthing*

9.3.2.53.1 The metal parts of electrical appliances in the cargo area which are not live as well as protective metal tubes or metal sheaths of cables in normal service shall be earthed, unless they are so arranged that they are automatically earthed by bonding to the metal structure of the vessel.

9.3.2.53.2 The provisions of 9.3.2.53.1 above apply also to equipment having service voltages of less than 50 V.

9.3.2.53.3 Independent cargo tanks, metal intermediate bulk containers and tank–containers shall be earthed.

9.3.2.53.4 Receptacles for residual products shall be capable of being earthed.

9.3.2.54 and 9.3.2.55 (*Reserved*)

9.3.2.56 *Electrical cables*

9.3.2.56.1 All cables in the cargo area shall have a metallic sheath.

9.3.2.56.2 Cables and sockets in the cargo area shall be protected against mechanical damage.

9.3.2.56.3 Movable cables are prohibited in the cargo area, except for intrinsically safe electric circuits or for the supply of signal lights and gangway lighting.

9.3.2.56.4 Cables of intrinsically safe circuits shall only be used for such circuits and shall be separated from other cables not intended for being used in such circuits (e.g. they shall not be installed together in the same string of cables and they shall not be fixed by the same cable clamps).

9.3.2.56.5 For movable cables intended for signal lights and gangway lighting, only sheathed cables of type H 07 RN-F in accordance with standard IEC 60 245-4:1994 or cables of at least equivalent design having conductors with a cross-section of not less than 1.5 mm^2 shall be used.

These cables shall be as short as possible and installed so that damage is not likely to occur.

9.3.2.56.6 The cables required for the electrical equipment referred to in 9.3.2.51.1 (b) and (c) are accepted in cofferdams, double-hull spaces, double bottoms, hold spaces and service spaces below deck.

9.3.2.57 to 9.3.2.59 (*Reserved*)

9.3.2.60   *Special equipment*

A shower and an eye and face bath shall be provided on the vessel at a location which is directly accessible from the cargo area.

9.3.2.61 to 9.3.2.70   *(Reserved)*

9.3.2.71   *Admittance on board*

The notice boards displaying the prohibition of admittance in accordance with 8.3.3 shall be clearly legible from either side of the vessel.

9.3.2.72 and 9.3.2.73   *(Reserved)*

9.3.2.74   *Prohibition of smoking, fire or naked light*

9.3.2.74.1   The notice boards displaying the prohibition of smoking in accordance with 8.3.4 shall be clearly legible from either side of the vessel.

9.3.2.74.2   Notice boards indicating the circumstances under which the prohibition is applicable shall be fitted near the entrances to the spaces where smoking or the use of fire or naked light is not always prohibited.

9.3.2.74.3   Ashtrays shall be provided close to each exit of the accommodation and the wheelhouse.

9.3.2.75 to 9.3.2.91   *(Reserved)*

9.3.2.92   *Emergency exit*

Spaces the entrances or exits of which are likely to become partly or completely immersed in the damaged condition shall have an emergency exit which is situated not less than 0.10 m above the damage waterline. This requirement does not apply to forepeak and afterpeak.

9.3.2.93 to 9.3.2.99   *(Reserved)*

9.3.3   **Rules for construction of type N tank vessels**

The rules for construction of 9.3.3.0 to 9.3.3.99 apply to type N tank vessels.

9.3.3.0   *Materials of construction*

9.3.3.0.1   (a)   The vessel's hull and the cargo tanks shall be constructed of shipbuilding steel or other at least equivalent metal.

The independent cargo tanks may also be constructed of other materials, provided these have at least equivalent mechanical properties and resistance against the effects of temperature and fire.

(b)   Every part of the vessel including any installation and equipment which may come into contact with the cargo shall consist of materials which can neither be dangerously affected by the cargo nor cause decomposition of the cargo or react with it so as to form harmful or hazardous products. In case it has not been possible to examine this during classification and inspection of the vessel a relevant reservation shall be entered in the vessel substance list according to 1.16.1.2.5.

(c)   Inside venting piping shall be protected against corrosion.

9.3.3.0.2   Except where explicitly permitted in 9.3.3.03 below or in the certificate of approval, the use of wood, aluminium alloys or plastic materials within the cargo area is prohibited.

9.3.3.0.3   (a)   The use of wood, aluminium alloys or plastic materials within the cargo area is only permitted for:

–   gangways and external ladders;

–   movable items of equipment (aluminium gauging rods are, however, permitted provided that they are fitted with brass feet or protected in another way to avoid sparking);

–   chocking of cargo tanks which are independent of the vessel's hull and chocking of installations and equipment;

–   masts and similar round timber;

–   engine parts;

–   parts of the electrical installation;

–   loading and unloading appliances;

–   lids of boxes which are placed on the deck.

(b)   The use of wood or plastic materials within the cargo area is only permitted for:

–   supports and stops of any kind.

(c)   The use of plastic materials or rubber within the cargo area is only permitted for:

–   coating of cargo tanks and of piping for loading and unloading;

–   all kinds of gaskets (e.g. for dome or hatch covers);

–   electric cables;

–   hose assemblies for loading and unloading;

–   insulation of cargo tanks and of piping for loading and unloading;

–   photo–optical copies of the certificate of approval according to 8.1.2.6 or 8.1.2.7.

(d)   All permanently fitted materials in the accommodation or wheelhouse, with the exception of furniture, shall not readily ignite. They shall not evolve fumes or toxic gases in dangerous quantities, if involved in a fire.

9.3.3.0.4   The paint used in the cargo area shall not be liable to produce sparks in case of impact.

9.3.3.0.5   The use of plastic material for vessel's boats is permitted only if the material does not readily ignite.

9.3.3.1   **Vessel record**

NOTE: For the purpose of this paragraph, the term "owner" has the same meaning as in 1.16.0.

The vessel record shall be retained by the owner who shall be able to provide this documentation at the request of the competent authority and the recognized classification society.

The vessel record shall be maintained and updated throughout the life of the vessel and shall be retained for 6 months after the vessel is taken out of service.

Should a change of owner occur during the life of the vessel the vessel record shall be transferred to the new owner.

Copies of the vessel record or all necessary documents shall be made available on request to the competent authority for the issuance of the certificate of approval and for the recognized classification society or inspection body for first inspection, periodic inspection, special inspection or exceptional checks.

9.3.3.2 to 9.3.3.7   (*Reserved*)

9.3.3.8   *Classification*

9.3.3.8.1   The tank vessel shall be built under the survey of a recognised classification society and be classed in its highest class.

The vessel's highest class shall be continued. This shall be confirmed by an appropriate certificate issued by the recognized classification society (certificate of class).

The design pressure and the test pressure of cargo tanks shall be entered in the certificate.

If a vessel has cargo tanks with different valve opening pressures, the design and test pressures of each tank shall be entered in the certificate.

The recognized classification society shall draw up a vessel substance list mentioning all the dangerous goods accepted for carriage by the tank vessel (see also 1.16.1.2.5).

9.3.3.8.2   The cargo pump-rooms shall be inspected by a recognised classification society whenever the certificate of approval has to be renewed as well as during the third year of validity of the certificate of approval. The inspection shall comprise at least:

–   an inspection of the whole system for its condition, for corrosion, leakage or conversion works which have not been approved;

–   a checking of the condition of the gas detection system in the cargo pump-rooms.

Inspection certificates signed by the recognised classification society with respect to the inspection of the cargo pump-rooms shall be kept on board. The inspection certificates shall at least include particulars of the above inspection and the results obtained as well as the date of the inspection.

9.3.3.8.3   The condition of the gas detection system referred to in 9.3.3.52.3 shall be checked by a recognised classification society whenever the certificate of approval has to be renewed and during the third year of validity of the certificate of approval. A certificate signed by the recognised classification society shall be kept on board.

9.3.3.8.4 9.3.3.8.2 and 9.3.3.8.3, checking of the condition of the gas detection system, do not apply to open type N.

9.3.3.9 *(Reserved)*

9.3.3.10 *Protection against the penetration of gases*

9.3.3.10.1 The vessel shall be designed so as to prevent gases from penetrating into the accommodation and the service spaces.

9.3.3.10.2 Outside the cargo area, the lower edges of door-openings in the sidewalls of superstructures and the coaming of access hatches to under-deck spaces shall have a height of not less than 0.50 m above the deck.

This requirement need not be complied with if the wall of the superstructures facing the cargo area extends from one side of the ship to the other and has doors the sills of which have a height of not less than 0.50 m above the deck. The height of this wall shall be not less than 2.00 m. In this case, the lower edges of door-openings in the sidewalls of superstructures and the coamings of access hatches behind this wall shall have a height of not less than 0.10 m above the deck. The sills of engine room doors and the coamings of its access hatches shall, however, always have a height of not less than 0.50 m.

9.3.3.10.3 In the cargo area, the lower edges of door-openings in the sidewalls of superstructures shall have a height of not less than 0.50 m above the deck and the sills of hatches and ventilation openings of premises located under the deck shall have a height of not less than 0.50 m above the deck. This requirement does not apply to access openings to double-hull and double bottom spaces.

9.3.3.10.4 The bulwarks, foot-rails etc. shall be provided with sufficiently large openings which are located directly above the deck.

9.3.3.10.5 9.3.3.10.1 to 9.3.3.10.4 above do not apply to open type N.

9.3.3.11 *Hold spaces and cargo tanks*

9.3.3.11.1 (a) The maximum permissible capacity of a cargo tank shall be determined in accordance with the following table:

$L \times B \times H$ (m³)	Maximum permissible capacity of a cargo tank (m³)
up to 600	$L \times B \times H \times 0.3$
600 to 3 750	$180 + (L \times B \times H - 600) \times 0.0635$
> 3 750	380

Alternative constructions in accordance with 9.3.4 are permitted.
In the table above $L \times B \times H$ is the product of the main dimensions of the tank vessel in metres (according to the measurement certificate), where:

L = overall length of the hull in m;
B = extreme breadth of the hull in m;
H = shortest vertical distance between the top of the keel and the lowest point of the deck at the side of the vessel (moulded depth) within the cargo area in m;

where:

For trunk vessels, H shall be replaced by H', where H' shall be obtained from the following formula:

$$H' = H + \left(ht \times \frac{bt}{B} \times \frac{lt}{L} \right)$$

where:

ht = trunk height (distance between trunk deck and main deck measured on trunk side at L/2) in m;

bt = trunk breadth in m;

lt = trunk length in m.

(b) The relative density of the substances to be carried shall be taken into consideration in the design of the cargo tanks. The maximum relative density shall be indicated in the certificate of approval.

(c) When the vessel is provided with pressure tanks, these tanks shall be designed for a working pressure of 400 kPa (4 bar).

(d) For vessels with a length of not more than 50.00 m, the length of a cargo tank shall not exceed 10.00 m; and

For vessels with a length of more than 50.00 m, the length of a cargo tank shall not exceed 0.20 L.

This provision does not apply to vessels with independent built–in cylindrical tanks having a length to diameter ratio ≤ 7.

9.3.3.11.2 (a) The cargo tanks independent of the vessel's hull shall be fixed so that they cannot float.

Refrigerated cargo tank fastenings shall meet the requirements of a recognised classification society.

(b) The capacity of a suction well shall be limited to not more than 0.10 m³.

9.3.3.11.3 (a) The cargo tanks shall be separated by cofferdams of at least 0.60 m in width from the accommodation, engine rooms and service spaces outside the cargo area below deck or, if there are no such accommodation, engine room and service spaces, from the vessel's ends. Where the cargo tanks are installed in a hold space, a space of not less than 0.50 m shall be provided between such tanks and the end bulkheads of the hold space. In this case an insulated end bulkhead meeting the definition for Class "A-60" according to SOLAS 74, Chapter II-2, Regulation 3, shall be deemed equivalent to a cofferdam. For pressure cargo tanks, the 0.50 m distance may be reduced to 0.20 m.

(b) Hold spaces, cofferdams and cargo tanks shall be capable of being inspected.

(c) All spaces in the cargo area shall be capable of being ventilated. Means for checking their gas-free condition shall be provided.

9.3.3.11.4 The bulkheads bounding the cargo tanks, cofferdams and hold spaces shall be watertight. The cargo tanks and the bulkheads bounding the cargo area shall have no openings or penetrations below deck.

The bulkhead between the engine room and the cofferdam or service space in the cargo area or between the engine room and a hold space may be fitted with penetrations provided that they conform to the provisions of 9.3.3.17.5.

The bulkhead between the cargo tank and the cargo pump-room below deck may be fitted with penetrations provided that they conform to the provisions of 9.3.3.17.6. The bulkheads between the cargo tanks may be fitted with penetrations provided that the loading and unloading piping are fitted with shut-off devices in the cargo tank from which they come. These pipes shall be fitted at least 0.60m above the bottom. The shut-off devices shall be capable of being activated from the deck.

9.3.3.11.5   Double-hull spaces and double bottoms in the cargo area shall be arranged for being filled with ballast water only. Double bottoms may, however, be used as oil fuel tanks, provided they comply with the provisions of 9.3.3.32.

9.3.3.11.6   (a)   A cofferdam, the centre part of a cofferdam or another space below deck in the cargo area may be arranged as a service space, provided the bulkheads bounding the service space extend vertically to the bottom. This service space shall only be accessible from the deck.

           (b)   The service space shall be watertight with the exception of its access hatches and ventilation inlets.

           (c)   No piping for loading and unloading shall be fitted within the service space referred to under 9.3.3.11.4 above.

           Piping for loading and unloading may be fitted in the cargo pump-rooms below deck only when they conform to the provisions of 9.3.3.17.6.

9.3.3.11.7   Where independent cargo tanks are used, or for double-hull construction where the cargo tanks are integrated in the vessel's structure, the space between the wall of the vessel and wall of the cargo tanks shall be not less than 0.60 m.

The space between the bottom of the vessel and the bottom of the cargo tanks shall be not less than 0.50 m. The space may be reduced to 0.40 m under the pump sumps.

The vertical space between the suction well of a cargo tank and the bottom structures shall be not less than 0.10 m.

When a hull is constructed in the cargo area as a double hull with independent cargo tanks located in hold spaces, the above values are applicable to the double hull. If in this case the minimum values for inspections of independent tanks referred to in 9.3.3.11.9 are not feasible, it must be possible to remove the cargo tanks easily for inspection.

9.3.3.11.8   Where service spaces are located in the cargo area under deck, they shall be arranged so as to be easily accessible and to permit persons wearing protective clothing and breathing apparatus to safely operate the service equipment contained therein. They shall be designed so as to allow injured or unconscious personnel to be removed from such spaces without difficulties, if necessary by means of fixed equipment.

9.3.3.11.9   Cofferdams, double-hull spaces, double bottoms, cargo tanks, hold spaces and other accessible spaces within the cargo area shall be arranged so that they may be completely inspected and cleaned. The dimensions of openings except for those of double-hull spaces and double bottoms which do not have a wall adjoining the cargo tanks shall be sufficient to allow a person wearing breathing apparatus to enter or leave the space without difficulties. These openings shall have a minimum cross-section of 0.36 m^2 and a minimum side length of 0.50 m. They shall be designed so as to allow injured or unconscious personnel to be removed from the bottom of such a space without difficulties, if necessary by means of fixed equipment. In these spaces the free penetration width shall not be less than 0.50 m in the sector intended for the penetration. In double bottoms this distance may be reduced to 0.45 m.

Cargo tanks may have circular openings with a diameter of not less than 0.68 m.

9.3.3.11.10   9.3.3.11.6 (c) above does not apply to open type N.

9.3.3.12   *Ventilation*

9.3.3.12.1   Each hold space shall have two openings the dimensions and location of which shall be such as to permit effective ventilation of any part of the hold space. If there are no such openings, it shall be possible to fill the hold spaces with inert gas or dry air.

9.3.3.12.2   Double-hull spaces and double bottoms within the cargo area which are not arranged for being filled with ballast water, hold spaces and cofferdams shall be provided with ventilation systems.

9.3.3.12.3   Any service spaces located in the cargo area below deck shall be provided with a system of forced ventilation with sufficient power for ensuring at least 20 changes of air per hour based on the volume of the space.

The ventilation exhaust ducts shall be located up to 50 mm above the bottom of the service space. The fresh air inlets shall be located in the upper part; they shall be not less than 2.00 m above the deck, not less than 2.00 m from the openings of the cargo tanks and not less than 6.00 m from the outlets of safety valves.

The extension pipes which may be necessary may be of the hinged type.

On board open type N vessels other suitable installations without ventilator fans shall be sufficient.

9.3.3.12.4   Ventilation of accommodation and service spaces shall be possible.

9.3.3.12.5   Ventilators used in the cargo area shall be designed so that no sparks may be emitted on contact of the impeller blades with the housing and no static electricity may be generated.

9.3.3.12.6   Notice boards shall be fitted at the ventilation inlets indicating the conditions under which they shall be closed. Any ventilation inlets of accommodation and service spaces leading outside shall be fitted with fire flaps. Such ventilation inlets shall be located not less than 2.00 m from the cargo area.

Ventilation inlets of service spaces in the cargo area below deck may be located within such area.

9.3.3.12.7   Flame-arresters prescribed in 9.3.3.20.4, 9.3.3.22.4, 9.3.3.22.5 and 9.3.3.26.4 shall be of a type approved for this purpose by the competent authority.

9.3.3.12.8   9.3.3.12.5, 9.3.3.12.6 and 9.3.3.12.7 above do not apply to open type N.

9.3.3.13   *Stability (general)*

9.3.3.13.1   Proof of sufficient stability shall be furnished. This proof is not required for single hull vessels with cargo tanks the width of which is not more than 0.70 B.

9.3.3.13.2   The basic values for the stability calculation – the vessel's lightweight and location of the centre of gravity – shall be determined either by means of an inclining experiment or by detailed mass and moment calculation. In the latter case the lightweight of the vessel shall be checked by means of a lightweight test with a tolerance limit of ± 5% between the mass determined by calculation and the displacement determined by the draught readings.

9.3.3.13.3   Proof of sufficient intact stability shall be furnished for all stages of loading and unloading and for the final loading condition for all the relative densities of the substances transported contained in the vessel substance list according to 1.16.1.2.5.

For every loading operation, taking account of the actual fillings and floating position of cargo tanks, ballast tanks and compartments, drinking water and sewage tanks and tanks containing products for the operation of the vessel, the vessel shall comply with the intact and damage stability requirements.

Intermediate stages during operations shall also be taken into consideration.

The proof of sufficient stability shall be shown for every operating, loading and ballast condition in the stability booklet, to be approved by the recognized classification society, which classes the vessel. If it is unpractical to pre-calculate the operating, loading and ballast conditions, a loading instrument approved by the recognised classification society which classes the vessel shall be installed and used which contains the contents of the stability booklet.

NOTE: A stability booklet shall be worded in a form comprehensible for the responsible master and containing the following details:

General description of the vessel:

— *General arrangement and capacity plans indicating the assigned use of compartments and spaces (cargo tanks, stores, accommodation, etc.);*

— *A sketch indicating the position of the draught marks referring to the vessel's perpendiculars;*

— *A scheme for ballast/bilge pumping and overflow prevention systems;*

— *Hydrostatic curves or tables corresponding to the design trim, and, if significant trim angles are foreseen during the normal operation of the vessel, curves or tables corresponding to such range of trim are to be introduced;*

— *Cross curves or tables of stability calculated on a free trimming basis, for the ranges of displacement and trim anticipated in normal operating conditions, with an indication of the volumes which have been considered buoyant;*

— *Tank sounding tables or curves showing capacities, centres of gravity, and free surface data for all cargo tanks, ballast tanks and compartments, drinking water and sewage water tanks and tanks containing products for the operation of the vessel;*

— *Lightship data (weight and centre of gravity) resulting from an inclining test or deadweight measurement in combination with a detailed mass balance or other acceptable measures. Where the above–mentioned information is derived from a sister vessel, the reference to this sister vessel shall be clearly indicated, and a copy of the approved inclining test report relevant to this sister vessel shall be included;*

— *A copy of the approved test report shall be included in the stability booklet;*

–   *Operating loading conditions with all relevant details, such as:*

–   *Lightship data, tank fillings, stores, crew and other relevant items on board (mass and centre of gravity for each item, free surface moments for liquid loads);*
–   *Draughts amidships and at perpendiculars;*
–   *Metacentric height corrected for free surfaces effect;*
–   *Righting lever values and curve;*
–   *Longitudinal bending moments and shear forces at read–out points;*
–   *Information about openings (location, type of tightness, means of closure); and*
–   *Information for the master;*

–   *Calculation of the influence of ballast water on stability with information on whether fixed level gauges for ballast tanks and compartments have to be installed or whether the ballast tanks or compartments shall only be completely full or completely empty when underway.*

9.3.3.13.4   Floatability after damage shall be proved for the most unfavourable loading condition. For this purpose, calculated proof of sufficient stability shall be established for critical intermediate stages of flooding and for the final stage of flooding.

9.3.3.14   *Stability (intact)*

9.3.3.14.1   For vessels with independent cargo tanks and for double-hull constructions with cargo tanks integrated in the frames of the vessel, the requirements for intact stability resulting from the damage stability calculation shall be fully complied with.

9.3.3.14.2   For vessels with cargo tanks of more than 0.70 B in width, proof shall be furnished that the following stability requirements have been complied with:

(a)   In the positive area of the righting lever curve up to immersion of the first non-watertight opening there shall be a righting lever (GZ) of not less than 0.10 m;

(b)   The surface of the positive area of the righting lever curve up to immersion of the first non-watertight opening and in any event up to an angle of heel $\leq 27°$ shall not be less than 0.024 m.rad;

(c)   The metacentric height (GM) shall be not less than 0.10 m.

These conditions shall be met bearing in mind the influence of all free surfaces in tanks for all stages of loading and unloading.

9.3.3.15   *Stability (damaged condition)*

9.3.3.15.1   For vessels with independent cargo tanks and for double-hull vessels with cargo tanks integrated in the construction of the vessel, the following assumptions shall be taken into consideration for the damaged condition:

(a)   The extent of side damage is as follows:

longitudinal extent: at least 0.10 L, but not less than 5.00 m;
transverse extent: 0.59 m inboard from the vessel's side at right angles to the centreline at the level corresponding to the maximum draught, or when applicable, the distance allowed by section 9.3.4, reduced by 0.01 m;
vertical extent: from the base line upwards without limit.

(b) The extent of bottom damage is as follows:

longitudinal extent:	at least 0.10 L, but not less than 5.00 m;
transverse extent:	3.00 m;
vertical extent:	from the base 0.49 m upwards, the sump excepted.

(c) Any bulkheads within the damaged area shall be assumed damaged, which means that the location of bulkheads shall be chosen so as to ensure that the vessel remains afloat after the flooding of two or more adjacent compartments in the longitudinal direction.

The following provisions are applicable:

- For bottom damage, adjacent athwartship compartments shall also be assumed as flooded;

- The lower edge of any non-watertight openings (e.g. doors, windows, access hatchways) shall, at the final stage of flooding, be not less than 0.10 m above the damage waterline;

- In general, permeability shall be assumed to be 95%. Where an average permeability of less than 95% is calculated for any compartment, this calculated value obtained may be used.

However, the following minimum values shall be used:

– engine rooms:	85%;
– accommodation:	95%;
– double bottoms, oil fuel tanks, ballast tanks, etc., depending on whether, according to their function, they have to be assumed as full or empty for the vessel floating at the maximum permissible draught:	0% or 95%.

For the main engine room only the one–compartment standard need be taken into account, i.e. the end bulkheads of the engine room shall be assumed as not damaged.

9.3.3.15.2 For the intermediate stage of flooding the following criteria have to be fulfilled:

GZ >= 0.03m

Range of positive GZ: 5°.

At the stage of equilibrium (final stage of flooding), the angle of heel shall not exceed 12°. Non-watertight openings shall not be flooded before reaching the stage of equilibrium. If such openings are immersed before that stage, the corresponding spaces shall be considered as flooded for the purpose of the stability calculation.

The positive range of the righting lever curve beyond the stage of equilibrium shall have a righting lever of ≥ 0.05 m in association with an area under the curve of ≥ 0.0065 m.rad. The minimum values of stability shall be satisfied up to immersion of the first non-watertight opening and in any event up to an angle of heel $\leq 27°$. If non-watertight openings are immersed before that stage, the corresponding spaces shall be considered as flooded for the purposes of stability calculation.

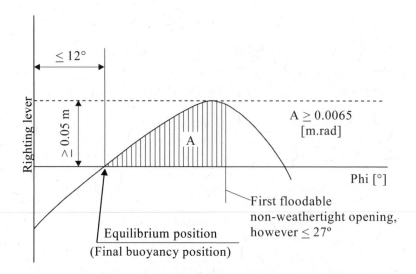

9.3.3.15.3   If openings through which undamaged compartments may additionally become flooded are capable of being closed watertight, the closing appliances shall be marked accordingly.

9.3.3.15.4   Where cross- or down-flooding openings are provided for reduction of unsymmetrical flooding, the time for equalization shall not exceed 15 minutes, if during the intermediate stages of flooding sufficient stability has been proved.

9.3.3.16   ***Engine rooms***

9.3.3.16.1   Internal combustion engines for the vessel's propulsion as well as internal combustion engines for auxiliary machinery shall be located outside the cargo area. Entrances and other openings of engine rooms shall be at a distance of not less than 2.00 m from the cargo area.

9.3.3.16.2   The engine rooms shall be accessible from the deck; the entrances shall not face the cargo area. Where the doors are not located in a recess whose depth is at least equal to the door width, the hinges shall face the cargo area.

9.3.3.16.3   The last sentence of 9.3.3.16.2 does not apply to oil separator or supply vessels.

9.3.3.17   ***Accommodation and service spaces***

9.3.3.17.1   Accommodation spaces and the wheelhouse shall be located outside the cargo area forward of the fore vertical plane or abaft the aft vertical plane bounding the part of the cargo area below deck. Windows of the wheelhouse which are located not less than 1.00 m above the bottom of the wheelhouse may tilt forward.

9.3.3.17.2   Entrances to spaces and openings of superstructures shall not face the cargo area. Doors opening outward and not located in a recess whose depth is at least equal to the width of the doors shall have their hinges face the cargo area.

9.3.3.17.3   Entrances from the deck and openings of spaces facing the weather shall be capable of being closed. The following instruction shall be displayed at the entrance of such spaces:

Do not open during loading, unloading and degassing
without the permission of the master.
Close immediately.

9.3.3.17.4   Entrances and windows of superstructures and accommodation spaces which can be opened as well as other openings of these spaces shall be located not less than 2.00 m from the cargo area. No wheelhouse doors and windows shall be located within 2.00 m from the cargo area, except where there is no direct connection between the wheelhouse and the accommodation.

9.3.3.17.5   (a)   Driving shafts of the bilge or ballast pumps may penetrate through the bulkhead between the service space and the engine room, provided the arrangement of the service space is in compliance with 9.3.3.11.6.

(b)   The penetration of the shaft through the bulkhead shall be gastight and shall have been approved by a recognised classification society.

(c)   The necessary operating instructions shall be displayed.

(d)   Penetrations through the bulkhead between the engine room and the service space in the cargo area and the bulkhead between the engine room and the hold spaces may be provided for electrical cables, hydraulic lines and piping for measuring, control and alarm systems, provided that the penetrations have been approved by a recognised classification society. The penetrations shall be gastight. Penetrations through a bulkhead with an "A-60" fire protection insulation according to SOLAS 74, Chapter II-2, Regulation 3, shall have an equivalent fire protection.

(e)   Pipes may penetrate the bulkhead between the engine room and the service space in the cargo area provided that these are pipes between the mechanical equipment in the engine room and the service space which do not have any openings within the service space and which are provided with shut-off devices at the bulkhead in the engine room.

(f)   Notwithstanding 9.3.3.11.4, pipes from the engine room may pass through the service space in the cargo area or a cofferdam or a hold space or a double-hull space to the outside provided that within the service space or cofferdam or hold space or double-hull space they are of the thick-walled type and have no flanges or openings.

(g)   Where a driving shaft of auxiliary machinery penetrates through a wall located above the deck the penetration shall be gastight.

9.3.3.17.6   A service space located within the cargo area below deck shall not be used as a cargo pump-room for the loading and unloading system, except where:

–   the cargo pump-room is separated from the engine room or from service spaces outside the cargo area by a cofferdam or a bulkhead with an "A-60" fire protection insulation according to SOLAS 74, Chapter II-2, Regulation 3, or by a service space or a hold space;

–   the "A-60" bulkhead required above does not include penetrations referred to in 9.3.3.17.5 (a);

–   ventilation exhaust outlets are located not less than 6.00 m from entrances and openings of the accommodation and service spaces outside the cargo area;

–   the access hatches and ventilation inlets can be closed from the outside;

–   all piping for loading and unloading as well as those of stripping systems are provided with shut-off devices at the pump suction side in the cargo pump-room immediately at the bulkhead. The necessary operation of the control devices in the pump-room, starting of pumps and necessary control of the liquid flow rate shall be effected from the deck;

–   the bilge of the cargo pump-room is equipped with a gauging device for measuring the filling level which activates a visual and audible alarm in the wheelhouse when liquid is accumulating in the cargo pump-room bilge;

–   the cargo pump-room is provided with a permanent gas detection system which automatically indicates the presence of explosive gases or lack of oxygen by means of direct-measuring sensors and which actuates a visual and audible alarm when the gas concentration has reached 20% of the lower explosive limit. The sensors of this system shall be placed at suitable positions at the bottom and directly below the deck.

Measurement shall be continuous.

The audible and visual alarms are installed in the wheelhouse and in the cargo pump-room and, when the alarm is actuated, the loading and unloading system is shut down. Failure of the gas detection system shall be immediately signalled in the wheelhouse and on deck by means of audible and visual alarms;

–   the ventilation system prescribed in 9.3.3.12.3 has a capacity of not less than 30 changes of air per hour based on the total volume of the service space.

9.3.3.17.7   The following instruction shall be displayed at the entrance of the cargo pump-room:

**Before entering the cargo pump-room check whether
it is free from gases and contains sufficient oxygen.
Do not open doors and entrance openings without
the permission of the master.
Leave immediately in the event of alarm.**

9.3.3.17.8   9.3.3.17.5 (g), 9.3.3.17.6 and 9.3.3.17.7 do not apply to open type N.

9.3.3.17.2, last sentence, 9.3.3.17.3, last sentence and 9.3.3.17.4 do not apply to oil separator and supply vessels.

9.3.3.18   *Inerting facility*

In cases in which inerting or blanketing of the cargo is prescribed, the vessel shall be equipped with an inerting system.

This system shall be capable of maintaining a permanent minimum pressure of 7 kPa (0.07 bar) in the spaces to be inerted. In addition, the inerting system shall not increase the pressure in the cargo tank to a pressure greater than that at which the pressure valve is regulated. The set pressure of the vacuum-relief valve shall be 3.5 kPa (0.035 bar).

A sufficient quantity of inert gas for loading or unloading shall be carried or produced on board if it is not possible to obtain it on shore. In addition, a sufficient quantity of inert gas to offset normal losses occurring during carriage shall be on board.

The premises to be inerted shall be equipped with connections for introducing the inert gas and monitoring systems so as to ensure the correct atmosphere on a permanent basis.

When the pressure or the concentration of inert gas in the gaseous phase falls below a given value, this monitoring system shall activate an audible and visible alarm in the wheelhouse. When the wheelhouse is unoccupied, the alarm shall also be perceptible in a location occupied by a crew member.

9.3.3.19   (*Reserved*)

9.3.3.20   *Arrangement of cofferdams*

9.3.3.20.1   Cofferdams or cofferdam compartments remaining once a service space has been arranged in accordance with 9.3.3.11.6 shall be accessible through an access hatch.

9.3.3.20.2   Cofferdams shall be capable of being filled with water and emptied by means of a pump. Filling shall be effected within 30 minutes. These requirements are not applicable when the bulkhead between the engine room and the cofferdam has an "A-16" fire protection insulation according to SOLAS 74, Chapter II-2, Regulation 3.

The cofferdams shall not be fitted with inlet valves.

9.3.3.20.3   No fixed pipe shall permit connection between a cofferdam and other piping of the vessel outside the cargo area.

9.3.3.20.4   When the list of substances on the vessel according to 1.16.1.2.5 contains substances for which protection against explosion is required in column (17) of Table C of Chapter 3.2, the ventilation openings of cofferdams shall be fitted with a flame-arrester withstanding a deflagration.

9.3.3.20.5   9.3.3.20.4 above does not apply to open type N.

9.3.3.20.2 above does not apply to oil separator and supply vessels.

9.3.3.21   *Safety and control installations*

9.3.3.21.1   Cargo tanks shall be provided with the following equipment:

(a)   a mark inside the tank indicating the liquid level of 97%;

(b)   a level gauge;

(c)   a level alarm device which is activated at the latest when a degree of filling of 90% is reached;

(d)   a high level sensor for actuating the facility against overflowing when a degree of filling of 97.5% is reached;

(e)   an instrument for measuring the pressure of the vapour phase inside the cargo tank;

(f)   an instrument for measuring the temperature of the cargo if in column (9) of Table C of Chapter 3.2 a heating installation is required or if in column (20) a possibility of heating the cargo is required or if a maximum temperature is indicated;

(g)   a connection for a closed-type or partly closed-type sampling device, and/or at least one sampling opening as required in column (13) of Table C of Chapter 3.2.

9.3.3.21.2   When the degree of filling in per cent is determined, an error of not more than 0.5% is permitted. It shall be calculated on the basis of the total cargo tank capacity including the expansion trunk.

9.3.3.21.3   The level gauge shall allow readings from the control position of the shut-off devices of the particular cargo tank. The permissible maximum filling levels of 95% and 97%, as given in the list of substances, shall be marked on each level gauge.

Permanent reading of the overpressure and vacuum shall be possible from a location from which loading or unloading operations may be interrupted. The permissible maximum overpressure and vacuum shall be marked on each level gauge.

Readings shall be possible in all weather conditions.

9.3.3.21.4   The level alarm device shall give a visual and audible warning on board when actuated. The level alarm device shall be independent of the level gauge.

9.3.3.21.5   (a)   The high level sensor referred to in 9.3.3.21.1 (d) above shall give a visual and audible alarm on board and at the same time actuate an electrical contact which in the form of a binary signal interrupts the electric current loop provided and fed by the shore facility, thus initiating measures at the shore facility against overflowing during loading operations. The signal shall be transmitted to the shore facility via a watertight two–pin plug of a connector device in accordance with standard EN 60309-2:1999 + A1:2007 + A2:2012 for direct current of 40 to 50 volts, identification colour white, position of the nose 10 h.

                The plug shall be permanently fitted to the vessel close to the shore connections of the loading and unloading piping.

                The high level sensor shall also be capable of switching off the vessel's own discharging pump.

                The high level sensor shall be independent of the level alarm device, but it may be connected to the level gauge.

  (b)   On board oil separator vessels the sensor referred to in 9.3.3.21.1 (d) shall activate a visual and audible alarm and switch off the pump used to evacuate bilge water.

  (c)   Supply vessels and other vessels which may be delivering products required for operation shall be equipped with a transshipment facility compatible with European standard EN 12827:1999 and a rapid closing device enabling refuelling to be interrupted. It shall be possible to actuate this rapid closing device by means of an electrical signal from the overflow prevention system. The electrical circuits actuating the rapid closing device shall be secured according to the quiescent current principle or other appropriate error detection measures. The state of operation of electrical circuits which cannot be controlled using the quiescent current principle shall be capable of being easily checked.

                It shall be possible to actuate the rapid closing device independently of the electrical signal.

                The rapid closing device shall actuate a visual and an audible alarm on board.

  (d)   During discharging by means of the on-board pump, it shall be possible for the shore facility to switch it off. For this purpose, an independent intrinsically safe power line, fed by the vessel, shall be switched off by the shore facility by means of an electrical contact.

                It shall be possible for the binary signal of the shore facility to be transmitted via a watertight two–pole socket or a connector device in accordance with standard EN 60309-2:1999 + A1:2007 + A2:2012, for direct current of 40 to 50 volts, identification colour white, position of the nose 10 h.

                This socket shall be permanently fitted to the vessel close to the shore connections of the unloading piping.

9.3.3.21.6   The visual and audible signals given by the level alarm device shall be clearly distinguishable from those of the high level sensor.

The visual alarm shall be visible at each control position on deck of the cargo tank stop valves. It shall be possible to easily check the functioning of the sensors and electric circuits or these shall be intrinsically safe apparatus.

9.3.3.21.7   When the pressure or temperature exceeds a set value, instruments for measuring the vacuum or overpressure of the gaseous phase in the cargo tank or the temperature of the cargo, shall activate a visual and audible alarm in the wheelhouse. When the wheelhouse is unoccupied, the alarm shall also be perceptible in a location occupied by a crew member.

When the pressure exceeds the set value during loading and unloading, the instrument for measuring the pressure shall, by means of the plug referred to in 9.3.3.21.5, initiate simultaneously an electrical contact which shall put into effect measures to interrupt the loading and unloading operation. If the vessel's own discharge pump is used, it shall be switched off automatically.

The instrument for measuring the overpressure or vacuum shall activate the alarm at latest when an overpressure is reached equal to 1.15 times the opening pressure of the pressure relief device, or a vacuum pressure equal to the construction vacuum pressure but not exceeding 5 kPa. The maximum allowable temperature is indicated in column (20) of Table C of Chapter 3.2. The sensors for the alarms mentioned in this paragraph may be connected to the alarm device of the sensor.

When it is prescribed in column (20) of Table C of Chapter 3.2 the instrument for measuring the overpressure of the gaseous phase shall activate a visible and audible alarm in the wheelhouse when the overpressure exceeds 40 kPa during the voyage. When the wheelhouse is unoccupied, the alarm shall also be perceptible in a location occupied by a crew member. It shall be possible to read the gauges in direct proximity to the control for the water spray system.

9.3.3.21.8   Where the control elements of the shut-off devices of the cargo tanks are located in a control room, it shall be possible to stop the loading pumps and read the level gauges in the control room, and the visual and audible warning given by the level alarm device, the high level sensor referred to in 9.3.3.21.1 (d) and the instruments for measuring the pressure and temperature of the cargo shall be noticeable in the control room and on deck.

Satisfactory monitoring of the cargo area shall be ensured from the control room.

9.3.3.21.9   9.3.3.21.1 (e), 9.3.3.21.7 as regards measuring the pressure, do not apply to open type N with flame-arrester and to open type N.

9.3.3.21.1 (b), (c) and (g), 9.3.3.21.3 and 9.3.3.21.4 do not apply to oil separator and supply vessels.

A flame arrester plate stack in sampling openings is not required on board open type N tank vessels.

9.3.3.21.1 (f) and 9.3.3.21.7 do not apply to supply vessels.

9.3.3.21.5 (a) does not apply to oil separator vessels.

9.3.3.21.10   When refrigerated substances are carried the opening pressure of the safety system shall be determined by the design of the cargo tanks. In the event of the transport of substances that must be carried in a refrigerated state the opening pressure of the safety system shall be not less than 25 kPa (0.25 bar) greater than the maximum pressure calculated according to 9.3.3.27.

9.3.3.22   *Cargo tank openings*

9.3.3.22.1   (a)   Cargo tank openings shall be located on deck in the cargo area.

(b)   Cargo tank openings with a cross-section of more than 0.10 m^2 and openings of safety devices for preventing overpressures shall be located not less than 0.50 m above deck.

9.3.3.22.2   Cargo tank openings shall be fitted with gastight closures capable of withstanding the test pressure in accordance with 9.3.3.23.2.

9.3.3.22.3   Closures which are normally used during loading or unloading operations shall not cause sparking when operated.

9.3.3.22.4   (a)   Each cargo tank or group of cargo tanks connected to a common venting piping shall be fitted with safety devices for preventing unacceptable overpressures or vacuums.

These safety devices shall be as follows:

for the open N type:

–   safety devices designed to prevent any accumulation of water and its penetration into the cargo tanks;

for the open N type with flame-arresters:

–   safety equipment fitted with flame-arresters capable of withstanding steady burning and designed to prevent any accumulation of water and its penetration into the cargo tank;

for the closed N type:

–   safety devices for preventing unacceptable overpressure or vacuum. Where anti-explosion protection is required in column (17) of Table C of Chapter 3.2, the vacuum valve shall be fitted with a flame arrester capable of withstanding a deflagration and the pressure relief valve with a high-velocity vent valve acting as a flame arrester capable of withstanding steady burning. Gases shall be discharged upwards. The opening pressure of the high-velocity vent valve and the opening pressure of the vacuum valve shall be permanently marked on the valves;

–   a connection for the safe return ashore of gases expelled during loading;

–   a device for the safe depressurization of the tanks. When the list of substances on the vessel according to 1.16.1.2.5 contains substances for which protection against explosion is required in column (17) of Table C of Chapter 3.2, this device shall include at least a fire–resistant flame arrester and a stop valve which clearly indicates whether it is open or shut.

(b)   The outlets of high-velocity vent valves shall be located not less than 2.00 m above the deck and at a distance of not less than 6.00 m from the accommodation and from the service spaces outside the cargo area. This height may be reduced when within a radius of 1.00 m round the outlet of the high-velocity vent valve, there is no equipment, no work is being carried out and signs indicate the area. The setting of the high-velocity vent valves shall be such that during the transport operation they do not blow off until the maximum permissible working pressure of the cargo tanks is reached.

9.3.3.22.5   (a)   Insofar as anti-explosion protection is prescribed in column (17) of Table C of Chapter 3.2, venting piping connecting two or more cargo tanks shall be fitted, at the connection to each cargo tank, with a flame arrester with a fixed or spring-loaded plate stack, capable of withstanding detonation. This equipment may consist of:

(i)   a flame arrester fitted with a fixed plate stack, where each cargo tank is fitted with a vacuum valve capable of withstanding a deflagration and a high-velocity vent valve capable of withstanding steady burning;

(ii) a flame arrester fitted with a spring-loaded plate stack, where each cargo tank is fitted with a vacuum valve capable of withstanding a deflagration;

(iii) a flame arrester with a fixed or spring-loaded plate stack;

(iv) a flame arrester with a fixed plate stack, where the pressure measurement device is fitted with an alarm system in accordance with 9.3.3.21.7;

(v) a flame arrester with a spring-loaded plate stack, where the pressure measurement device is fitted with an alarm system in accordance with 9.3.3.21.7.

Only substances which do not mix and which do not react dangerously with each other may be carried simultaneously in cargo tanks connected to a common venting piping;

or

(b) Insofar as anti-explosion protection is prescribed in column (17) of Table C of Chapter 3.2, venting piping connecting two or more cargo tanks shall be fitted, at the connection to each cargo tank, with a pressure/vacuum valve incorporating a flame arrester capable of withstanding a detonation/deflagration so that any gas released is removed by the venting piping;

Only substances which do not mix and which do not react dangerously with each other may be carried simultaneously in cargo tanks connected to a common venting piping;

or

(c) Insofar as anti-explosion protection is prescribed in column (17) of Table C of Chapter 3.2, an independent venting piping for each cargo tank, fitted with a vacuum valve incorporating a flame arrester capable of withstanding a deflagration and a high-velocity vent valve incorporating a flame arrester capable of withstanding steady burning. Several different substances may be carried simultaneously;

or

(d) Insofar as anti-explosion protection is prescribed in column (17) of Table C of Chapter 3.2, venting piping connecting two or more cargo tanks shall be fitted, at the connection to each cargo tank, with a shut-off device capable of withstanding a detonation, where each cargo tank is fitted with a vacuum valve capable of withstanding a deflagration and a high-velocity vent valve capable of withstanding steady burning;

Only substances which do not mix and which do not react dangerously with each other may be carried simultaneously in cargo tanks connected to a common venting piping.

9.3.3.22.6   9.3.3.22.2, 9.3.3.22.4 (b) and 9.3.3.22.5 do not apply to open type N with flame-arrester and to open type N.

9.3.3.22.3 does not apply to open type N.

9.3.3.23   *Pressure tests*

9.3.3.23.1   The cargo tanks, residual cargo tanks, cofferdams, piping for loading and unloading, with the exception of discharge hoses shall be subjected to initial tests before being put into service and thereafter at prescribed intervals.

Where a heating system is provided inside the cargo tanks, the heating coils shall be subjected to initial tests before being put into service and thereafter at prescribed intervals.

9.3.3.23.2 The test pressure for the cargo tanks and residual cargo tanks shall be not less than 1.3 times the design pressure. The test pressure for the cofferdams and open cargo tanks shall be not less than 10 kPa (0.10 bar) gauge pressure.

9.3.3.23.3 The test pressure for piping for loading and unloading shall be not less than 1,000 kPa (10 bar) gauge pressure.

9.3.3.23.4 The maximum intervals for the periodic tests shall be 11 years.

9.3.3.23.5 The procedure for pressure tests shall comply with the provisions established by the competent authority or a recognised classification society.

9.3.3.24 *Regulation of cargo pressure and temperature*

9.3.3.24.1 Unless the entire cargo system is designed to resist the full effective vapour pressure of the cargo at the upper limits of the ambient design temperatures, the pressure of the tanks shall be kept below the permissible maximum set pressure of the safety valves, by one or more of the following means:

(a) a system for the regulation of cargo tank pressure using mechanical refrigeration;

(b) a system ensuring safety in the event of the heating or increase in pressure of the cargo. The insulation or the design pressure of the cargo tank, or the combination of these two elements, shall be such as to leave an adequate margin for the operating period and the temperatures expected; in each case the system shall be deemed acceptable by a recognised classification society and shall ensure safety for a minimum time of three times the operation period;

(c) other systems deemed acceptable by a recognised classification society.

9.3.3.24.2 The systems prescribed in 9.3.3.24.1 shall be constructed, installed and tested to the satisfaction of the recognised classification society. The materials used in their construction shall be compatible with the cargoes to be carried. For normal service, the upper ambient design temperature limits shall be:

air: +30° C;
water: +20° C.

9.3.3.24.3 The cargo storage system shall be capable of resisting the full vapour pressure of the cargo at the upper limits of the ambient design temperatures, whatever the system adopted to deal with the boil-off gas. This requirement is indicated by remark 37 in column (20) of Table C of Chapter 3.2.

9.3.3.25 *Pumps and piping*

9.3.3.25.1 (a) Pumps and accessory loading and unloading piping shall be located in the cargo area;

(b) Cargo pumps shall be capable of being shut down from the cargo area and from a position outside the cargo area;

(c) Cargo pumps situated on deck shall be located not less than 6.00 m from entrances to, or openings of, the accommodation and service spaces outside the cargo area.

9.3.3.25.2 (a) Piping for loading and unloading shall be independent of any other piping of the vessel. No cargo piping shall be located below deck, except those inside the cargo tanks and inside the cargo pump-room;

(b)   The piping for loading and unloading shall be arranged so that, after loading or unloading operations, the liquid remaining in these pipes may be safely removed and may flow either into the vessel's cargo tanks or the tanks ashore;

(c)   Piping for loading and unloading shall be clearly distinguishable from other piping, e.g. by means of colour marking;

(d)   *(Reserved);*

(e)   The shore connections shall be located not less than 6.00 m from the entrances to, or openings of, the accommodation and service spaces outside the cargo area;

(f)   Each shore connection of the venting piping and shore connections of the piping for loading and unloading, through which the loading or unloading operation is carried out, shall be fitted with a shut-off device. However, each shore connection shall be fitted with a blind flange when it is not in operation;

(g)   *(Deleted);*

(h)   Piping for loading and unloading, and venting piping, shall not have flexible connections fitted with sliding seals.

9.3.3.25.3   The distance referred to in 9.3.3.25.1 (c) and 9.3.3.25.2 (e) may be reduced to 3.00 m if a transverse bulkhead complying with 9.3.3.10.2 is situated at the end of the cargo area. The openings shall be provided with doors.

The following notice shall be displayed on the doors:

Do not open during loading and unloading without
the permission of the master.
Close immediately.

9.3.3.25.4   (a)   Every component of the piping for loading and unloading shall be electrically connected to the hull;

(b)   The piping for loading shall extend down to the bottom of the cargo tanks.

9.3.3.25.5   The stop valves or other shut-off devices of the piping for loading and unloading shall indicate whether they are open or shut.

9.3.3.25.6   The piping for loading and unloading shall have, at the test pressure, the required elasticity, leakproofness and resistance to pressure.

9.3.3.25.7   The piping for loading and unloading shall be fitted with pressure gauges at the outlet of the pumps. The permissible maximum overpressure or vacuum value shall be indicated on each measuring device. Readings shall be possible in all weather conditions.

9.3.3.25.8   (a)   When piping for loading and unloading are used for supplying the cargo tanks with washing or ballast water, the suctions of these pipes shall be located within the cargo area but outside the cargo tanks;

Pumps for tank washing systems with associated connections may be located outside the cargo area, provided the discharge side of the system is arranged in such a way that suction is not possible through that part;

A spring-loaded non-return valve shall be provided to prevent any gases from being expelled from the cargo area through the tank washing system.

(b)　A non-return valve shall be fitted at the junction between the water suction pipe and the cargo loading pipe.

9.3.3.25.9　The permissible loading and unloading flows shall be calculated. For open type N with flame-arrester and open type N the loading and unloading flows depend on the total cross-section of the exhaust ducts.

Calculations concerning the permissible maximum loading and unloading flows for each cargo tank or each group of cargo tanks, taking into account the design of the ventilation system. These calculations shall take into consideration the fact that in the event of an unforeseen cut–off of the vapour return piping of the shore facility, the safety devices of the cargo tanks will prevent pressure in the cargo tanks from exceeding the following values:

over pressure:　　　115% of the opening pressure of the high velocity vent valve.
vacuum pressure:　　not more than the construction vacuum pressure but not exceeding 5 kPa (0.05 bar).

The main factors to be considered are the following:

1.　Dimensions of the ventilation system of the cargo tanks;

2.　Gas formation during loading: multiply the largest loading flow by a factor of not less than 1.25;

3.　Density of the vapour mixture of the cargo based on 50% volume vapour of 50% volume air;

4.　Loss of pressure through ventilation pipes, valves and fittings. Account will be taken of a 30% clogging of the mesh of the flame-arrester;

5.　Chocking pressure of the safety valves.

The permissible maximum loading and unloading flows for each cargo tank or for each group of cargo tanks shall be given in an on-board instruction.

9.3.3.25.10　Compressed air generated outside the cargo area or wheelhouse can be used in the cargo area subject to the installation of a spring-loaded non-return valve to ensure that no gases can escape from the cargo area through the compressed air system into accommodation or service spaces outside the cargo area.

9.3.3.25.11　If the vessel is carrying several dangerous substances liable to react dangerously with each other, a separate pump with its own piping for loading and unloading shall be installed for each substance. The piping shall not pass through a cargo tank containing dangerous substances with which the substance in question is liable to react.

9.3.3.25.12　9.3.3.25.1 (a) and (c), 9.3.3.25.2 (a), last sentence and (e), 9.3.3.25.3 and 9.3.3.25.4 (a) do not apply to type N open unless the substance carried has corrosive properties (see column (5) of Table C of Chapter 3.2, hazard 8).

9.3.3.25.4 (b) does not apply to open type N.

9.3.3.25.2 (f), last sentence, 9.3.3.25.2 (g), 9.3.3.25.8 (a), last sentence and 9.3.3.25.10 do not apply to oil separator and supply vessels.

9.3.3.25.9 does not apply to oil separator vessels.

9.3.3.25.2 (h) does not apply to supply vessels.

9.3.3.26   *Receptacles for residual products and receptacles for slops*

9.3.3.26.1   If vessels are provided with a tank for residual products, it shall comply with the provisions of 9.3.3.26.3 and 9.3.3.26.4. Receptacles for residual products and receptacles for slops shall be located only in the cargo area. During filling of receptacles for residual products, means for collecting any leakage shall be placed under the filling connections.

9.3.3.26.2   Receptacles for slops shall be fire resistant and shall be capable of being closed with lids (drums with removable heads, code 1A2, ADR). The receptacles for slops shall be marked and easy to handle.

9.3.3.26.3   The maximum capacity of a tank for residual products is 30 m^3.

9.3.3.26.4   The tank for residual products shall be equipped with:

–   in the case of an open system:

  –   a device for ensuring pressure equilibrium;
  –   an ullage opening;
  –   connections, with stop valves, for pipes and hose assemblies;

–   in the case of a protected system:

  –   a device for ensuring pressure equilibrium, fitted with a flame-arrester capable of withstanding steady burning;
  –   an ullage opening;
  –   connections, with stop valves, for pipes and hose assemblies;

–   in the case of a closed system:

  –   a vacuum valve and a high-velocity vent valve.

    The high-velocity vent valve shall be so regulated that it does not open during carriage. This condition is met when the opening pressure of the valve meets the conditions required in column (10) of Table C of Chapter 3.2 for the substance to be carried. When anti-explosion protection is required in column (17) of Table C of Chapter 3.2, the vacuum valve shall be capable of withstanding deflagrations and the high-velocity vent valve steady burning;

–   a device for measuring the degree of filling;

–   connections, with stop valves, for pipes and hose assemblies.

Receptacles for residual products shall be equipped with:

  –   a connection enabling gases released during filling to be evacuated safely;
  –   a possibility of indicating the degree of filling;
  –   connections with shut-off devices, for pipes and hose assemblies.

Receptacles for residual products shall be connected to the venting piping of cargo tanks only for the time necessary to fill them in accordance with 7.2.4.15.2.

Receptacles for residual products and receptacles for slops placed on the deck shall be located at a minimum distance from the hull equal to one quarter of the vessel's breadth.

9.3.3.26.5   Paragraphs 9.3.3.26.1, 9.3.3.26.3 and 9.3.3.26.4 above do not apply to oil separator vessels.

9.3.3.27 *Refrigeration system*

9.3.3.27.1 The refrigeration system referred to in 9.3.3.24.1 (a) shall be composed of one or more units capable of keeping the pressure and temperature of the cargo at the upper limits of the ambient design temperatures at the prescribed level. Unless another means of regulating cargo pressure and temperature deemed satisfactory by a recognised classification society is provided, provision shall be made for one or more stand-by units with an output at least equal to that of the largest prescribed unit. A stand-by unit shall include a compressor, its engine, its control system and all necessary accessories to enable it to operate independently of the units normally used. Provision shall be made for a stand-by heat-exchanger unless the system's normal heat-exchanger has a surplus capacity equal to at least 25% of the largest prescribed capacity. It is not necessary to make provision for separate piping.

Cargo tanks, piping and accessories shall be insulated so that, in the event of a failure of all cargo refrigeration systems, the entire cargo remains for at least 52 hours in a condition not causing the safety valves to open.

9.3.3.27.2 The security devices and the connecting lines from the refrigeration system shall be connected to the cargo tanks above the liquid phase of the cargo when the tanks are filled to their maximum permissible degree of filling. They shall remain within the gaseous phase, even if the vessel has a list up to 12 degrees.

9.3.3.27.3 When several refrigerated cargoes with a potentially dangerous chemical reaction are carried simultaneously, particular care shall be given to the refrigeration systems so as to prevent any mixing of the cargoes. For the carriage of such cargoes, separate refrigeration systems, each including the full stand-by unit referred to in 9.3.3.27.1, shall be provided for each cargo. When, however, refrigeration is ensured by an indirect or combined system and no leak in the heat exchangers can under any foreseeable circumstances lead to the mixing of cargoes, no provision need be made for separate refrigeration units for the different cargoes.

9.3.3.27.4 When several refrigerated cargoes are not soluble in each other under conditions of carriage such that their vapour pressures are added together in the event of mixing, particular care shall be given to the refrigeration systems to prevent any mixing of the cargoes.

9.3.3.27.5 When the refrigeration systems require water for cooling, a sufficient quantity shall be supplied by a pump or pumps used exclusively for the purpose. This pump or pumps shall have at least two suction pipes, leading from two water intakes, one to port, the other to starboard. Provision shall be made for a stand-by pump with a satisfactory flow; this may be a pump used for other purposes provided that its use for supplying water for cooling does not impair any other essential service.

9.3.3.27.6 The refrigeration system may take one of the following forms:

(a) Direct system: the cargo vapours are compressed, condensed and returned to the cargo tanks. This system shall not be used for certain cargoes specified in Table C of Chapter 3.2. This requirement is indicated by remark 35 in column (20) of Table C of Chapter 3.2;

(b) Indirect system: the cargo or the cargo vapours are cooled or condensed by means of a coolant without being compressed;

(c) Combined system: the cargo vapours are compressed and condensed in a cargo/coolant heat-exchanger and returned to the cargo tanks. This system shall not be used for certain cargoes specified in Table C of Chapter 3.2. This requirement is indicated by remark 36 in column (20) of Table C of Chapter 3.2.

9.3.3.27.7　　All primary and secondary coolant fluids shall be compatible with each other and with the cargo with which they may come into contact. Heat exchange may take place either at a distance from the cargo tank, or by using cooling coils attached to the inside or the outside of the cargo tank.

9.3.3.27.8　　When the refrigeration system is installed in a separate service space, this service space shall meet the requirements of 9.3.3.17.6.

9.3.3.27.9　　For all cargo systems, the heat transmission coefficient as used for the determination of the holding time (7.2.4.16.16 and 7.2.4.16.17) shall be determined by calculation. Upon completion of the vessel, the correctness of the calculation shall be checked by means of a heat balance test. The calculation and test shall be performed under supervision by the recognized classification society which classified the vessel.

The heat transmission coefficient shall be documented and kept on board. The heat transmission coefficient shall be verified at every renewal of the certificate of approval.

9.3.3.27.10　　A certificate from a recognised classification society stating that 9.3.3.24.1 to 9.3.3.24.3, 9.3.3.27.1 and 9.3.3.27.4 above have been complied with shall be submitted together with the application for issue or renewal of the certificate of approval.

9.3.3.28　　*Water-spray system*

When water-spraying is required in column (9) of Table C of Chapter 3.2, a water-spray system shall be installed in the cargo area on deck for the purpose of cooling the tops of cargo tanks by spraying water over the whole surface so as to avoid safely the activation of the high-velocity vent valve at 10 kPa or as regulated.

The spray nozzles shall be so installed that the entire cargo deck area is covered and the gases released are precipitated safely.

The system shall be capable of being put into operation from the wheelhouse and from the deck. Its capacity shall be such that when all the spray nozzles are in operation, the outflow is not less than 50 litres per square metre of deck area and per hour.

9.3.3.29 and 9.3.3.30　　(*Reserved*)

9.3.3.31　　*Engines*

9.3.3.31.1　　Only internal combustion engines running on fuel with a flashpoint of more than 55 °C are allowed.

9.3.3.31.2　　Ventilation inlets of the engine room and, when the engines do not take in air directly from the engine room, air intakes of the engines shall be located not less than 2.00 m from the cargo area.

9.3.3.31.3　　Sparking shall not be possible within the cargo area.

9.3.3.31.4　　The surface temperature of the outer parts of engines used during loading or unloading operations, as well as that of their air inlets and exhaust ducts shall not exceed the allowable temperature according to the temperature class of the substances carried. This provision does not apply to engines installed in service spaces provided the provisions of 9.3.3.52.3 are fully complied with.

9.3.3.31.5　　The ventilation in the closed engine room shall be designed so that, at an ambient temperature of 20 °C, the average temperature in the engine room does not exceed 40 °C.

9.3.3.31.6　　9.3.3.31.2 above does not apply to oil separator or supply vessels.

9.3.3.32   *Oil fuel tanks*

9.3.3.32.1   Where the vessel is provided with hold spaces, the double bottoms within these spaces may be arranged as oil fuel tanks, provided their depth is not less than 0.6 m.

Oil fuel pipes and openings of such tanks are not permitted in the hold space.

9.3.3.32.2   The open ends of the air pipes of each oil fuel tank shall extend to 0.5 m above the open deck. These open ends and the open ends of overflow pipes leading to the deck shall be provided with a protective device consisting of a gauze diaphragm or a perforated plate.

9.3.3.33   (*Reserved*)

9.3.3.34   *Exhaust pipes*

9.3.3.34.1   Exhaust shall be evacuated from the vessel into the open air either upwards through an exhaust pipe or through the shell plating. The exhaust outlet shall be located not less than 2.00 m from the cargo area. The exhaust pipes of engines shall be arranged so that the exhausts are led away from the vessel. The exhaust pipes shall not be located within the cargo area.

9.3.3.34.2   Exhaust pipes shall be provided with a device preventing the escape of sparks, e.g. spark arresters.

9.3.3.34.3   The distance prescribed in 9.3.3.34.1 above does not apply to oil separator or supply vessels.

9.3.3.35   *Bilge pumping and ballasting arrangements*

9.3.3.35.1   Bilge and ballast pumps for spaces within the cargo area shall be installed within such area.

This provision does not apply to:

–   double-hull spaces and double bottoms which do not have a common boundary wall with the cargo tanks;

–   cofferdams, double-hull, double bottom and hold spaces where ballasting is carried out using the piping of the fire-fighting system in the cargo area and bilge–pumping is performed using eductors.

9.3.3.35.2   Where the double bottom is used as a liquid oil fuel tank, it shall not be connected to the bilge piping system.

9.3.3.35.3   Where the ballast pump is installed in the cargo area, the standpipe and its outboard connection for suction of ballast water shall be located within the cargo area but outside the cargo tanks.

9.3.3.35.4   A cargo pump-room below deck shall be capable of being drained in an emergency by an installation located in the cargo area and independent from any other installation. The installation shall be provided outside the cargo pump-room.

9.3.3.36 to 9.3.3.39   (*Reserved*)

9.3.3.40   *Fire-extinguishing arrangements*

9.3.3.40.1   A fire-extinguishing system shall be installed on the vessel. This system shall comply with the following requirements:

–   It shall be supplied by two independent fire or ballast pumps, one of which shall be ready for use at any time. These pumps and their means of propulsion and electrical equipment shall not be installed in the same space;

–   It shall be provided with a water main fitted with at least three hydrants in the cargo area above deck. Three suitable and sufficiently long hoses with jet/spray nozzles having a diameter of not less than 12 mm shall be provided. Alternatively one or more of the hose assemblies may be substituted by directable jet/spray nozzles having a diameter of not less than 12 mm. It shall be possible to reach any point of the deck in the cargo area simultaneously with at least two jets of water which do not emanate from the same hydrant;

A spring-loaded non-return valve shall be fitted to ensure that no gases can escape through the fire-extinguishing system into the accommodation or service spaces outside the cargo area or wheelhouse;

–   The capacity of the system shall be at least sufficient for a jet of water to have a minimum reach of not less than the vessel's breadth from any location on board with two spray nozzles being used at the same time;

–   The water supply system shall be capable of being put into operation from the wheelhouse and from the deck;

–   Measures shall be taken to prevent the freezing of fire-mains and hydrants.

9.3.3.40.2   In addition the engine room, the pump-room and all spaces containing essential equipment (switchboards, compressors, etc.) for the refrigeration equipment, if any, shall be provided with a fixed fire-extinguishing system meeting the following requirements:

9.3.3.40.2.1   *Extinguishing agents*

For the protection of spaces in engine rooms, boiler rooms and pump rooms, only permanently fixed fire-extinguishing systems using the following extinguishing agents are permitted:

(a) CO_2 (carbon dioxide);

(b)   HFC 227 ea (heptafluoropropane);

(c)   IG-541 (52% nitrogen, 40% argon, 8% carbon dioxide);

(d)   FK-5-1-12 (dodecafluoro 2–methylpentane–3–one).

Other extinguishing agents are permitted only on the basis of recommendations by the Administrative Committee.

9.3.3.40.2.2   *Ventilation, air extraction*

(a)   The combustion air required by the combustion engines which ensure propulsion should not come from spaces protected by permanently fixed fire-extinguishing systems. This requirement is not mandatory if the vessel has two independent main engine rooms with a gastight separation or if, in addition to the main engine room, there is a separate engine room installed with a bow thruster that can independently ensure propulsion in the event of a fire in the main engine room;

(b)   All forced ventilation systems in the space to be protected shall be shut down automatically as soon as the fire-extinguishing system is activated;

(c)   All openings in the space to be protected which permit air to enter or gas to escape shall be fitted with devices enabling them to be closed rapidly. It shall be clear whether they are open or closed;

(d)   Air escaping from the pressure–relief valves of the pressurised air tanks installed in the engine rooms shall be evacuated to the open air;

(e)   Overpressure or negative pressure caused by the diffusion of the extinguishing agent shall not destroy the constituent elements of the space to be protected. It shall be possible to ensure the safe equalisation of pressure;

(f)   Protected spaces shall be provided with a means of extracting the extinguishing agent. If extraction devices are installed, it shall not be possible to start them up during extinguishing.

9.3.3.40.2.3   *Fire alarm system*

The space to be protected shall be monitored by an appropriate fire alarm system. The alarm signal shall be audible in the wheelhouse, the accommodation and the space to be protected.

9.3.3.40.2.4   *Piping system*

(a)   The extinguishing agent shall be routed to and distributed in the space to be protected by means of a permanent piping system. Piping installed in the space to be protected and their fittings shall be made of steel. This shall not apply to the connecting nozzles of tanks and compensators provided that the materials used have equivalent fire–retardant properties. Piping shall be protected against corrosion both internally and externally;

(b)   The discharge nozzles shall be so arranged as to ensure the regular diffusion of the extinguishing agent. In particular, the extinguishing agent must also be effective beneath the floor.

9.3.3.40.2.5   *Triggering device*

(a)   Automatically activated fire-extinguishing systems are not permitted;

(b)   It shall be possible to activate the fire-extinguishing system from a suitable point located outside the space to be protected;

(c)   Triggering devices shall be so installed that they can be activated in the event of a fire and so that the risk of their breakdown in the event of a fire or an explosion in the space to be protected is reduced as far as possible;

Systems which are not mechanically activated shall be supplied from two energy sources independent of each other. These energy sources shall be located outside the space to be protected. The control lines located in the space to be protected shall be so designed as to remain capable of operating in the event of a fire for a minimum of 30 minutes. The electrical installations are deemed to meet this requirement if they conform to the IEC 60331–21:1999 standard;

When the triggering devices are so placed as not to be visible, the component concealing them shall carry the "Fire-fighting system" symbol, each side being not less than 10 cm in length, with the following text in red letters on a white ground:

Fire-extinguishing system

(d)   If the fire-extinguishing system is intended to protect several spaces, it shall comprise a separate and clearly–marked triggering device for each space;

(e)   The instructions shall be posted alongside all triggering devices and shall be clearly visible and indelible. The instructions shall be in a language the master can read and understand and if this language is not English, French or German, they shall be in English, French or German. They shall include information concerning:

    (i)   the activation of the fire-extinguishing system;

    (ii)   the need to ensure that all persons have left the space to be protected;

    (iii)   the correct behaviour of the crew in the event of activation and when accessing the space to be protected following activation or diffusion, in particular in respect of the possible presence of dangerous substances;

    (iv)   the correct behaviour of the crew in the event of the failure of the fire-extinguishing system to function properly.

(f)   The instructions shall mention that prior to the activation of the fire-extinguishing system, combustion engines installed in the space and aspirating air from the space to be protected, shall be shut down.

9.3.3.40.2.6   *Alarm device*

(a)   Permanently fixed fire-extinguishing systems shall be fitted with an audible and visual alarm device;

(b)   The alarm device shall be set off automatically as soon as the fire-extinguishing system is first activated. The alarm device shall function for an appropriate period of time before the extinguishing agent is released; it shall not be possible to turn it off;

(c)   Alarm signals shall be clearly visible in the spaces to be protected and their access points and be clearly audible under operating conditions corresponding to the highest possible sound level. It shall be possible to distinguish them clearly from all other sound and visual signals in the space to be protected;

(d)   Sound alarms shall also be clearly audible in adjoining spaces, with the communicating doors shut, and under operating conditions corresponding to the highest possible sound level;

(e)   If the alarm device is not intrinsically protected against short circuits, broken wires and drops in voltage, it shall be possible to monitor its operation;

(f)   A sign with the following text in red letters on a white ground shall be clearly posted at the entrance to any space the extinguishing agent may reach:

Warning, fire-extinguishing system!
Leave this space immediately when the ... (description)
alarm is activated!

9.3.3.40.2.7   *Pressurised tanks, fittings and piping*

(a)   Pressurised tanks, fittings and piping shall conform to the requirements of the competent authority or, if there are no such requirements, to those of a recognized classification society;

(b)   Pressurised tanks shall be installed in accordance with the manufacturer's instructions;

(c) Pressurised tanks, fittings and piping shall not be installed in the accommodation;

(d) The temperature of cabinets and storage spaces for pressurised tanks shall not exceed 50 °C;

(e) Cabinets or storage spaces on deck shall be securely stowed and shall have vents so placed that in the event of a pressurised tank not being gastight, the escaping gas cannot penetrate into the vessel. Direct connections with other spaces are not permitted.

9.3.3.40.2.8   *Quantity of extinguishing agent*

If the quantity of extinguishing agent is intended for more than one space, the quantity of extinguishing agent available does not need to be greater than the quantity required for the largest of the spaces thus protected.

9.3.3.40.2.9   *Installation, maintenance, monitoring and documents*

(a) The mounting or modification of the system shall only be performed by a company specialised in fire-extinguishing systems. The instructions (product data sheet, safety data sheet) provided by the manufacturer of the extinguishing agent or the system shall be followed;

(b) The system shall be inspected by an expert:

(i) before being brought into service;

(ii) each time it is put back into service after activation;

(iii) after every modification or repair;

(iv) regularly, not less than every two years.

(c) During the inspection, the expert is required to check that the system conforms to the requirements of 9.3.3.40.2;

(d) The inspection shall include, as a minimum:

(i) an external inspection of the entire system;

(ii) an inspection to ensure that the piping is leakproof;

(iii) an inspection to ensure that the control and activation systems are in good working order;

(iv) an inspection of the pressure and contents of tanks;

(v) an inspection to ensure that the means of closing the space to be protected are leakproof;

(vi) an inspection of the fire alarm system;

(vii) an inspection of the alarm device.

(e) The person performing the inspection shall establish, sign and date a certificate of inspection;

(f) The number of permanently fixed fire-extinguishing systems shall be mentioned in the vessel certificate.

9.3.3.40.2.10 *Fire-extinguishing system operating with CO₂*

In addition to the requirements contained in 9.3.3.40.2.1 to 9.3.3.40.2.9, fire-extinguishing systems using CO_2 as an extinguishing agent shall conform to the following provisions:

(a) Tanks of CO_2 shall be placed in a gastight space or cabinet separated from other spaces. The doors of such storage spaces and cabinets shall open outwards; they shall be capable of being locked and shall carry on the outside the symbol "Warning: danger", not less than 5 cm high and "CO_2" in the same colours and the same size;

(b) Storage cabinets or spaces for CO_2 tanks located below deck shall only be accessible from the outside. These spaces shall have an artificial ventilation system with extractor hoods and shall be completely independent of the other ventilation systems on board;

(c) The level of filling of CO_2 tanks shall not exceed 0.75 kg/l. The volume of depressurised CO_2 shall be taken to be 0.56 m³/kg;

(d) The concentration of CO_2 in the space to be protected shall be not less than 40% of the gross volume of the space. This quantity shall be released within 120 seconds. It shall be possible to monitor whether diffusion is proceeding correctly;

(e) The opening of the tank valves and the control of the diffusing valve shall correspond to two different operations;

(f) The appropriate period of time mentioned in 9.3.3.40.2.6 (b) shall be not less than 20 seconds. A reliable installation shall ensure the timing of the diffusion of CO_2.

9.3.3.40.2.11 *Fire-extinguishing system operating with HFC-227 ea (heptafluoropropane)*

In addition to the requirements of 9.3.3.40.2.1 to 9.3.3.40.2.9, fire-extinguishing systems using HFC-227 ea as an extinguishing agent shall conform to the following provisions:

(a) Where there are several spaces with different gross volumes, each space shall be equipped with its own fire-extinguishing system;

(b) Every tank containing HFC-227 ea placed in the space to be protected shall be fitted with a device to prevent overpressure. This device shall ensure that the contents of the tank are safely diffused in the space to be protected if the tank is subjected to fire, when the fire-extinguishing system has not been brought into service;

(c) Every tank shall be fitted with a device permitting control of the gas pressure;

(d) The level of filling of tanks shall not exceed 1.15 kg/l. The specific volume of depressurised HFC-227 ea shall be taken to be 0.1374 m³/kg;

(e) The concentration of HFC-227 ea in the space to be protected shall be not less than 8% of the gross volume of the space. This quantity shall be released within 10 seconds;

(f) Tanks of HFC-227 ea shall be fitted with a pressure monitoring device which triggers an audible and visual alarm in the wheelhouse in the event of an unscheduled loss of propellant gas. Where there is no wheelhouse, the alarm shall be triggered outside the space to be protected;

(g) After discharge, the concentration in the space to be protected shall not exceed 10.5% (volume);

(h) The fire-extinguishing system shall not comprise aluminium parts.

9.3.3.40.2.12 *Fire-extinguishing system operating with IG-541*

In addition to the requirements of 9.3.3.40.2.1 to 9.3.3.40.2.9, fire-extinguishing systems using IG-541 as an extinguishing agent shall conform to the following provisions:

(a) Where there are several spaces with different gross volumes, every space shall be equipped with its own fire-extinguishing system;

(b) Every tank containing IG-541 placed in the space to be protected shall be fitted with a device to prevent overpressure. This device shall ensure that the contents of the tank are safely diffused in the space to be protected if the tank is subjected to fire, when the fire-extinguishing system has not been brought into service;

(c) Each tank shall be fitted with a device for checking the contents;

(d) The filling pressure of the tanks shall not exceed 200 bar at a temperature of +15 °C;

(e) The concentration of IG-541 in the space to be protected shall be not less than 44% and not more than 50% of the gross volume of the space. This quantity shall be released within 120 seconds.

9.3.3.40.2.13 *Fire-extinguishing system operating with FK-5-1-12*

In addition to the requirements of 9.3.3.40.2.1 to 9.3.3.40.2.9, fire-extinguishing systems using FK-5-1-12 as an extinguishing agent shall comply with the following provisions:

(a) Where there are several spaces with different gross volumes, every space shall be equipped with its own fire-extinguishing system;

(b) Every tank containing FK-5-1-12 placed in the space to be protected shall be fitted with a device to prevent overpressure. This device shall ensure that the contents of the tank are safely diffused in the space to be protected if the tank is subjected to fire, when the fire-extinguishing system has not been brought into service;

(c) Every tank shall be fitted with a device permitting control of the gas pressure;

(d) The level of filling of tanks shall not exceed 1.00 kg/l. The specific volume of depressurized FK-5-1-12 shall be taken to be 0.0719 m³/kg;

(e) The volume of FK-5-1-12 in the space to be protected shall be not less than 5.5% of the gross volume of the space. This quantity shall be released within 10 seconds;

(f) Tanks of FK-5-1-12 shall be fitted with a pressure monitoring device which triggers an audible and visual alarm in the wheelhouse in the event of an unscheduled loss of extinguishing agent. Where there is no wheelhouse, the alarm shall be triggered outside the space to be protected;

(g) After discharge, the concentration in the space to be protected shall not exceed 10.0%.

9.3.3.40.2.14   *Fixed fire-extinguishing system for physical protection*

In order to ensure physical protection in the engine rooms, boiler rooms and pump rooms, permanently fixed fire-extinguishing systems are accepted solely on the basis of recommendations by the Administrative Committee.

9.3.3.40.3   The two hand fire–extinguishers referred to in 8.1.4 shall be located in the cargo area.

9.3.3.40.4   The fire-extinguishing agent and the quantity contained in the permanently fixed fire-extinguishing system shall be suitable and sufficient for fighting fires.

9.3.3.40.5   9.3.3.40.1 and 9.3.3.40.2 above do not apply to oil separator or supply vessels.

9.3.3.41   ***Fire and naked light***

9.3.3.41.1   The outlets of funnels shall be located not less than 2.00 m from the cargo area. Arrangements shall be provided to prevent the escape of sparks and the entry of water.

9.3.3.41.2   Heating, cooking and refrigerating appliances shall not be fuelled with liquid fuels, liquid gas or solid fuels.

The installation in the engine room or in another separate space of heating appliances fuelled with liquid fuel having a flashpoint above 55 °C is, however, permitted.

Cooking and refrigerating appliances are permitted only in the accommodation.

9.3.3.41.3   Only electrical lighting appliances are permitted.

9.3.3.42   ***Cargo heating system***

9.3.3.42.1   Boilers which are used for heating the cargo shall be fuelled with a liquid fuel having a flashpoint of more than 55 °C. They shall be placed either in the engine room or in another separate space below deck and outside the cargo area, which is accessible from the deck or from the engine room.

9.3.3.42.2   The cargo heating system shall be designed so that the cargo cannot penetrate into the boiler in the case of a leak in the heating coils. A cargo heating system with artificial draught shall be ignited electrically.

9.3.3.42.3   The ventilation system of the engine room shall be designed taking into account the air required for the boiler.

9.3.3.42.4   Where the cargo heating system is used during loading, unloading or gas-freeing, the service space which contains this system shall fully comply with the requirements of 9.3.3.52.3. This requirement does not apply to the inlets of the ventilation system. These inlets shall be located at a minimum distance of 2 m from the cargo area and 6 m from the openings of cargo tanks or residual cargo tanks, loading pumps situated on deck, openings of high-velocity vent valves, pressure relief devices and shore connections of loading and unloading piping and must be located not less than 2 m above the deck.

The requirements of 9.3.3.52.3 are not applicable to the unloading of substances having a flashpoint of 60 °C or more when the temperature of the product is at least 15 K lower at the flashpoint.

9.3.3.43 to 9.3.3.49   (*Reserved*)

9.3.3.50   *Documents concerning electrical installations*

9.3.3.50.1   In addition to the documents required in accordance with the Regulations referred to in 1.1.4.6, the following documents shall be on board:

(a)   a drawing indicating the boundaries of the cargo area and the location of the electrical equipment installed in this area;

(b)   a list of the electrical equipment referred to in (a) above including the following particulars:

machine or appliance, location, type of protection, type of protection against explosion, testing body and approval number;

(c)   a list of or general plan indicating the electrical equipment outside the cargo area which may be operated during loading, unloading or gas-freeing. All other electrical equipment shall be marked in red. See 9.3.3.52.3 and 9.3.3.52.4.

9.3.3.50.2   The documents listed above shall bear the stamp of the competent authority issuing the certificate of approval.

9.3.3.51   *Electrical installations*

9.3.3.51.1   Only distribution systems without return connection to the hull are permitted.

This provision does not apply to:

–   active cathodic corrosion protection;

–   certain limited sections of the installations situated outside the cargo area (e.g. connections of starters of diesel engines);

–   the device for checking the insulation level referred to in 9.3.3.51.2 below.

9.3.3.51.2   Every insulated distribution network shall be fitted with an automatic device with a visual and audible alarm for checking the insulation level.

9.3.3.51.3   For the selection of electrical equipment to be used in zones presenting an explosion risk, the explosion groups and temperature classes assigned to the substances carried in columns (15) and (16) of Table C of Chapter 3.2 shall be taken into consideration.

9.3.3.52   *Type and location of electrical equipment*

9.3.3.52.1   (a)   Only the following equipment may be installed in cargo tanks, residual cargo tanks, and piping for loading and unloading (comparable to zone 0):

–   measuring, regulation and alarm devices of the EEx (ia) type of protection.

(b)   Only the following equipment may be installed in the cofferdams, double-hull spaces, double bottoms and hold spaces (comparable to zone 1):

–   measuring, regulation and alarm devices of the certified safe type;

–   lighting appliances of the "flame-proof enclosure" or "apparatus protected by pressurization" type of protection;

–   hermetically sealed echo sounding devices the cables of which are led through thick-walled steel tubes with gastight connections up to the main deck;

         –   cables for the active cathodic protection of the shell plating in protective steel tubes such as those provided for echo sounding devices.

The following equipment may be installed only in double-hull spaces and double bottoms if used for ballasting:

         –   Permanently fixed submerged pumps with temperature monitoring, of the certified safe type.

(c)   Only the following equipment may be installed in the service spaces in the cargo area below deck (comparable to zone 1):

         –   measuring, regulation and alarm devices of the certified safe type;

         –   lighting appliances of the "flame-proof enclosure" or "apparatus protected by pressurization" type of protection;

         –   motors driving essential equipment such as ballast pumps with temperature monitoring; they shall be of the certified safe type.

(d)   The control and protective equipment of the electrical equipment referred to in paragraphs (a), (b) and (c) above shall be located outside the cargo area if they are not intrinsically safe.

(e)   The electrical equipment in the cargo area on deck (comparable to zone 1) shall be of the certified safe type.

9.3.3.52.2   Accumulators shall be located outside the cargo area.

9.3.3.52.3   (a)   Electrical equipment used during loading, unloading and gas-freeing during berthing and which are located outside the cargo area shall (comparable to zone 2) be at least of the "limited explosion risk" type.

(b)   This provision does not apply to:

    (i)   lighting installations in the accommodation, except for switches near entrances to accommodation;

    (ii)   radiotelephone installations in the accommodation or the wheelhouse;

    (iii)   mobile and fixed telephone installations in the accommodation or the wheelhouse;

    (iv)   electrical installations in the accommodation, the wheelhouse or the service spaces outside the cargo areas if:

        1.   These spaces are fitted with a ventilation system ensuring an overpressure of 0.1 kPa (0.001 bar) and none of the windows is capable of being opened; the air intakes of the ventilation system shall be located as far away as possible, however, not less than 6.00 m from the cargo area and not less than 2.00 m above the deck;

        2.   The spaces are fitted with a gas detection system with sensors:

            –   at the suction inlets of the ventilation system;

            –   directly at the top edge of the sill of the entrance doors of the accommodation and service spaces;

        3.   The gas concentration measurement is continuous;

4.   When the gas concentration reaches 20% of the lower explosive limit, the ventilators are switched off. In such a case and when the overpressure is not maintained or in the event of failure of the gas detection system, the electrical installations which do not comply with (a) above, shall be switched off. These operations shall be performed immediately and automatically and activate the emergency lighting in the accommodation, the wheelhouse and the service spaces, which shall comply at least with the "limited explosion risk" type. The switching-off shall be indicated in the accommodation and wheelhouse by visual and audible signals;

5.   The ventilation system, the gas detection system and the alarm of the switch-off device fully comply with the requirements of (a) above;

6.   The automatic switch-off device is set so that no automatic switching-off may occur while the vessel is under way.

(v)   Inland AIS (automatic identification systems) stations in the accommodation and in the wheelhouse if no part of an aerial for electronic apparatus is situated above the cargo area and if no part of a VHF antenna for AIS stations is situated within 2 m from the cargo area.

9.3.3.52.4   The electrical equipment which does not meet the requirements set out in 9.3.3.52.3 above together with its switches shall be marked in red. The disconnection of such equipment shall be operated from a centralised location on board.

9.3.3.52.5   An electric generator which is permanently driven by an engine and which does not meet the requirements of 9.3.3.52.3 above, shall be fitted with a switch capable of shutting down the excitation of the generator. A notice board with the operating instructions shall be displayed near the switch.

9.3.3.52.6   Sockets for the connection of signal lights and gangway lighting shall be permanently fitted to the vessel close to the signal mast or the gangway. Connecting and disconnecting shall not be possible except when the sockets are not live.

9.3.3.52.7   The failure of the power supply for the safety and control equipment shall be immediately indicated by visual and audible signals at the locations where the alarms are usually actuated.

9.3.3.53   *Earthing*

9.3.3.53.1   The metal parts of electrical appliances in the cargo area which are not live as well as protective metal tubes or metal sheaths of cables in normal service shall be earthed, unless they are so arranged that they are automatically earthed by bonding to the metal structure of the vessel.

9.3.3.53.2   The provisions of 9.3.3.53.1 above apply also to equipment having service voltages of less than 50 V.

9.3.3.53.3   Independent cargo tanks shall be earthed.

9.3.3.53.4   Receptacles for residual products shall be capable of being earthed.

9.3.3.54 and 9.3.3.55   (*Reserved*)

9.3.3.56   *Electrical cables*

9.3.3.56.1   All cables in the cargo area shall have a metallic sheath.

9.3.3.56.2   Cables and sockets in the cargo area shall be protected against mechanical damage.

9.3.3.56.3   Movable cables are prohibited in the cargo area, except for intrinsically safe electric circuits or for the supply of signal lights, gangway lighting and submerged pumps on board oil separator vessels.

9.3.3.56.4   Cables of intrinsically safe circuits shall only be used for such circuits and shall be separated from other cables not intended for being used in such circuits (e.g. they shall not be installed together in the same string of cables and they shall not be fixed by the same cable clamps).

9.3.3.56.5   For movable cables intended for signal lights, gangway lighting, and submerged pumps on board oil separator vessels, only sheathed cables of type H 07 RN-F in accordance with IEC publication-60 245-4 (1994) or cables of at least equivalent design having conductors with a cross-section of not less than 1.5 mm^2 shall be used.

These cables shall be as short as possible and installed so that damage is not likely to occur.

9.3.3.56.6   The cables required for the electrical equipment referred to in 9.3.3.52.1 (b) and (c) are accepted in cofferdams, double-hull spaces, double bottoms, hold spaces and service spaces below deck. When the vessel is only authorized to carry substances for which no anti-explosion protection is required in column (17) of Table C in Chapter 3.2, cable penetration is permitted in the hold spaces.

9.3.3.57 to 9.3.3.59   (*Reserved*)

9.3.3.60   *Special equipment*

A shower and an eye and face bath shall be provided on the vessel at a location which is directly accessible from the cargo area.

This requirement does not apply to oil separator and supply vessels.

9.3.3.61 to 9.3.3.70   (*Reserved*)

9.3.3.71   *Admittance on board*

The notice boards displaying the prohibition of admittance in accordance with 8.3.3 shall be clearly legible from either side of the vessel.

9.3.3.72 and 9.3.3.73   (*Reserved*)

9.3.3.74   *Prohibition of smoking, fire or naked light*

9.3.3.74.1   The notice boards displaying the prohibition of smoking in accordance with 8.3.4 shall be clearly legible from either side of the vessel.

9.3.3.74.2   Notice boards indicating the circumstances under which the prohibition is applicable shall be fitted near the entrances to the spaces where smoking or the use of fire or naked light is not always prohibited.

9.3.3.74.3   Ashtrays shall be provided close to each exit in the accommodation and the wheelhouse.

9.3.3.75 to 9.3.3.91   (*Reserved*)

9.3.3.92   On board of tank vessels referred to in 9.3.3.11.7, spaces the entrances or exits of which are likely to become partly or completely immersed in the damaged condition shall have an emergency exit which is situated not less than 0.10 m above the damage waterline. This requirement does not apply to forepeak and afterpeak.

9.3.3.93 to 9.3.3.99   (*Reserved*)

9.3.4   **Alternative constructions**

9.3.4.1   *General*

9.3.4.1.1   The maximum permissible capacity and length of a cargo tank in accordance with 9.3.1.11.1, 9.3.2.11.1 and 9.3.3.11.1 may be exceeded and the minimum distances in accordance with 9.3.1.11.2 a) and 9.3.2.11.7 may be deviated from provided that the provisions of this section are complied with. The capacity of a cargo tank shall not exceed 1000 m^3.

9.3.4.1.2   Tank vessels whose cargo tanks exceed the maximum allowable capacity or where the distance between the side wall and the cargo tank is smaller than required, shall be protected through a more crashworthy side structure. This shall be proved by comparing the risk of a conventional construction (reference construction), complying with the ADN regulations with the risk of a crashworthy construction (alternative construction).

9.3.4.1.3   When the risk of the more crashworthy construction is equal to or lower than the risk of the conventional construction, equivalent or higher safety is proven. The equivalent or higher safety shall be proven in accordance with 9.3.4.3.

9.3.4.1.4   When a vessel is built in compliance with this section, a recognised classification society shall document the application of the calculation procedure in accordance with 9.3.4.3 and shall submit its conclusions to the competent authority for approval.

                 The competent authority may request additional calculations and proof.

9.3.4.1.5   The competent authority shall include this construction in the certificate of approval in accordance with 8.6.1.

9.3.4.2   *Approach*

9.3.4.2.1   The probability of cargo tank rupture due to a collision and the area around the vessel affected by the cargo outflow as a result thereof are the governing parameters. The risk is described by the following formula:

$$R = P \cdot C$$

Wherein: R risk [m^2],

                  P probability of cargo tank rupture [ ],

                  C consequence (measure of damage) of cargo tank rupture [m^2].

9.3.4.2.2   The probability P of cargo tank rupture depends on the probability distribution of the available collision energy represented by vessels, which the victim is likely to encounter in a collision, and the capability of the struck vessel to absorb collision energy without cargo tank rupture. A decrease of this probability can be achieved by means of a more crashworthy side structure.

                 The consequence C of cargo spillage resulting from cargo tank rupture is expressed as an affected area around the struck vessel.

9.3.4.2.3   The procedure according to 9.3.4.3 shows how tank rupture probabilities shall be calculated as well as how the collision energy absorbing capacity of side structure and a consequence increase shall be determined.

9.3.4.3   *Calculation procedure*

9.3.4.3.1   The calculation procedure shall follow 13 basic steps. Steps 2 through 10 shall be carried out for both the alternative design and the reference design. The following table shows the calculation of the weighted probability of cargo tank rupture:

A	B	C	D	E	F	G	H	I	J	K	L	M	N	O
							F x G			I x J			L x M	
Identify collision locations and associated weighting factors, Collision scenario I	Loc1	Finite element analysis	Eloc1	Calculate probability with CPDF 50%	P50%	wf 50%	Pw50%							
				Calculate probability with CPDF 66%	P66%	wf 66%	Pw66%							
				Calculate probability with CPDF 100%	P100%	wf 100%	Pw100%	+						
							sum	Ploc. 1	wf loc 1	Pwloc. 1				
	Loci	Finite element analysis	Eloci	Calculate probability with CPDF 50%	P50%	wf 50%	Pw50%							
				Calculate probability with CPDF 66%	P66%	wf 66%	Pw66%							
				Calculate probability with CPDF 100%	P100%	wf 100%	Pw100%	+						
							sum	Ploci	wf loci	Pwloci				
	Locn	Finite element analysis	Elocn	Calculate probability with CPDF 50%	P50%	wf 50%	Pw50%							
				Calculate probability with CPDF 66%	P66%	wf 66%	Pw66%							
				Calculate probability with CPDF 100%	P100%	wf 100%	Pw100%	+						
							sum	Plocn	wf locn	Pwlocn	+			
										sum	PscenI	wfscenI	PwscenI	
Identify collision locations and associated weighting factors, Collision scenario II	Loc1	Finite element analysis	Eloc1	Calculate probability with CPDF 30%	P30%	wf 30%	Pw30%							
				Calculate probability with CPDF 100%	P100%	wf 100%	Pw100%	+						
							sum	Ploc 1	wf loc 1	Pwloc 1				
	Locn	Finite element analysis	Elocn	Calculate probability with CPDF 30%	P30%	wf 30%	Pw30%							
				Calculate probability with CPDF 100%	P100%	wf 100%	Pw100%	+						
							sum	Plocn	wf locn	Pwlocn	+			
										sum	PscenII	wfscenII	PwscenII	+
													sum	Pw

CPDF: Cumulative probability density function

9.3.4.3.1.1   *Step 1*

Besides the alternative design, which is used for cargo tanks exceeding the maximum allowable capacity or a reduced distance between the side wall and the cargo tank as well as a more crashworthy side structure, a reference design with at least the same dimensions (length, width, depth, displacement) shall be drawn up. This reference design shall fulfil the requirements specified in section 9.3.1 (Type G), 9.3.2 (Type C) or 9.3.3 (Type N) and shall comply with the minimum requirements of a recognised classification society.

9.3.4.3.1.2   *Step 2*

9.3.4.3.1.2.1   The relevant typical collision locations i=1 through n shall be determined. The table in 9.3.4.3.1 depicts the general case where there are 'n' typical collision locations.

The number of typical collision locations depends on the vessel design. The choice of the collision locations shall be accepted by the recognised classification society.

9.3.4.3.1.2.2   *Vertical collision locations*

9.3.4.3.1.2.2.1   *Tank vessel type C and N*

9.3.4.3.1.2.2.1.1 The determination of the collision locations in the vertical direction depends on the draught differences between striking and struck vessel, which is limited by the maximum and minimum draughts of both vessels and the construction of the struck vessel. This can be depicted graphically through a rectangular area which is enclosed by the values of the maximum and minimum draught of both striking and struck vessel (see following figure).

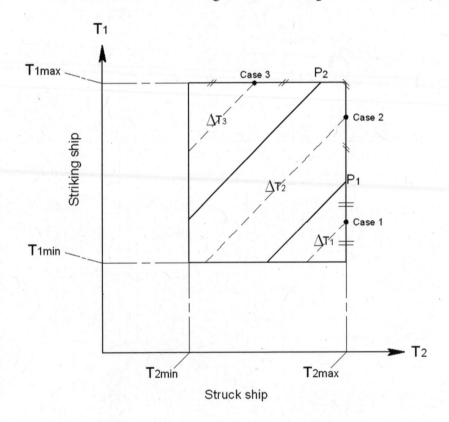

Definition of vertical striking locations

9.3.4.3.1.2.2.1.2 Each point in this area represents a possible draught combination. T_{1max} is the maximum draught and T_{1min} is the minimum draught of the striking vessel, while T_{2max} and T_{2min} are the corresponding minimum and maximum draughts of the struck vessel. Each draught combination has an equal probability of occurrence.

9.3.4.3.1.2.2.1.3 Points on each inclined line in the figure in 9.3.4.3.1.2.2.1.1 indicate the same draught difference. Each of these lines reflects a vertical collision location. In the example in the figure in 9.3.4.3.1.2.2.1.1 three vertical collision locations are defined, depicted by three areas. Point P_1 is the point where the lower edge of the vertical part of the push barge or V–bow strikes at deck level of the struck vessel. The triangular area for collision case 1 is bordered by point P_1. This corresponds to the vertical collision location "collision at deck level". The triangular upper left area of the rectangle corresponds to the vertical collision location "collision below deck". The draught difference ΔT_i, i=1,2,3 shall be used in the collision calculations (see following figure).

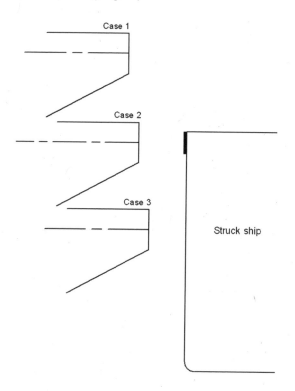

Example of vertical collision locations

9.3.4.3.1.2.2.1.4 For the calculation of the collision energies the maximum masses of both striking vessel and struck vessel must be used (highest point on each respective diagonal ΔT_i).

9.3.4.3.1.2.2.1.5 Depending on the vessel design, the recognised classification society may require additional collision locations.

9.3.4.3.1.2.2.2 *Tank vessel type G*

For a tank vessel type G a collision at half tank height shall be assumed. The recognised classification society may require additional collision locations at other heights. This shall be agreed with the recognised classification society.

9.3.4.3.1.2.3 *Longitudinal collision location*

9.3.4.3.1.2.3.1 *Tank vessels type C and N*

At least the following three typical collision locations shall be considered:

–   at bulkhead,

–   between webs and

–   at web.

9.3.4.3.1.2.3.2 *Tank vessel type G*

For a tank vessel type G at least the following three typical collision locations shall be considered:

- at cargo tank end,
- between webs and
- at web.

9.3.4.3.1.2.4 Number of collision locations

9.3.4.3.1.2.4.1 Tank vessel type C and N

The combination of vertical and longitudinal collision locations in the example mentioned in 9.3.4.3.1.2.2.1.3 and 9.3.4.3.1.2.3.1 results in 3 • 3 = 9 collision locations.

9.3.4.3.1.2.4.2 Tank vessel type G

The combination of vertical and longitudinal collision locations in the example mentioned in 9.3.4.3.1.2.2.2 and 9.3.4.3.1.2.3.2 results in 1 • 3 = 3 collision locations.

9.3.4.3.1.2.4.3 Additional examinations for tank vessels type G, C and N with independent cargo tanks

As proof that the tank seatings and the buoyancy restraints do not cause any premature tank rupture, additional calculations shall be carried out. The additional collision locations for this purpose shall be agreed with the recognised classification society.

9.3.4.3.1.3 Step 3

9.3.4.3.1.3.1 For each typical collision location a weighting factor which indicates the relative probability that such a typical collision location will be struck shall be determined. In the table in 9.3.4.3.1 these factors are named $wf_{loc(i)}$ (column J). The assumptions shall be agreed with the recognised classification society.

The weighting factor for each collision location is the product of the factor for the vertical collision location by the factor for the longitudinal collision location.

9.3.4.3.1.3.2 Vertical collision locations

9.3.4.3.1.3.2.1 Tank vessel type C and N

The weighting factors for the various vertical collision locations are in each case defined by the ratio between the partial area for the corresponding collision case and the total area of the rectangle shown in the Figure in 9.3.4.3.1.2.2.1.1.

For example, for collision case 1 (see figure in 9.3.4.3.1.2.2.1.3) the weighting factor equals the ratio between the triangular lower right area of the rectangle, and the area of the rectangle between minimum and maximum draughts of striking and struck vessels.

9.3.4.3.1.3.2.2 Tank vessel type G

The weighting factor for the vertical collision location has the value 1.0, if only one collision location is assumed. When the recognised classification society requires additional collision locations, the weighting factor shall be determined analogous to the procedure for tank vessels type C and N.

9.3.4.3.1.3.3   Longitudinal collision locations

9.3.4.3.1.3.3.1   Tank vessel type C and N

The weighting factor for each longitudinal collision location is the ratio between the "calculational span length" and the tank length.

The calculational span length shall be calculated as follows:

(a)   collision on bulkhead:

    0.2 • distance between web frame and bulkhead, but not larger than 450 mm,

(b)   collision on web frame:

    sum of 0.2 • web frame spacing forward of the web frame, but not larger than 450 mm, and 0.2 • web frame spacing aft of the web frame, but not larger than 450 mm, and

(c)   collision between web frames:

    cargo tank length minus the length "collision at bulkhead" and minus the length "collision at web frame".

9.3.4.3.1.3.3.2   Tank vessel type G

The weighting factor for each longitudinal collision location is the ratio between the "calculational span length" and the length of the hold space.

The calculational span length shall be calculated as follows:

(a)   collision at cargo tank end:

    distance between bulkhead and the start of the cylindrical part of the cargo tank,

(b)   collision on web frame:

    sum of 0.2 • web frame spacing forward of the web frame, but not larger than 450 mm, and 0.2 • web frame spacing aft of the web frame, but not larger than 450 mm, and

(c)   collision between web frames:

    cargo tank length minus the length "collision at cargo tank end" and minus the length "collision at web frame".

9.3.4.3.1.4   *Step 4*

9.3.4.3.1.4.1   For each collision location the collision energy absorbing capacity shall be calculated. For that matter the collision energy absorbing capacity is the amount of collision energy absorbed by the vessel structure up to initial rupture of the cargo tank (see the table in 9.3.4.3.1, column D: $E_{loc(i)}$). For this purpose a finite element analysis in accordance with 9.3.4.4.2 shall be used.

These calculations shall be done for two collision scenarios according to the following table. Collision scenario I shall be analysed under the assumption of a push barge bow shape. Collision scenario II shall be analysed under the assumption of a V–shaped bow.

These bow shapes are defined in 9.3.4.4.8.

Table : Speed reduction factors for scenario I or scenario II with weighting factors

				Causes		
				Communication error and poor visibility	Technical error	Human error
				0,50	0,20	0,30
Worst case scenarios	I		Push barge–bow, striking angle 55° 0,80	0.66	0.50	1.00
	II		V-shaped-bow, striking angle 90° 0,20	0.30		1.00

9.3.4.3.1.5 *Step 5*

9.3.4.3.1.5.1 For each collision energy absorption capacity $E_{loc(i)}$, the associated probability of exceedance is to be calculated, i.e. the probability of cargo tank rupture. For this purpose, the formula for the cumulative probability density functions (CPDF) below shall be used. The appropriate coefficients shall be selected from the Table in 9.3.4.3.1.5.6 for the effective mass of the struck vessel.

$$P_{x\%} = C_1 (E_{loc(i)})^3 + C_2 (E_{loc(i)})^2 + C_3 E_{loc(i)} + C_4$$

with: $P_{x\%}$ probability of tank rupture,

C_{1-4} coefficients from table in 9.3.4.3.1.5.6,

$E_{loc(i)}$ collision energy absorbing capacity.

9.3.4.3.1.5.2 The effective mass shall be equal to the maximum displacement of the vessel multiplied by a factor of 1.4. Both collision scenarios (9.3.4.3.1.4.2) shall be considered.

9.3.4.3.1.5.3 In the case of collision scenario I (push barge bow at 55°), three CPDF formulas shall be used:

CPDF 50% (velocity 0.5 V_{max}),

CPDF 66% (velocity 2/3 V_{max}) and

CPDF 100% (velocity V_{max}).

9.3.4.3.1.5.4 In the case of scenario II (V–shaped bow at 90°), the following two CPDF formulas shall be used:

CPDF 30% (velocity 0.3 V_{max}) and

CPDF 100% (velocity V_{max}).

9.3.4.3.1.5.5 In the table in 9.3.4.3.1, column F, these probabilities are called *P50%*, *P66%*, *P100%* and *P30%*, *P100%* respectively.

9.3.4.3.1.5.6 Table: Coefficients for the CPDF formulas

| Effective mass of struck vessel in tonnes | velocity = 1 x V_{max} | | | | |
| | coefficients | | | | |
	C_1	C_2	C_3	C_4	range
14000	4.106E–05	–2.507E–03	9.727E–03	9.983E–01	4<E_{loc}<39
12000	4.609E–05	–2.761E–03	1.215E–02	9.926E–01	4<E_{loc}<36
10000	5.327E–05	–3.125E–03	1.569E–02	9.839E–01	4<E_{loc}<33
8000	6.458E–05	–3.691E–03	2.108E–02	9.715E–01	4<E_{loc}<31
6000	7.902E–05	–4.431E–03	2.719E–02	9.590E–01	4<E_{loc}<27
4500	8.823E–05	–5.152E–03	3.285E–02	9.482E–01	4<E_{loc}<24
3000	2.144E–05	–4.607E–03	2.921E–02	9.555E–01	2<E_{loc}<19
1500	– 2.071E–03	2.704E–02	–1.245E–01	1.169E+00	2<E_{loc}<12

| Effective mass of struck vessel in tonnes | velocity = 0.66 x V_{max} | | | | |
| | coefficients | | | | |
	C_1	C_2	C_3	C_4	range
14000	4.638E–04	–1.254E–02	2.041E–02	1.000E+00	2<E_{loc}<17
12000	5.377E–04	–1.427E–02	2.897E–02	9.908E–01	2<E_{loc}<17
10000	6.262E–04	–1.631E–02	3.849E–02	9.805E–01	2<E_{loc}<15
8000	7.363E–04	–1.861E–02	4.646E–02	9.729E–01	2<E_{loc}<13
6000	9.115E–04	–2.269E–02	6.285E–02	9.573E–01	2<E_{loc}<12
4500	1.071E–03	–2.705E–02	7.738E–02	9.455E–01	1<E_{loc}<11
3000	–1.709E–05	–1.952E–02	5.123E–02	9.682E–01	1<E_{loc}<8
1500	–2.479E–02	1.500E–01	–3.218E–01	1.204E+00	1<E_{loc}<5

| Effective mass of struck vessel in tonnes | velocity = 0.5 x V_{max} | | | | |
| | coefficients | | | | |
	C_1	C_2	C_3	C_4	range
14000	2.621E–03	–3.978E–02	3.363E–02	1.000E+00	1<E_{loc}<10
12000	2.947E–03	–4.404E–02	4.759E–02	9.932E–01	1<E_{loc}<9
10000	3.317E–03	–4.873E–02	5.843E–02	9.878E–01	2<E_{loc}<8
8000	3.963E–03	–5.723E–02	7.945E–02	9.739E–01	2<E_{loc}<7
6000	5.349E–03	–7.407E–02	1.186E–01	9.517E–01	1<E_{loc}<6
4500	6.303E–03	–8.713E–02	1.393E–01	9.440E–01	1<E_{loc}<6
3000	2.628E–03	–8.504E–02	1.447E–01	9.408E–01	1<E_{loc}<5
1500	–1.566E–01	5.419E–01	–6.348E–01	1.209E+00	1<E_{loc}<3

Effective mass of struck vessel in tonnes	velocity = 0.3 x V_{max}				
	coefficients				
	C_1	C_2	C_3	C_4	range
14000	5.628E–02	–3.081E–01	1.036E–01	9.991E–01	$1<E_{loc}<3$
12000	5.997E–02	–3.212E–01	1.029E–01	1.002E+00	$1<E_{loc}<3$
10000	7.477E–02	–3.949E–01	1.875E–01	9.816E–01	$1<E_{loc}<3$
8000	1.021E–02	–5.143E–01	2.983E–01	9.593E–01	$1<E_{loc}<2$
6000	9.145E–02	–4.814E–01	2.421E–01	9.694E–01	$1<E_{loc}<2$
4500	1.180E–01	–6.267E–01	3.542E–01	9.521E–01	$1<E_{loc}<2$
3000	7.902E–02	–7.546E–01	5.079E–01	9.218E–01	$1<E_{loc}<2$
1500	–1.031E+00	2.214E–01	1.891E–01	9.554E–01	$0.5<E_{loc}<1$

The range where the formula is valid is given in column 6. In case of an E_{loc} value below the range the probability equals $P_{x\%} = 1.0$. In case of a value above the range $P_{x\%}$ equals 0.

9.3.4.3.1.6 *Step 6*

The weighted probabilities of cargo tank rupture $P_{wx\%}$ (table in 9.3.4.3.1, column H) shall be calculated by multiplying each cargo tank rupture probability $P_{x\%}$ (table in 9.3.4.3.1, column F) by the weighting factors $wf_{x\%}$ according to the following table:

Table: Weighting factors for each characteristic collision speed

			weighting factor
Scenario I	CPDF 50%	wf50%	0.2
	CPDF 66%	wf66%	0.5
	CPDF 100%	wf100%	0.3
Scenario II	CPDF 30%	wf30%	0.7
	CPDF 100%	wf100%	0.3

9.3.4.3.1.7 *Step 7*

The total probabilities of cargo tank rupture $P_{loc(i)}$ (table in 9.3.4.3.1, column I) resulting from 9.3.4.3.1.6 (step 6) shall be calculated as the sum of all weighted cargo tank rupture probabilities $P_{wx\%}$ (table in 9.3.4.3.1, column H) for each collision location considered.

9.3.4.3.1.8 *Step 8*

For both collision scenarios the weighted total probabilities of cargo tank rupture $P_{wloc(i)}$ shall, in each case, be calculated by multiplying the total tank probabilities of cargo tank rupture $P_{loc(i)}$ for each collision location, by the weighting factors $wf_{loc(i)}$ corresponding to the respective collision location (see 9.3.4.3.1.3 (step 3) and table in 9.3.4.3.1, column J).

9.3.4.3.1.9 *Step 9*

Through the addition of the weighted total probabilities of cargo tank rupture $P_{wloc(i)}$, the scenario specific total probabilities of cargo tank rupture P_{scenI} and P_{scenII} (table in 9.3.4.3.1, column L) shall be calculated, for each collision scenario I and II separately.

9.3.4.3.1.10 *Step 10*

Finally the weighted value of the overall total probability of cargo tank rupture P_w shall be calculated by the formula below (table in 9.3.4.3.1, column O):

$$P_w = 0.8 \cdot P_{scenI} + 0.2 \cdot P_{scenII}$$

9.3.4.3.1.11   *Step 11*

The overall total probability of cargo tank rupture P_w for the alternative design is called P_n. The overall total probability of cargo tank rupture P_w for the reference design is called P_r.

9.3.4.3.1.12   *Step 12*

9.3.4.3.1.12.1   The ratio (C_n/C_r) between the consequence (measure of damage) C_n of a cargo tank rupture of the alternative design and the consequence C_r of a cargo tank rupture of the reference design shall be determined with the following formula:

$$C_n/C_r = V_n / V_r$$

With C_n/C_r the ratio between the consequence related to the alternative design, and the consequence related to the reference design,

V_n maximum capacity of the largest cargo tank in the alternative design,

V_r maximum capacity of the largest cargo tank reference design.

9.3.4.3.1.12.2   This formula was derived for characteristic cargoes as listed in the following table.

Table: Characteristic cargoes

	UN No.	Description
Benzene	1114	Flammable liquid Packing group II Hazardous to health
Acrylonitrile Stabilised ACN	1093	Flammable liquid Packing group I Toxic, stabilised
n–Hexane	1208	Flammable liquid Packing group II
Nonane	1920	Flammable liquid Packing group III
Ammonia	1005	Toxic, corrosive gas Liquefied under pressure
Propane	1978	Flammable gas Liquefied under pressure

9.3.4.3.1.12.3   For cargo tanks with capacities between 380 m^3 and 1000 m^3 containing flammable, toxic and acid liquids or gases it shall be assumed that the effect increase relates linearly to the increased tank capacity (proportionality factor 1.0).

9.3.4.3.1.12.4   If substances are to be carried in tank vessels, which have been analysed according to this calculation procedure, where the proportionality factor between the total cargo tank capacity and the affected area is expected to be larger than 1.0, as assumed in the previous paragraph, the affected area shall be determined through a separate calculation. In this case the comparison as described in 9.3.4.3.1.13 (step 13) shall be carried out with this different value for the size of the affected area, t.

9.3.4.3.1.13   *Step 13*

Finally the ratio $\dfrac{P_r}{P_n}$ between the overall total probability of cargo tank rupture P_r for the reference design and the overall total probability of cargo tank rupture P_n for the alternative design shall be compared with the ratio $\dfrac{C_n}{C_r}$ between the consequence related to the alternative design, and the consequence related to the reference design.

When $\dfrac{C_n}{C_r} \leq \dfrac{P_r}{P_n}$ is fulfilled, the evidence according to 9.3.4.1.3 for the alternative design is provided.

9.3.4.4   *Determination of the collision energy absorbing capacity*

9.3.4.4.1   *General*

9.3.4.4.1.1   The determination of the collision energy absorbing capacity shall be carried out by means of a finite element analysis (FEA). The analysis shall be carried out using a customary finite element code (e.g. LS–DYNA[2], PAM–CRASH[3], ABAQUS[4] etc.) capable of dealing with both geometrical and material non–linear effects. The code shall also be able to simulate rupture realistically.

9.3.4.4.1.2   The program actually used and the level of detail of the calculations shall be agreed upon with a recognised classification society.

9.3.4.4.2   *Creating the finite element models (FE models)*

9.3.4.4.2.1   First of all, FE models for the more crashworthy design and one for the reference design shall be generated. Each FE model shall describe all plastic deformations relevant for all collision cases considered. The section of the cargo area to be modelled shall be agreed upon with a recognised classification society.

9.3.4.4.2.2   At both ends of the section to be modelled all three translational degrees of freedom are to be restrained. Because in most collision cases the global horizontal hull girder bending of the vessel is not of significant relevance for the evaluation of plastic deformation energy it is sufficient that only half beam of the vessel needs to be considered. In these cases the transverse displacements at the centre line (CL) shall be constrained. After generating the FE model, a trial collision calculation shall be carried out to ensure that there is no occurrence of plastic deformations near the constraint boundaries. Otherwise the FE modelled area has to be extended.

9.3.4.4.2.3   Structural areas affected during collisions shall be sufficiently finely idealized, while other parts may be modelled more coarsely. The fineness of the element mesh shall be suitable for an adequate description of local folding deformations and for determination of realistic rupture of elements.

---

[2]   *LSTC, 7374 Las Positas Rd, Livermore, CA 94551, USA Tel : +1 925 245–4500.*
[3]   *ESI Group, 8, Rue Christophe Colomb, 75008 Paris, France*
Tel: +33 (0)1 53 65 14 14, Fax: +33 (0)1 53 65 14 12, E–mail: info@esi–group.com.
[4]   *SIMULIA, Rising Sun Mills, 166 Valley Street, Providence, RI 02909–2499 USA*
Tel: +1 401 276–4400, Fax: +1 401 276–4408, E–mail: info@simulia.com.

9.3.4.4.2.4   The calculation of rupture initiation must be based on fracture criteria which are suitable for the elements used. The maximum element size shall be less than 200 mm in the collision areas. The ratio between the longer and the shorter shell element edge shall not exceed the value of three. The element length L for a shell element is defined as the longer length of both sides of the element. The ratio between element length and element thickness shall be larger than five. Other values shall be agreed upon with the recognised classification society.

9.3.4.4.2.5   Plate structures, such as shell, inner hull (tank shell in the case of gas tanks), webs as well as stringers can be modelled as shell elements and stiffeners as beam elements. While modelling, cut outs and manholes in collision areas shall be taken into account.

9.3.4.4.2.6   In the FE calculation the 'node on segment penalty' method shall be used for the contact option. For this purpose the following options shall be activated in the codes mentioned:

-     "contact_automatic_single_surface" in LS–DYNA,

-     "self impacting" in PAMCRASH, and

-     similar contact types in other FE–programs.

9.3.4.4.3   Material properties

9.3.4.4.3.1   Because of the extreme behaviour of material and structure during a collision, with both geometrical and material non–linear effects, true stress–strain relations shall be used:

$$\sigma = C \cdot \varepsilon^n,$$

where

$$n = \ln(1 + A_g),$$

$$C = R_m \cdot \left(\frac{e}{n}\right)^n,$$

Ag = the maximum uniform strain related to the ultimate tensile stress Rm and

e = the natural logarithmic constant.

9.3.4.4.3.2   The values Ag and Rm shall be determined through tensile tests.

9.3.4.4.3.3   If only the ultimate tensile stress Rm is available, for shipbuilding steel with a yield stress ReH of not more than 355 N/mm² the following approximation shall be used in order to obtain the Ag value from a known Rm [N/mm²] value:

$$A_g = \frac{1}{0.24 + 0.01395 \cdot R_m}$$

9.3.4.4.3.4   If the material properties from tensile tests are not available when starting the calculations, minimum values of A_g and R_m, as defined in the rules of the recognised classification society, shall be used instead. For shipbuilding steel with a yield stress higher than 355 N/mm² or materials other than shipbuilding steel, material properties shall be agreed upon with a recognised classification society.

9.3.4.4.4   Rupture criteria

9.3.4.4.4.1   The first rupture of an element in a FEA is defined by the failure strain value. If the calculated strain, such as plastic effective strain, principal strain or, for shell elements, the strain in the thickness direction of this element exceeds its defined failure strain value, the element shall be deleted from the FE model and the deformation energy in this element will no longer change in the following calculation steps.

9.3.4.4.4.2   The following formula shall be used for the calculation of rupture strain:

$$\varepsilon_f(l_e) = \varepsilon_g + \varepsilon_e \cdot \frac{t}{l_e}$$

where
ε_g = *uniform strain*
ε_e = *necking*
t = *plate thickness*
l_e = *individual element length.*

9.3.4.4.4.3   The values of uniform strain and the necking for shipbuilding steel with a yield stress R_{eH} of not more than 355 N/mm² shall be taken from the following table:

Table

stress states	1–D	2–D
ε_g	0.079	0.056
ε_e	0.76	0.54
element type	truss beam	shell plate

9.3.4.4.4.4   Other ε_g and ε_e values taken from thickness measurements of exemplary damage cases and experiments may be used in agreement with the recognised classification society.

9.3.4.4.4.5   Other rupture criteria may be accepted by the recognised classification society if proof from adequate tests is provided.

9.3.4.4.4.6   *Tank vessel type G*

For a tank vessel type G the rupture criterion for the pressure tank shall be based on equivalent plastic strain. The value to be used while applying the rupture criterion shall be agreed upon with the recognised classification society. Equivalent plastic strains associated with compressions shall be ignored.

9.3.4.4.5   *Calculation of the collision energy absorbing capacity*

9.3.4.4.5.1   The collision energy absorbing capacity is the summation of internal energy (energy associated with deformation of structural elements) and friction energy.

The friction coefficient μ_c is defined as:

$$\mu_c = FD + \left(FS - FD\right) \cdot e^{-DC|v_{rel}|},$$

with FD = 0.1,
$\quad\quad FS$ = 0.3,
$\quad\quad DC$ = 0.01
$\quad\quad |v_{rel}|$ = relative friction velocity.

NOTE: *Values are default for shipbuilding steel.*

9.3.4.4.5.2   The force penetration curves resulting from the FE model calculation shall be submitted to the recognised classification society.

9.3.4.4.5.3   *Tank vessel type G*

9.3.4.4.5.3.1   In order to obtain the total energy absorbing capacity of a tank vessel type G the energy absorbed through compression of the vapour during the collision shall be calculated.

9.3.4.4.5.3.2   The energy E absorbed by the vapour shall be calculated as follows:

$$E = \frac{p_1 \cdot V_1 - p_0 \cdot V_0}{1 - \gamma}$$

with:

γ 1.4
(Note: The value 1.4 is the default value c_p/c_v with, in principle:
c_p = specific heat at constant pressure [J/(kgK)]
c_v = specific heat at constant volume [J/(kgK)])
p_0 = pressure at start of compression [Pa]
p_1 = pressure at end of compression [Pa]
V_0 = volume at start of compression [m^3]
V_1 = volume at end of compression [m^3]

9.3.4.4.6   Definition of striking vessel and striking bow

9.3.4.4.6.1   At least two types of bow shapes of the striking vessel shall be used for calculating the collision energy absorbing capacities:

- bow shape I: push barge bow (see 9.3.4.4.8),

- bow shape II: V–shape bow without bulb (see 9.3.4.4.8).

9.3.4.4.6.2   Because in most collision cases the bow of the striking vessel shows only slight deformations compared to the side structure of the struck vessel, a striking bow will be defined as rigid. Only for special situations, where the struck vessel has an extremely strong side structure compared to the striking bow and the structural behaviour of the struck vessel is influenced by the plastic deformation of the striking bow, the striking bow shall be considered as deformable. In this case the structure of the striking bow should also be modelled. This shall be agreed upon with the recognised classification society.

9.3.4.4.7   Assumptions for collision cases

For the collision cases the following shall be assumed:

As collision angle between striking and struck vessel 90° shall be taken in case of a V–shaped bow and 55° in case of a push barge bow; and

The struck vessel has zero speed, while the striking vessel runs into the side of the struck ship with a constant speed of 10 m/s.

The collision velocity of 10 m/s is an assumed value to be used in the FE analysis.

9.3.4.4.8   Types of bow shapes

9.3.4.4.8.1   Push barge bow

Characteristic dimensions shall be taken from the table below:

| | half breadths | | | | heights | | | |
fr	Knuckle 1	Knuckle 2	deck	stem	Knuckle 1	Knuckle 2	deck
145	4.173	5.730	5.730	0.769	1.773	2.882	5.084
146	4.100	5.730	5.730	0.993	2.022	3.074	5.116
147	4.028	5.730	5.730	1.255	2.289	3.266	5.149
148	3.955	5.711	5.711	1.559	2.576	3.449	5.181
149	3.883	5.653	5.653	1.932	2.883	3.621	5.214
150	3.810	5.555	5.555	2.435	3.212	3.797	5.246
151	3.738	5.415	5.415	3.043	3.536	3.987	5.278
152	3.665	5.230	5.230	3.652	3.939	4.185	5.315
transom	3.600	4.642	4.642	4.200	4.300	4.351	5.340

The following figures are intended to provide illustration.

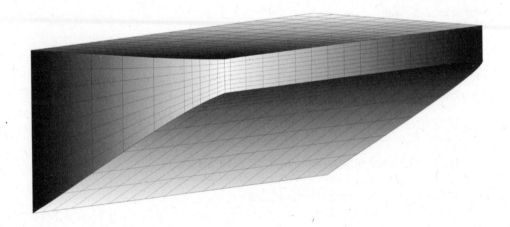

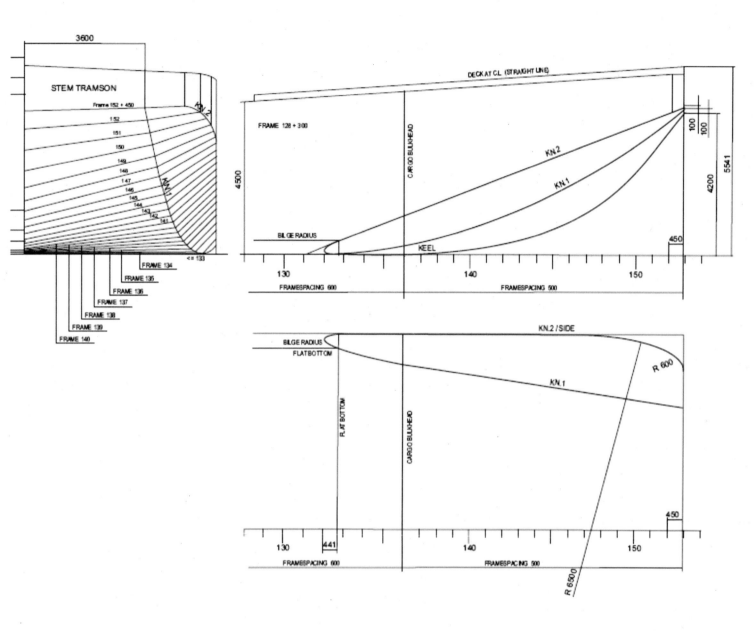

9.3.4.4.8.2   V–bow

Characteristic dimensions shall be taken from the table below:

Reference number	x	y	z
1	0.000	3.923	4.459
2	0.000	3.923	4.852
11	0.000	3.000	2.596
12	0.652	3.000	3.507
13	1.296	3.000	4.535
14	1.296	3.000	4.910
21	0.000	2.000	0.947
22	1.197	2.000	2.498
23	2.346	2.000	4.589
24	2.346	2.000	4.955
31	0.000	1.000	0.085
32	0.420	1.000	0.255
33	0.777	1.000	0.509
34	1.894	1.000	1.997
35	3.123	1.000	4.624
36	3.123	1.000	4.986
41	1.765	0.053	0.424
42	2.131	0.120	1.005
43	2.471	0.272	1.997
44	2.618	0.357	2.493
45	2.895	0.588	3.503
46	3.159	0.949	4.629
47	3.159	0.949	4.991
51	0.000	0.000	0.000
52	0.795	0.000	0.000
53	2.212	0.000	1.005
54	3.481	0.000	4.651
55	3.485	0.000	5.004

The following figures are intended to provide illustration.